Automatic Transmissions

SOUTH-WESTERN AUTO-TECH SERIES

Automatic Transmissions

Robert T. Kovacik

Saverio G. Bono

Published by

IE06 SOUTH-WESTERN PUBLISHING CO.

CINCINNATI WEST CHICAGO, ILL. DALLAS PELHAM MANOR, N.Y. PALO ALTO, CALIF.

South-Western Auto-Tech Series:

Auto Engine Repair

Manual Transmissions and Drivetrains

Automatic Transmissions

ISBN: 0-538-33060-0

Library of Congress Catalog Card Number: 83-50873

2 3 4 5 6 7 H 9 8 7 6 5 4

Printed in the United States of America

Table of Contents

Preface ix

PART I—THE DRIVETRAIN SYSTEM

Unit 1: Parts of the Drivetrain 1
Certification Practice, Test 1 15

Unit 2: Servicing the Automatic Transmission 17
Job Sheet 2-1 21
Job Sheet 2-2 29
Certification Practice, Test 2 33

Unit 3: Shift Linkage and Adjustment 35
Job Sheet 3-1 41
Job Sheet 3-2 45
Job Sheet 3-3 51
Job Sheet 3-4 53
Job Sheet 3-5 55
Job Sheet 3-6 57
Certification Practice, Test 3 59

PART II—TORQUE CONVERTERS

Unit 4: Torque Converter Systems 61
Certification Practice, Test 4 67

Unit 5: Checking the Torque Converter 69
Job Sheet 5-1 71
Job Sheet 5-2 75
Job Sheet 5-3 81
Job Sheet 5-4 85
Certification Practice, Test 5 87

Unit 6: Transmission Cooling Systems and Repairs 89
Job Sheet 6-1 93
Job Sheet 6-2 97
Certification Practice, Test 6 101

PART III—PLANETARY GEARSETS AND APPLY DEVICES

Unit 7: Planetary Gearsets and Apply Devices 103
Certification Practice, Test 7 111

Unit 8: Repairing Planetary Gearsets and Apply Devices 113
Job Sheet 8-1 117
Job Sheet 8-1A 119
Job Sheet 8-1B 121
Job Sheet 8-2 127
Job Sheet 8-3 135
Certification Practice, Test 8 139

PART IV—HYDRAULIC SYSTEMS

Unit 9: Hydraulic Principles 141
Certification Practice, Test 9 151

Unit 10: Hydraulic Systems 153
Certification Practice, Test 10 165

Unit 11: Reading Flow Charts 167
Certification Practice, Test 11 187

Unit 12: Hydraulic System Repairs 189
Job Sheet 12-1 195
Certification Practice, Test 12 199

PART V—TROUBLESHOOTING AND DIAGNOSIS

Unit 13: Troubleshooting and Diagnosis 201
Job Sheet 13-1 209
Job Sheet 13-2 217
Job Sheet 13-3 223
Job Sheet 13-4 227
Certification Practice, Test 13 237

PART VI—AUTOMATIC TRANSMISSION OVERHAUL

Unit 14: Removing the Transmission 239
Job Sheet 14-1 243
Certification Practice, Test 14 247

Unit 15: Overhauling the Chrysler TorqueFlite A-727 249
Job Sheet 15-1 273
Certification Practice, Test 15 279

Unit 16: Overhauling the Ford AOD 281
Job Sheet 16-1 303
Certification Practice, Test 16 309

Unit 17: Installing the Transmission 311
Job Sheet 17-1 315
Certification Practice, Test 17 319

PART VII—AUTOMATIC TRANSMISSION DESCRIPTIONS

Unit 18: Turbo Hydra-Matic 200, 200C, 200-4R, 325, 325-4L 321
Certification Practice, Test 18 327

Unit 19: Turbo Hydra-Matic 250, 250C, 350, 350C, 375B 329
Certification Practice, Test 19 337

Unit 20: Turbo Hydra-Matic 375, 400, 425 339
Certification Practice, Test 20 343

Unit 21: Ford C-3, C-4, C-5 345
Certification Practice, Test 21 351

Unit 22: Ford C-6 353
Certification Practice, Test 22 357

Unit 23: Ford Jatco 359
Certification Practice, Test 23 363

Unit 24: Borg-Warner A40 365
Certification Practice, Test 24 369

PART VIII—AUTOMATIC TRANSAXLES

Unit 25: Transaxle Differences 371
Certification Practice, Test 25 377

Unit 26: Overhauling the Turbo Hydra-Matic 125/125C 379
Job Sheet 26-1 385
Job Sheet 26-2 409
Job Sheet 26-3 417
Certification Practice, Test 26 421

Glossary 423

Index 432

Preface

This is both a text and a shop manual to guide students through hands-on experience in disassembly, troubleshooting, and repair of *Automatic Transmissions.* As a text, this book provides knowledge-building units aimed at assuring student understanding about automatic transmissions. The knowledge acquired by students, in turn, serves as the basis for performance of practical, mechanical tasks presented in skill-building units. Where appropriate, knowledge- and skill-building units are combined. These units provide instructions and guidance in a series of tasks that equip students for job qualifications in diagnosis and repair of automatic transmissions.

In addition, a series of review and practice tests serve to prepare students who will go on to seek certification through completion of the tests administered by the National Institute for Automotive Service Excellence (NIASE).

ORGANIZATION

This text/shop manual is organized into units that establish natural learning experiences for students. Each unit begins with a section, entitled *Your Learning Job*, that presents an overview of content and a list of achievement objectives that should be mastered. This listing serves as both a focus for the study and practice that follow, and a reference by which both students and instructors can measure knowledge- and skill-building progress.

Both textual and hands-on instruction are presented in readable and understandable format. Illustrations reinforce the instructions and guide work performance. Checkpoints are incorporated following each significant topic in knowledge-building units. These provide guideposts for self-checking of progress and reassure instructors about student progress.

For skill-building activities, job sheets are provided that provide detailed performance descriptions. These instructions guide students and provide built-in checkpoints for monitoring of shop work by instructors.

Practice tests are provided at the end of each key learning segment of the book. These are patterned specifically as preparation for the certification tests administered by NIASE. Thus, a student who finishes work in this book should be prepared—in terms of both format and content—to deal with certification tests.

FEATURES

Objectives present clear statements of the knowledge and skill achievements expected of students.

Illustrations are positioned in direct relationship to the text content to which they apply. All illustrations are cited in the text to establish solid learning reinforcement.

Safety is emphasized throughout the text/shop manual, with care taken to relate safety measures to the use of tools and equipment.

An *Instructor's Manual and Key* contains an outline of each unit, teaching suggestions, and a key to checkpoints and the Certification Practice Tests.

This text/shop manual is part of the South-Western Auto Tech Series. This series is a comprehensive teaching/learning system developed to prepare today's students for the fast-growing automotive service field.

ACKNOWLEDGEMENTS

Photography for the step-by-step transmission overhaul procedures (Units 15, 16, and 26) was done by Kalton C. Lahue.

The authors wish to thank Dave Watts, technical consultant, and the following companies, which supplied photos and illustrations used throughout this book: Chevrolet Motor Division, Pontiac Motor Division, Oldsmobile Division, Buick Motor Division, and Cadillac Motor Division, General Motors Corporation; Hydra-Matic, Division of General Motors Corporation; Chrysler Corporation; Ford Motor Company; American Motors Company; Toyota Motor Sales, U.S.A., Inc.; Mazda (North America), Inc. (Toyo Kogo Co., Ltd.); Volvo of America Corporation; Nissan Motor Corporation in U.S.A.

I

The Drivetrain System

1 Parts of the Drivetrain

YOUR LEARNING JOB

It is important to understand how the drivetrain works and the role of the automatic transmission in the drivetrain system. This unit describes how a drivetrain functions. It also explains the operation of the drivetrain parts and how to identify them.

JOB COMPETENCY OBJECTIVES

This is a knowledge-building unit. When you have completed your assignments and exercises in this unit, you should be able to:

- ☐ Explain what the drivetrain is and what it does.
- ☐ Describe the operation of the automatic transmission.
- ☐ Describe the operation of the driveline.
- ☐ Explain the function of the differential and driving axles.
- ☐ Identify the parts of the drivetrain.

1.1 PARTS OF THE DRIVETRAIN

The drivetrain takes power from the engine. This power is controlled and transferred by the drivetrain to the driving wheels, causing the wheels to turn. The driving wheels may be either the rear wheels or the front wheels. Rear-wheel-drive drivetrains are discussed in this unit. Front-wheel-drive drivetrains are covered in Unit 25. Parts of the rear-wheel-drive drivetrain are:

- Automatic transmission
- Driveline
- Differential
- Driving axles

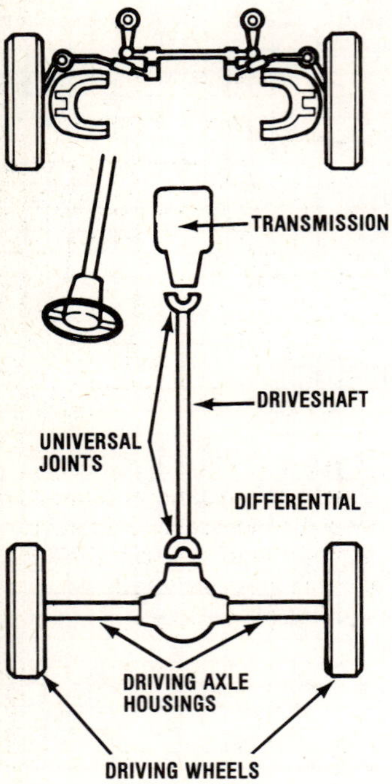

Figure 1-1. A typical rear-wheel drivetrain. (Chevrolet Motor Division-GMC)

A rear-wheel-drive drivetrain system is shown in Figure 1-1.

When a manual transmission is part of the drivetrain, a clutch is used. The clutch is located between the engine and the manual transmission. Manual transmissions are covered in a separate book, in the Auto Tech series, *Manual Transmissions and Drivetrains*.

Another term frequently used is powertrain. The powertrain includes the engine *and* drivetrain. The engine powers the drivetrain. For this reason, the engine and drivetrain together are referred to as the powertrain.

1.2 AUTOMATIC TRANSMISSION

The first part of the drivetrain system is the automatic transmission. The automatic transmission connects and disconnects engine power from the rest of the drivetrain. The automatic transmission also controls the torque, or twisting force, from the engine.

The automatic transmission works both hydraulically and mechanically. Hydraulics is a branch of science that deals with applied fluid power or the practical applications of confined fluids. Hydraulic principles can be used for operating and controlling parts. Hydraulics is covered in greater detail in Unit 9. The main mechanical parts in an automatic transmission are gears, shafts, and clutches. The major parts of an automatic transmission are:

- Torque converter
- Planetary gearset
- Control valve body
- Clutches.

CHECKPOINT

1. **The ____________________ takes power from the engine and delivers it to the vehicle's driving wheels.**
2. **A drivetrain consists of ____________________, ____________________, ____________________, and ____________________.**
3. **The engine and drivetrain together are referred to as the ____________________.**
4. **____________________ principles can be used for controlling and operating parts.**
5. **The main parts of an automatic transmission are: ____________________, ____________________, ____________________, and ____________________.**

1.3 TORQUE CONVERTER

The automatic transmission takes power from the engine through a fluid connection, or coupling. The fluid used in automatic transmissions is a special oil called automatic transmission fluid, or ATF for short. The fluid coupling consists of two main parts: an impeller and a turbine. See Figure 1-2.

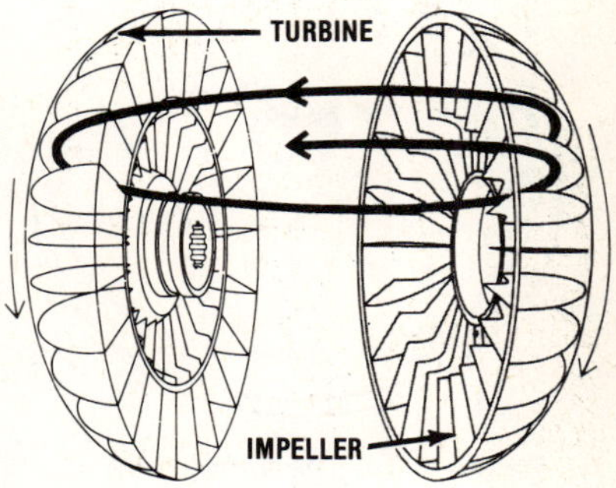

Figure 1-2. A fluid coupling consists of an impeller (turned by engine torque) and a turbine (connected to the input shaft of the transmission), inside a liquid-filled housing.

The impeller is bolted to the engine flywheel. The turbine is placed next to the impeller and drives the transmission input shaft. The transmission input shaft connects the turbine to the gears.

The impeller and turbine resemble two fans placed so that the blades, or vanes, face each other. When one fan (the impeller) turns, it forces air through the vanes of the second fan (the turbine). The air passing through the vanes of the second fan cause it to turn.

In a fluid coupling, an impeller forces liquid, instead of air, through the vanes of the turbine.

The impeller turns at engine speed and forces liquid into the vanes of the turbine. As the turbine turns, it turns the transmission input shaft. The input shaft turns the gears, and the gears turn the output shaft. The output shaft turns the driveline.

In a fluid coupling, a large amount of power is lost, because the impeller can turn faster than the turbine. This condition is called slippage, or slip. To help eliminate much of this slip, a smaller turbine-like part is placed between the impeller and the turbine. This third part is a stator.

As engine rotation directs fluid flow from the impeller blades to the turbine blades, fluid flow is reversed. This reversed fluid flow works against the rotation of the impeller and decreases impeller efficiency.

To correct this reversed fluid flow condition, a stator (Figure 1-3) is placed between the turbine and the impeller. The stator reverses the return flow of the turbine fluid so that the flow matches the impeller's direction of rotation. This is what causes torque multiplication, discussed in 1.4.

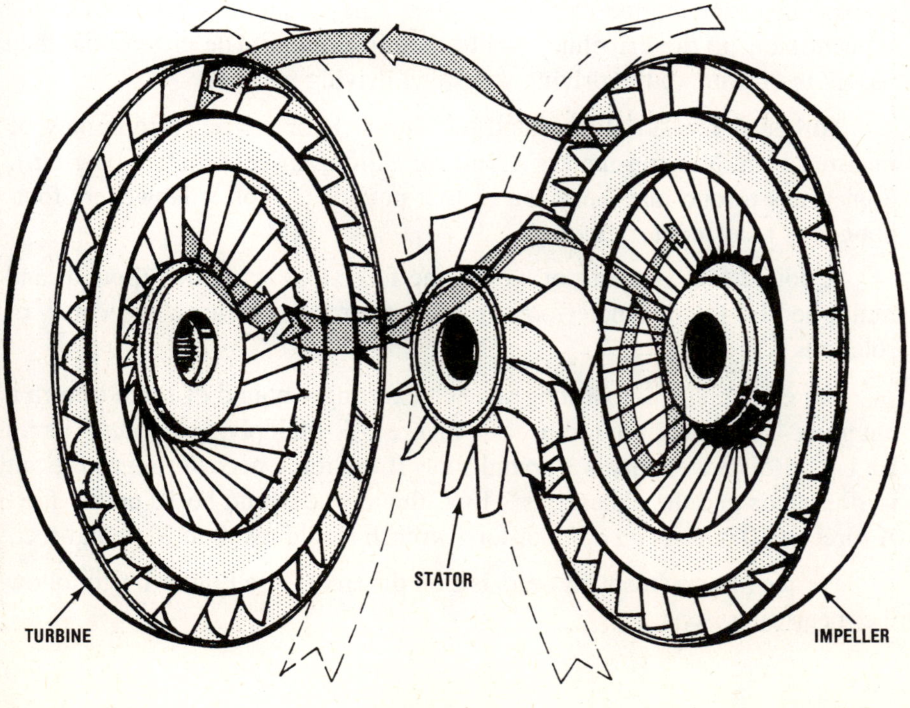

Figure 1-3. The stator, positioned between the impeller and the turbine, redirects fluid flow for more efficient coupling action. (American Motors Corporation)

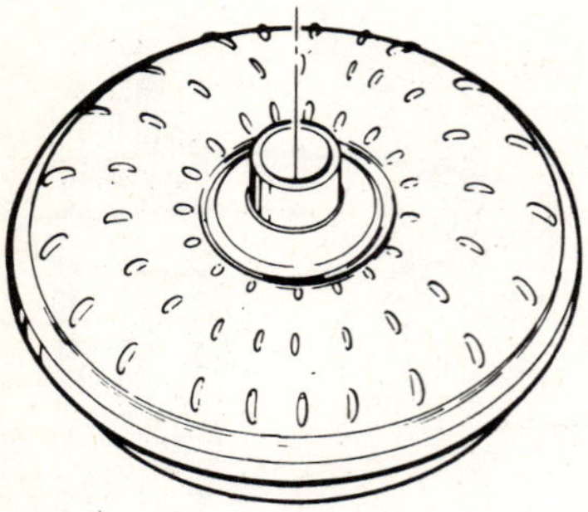

Figure 1-4. The torque converter housing is made of pressed sheet metal, welded at the seams. (Chevrolet Motor Division-GMC)

The impeller, stator, and turbine are enclosed in a housing, called the torque-converter housing. See Figure 1-4. The torque-converter housing also acts as a reservoir to store automatic transmission fluid. The fluid reservoir is necessary to make sure there is always enough ATF available for proper coupling.

All of these parts are referred to as the torque converter. The torque converter reduces slip and multiplies torque.

CHECKPOINT

1. **The automatic transmission takes power from the engine by using a fluid ________________.**
2. **A fluid coupling consists of two parts: ________________ and ________________.**
3. **The ________________ is bolted to the engine flywheel.**
4. **As the ________________ turns, it turns the transmission input shaft.**
5. **The ________________ redirects fluid flow from the impeller to the turbine.**
6. **A torque converter ________________ slip and ________________ torque.**

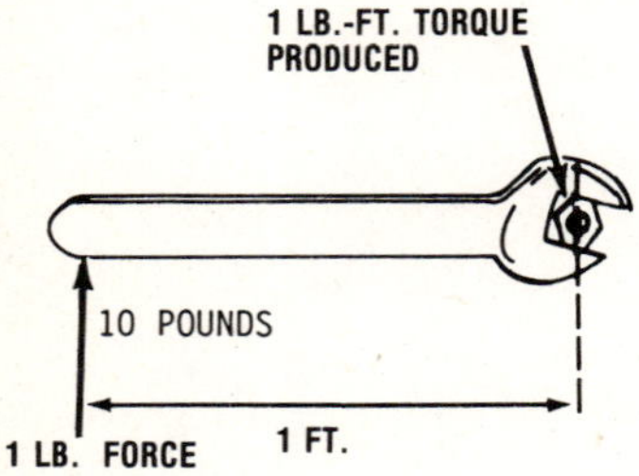

Figure 1-5. A force of 1 pound exerted on a wrench 1 foot long produces a torque of 1 pound-foot (lb.-ft.) at the axis (center) of the nut.

1.4 TORQUE MULTIPLICATION

Any shaft or gear that is turned has torque applied to it. Torque is a turning or twisting force. The term *torque* is commonly used throughout the automotive powertrain system. Pistons and connecting rods acting on the crankshaft in the engine produce torque. This torque is transferred to the transmission, to the driveline, and to the differential. The differential then passes the torque to the driving axles and driving wheels.

Torque is measured in pound-feet (lb.-ft.). In metric, torque may be measured in Newton-meters (Nm) or kilogram-centimeters (kg-cm). Sometimes torque may be given in foot-pounds (ft.-lbs.). However, foot-pounds is incorrect terminology.

A pound-foot is a unit of work—the force that is applied (pound) and multiplied by the distance (foot) from that force to the axis, or center, of rotation.

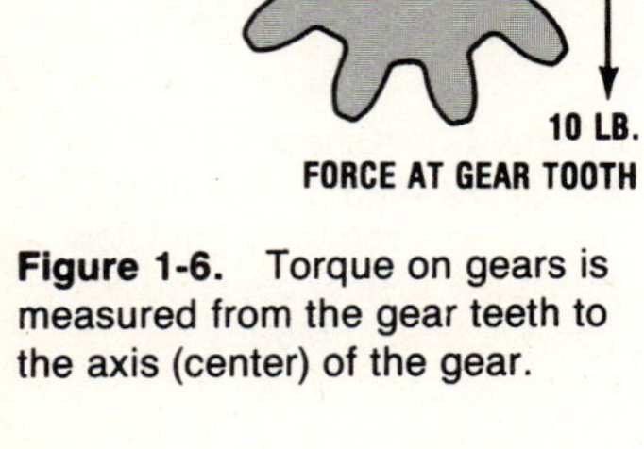

Figure 1-6. Torque on gears is measured from the gear teeth to the axis (center) of the gear.

For example, a 1-foot-long wrench exerting 1 pound of force to turn a nut gives 1 pound-foot of torque. See Figure 1-5. Multiplying 1 pound (force) by 1 foot (distance) equals 1 pound-foot. If the force needed to turn the nut is 10 pounds and the distance is 1 foot, the torque would be 10 pound-feet. A force of 10 pounds on a 2-foot-long wrench would result in 20 pound-feet.

Torque measurement on gears is a bit different. Use Figure 1-6 to follow the measuring steps.

The gear tooth can be likened to the handle of the wrench discussed earlier. Force is exerted at the gear tooth. The center of the gear, or axis, is where the nut would be located. A force of 10 pounds on the gear tooth (wrench handle) multiplied by the distance of one foot from the axis (nut) equals 10 pound-feet.

Consider what happens when a gear is meshed with a larger one. Remember, force occurs where the gears mesh, and distance is measured in a straight line from one gear axis to the other gear axis. The 10 pound-feet torque of the smaller gear is applied to the teeth on the larger gear. Figure 1-7 shows the measurements of the two gears. However, the larger gear has a distance to the axis of 2 feet. By multiplying the force (10 pounds) by the distance (2 feet), the torque has been increased, or multiplied, to 20 lb.-ft.

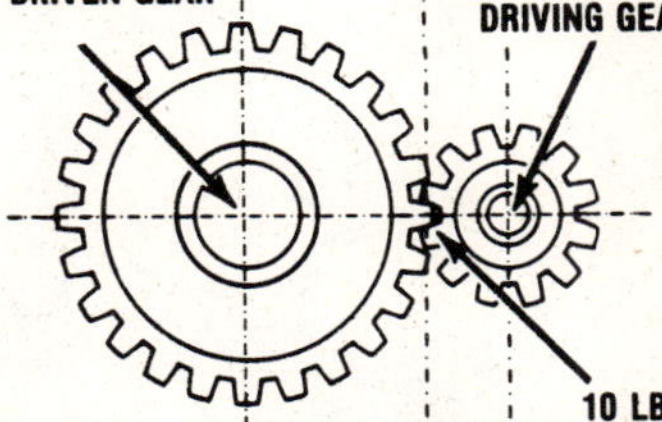

Figure 1-7. The distance from the gear teeth to the axis of the gear is 2 feet for the larger gear, 1 foot for the smaller gear.

When a smaller gear drives a larger gear, the larger gear turns slower. This is known as speed, or gear, reduction. The larger gear slows, or reduces, the original rotating speed.

A gear reduction occurs in the transmission and in the differential. Any time a smaller gear, called a pinion, drives a larger gear, gear reduction and torque multiplication occur.

To multiply torque hydraulically, fluid flow is increased from the drive member to the driven member.

The drive member is the part that applies power. The driven member is the part that accepts power. In the case of a torque converter, the impeller is the driving member, and the turbine is the driven member.

CHECKPOINT

1. **Torque is measured in ____________________.**
2. **If the force needed to turn a nut is 10 pounds and the distance is 1 foot, the torque would be ________________ pound-feet.**
3. **Any time a smaller gear drives a larger gear, __________________ reduction and ________________ multiplication occur.**
4. **To multiply torque hydraulically, ____________________ is increased.**
5. **The ____________________ member is the part that applies power.**

1.5 PLANETARY GEARSET

In a manual transmission, a clutch is used to disengage power before the gears can be shifted properly. An automatic transmission is shifted without disconnecting engine power from the transmission. A special gear system is used that does not require manual shifting to change gears. To do this, automatic transmissions use a planetary gearset. A gearset is a series of gears.

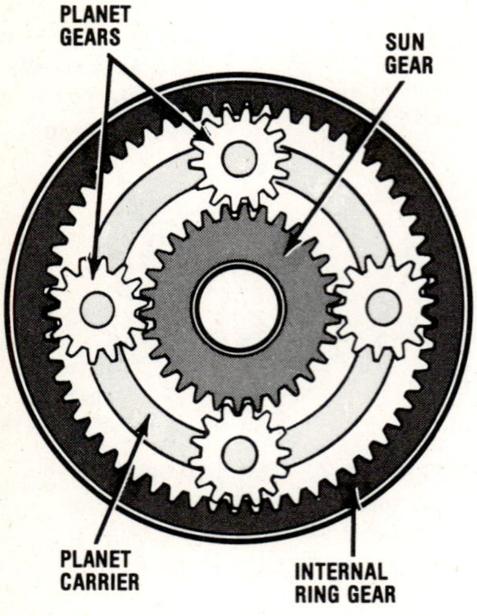

Figure 1-8. A simple planetary gearset.

A simple planetary gearset, shown in Figure 1-8, consists of the following parts:

- Sun gear
- Ring gear with internal teeth
- Planet gears
- Planet carrier.

A planetary gearset works much like the solar system. Planet gears orbit, or circle around, the sun gear. The planet gears are held in place by the planet carrier and the ring gear. The ring gear surrounds, and meshes with, the planet gears. When gears mesh, the teeth interlock so that they will turn together.

In a planetary gear system, the teeth of each gear are always meshed with the teeth of another gear. So, whenever one gear or part is turned, all other gears and parts in the system are affected.

To transmit power through a planetary gearset, one gear or part turns as the drive member. The drive member is turned by torque from the input shaft. Another part, called the reaction member, is held and prevented from moving. The reaction member can be locked in place by a brake band, or a clutch. A third part becomes the driven, or output, member. The driven member transmits torque to the driveline.

Any part or gear of the planetary gearset can be the drive, or input, member.

The reaction member does not turn. When the planet carrier is the reaction member, the planet gears turn on their shafts, but do not orbit around the sun and ring gears. The sun gear can be locked in place by a clutch. The ring gear can be held by a brake band that surrounds the ring gear and squeezes against the outside of the gear to stop it from moving.

A variety of gear reductions, or ratios, is possible, depending on which gear is the reaction gear. A choice of transmission gear ratios allows the vehicle's engine to run efficiently at different road speeds.

Two simple planetary gearsets can also be connected together for additional gear ratios. When two planetary gearsets are connected together, the sun gear is usually a common, or shared, part. The combination of two planetary gearsets coupled, or connected, with a common sun gear is called a compound planetary gearset.

A detailed description of planetary gearset operation is discussed in Unit 7.

1.6 GEAR RATIOS

The relative number of turns and rotating speeds of two meshing gears is known as the gear ratio. Gear ratios are found in many different parts of vehicles. Examples include a steering ratio, a transmission ratio, and a differential ratio.

Two gears that are the same size, and have the same number of teeth, have a ratio of 1 to 1. This ratio is expressed as 1:1. A 1:1 ratio indicates that both gears turn at the same speed.

If a larger driven gear has twice as many teeth as the smaller driving gear, the ratio is 2:1. If the larger gear has three times as many teeth, the ratio is 3:1.

A gear ratio is determined by counting the number of teeth on the drive and driven gears. The number of teeth on the driving gear is then divided into the number of teeth on the driven gear. Suppose the smaller, or pinion, gear has 9 teeth and the larger gear has 37 teeth. Nine is divided into 37. The answer—4.11:1—is the gear ratio. Most gear ratios appear with decimals, such as: 2.43:1, 3.08:1, or 4.11:1.

A high gear-ratio number means the vehicle will have better acceleration and pulling power. A low gear-ratio number means the engine will run slower, which results in better fuel economy.

A transmission gear ratio of 1:1 is called direct drive. This means that the transmission output shaft makes one revolution for each revolution of the engine.

Some vehicles have an overdrive ratio. A transmission with an overdrive gear ratio improves highway fuel economy, but decreases pulling power. Overdrive gear ratios have become increasingly popular in recent years. A typical overdrive ratio might be 0.85:1. This means that the engine crankshaft will turn only 0.85 of a turn to cause the output shaft of the transmission to turn 1 full turn. The result is slower engine speeds for a given road speed. When the engine runs slower, fuel economy increases. Overdrive gear ratios are used to improve fuel economy at highway speeds.

CHECKPOINT

1. **An automatic transmission must be shifted without ________________ engine power from the transmission.**
2. **A planetary gearset consists of: ________________, ________________, ________________, and ________________.**
3. **In a planetary gearset, the teeth of each gear are always ________________ with the teeth of another gear.**
4. **To transmit power through a planetary gearset, one part or gear is the ________________ member, another part or gear is the ________________, and the third part or gear is the ________________ member.**
5. **The relative number of turns and rotating speeds of two meshing gears is known as the ________________ ratio.**
6. **________________ ________________ is the term for a transmission gear ratio of 1:1.**

1.7 CONTROLLING GEAR SHIFTS

To control gear shifts automatically, some sort of ''brain'' is necessary. The ''brain'' of the automatic transmission is the control valve body, often called, simply, the valve body. The control valve body contains numerous passages, valves, springs, pistons, accumulators, and servos. See Figure 1-9. A servo is a mechanism that converts hydraulic pressure to mechanical movement.

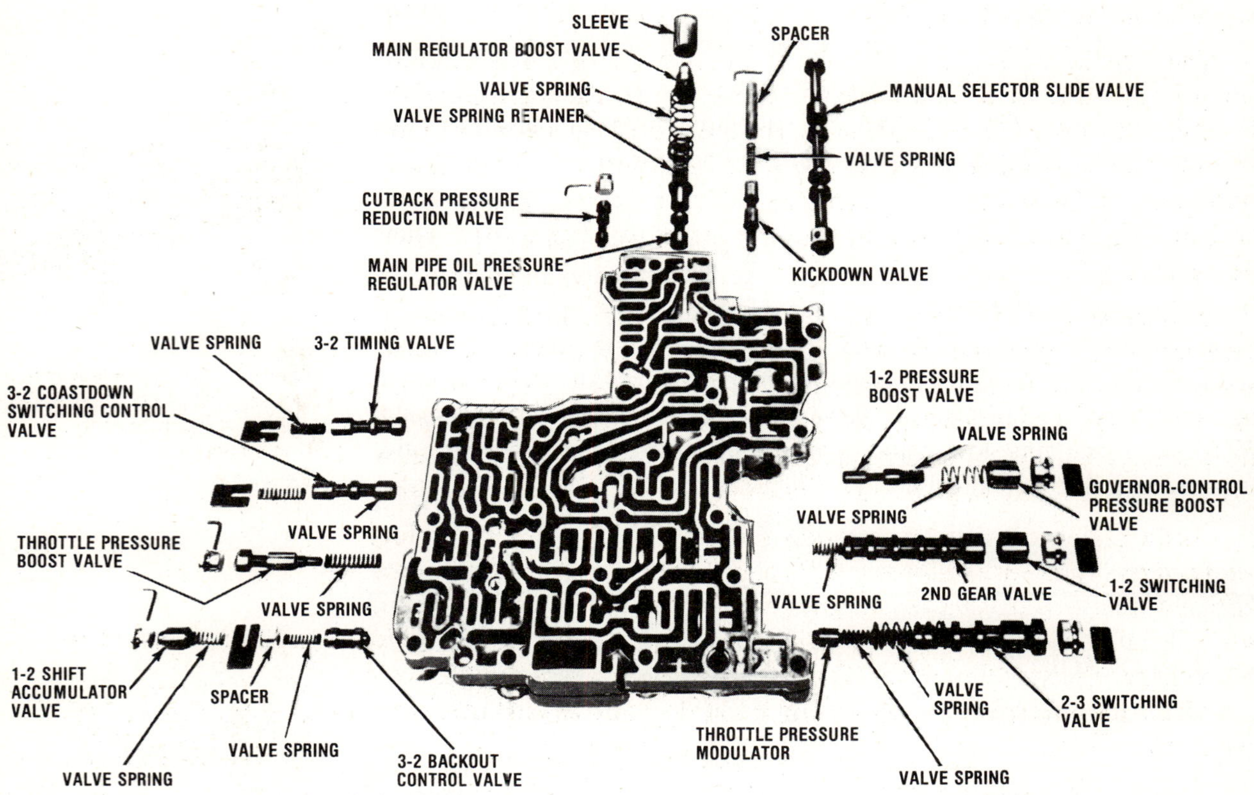

Figure 1-9. The control valve body contains many passages to direct fluid flow. (Ford Motor Company)

The control valve body receives signals on vehicle speed, throttle position, engine load, and the gear selected by the driver. Using these signals, the control valve body directs fluid pressure to the appropriate servos. The servos engage the proper gear for the situation.

When the shift lever is in the drive (D) position, a valve directs fluid pressure to the clutches that engage first gear. Other valves automatically control shifts as speed and load requirements dictate.

More detailed information on automatic transmission control systems is discussed in Parts III and IV.

1.8 DRIVELINE

The driveline consists of a driveshaft and two universal joints, or U-joints. The purpose of the driveline is to carry torque from the transmission output shaft to the differential. A driveline is used only on front-engine/rear-drive vehicles. A driveline is shown in Figure 1-10.

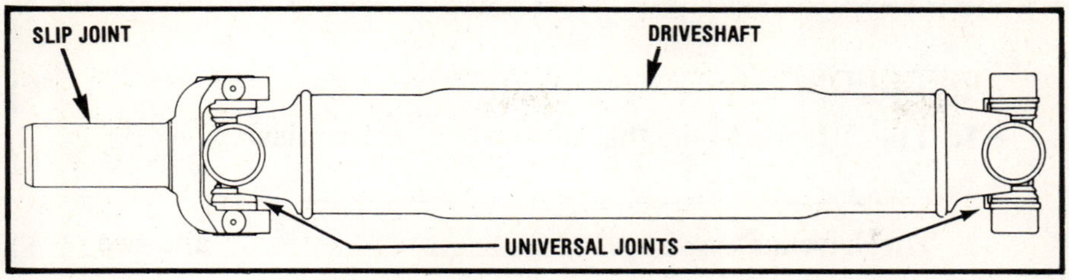

Figure 1-10. A typical driveline for a rear-wheel drivetrain. (Oldsmobile Division-GMC)

The driveshaft is a steel tube that is balanced to rotate smoothly. Balancing is done at the factory, usually by welding necessary weights to the driveshaft.

The driveline is connected to the transmission at one end. The transmission is firmly mounted to the vehicle so it cannot move. The other end of the driveline is connected to the differential. The differential moves up and down with the rear suspension of the vehicle. The movement of the differential causes the driveline to constantly change angles, up and down, while the vehicle is moving. To maintain a constant power flow to the differential, U-joints are necessary.

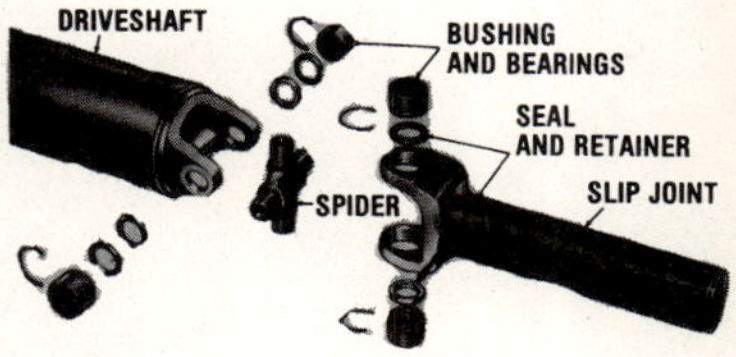

Figure 1-11. A universal joint can transmit power at an angle while rotating. (Chrysler Corporation)

Numerous types of U-joints are used, but they all work in the same way. A universal joint consists of two U-shaped yokes that hold a cross, or spider, as shown in Figure 1-11. Two cross arms, or trunnions, connect to one yoke. The remaining trunnions connect to the other yoke. The end of each trunnion has bearings that allow the spider to rotate inside the yoke holes. The U-joints adjust to accommodate angle changes, while allowing a constant power flow.

As the driveshaft moves up and down, it also moves in and out, or forward and rearward. To compensate for this movement, the yoke that fits onto the transmission output shaft is designed to slip in and out. For this reason, it is called the slip yoke, or slip joint. The slip joint is splined on the inside to match the splines on the output shaft. See Figure 1-12. The slip joint moves in and out on the splines as the driveline shortens or lengthens during movement.

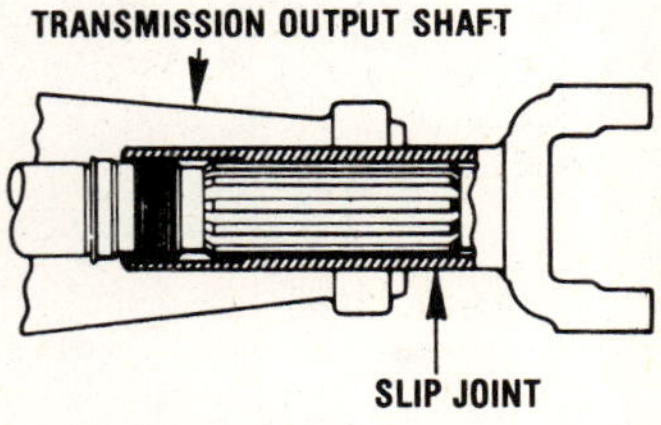

Figure 1-12. A slip joint can transmit power to the driveline while moving back and forth. (Oldsmobile Division-GMC)

Another U-joint is used at the differential end of the driveshaft to adjust up-and-down movement of the driveshaft. This U-joint is connected by a flange assembly, called a companion flange, as shown in Figure 1-13.

Detailed driveline description is covered in *Manual Transmissions and Drivetrains*, another manual in the Auto Tech series.

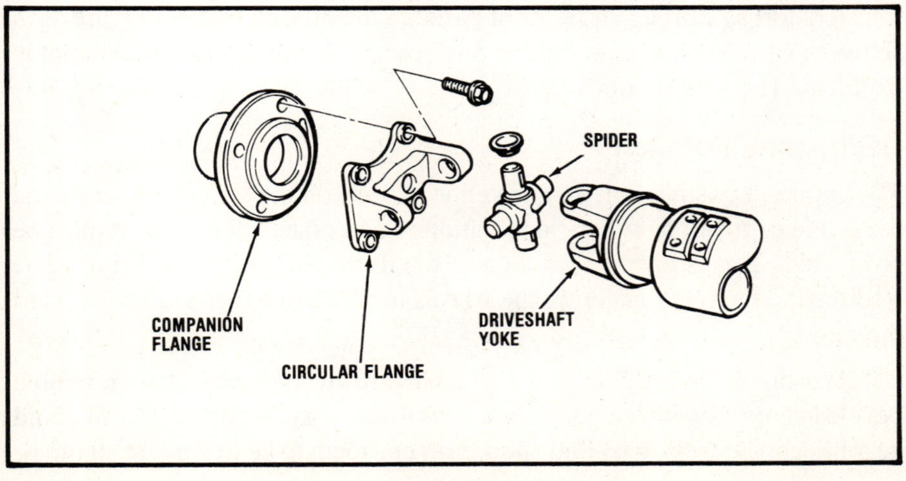

Figure 1-13. The part of the differential assembly that attaches to the rear universal joint is called the companion flange. (Ford Motor Company)

CHECKPOINT

1. The "brain" of the automatic transmission is the ________ ________.
2. The driveline consists of a ________ and two ________ ________.
3. The driveline connects the ________ in the front to the ________ in the rear of the vehicle.
4. A U-joint consists of two ________ that hold a ________.
5. The yoke that fits onto the transmission output shaft is called a ________ ________.

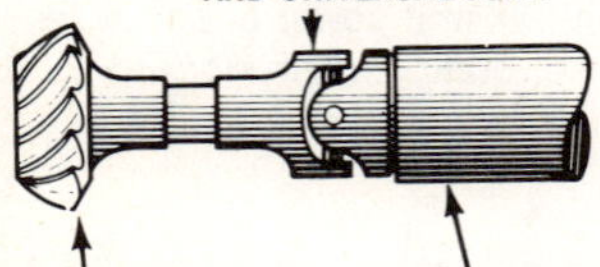

Figure 1-14. The differential drive pinion gear is attached to the rear yoke of the universal joint.

1.9 DIFFERENTIAL

The differential is located just behind the driveline and between the two rear driving axles. The purpose of the differential is to:

- Accept torque from the driveline
- Redirect, or turn, driveshaft rotation at right angles so it can turn the driving wheels
- Increase torque and divide it equally to the driving wheels
- Allow the rear wheels to turn at different speeds when the vehicle rounds a turn.

Accepting Driveline Torque

Driveline torque enters the differential through the drive pinion gear. The drive pinion gear attaches to the rear U-joint yoke, as shown in Figure 1-14. When the driveline turns, the drive pinion gear turns.

The drive pinion gear meshes with a larger gear that is inside of the differential housing. This larger gear is called the ring gear. The ring gear transfers torque, through a series of parts, to the driving wheels. Figure 1-15 shows the drive pinion gear and the ring gear. This pair of gears is commonly called the ring and pinion.

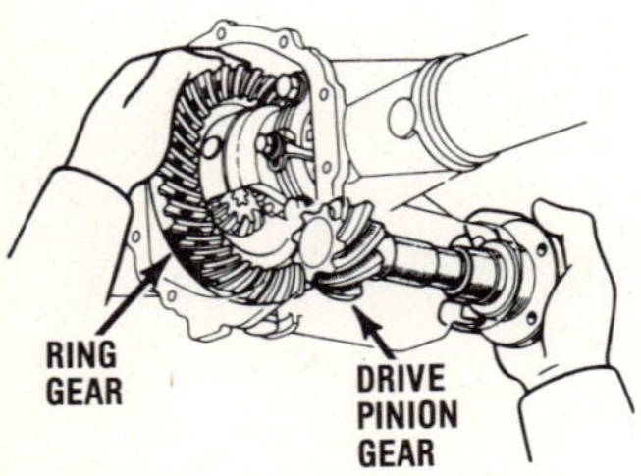

Figure 1-15. The gearset that redirects rotation 90 degrees in a differential is called a ring and pinion. (Toyota Motor Sales U.S.A., Inc.)

Redirecting Rotation

The engine, transmission, and driveline all turn on the same axis, or on the same line of rotation. When this rotation reaches the differential, it must be redirected. This is another function of the drive pinion gear and ring gear. Redirecting the rotation allows the wheels to rotate in a forward or rearward direction.

The drive pinion gear and ring gear are hypoid gears that resemble beveled gears. A beveled gear is one that is cut at a slant or angle. A hypoid gear has curved teeth that allow more than one tooth to be in mesh at all times.

The pinion gear of a hypoid gearset is mounted below the centerline of the ring gear so that the driveline can be lowered. A lowered driveline gives more interior room in the vehicle. A hypoid gear assembly is shown in Figure 1-16. The angle of the rotating ring and pinion gears changes the direction of rotation by 90 degrees.

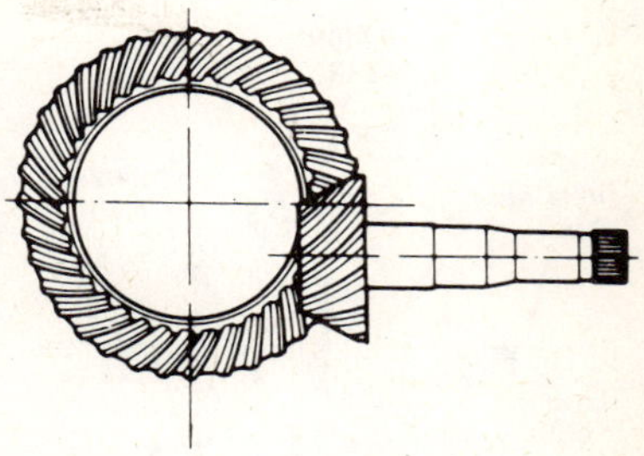

Figure 1-16. The differential drive pinion and ring gear teeth are specially made spiral bevel gear teeth, called hypoid gear teeth. Notice how the centerlines of the two gears are not aligned. (Nissan Motor Corporation)

Increasing Torque

Torque is increased, or multiplied, through gear reduction. The driving gear (pinion) acts on the driven gear (ring). The gear ratio between these two gears is called the axle ratio. Refer to Topics 1.4 and 1.6.

Differing Rear Wheel Speed

The differential gets its name by allowing each driving wheel to turn at a different speed during cornering. The difference in speed is necessary because the two rear wheels have different distances to travel when the vehicle turns. If both driving wheels were to turn at the same speed during cornering, one wheel would spin or skid, making control difficult. This skidding action would also wear out tires and suspension parts quickly.

If the vehicle only traveled in a straight line, a differential would not be needed. Since the vehicle must turn, however, a series of parts are necessary to rotate the driving wheels at different speeds.

Parts of the Differential

The differential consists of these basic parts, shown in Figure 1-17:

- Ring gear
- Differential case, or carrier
- Pinion shaft
- Differential pinion gears
- Differential side gears.

The ring gear is bolted to the differential case, or carrier. As the ring gear rotates, the carrier rotates with it. The rest of the differential parts are located inside of the carrier.

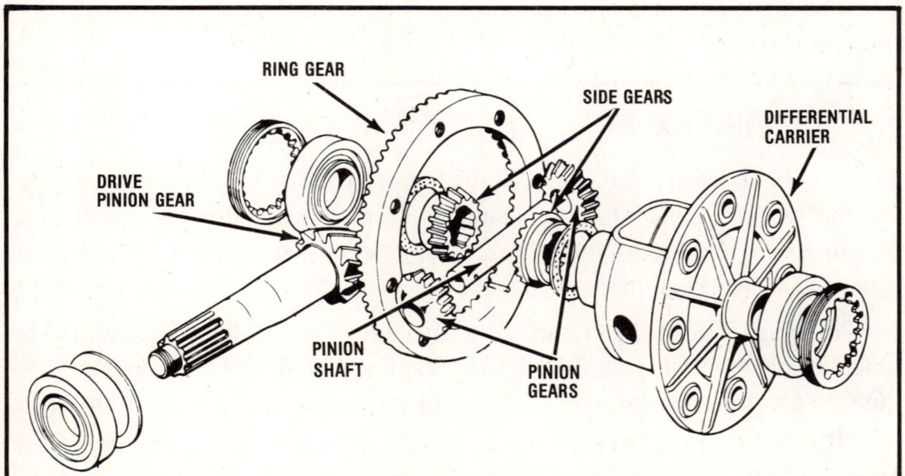

Figure 1-17. Parts of a differential assembly. (Toyo Kogyo Co., Ltd.)

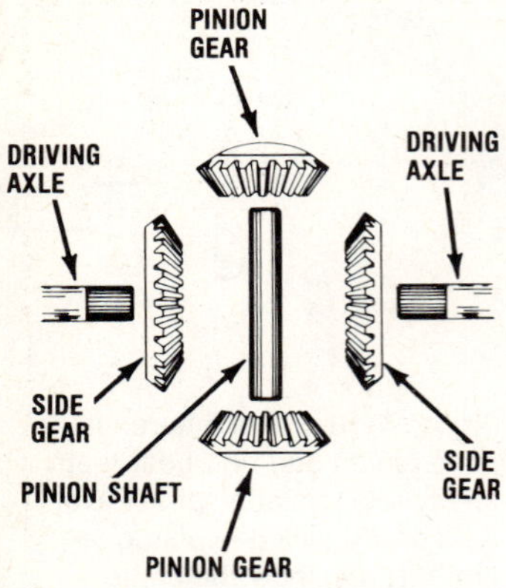

Figure 1-18. Notice the positions of the differential pinion gears and side gears in this differential.

The pinion shaft is mounted in holes in the carrier. Two differential pinion gears are used. Each pinion gear is mounted on the pinion shaft so that it meshes with the side gears. Both pinion gears and side gears are beveled, and turn in different directions. A differential gear layout is shown in Figure 1-18.

When driving in a straight line, the pinion gears do not rotate. The pinion shaft is mounted rigidly to the carrier. The carrier rotates, carrying the pinion shaft around with it, and the pinion gears remain stationary. As the pinion shaft moves with the carrier, the pinion gears mesh with, and exert pressure on, the side gears. The side gears now turn at the same speed as the ring gear, driving both wheels at the same speed.

When the vehicle rounds a corner, or curve in the road, the outer wheel turns faster than the inner wheel. To allow this to happen, the two pinion gears rotate on the pinion shaft. As the pinion gears rotate, they transmit more speed to the outer side. This allows the outside driving axle to turn faster than the inside driving axle.

The differential is described in more detail in *Manual Transmissions and Drivetrains*, another manual in the Auto Tech series.

CHECKPOINT

1. **The torque from the driveline enters the differential through the ____________ ____________ gear.**
2. **The drive pinion gear meshes with a larger gear in the differential called the ____________ gear.**
3. **The differential consists of the following basic parts: ____________, ____________, ____________, and ____________.**
4. **When driving in a straight line, the pinion gears do not ____________ on the pinion shaft.**
5. **When the vehicle rounds a corner, the outer wheels turn ____________ than the inner wheels.**

1.10 DRIVING AXLES

The driving axles are solid steel shafts that transfer differential torque to the driving wheels. A separate axle shaft is used for each driving wheel. The driving axles, and part of the differential, are enclosed in an axle housing that protects and supports these parts. Figure 1-19 shows a housing and axle setup.

Each driving axle is connected to the side gears in the differential. The inner, or differential, end of each axle is splined to fit into the side gears. As the side gears are turned, the axles to which they are splined turn at the same speed.

At the outer, or wheel, end, each axle is attached to the driving wheels. For attachment to a wheel, the outer end of each axle has a flange mounted

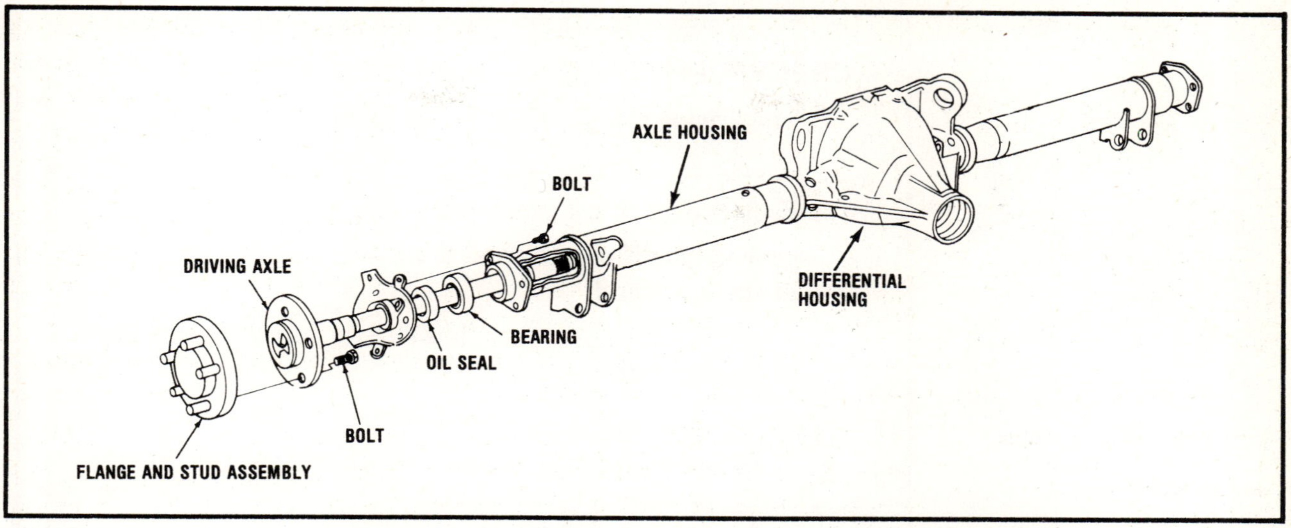

Figure 1-19. The driving axles are enclosed in protective tubes, called the axle housings. (Ford Motor Company)

to it. A flange is a rim for attaching one part to another part. To hold the wheel in place against the flange, studs are used. Studs are threaded shafts. One end of the stud is pressed into the flange. The wheel fits over the studs, and a nut, called a lug nut, is tightened over the open end of each stud. This holds the wheel in place, as shown in Figure 1-20.

The inner end of each axle shaft is supported by the differential carrier and side gear. The outer end of each axle shaft is supported by a bearing inside the axle housing. A bearing supports and holds a rotating part in place. This bearing—called an axle bearing—allows the axle to rotate smoothly inside of the axle housing.

More detailed information on driving axles is covered in a separate manual, *Manual Transmissions and Drivetrains*, another book in the Auto Tech series.

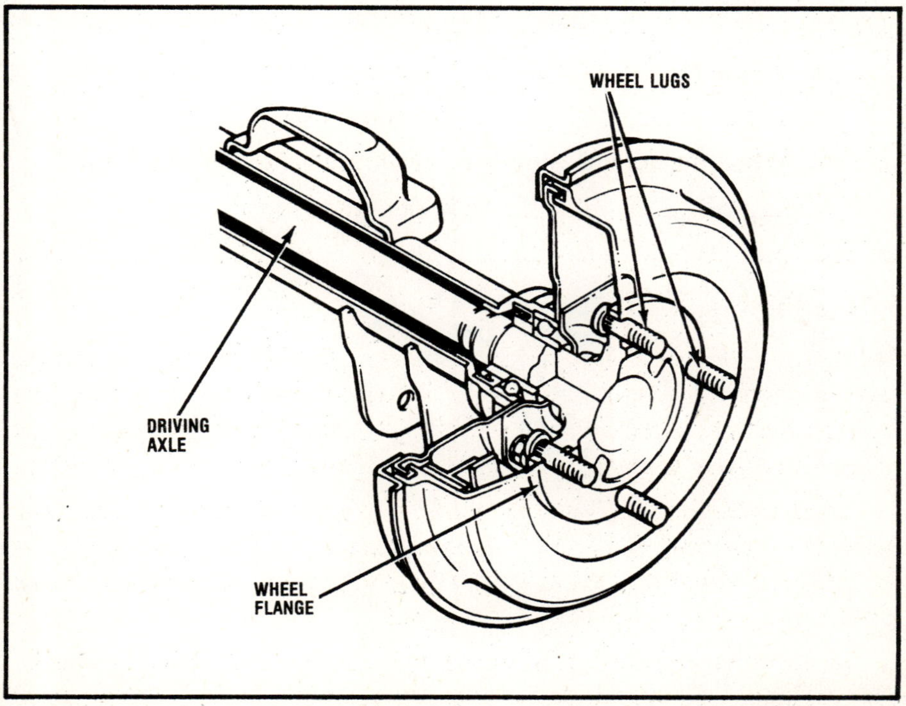

Figure 1-20. The end of the driving axle has a flange for attaching the driving wheels. Lug nuts or bolts hold the driving wheel to the flange. (Chrysler Corporation)

CHECKPOINT

1. **The driving axles transfer torque to the ________________ ________________.**
2. **Each driving axle is connected to a ________________ gear in the differential and to the ________________ at the outer end.**
3. **A ________________ is a rim for attaching one part to another part.**
4. **________________ are threaded shafts.**
5. **Each axle shaft is supported on the inside by the ________________ ________________.**

NAME ______________ SECTION ______________ DATE ______________

Certification Practice

Test 1

DIRECTIONS:

The following questions are similar to those used on mechanic certification tests. Answer each question by circling the letter of the correct choice.

1. All of the following are parts of the drivetrain EXCEPT
   A. the engine.
   B. the transmission.
   C. the differential.
   D. the driving axles.

2. The turbine transfers engine torque to the transmission input shaft.
   Mechanic A says that the impeller causes the turbine to rotate.
   Mechanic B says that the flywheel causes the turbine to rotate.
   Who is correct?
   A. A only B. B only C. Both A and B D. Neither A nor B

3. Which of the following statements is correct?
   I. In a normal fluid coupling, a lot of fluid is lost, causing slip.
   II. The stator redirects fluid from the impeller to the turbine.
   A. I only B. II only C. Both I and II D. Neither I nor II

4. The following statements apply to planetary gearsets EXCEPT
   A. The planet gears orbit the sun gear.
   B. The ring gear has internal teeth.
   C. The planet carrier is the housing for the planetary gearset.
   D. Of the three members in a planetary gearset, one member is driving, one member is driven, and one member is held.

5. The control valve body consists of numerous passages, valves, springs, pistons, and servos.
   Mechanic A says that the control valve body directs fluid to the servo to engage first gear.
   Mechanic B says that the control valve body directs fluid to the torque converter to increase power flow.
   Who is correct?
   A. A only B. B only C. Both A and B D. Neither A nor B

6. All of the following are true EXCEPT
   A. The driveline usually consists of a driveshaft and two universal joints.
   B. The driveshaft is balanced at the factory.
   C. Both the transmission and the differential move up and down with the vehicle's suspension, changing the angle of the driveshaft.
   D. To maintain a constant power flow as the driveshaft changes angles, universal joints are necessary.

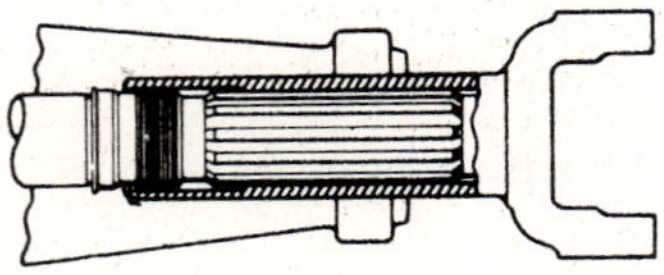

Question 7.

7. Mechanic A says that the part shown is a companion flange.
   Mechanic B says that it is a slip joint.
   Who is correct?
   A. A only B. B only C. Both A and B D. Neither A nor B
8. Which of the following is not part of the differential?
   A. Ring gear
   B. Differential case, or carrier
   C. Pinion shaft and differential pinion gears
   D. Driving axles
9. The differential allows the driving wheels to turn at different speeds when the vehicle turns a corner.
   Mechanic A says that, when driving in a straight line, the carrier rotates and the pinion shaft moves around with the carrier, but the pinion gears do not rotate.
   Mechanic B says that, when the vehicle rounds a corner, the pinion gears rotate on the pinion shaft and transfer more speed to the outer side.
   Who is correct?
   A. A only B. B only C. Both A and B D. Neither A nor B
10. All of the following are true EXCEPT
    A. The driving axles are solid steel shafts that transfer torque to the driving wheels.
    B. Each driving axle is splined and connected to a side gear in the differential.
    C. At their outer ends, the driving axles are attached to the driving wheels.
    D. Each axle shaft is supported at both ends by bearings.

2 Servicing the Automatic Transmission

YOUR LEARNING JOB

The ability to visually locate and identify problems is one of the first skills a mechanic must learn. As with any other major part of the automobile, routine service procedures for the automatic transmission must also be learned before tackling any major repair work. This unit describes the procedures for checking and servicing the automatic transmission.

JOB COMPETENCY OBJECTIVES

This is a skill-building unit. When you have completed your assignments and exercises in this unit, you should be able to:

- ☐ Identify leaks and locate external transmission damage.
- ☐ Diagnose the condition of the automatic transmission fluid.
- ☐ Drain the transmission and torque converter.
- ☐ Replace the transmission filter or service the filtering screen.

2.1 VISUAL CHECK OF THE AUTOMATIC TRANSMISSION

During the visual check, the vehicle should be raised on a hoist. Use a droplight to look around the transmission housing, various linkages, and controls. Many obvious problems can be overlooked when the light is poor.

SAFETY CAUTION: Be sure the hoist is properly secured before beginning any work. Always use safety glasses when working under the vehicle. The vehicle *must* be sufficiently cooled down before inspection. You will be using your bare hands, and serious burns can result if the vehicle is *not* cooled down. Move your hand slowly over the transmission. The automatic transmission and its various controls may have sharp edges that will cut. If you move slowly, you will be able to feel the danger in time to avoid injury.

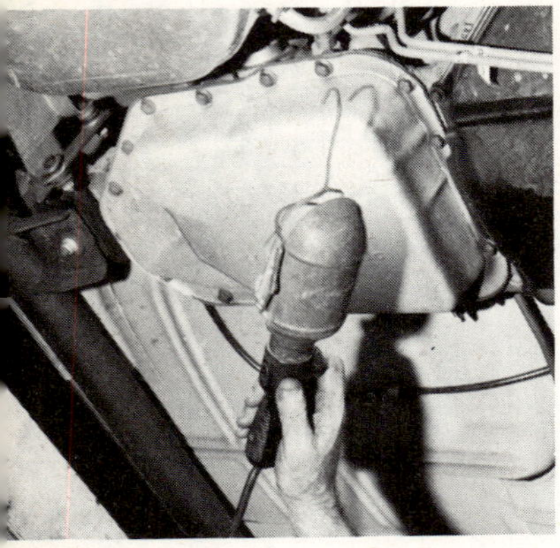

Figure 2-1. Be careful of sharp surfaces when checking the transmission housing.

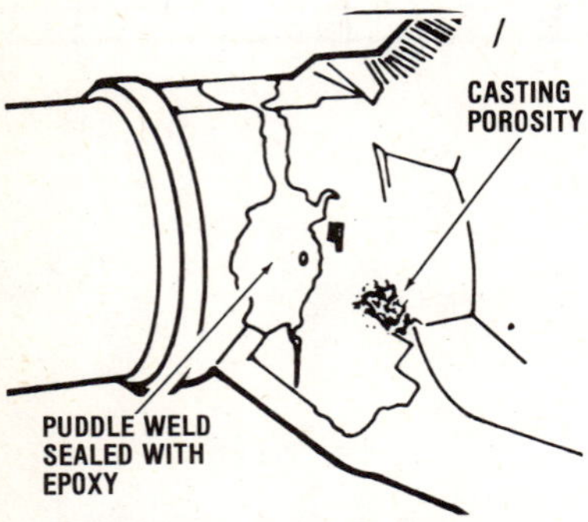

Figure 2-2. Casting porosity is caused by air bubbles in molten metal during the casting process. (Ford Motor Company)

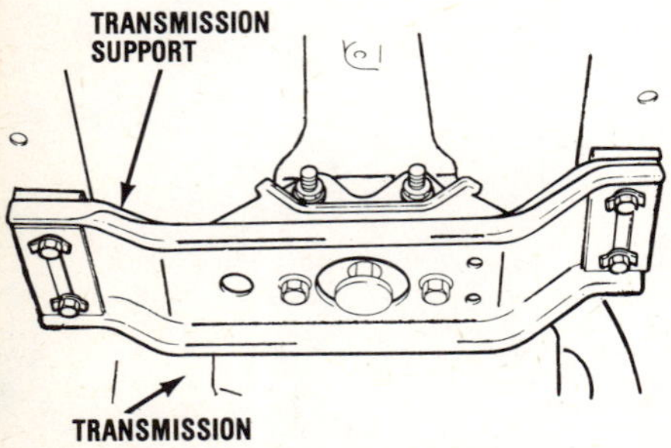

Figure 2-3. The transmission is supported by a rubber-and-metal transmission mount at the rear. The arrows indicate mounting bolts. (Chrysler Corporation)

Use a droplight and look for things that don't look or feel right. You will be looking for leaks, for looseness, and for damaged or missing parts. Use your hands in areas where your sight is obstructed.

To check shift linkage, it may be necessary to lower the vehicle and look under the hood. The main areas to check for leaks or damage are:

- Transmission housing
- Linkage systems.

Checking The Transmission Housing

Start at the front of the transmission housing where it connects to the engine. Use a droplight (Figure 2-1) to check for leaks. Reach with a bare hand over the top of the housing to feel for dampness or damage. Use your fingertips lightly, because there might be sharp edges where you can't see them. Check the entire transmission housing for cracks or damage.

Check where the slip joint enters the transmission housing for leaks. The rear portion of the transmission is called the extension housing. A leak in this area indicates a worn extension housing seal or bushing.

Sometimes a leak may appear in the transmission housing where it does not connect to a separate part. When there is a leak, or hole, in the housing, it can be the result of casting porosity. Casting porosity, shown in Figure 2-2, is the result of improper manufacturing procedures. Casting porosity is repairable, as discussed in 2.2.

There should be no leaks around any bolt hole or screw-in control device.

Shake the transmission hard. It should not move. Inspect the transmission mounts (Figure 2-3) for loose bolts or damage.

Checking Transmission Linkage

The transmission linkage may be controlled by a floor shifter or by a column shifter. See Figures 2-4 and 2-5. The linkage can be either of the rod-and-lever type or a cable.

Always wipe off the linkage with a rag to remove road dirt. This allows you a good look at the linkage. Be sure safety glasses are worn when removing the road dirt.

Floor shift linkage must be checked from under the vehicle. This linkage usually consists of one rod and two levers that mount to the left side of the transmission housing. Shake the linkage gently to be sure it is not excessively loose. Look at the places where the linkage pivots against another part. Small, synthetic bushings are usually located at the pivot points. The bushings, or washers, are designed to wear before the shift linkage does. The pivot points also require periodic lubrication with either white grease or a recommended lubricant. Also be sure that no retaining clips or nuts are missing from the linkage. Always use the proper service manual as a guide if you are uncertain.

NOTE: The right and left sides of a vehicle are determined from the driver's seat. As the driver sits behind the steering wheel, the right side of the vehicle is to the driver's right. The left side of the vehicle is to the driver's left.

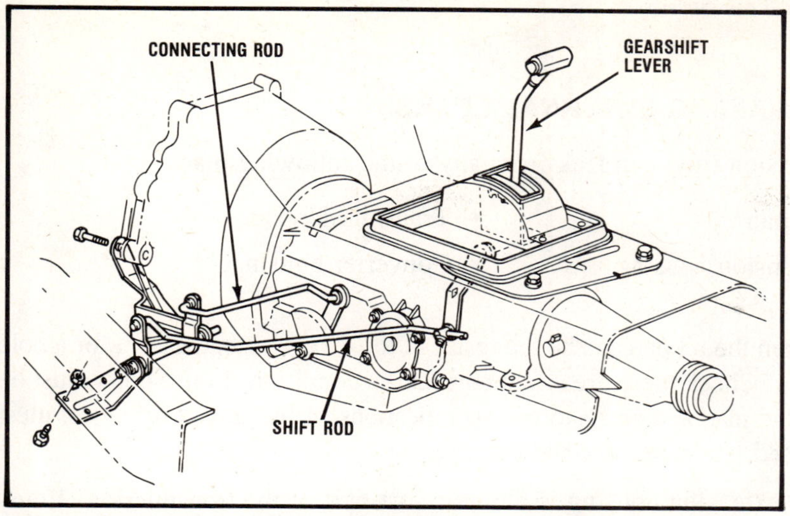

Figure 2-4. Rod-and-lever linkage can be used to operate the automatic transmission. (Ford Motor Company)

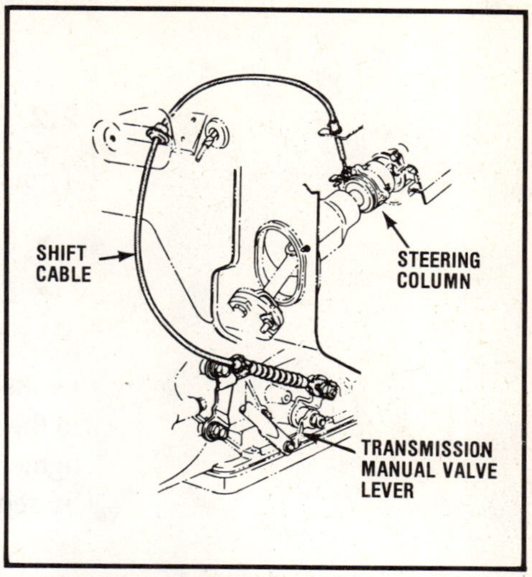

Figure 2-5. Cable linkage is used on many front-wheel-drive vehicles to operate the automatic transmission. (Ford Motor Company)

Column shift linkage is usually checked by opening the hood. The shift linkage will be located on the left side of the firewall. The firewall is the partition that separates the passenger compartment from the engine compartment. Column shift linkages are often controlled by cables.

Make sure the cable is not kinked. A cable that is kinked will not operate properly and must usually be replaced. Check the pivot points of the cable linkage where it connects to the shift lever and to the transmission. Look for missing parts and proper linkage lubrication as discussed for floor shift linkages.

The shift linkage at the transmission may also be connected to the carburetor by a cable or rod. This linkage allows the driver to engage a lower gear when the accelerator pedal is depressed. Figure 2-6 shows this type of linkage, also called kickdown linkage. Check this linkage carefully from under the hood.

More detailed information on automatic transmission linkage systems is covered in Part III.

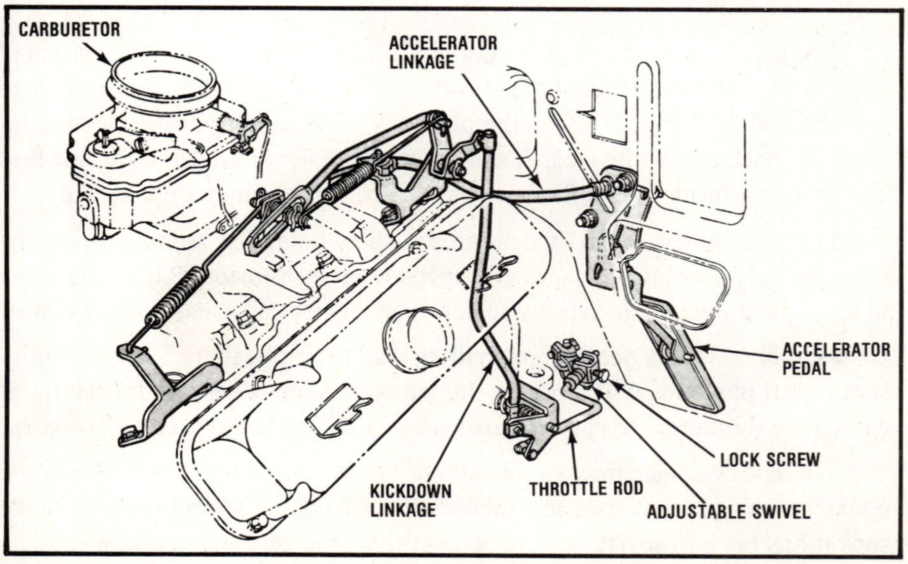

Figure 2-6. Kickdown linkage shifts the automatic transmission into a lower gear when the throttle is suddenly depressed. (Chrysler Corporation)

2.2 REPAIRING EXTERNAL LEAKS

Transmission fluid can leak from any of the following places:

- Oil pan
- Extension housing
- Transmission case
- Converter housing.

Leaks from the oil pan can be the result of loose bolts, a bad gasket, or a hole in the pan. Check these areas and correct the condition. Loose bolts must be tightened to manufacturer's torque specifications. A bad gasket or oil pan must be replaced (refer to 2.5 and 2.8).

The extension housing is the rearmost part of the transmission. If any leaks are evident at the extension housing, they may be caused by a bad seal, gasket, or bushing. The different types of seals used in automatic transmissions are covered in 8.13. Replacement of seals, gaskets, and bushings is discussed in the overhaul sections: Units 15, 16, and 26. Leaking seals can be found anywhere around the transmission, including at the transmission case and converter housing. Many transmissions do not have a separate converter housing, or bellhousing. In many instances, the transmission case and converter housing are one piece.

Check all parts of the transmission for loose bolts. Use a torque wrench to tighten the bolts to the manufacturer's recommendations.

Casting porosity is also a source of transmission leakage. Casting porosity is caused during manufacturing when air bubbles are trapped in the hot metal as it cools. The next section describes how to repair casting porosity. See Figure 2-2.

Repairing Casting Porosity

Place the shift lever in park (P), set the emergency brake, and bring the transmission to operating temperature. With the vehicle on a hoist, locate the leak. Stop the engine and wipe off the leaking area with a rag and solvent. Let the solvent air-dry.

Two methods can be used to correct casting porosity, depending on the size of the hole.

Body lead and an epoxy sealer will close small holes. Place a small amount of body lead in the hole. With a punch and hammer, force the lead into the hole by hammering on the punch. This is called *peening*, which indents, or pockmarks, the lead. A special epoxy sealer is used to cover the lead and ensure a proper seal. Some manufacturers recommend epoxy only.

For large holes, begin by drilling a shallow hole. Use a tap to thread the hole. A tap is a tool that cuts into the metal to create threads. Place a setscrew in the tapped hole. Place epoxy sealer over the setscrew and hole to seal it.

CAUTION: When peening or drilling and tapping, there is a possibility that metal particles will fall into the transmission. These metal particles can cause damage. Be very careful when making these types of repairs.

The epoxy sealer may require three hours or more to cure. When the epoxy sealer is cured, road-test the vehicle. Check the leakage area to make sure it has been properly sealed. Repeat the procedure, if necessary.

NAME ______________________ SECTION ____________ DATE ____________

Job Sheet 2-1

Visual Check of the Automatic Transmission

Before you begin:

Read Topics 2.1 through 2.2.

Vehicle Make ____________________ Model ____________ Year ____________

Time Started ____________________ Time Finished ____________________

Job Completion Time ______________ Flat Rate Time ____________________

Tools Required

- Hoist
- Droplight
- Torque wrench
- Safety glasses

References

Manufacturer's service manual ______________________________

Repair manual ______________________________

Service bulletin ______________________________

Procedure

SAFETY CAUTION: Be sure the vehicle to be inspected is cooled down. You will be using your bare hands, and you want to avoid getting burned on a hot part. Move your hand slowly over the surfaces. Many drivetrain parts have sharp edges that will cut. If you move slowly and carefully, you will be able to feel the danger in time to avoid injury.

1. Check the transmission housing for damage or cracks.
2. Check the slip joint area in the transmission extension housing for leaks.
3. Check for leaks in the transmission housing.
4. Check for leaks from any bolt-on or screw-in control device.
5. Check the transmission mounts for looseness or damage. See Figure 2-3.
6. Check the transmission linkage for looseness or damage.
7. Inspect the transmission linkage bushings for damage, and check for proper lubrication.
8. Check any shift linkage cables for kinking or damage.
9. Inspect the kickdown linkage for damage, and for proper connection to the transmission and carburetor linkage. See Figure 2-6.

Instructor Check: ____________________

Job Completion Report

1. What problems, if any, were noted during your inspection? ______________

____

____

____

2. Was the transmission linkage of the floor-shift or column-shift type?

____

3. What type of shift linkage (rod-and-lever or cable) did you inspect?

____

INSTRUCTOR SIGNOFF ______________________ **DATE** ______________

2.3 CHECKING TRANSMISSION FLUID LEVEL

Automatic transmission fluid level is checked from under the hood of the vehicle. A dipstick, similar to an engine oil dipstick, is located near the firewall. Be sure the dipstick hole leads to the automatic transmission. Some engine oil dipsticks may also be located at the rear of the engine compartment.

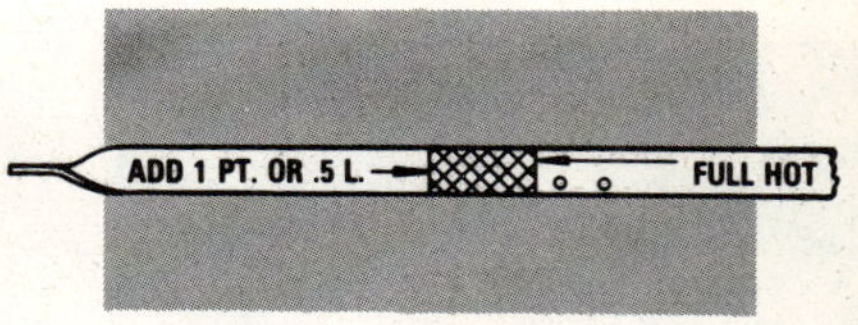

Figure 2-7. A typical automatic transmission dipstick has indications for full, safe amount of fluid, and add. (Pontiac Motor Division-GMC)

Automatic transmission fluid (ATF) levels must be checked at regular mileage or time intervals. Refer to the proper service manual. When the vehicle is operated in heavy stop-and-go traffic, under high load conditions, or in hot weather, fluid level should be checked more frequently.

In older vehicles, ATF level must be checked only when the engine and transmission are at operating temperatures. The engine must be running and the shift lever placed in park (P) or neutral (N) with the emergency brake on. TorqueFlite transmissions do not pick up fluid in P, so the shift lever must be placed in N.

If the engine and transmission are cold, start the engine and let it run until the transmission reaches operating temperature. Set the emergency brake and place one foot on the brake pedal. Place the shift lever into each lever position. Be sure the transmission engages in each lever position before moving it. Moving the shift lever circulates the fluid throughout the transmission.

CAUTION: The vehicle must be level when checking the ATF. Improper readings can lead to serious transmission damage. If the transmission is underfilled or overfilled, aeration, or bubbles, develop in the fluid. Aeration causes slip and can damage the transmission parts.

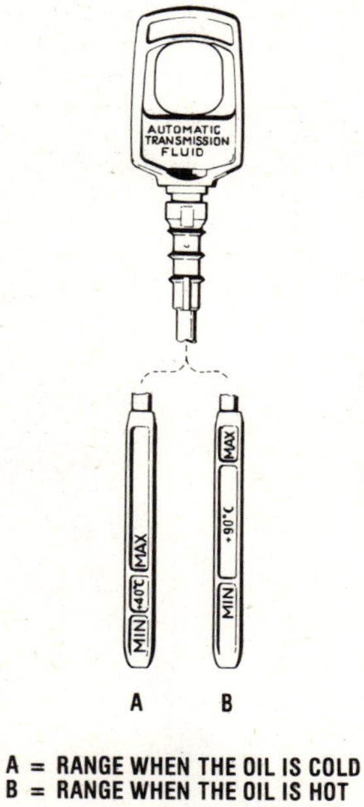

Figure 2-8. Some automatic transmission dipsticks indicate the correct readings for cold fluid on one side and hot fluid on the other side. (Volvo of America Corporation)

Fluid level can drop almost an inch when the transmission is cold. So, reading the dipstick at operating temperature is critical.

Pull out the transmission dipstick and wipe it clean with a lint-free cloth or paper towel. Reinsert the dipstick in the dipstick hole. Remove the dipstick and note the reading. A typical automatic transmission dipstick is shown in Figure 2-7.

If fluid level is between ADD and FULL, the fluid level is good. If fluid level reading is below ADD, at least 1 pint of fluid should be added. In most cases, the distance between the ADD and FULL marks on the dipstick indicates a 1 pint difference.

On most newer vehicles, the dipstick may have two or three different readings.

One dipstick, shown in Figure 2-8, can be read when the vehicle is either hot or cold. One side of the dipstick indicates the hot fluid level. The opposite side of the dipstick indicates the cold fluid level.

Another type of dipstick (Figure 2-9) can be read whether the fluid is cool, warm, or hot. To determine fluid temperature, touch the end of the dipstick where it was submerged in the ATF. Be careful, ATF can be very hot. When the temperature of the ATF has been determined, take the appropriate reading.

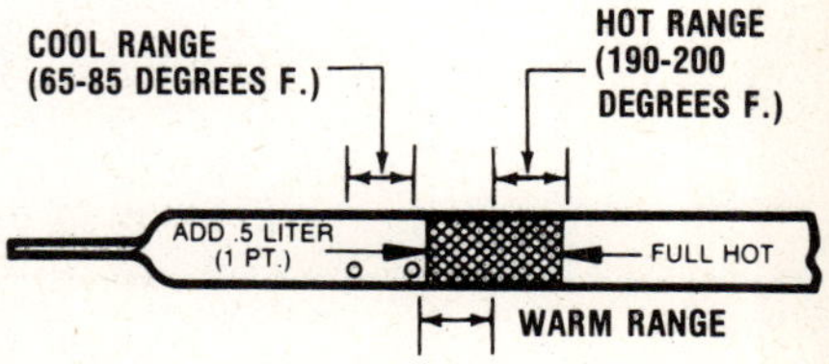

Figure 2-9. By referring to the proper service manual, correct fluid levels cool, warm, and hot can be determined. (Chevrolet Motor Division-GMC)

NOTE: If there is any uncertainty about the temperature of the ATF, it is recommended that the transmission be brought to operating temperature.

Figure 2-10. Use a funnel to add fluid, if necessary, through the dipstick tube.

Before adding ATF, first be sure the proper lubricant is used. Manufacturers differ as to the type of ATF that is required. Different types of ATF are discussed in 10.1.

Fluid is added through the dipstick hole, as shown in Figure 2-10. A special spout is used so that ATF is not spilled over the engine and transmission. No more than 1 pint of ATF should be added at any particular time. After several seconds, recheck fluid level. It should be between the appropriate ADD and FULL marks. If more fluid is needed, add a little at a time until the level is right. *Never* overfill the transmission. If too much fluid has been added, use a suction gun to remove the excess.

2.4 DIAGNOSING TRANSMISSION FLUID

Inspecting the ATF should be routine whenever fluid level is checked. Automatic transmission fluid is usually red in color. A special dye is added to the ATF so that transmission leaks can be easily identified. The two basic ways of telling the condition of ATF is by sight and smell.

If the ATF is dark brown in color, it may indicate overheating. Some fluids turn dark with age, so smell the fluid to determine if it has overheated. If it smells like burned food, the fluid has been overheated. An oil and filter change are necessary.

When a preliminary check for overheated fluid indicates a problem, the ATF should be checked more closely.

Wipe the dipstick on absorbent white paper. Look at the fluid stain, shown in Figure 2-11. Dark particles in the fluid indicate band or clutch material. Silvery particles indicate bits of metal that may be caused by excessive wear.

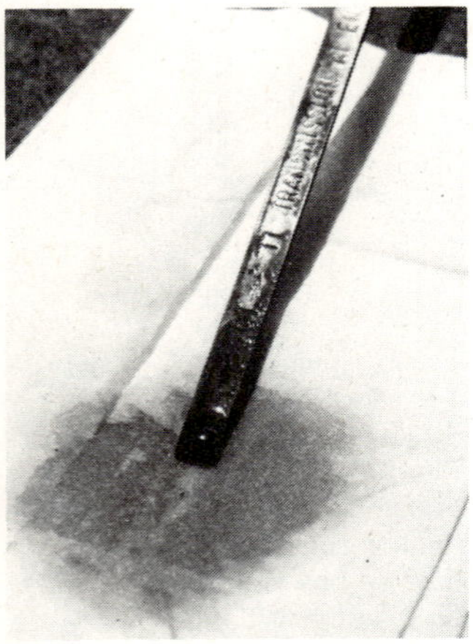

Figure 2-11. Check the automatic transmission fluid for dark color or particles by wiping the dipstick on absorbent white paper.

Check the cleaned dipstick. Any varnish or gum means that the ATF needs to be replaced, along with the filter.

If the ATF is pink or milky in appearance, it indicates that engine coolant has leaked into the transmission. This happens when there is a leak in the transmission cooler located in the radiator. Part 6 discusses transmission cooler operation and repair.

Bubbles appearing on the dipstick indicate that air has entered the transmission. If the fluid level is correct, this is usually caused by a high-pressure leak. Oil, under pressure, is sprayed into the oil in the reservoir. High-pressure oil aerates, or combines with air, to cause bubbles.

2.5 REMOVING THE OIL PAN

Most modern automatic transmissions do not have a drain plug. To drain the transmission, the oil pan must be removed.

SAFETY CAUTION: Transmission fluid should be drained when the ATF is warm. Contaminants, such as varnish or sludge, flow out more easily when the fluid is warm. Be careful when draining the ATF, because it may still be hot enough to cause serious burns.

Draining the ATF can be very sloppy when not done properly. First, place a large oil catchpan under the oil pan. Use the proper socket wrench to loosen the bolts at a front corner of the oil pan (Figure 2-12). Then loosen one or two bolts on each side of the corner. The oil pan will tilt forward, allowing the fluid to begin spilling out from one corner of the pan and in one continuous flow. When the fluid is barely dripping from the oil pan, support the pan, remove the remaining bolts, and lower the oil pan. The transmission will continue to drip ATF, so leave the oil catchpan in place.

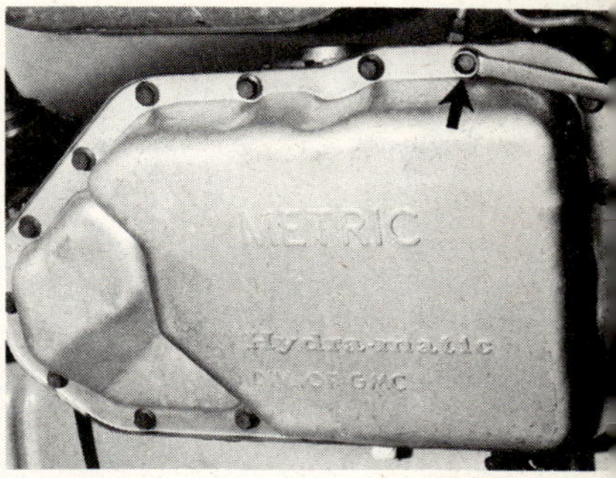

Figure 2-12. To drain an automatic transmission without a drain plug, begin by loosening the oil pan bolts at one corner. Loosen the bolts on each side away from the corner so that the pan will tilt down for draining.

Place the oil pan on a workbench. Inspect the inside of the oil pan for particles that may indicate damage. Some particles will always appear, which is caused by normal parts wear. If the particles are large and cover a large portion of the oil pan, it indicates a transmission problem.

If a small plastic plug is found in the bottom of the oil pan, this is not unusual. Some manufacturers place a plug in the dipstick hole to seal the transmission during shipment of a new vehicle. When the dipstick is inserted, it pushes this plug into the oil pan.

On transmissions with drain plugs, remove the drain plug with the appropriate socket or wrench. See Figure 2-13. Then remove the oil pan.

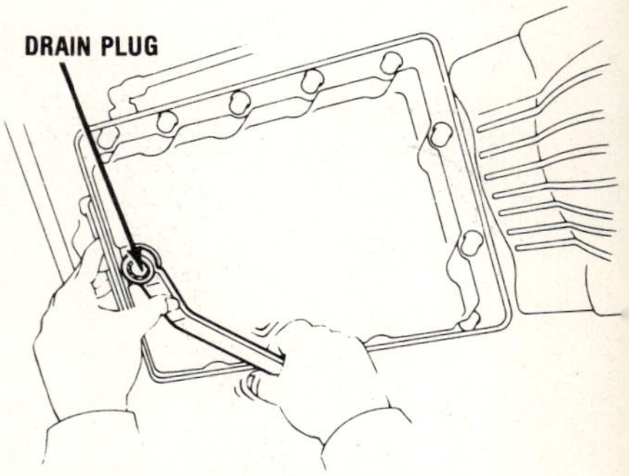

Figure 2-13. Some automatic transmissions have a drain plug in the oil pan to allow easier draining. (Toyota Motor Sales U.S.A., Inc.)

Clean the oil pan thoroughly with solvent. Remove any gasket material from the flange area with a putty knife or wide-blade screwdriver. Blow-dry the oil pan with low-pressure air. Also, carefully remove gasket material from the aluminum transmission case, taking care not to scratch the gasket-sealing surface.

CAUTION: ***Never* use a shop towel to wipe out the oil pan. Lint from the towel can lodge inside the transmission and cause improper operation.**

2.6 DRAINING THE TORQUE CONVERTER

All of the ATF does not drain out when the transmission oil pan is removed. Some ATF remains in the torque converter. In some vehicles, a drain plug is used to empty the torque converter. Most vehicles, however, have no drain plug. Check the proper service manual for the correct procedure to drain the torque converter.

If a drain plug is used, an access plate is usually located at the bottom of the converter housing. See Figure 2-14. Remove the access cover bolts, or screws, with the appropriate socket and wrench. Chances are, the drain plug will not be visible. Be sure the car is in park (P) and the emergency brake is applied. Have a helper inside of the car turn the starter switch in short bursts until the drain plug is in position to be removed.

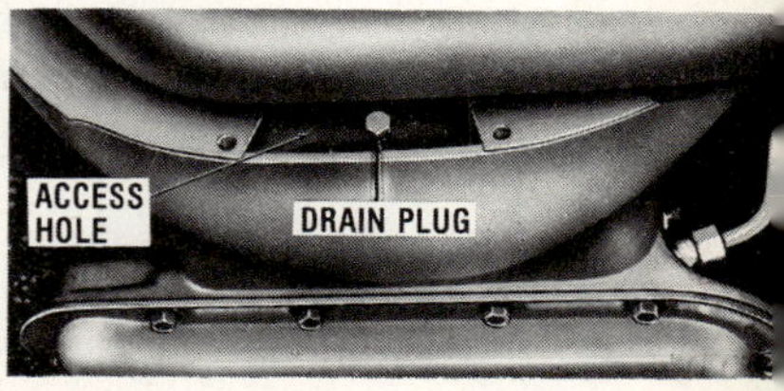

Figure 2-14. By removing a few bolts and an access cover, the torque converter drain plug can be reached for draining. (Chrysler Corporation)

Place an oil catchpan under the torque converter drain plug. Loosen and remove the drain plug. When the fluid has drained out, replace the drain plug and tighten it with a torque wrench to the manufacturer's specifications.

If no drain plug is furnished, and the ATF is badly contaminated, a different procedure must be used to replace the old fluid. Drain and refill the transmission, then drive five to ten miles. Repeat this procedure several times. This mixes the old and new ATF in the transmission and torque converter. Each time the transmission is drained, more impurities are removed. If the fluid still contains large amounts of impurities after two or three drainings, the torque converter must be replaced. Replacing the torque converter is covered in Unit 5.

2.7 FILTER AND SCREEN SERVICE

Transmission fluid must be filtered to remove particles from the fluid. Otherwise, damage will result. To remove harmful particles, either a disposable filter or reusable screen is used. The filter must be replaced and the screen cleaned or replaced at regular service intervals, according to the manufacturer's recommendations.

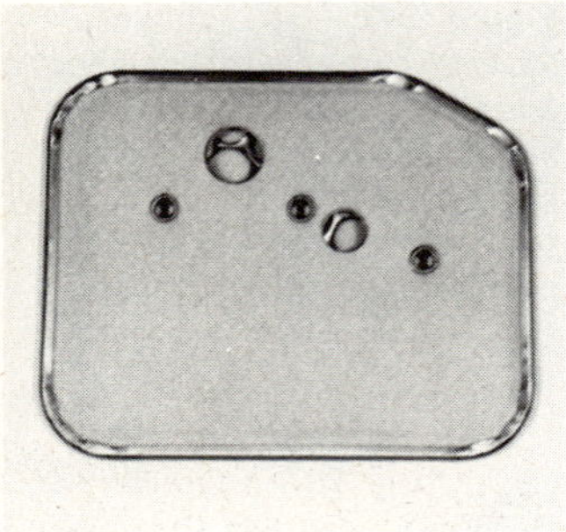

Figure 2-15. Most automatic transmissions use a disposable and replaceable filter to remove impurities from the circulating automatic transmission fluid.

The filter or screen can be any shape or size, depending on the type of automatic transmission. The filter or screen is located just beneath the valve body and just before the fluid can enter the oil pump. In this way, clean fluid enters the system.

Sometimes more than one filter or screen are used. Smaller filters are often found around the oil pump and valve body.

Filters must be replaced. They are made of paper or fabric. A filter may be held in place by screws, clips, or bolts. An O-ring, or circular rubber gasket, is used to ensure a leakproof seal between the filter and valve body, transmission case, or pickup tube. A pickup tube is a tube at the end of the oil pump through which clean ATF is drawn up by suction. A typical replaceable filter assembly is shown in Figure 2-15.

Remove the fasteners that hold the filter in place. Remove the old filter and O-ring, and discard them. A filter kit will include new O-rings and gaskets. Always replace the O-ring and gasket with new ones.

Install the new O-ring, filter, and any gaskets. Some filters and screens will be marked ''front'' to avoid improper installation. Be sure all parts fit together snugly, otherwise dirty oil will bypass the filter and enter the system. Some manufacturers require that the filter fasteners be tightened with a torque wrench to certain specifications. Check the proper service manual.

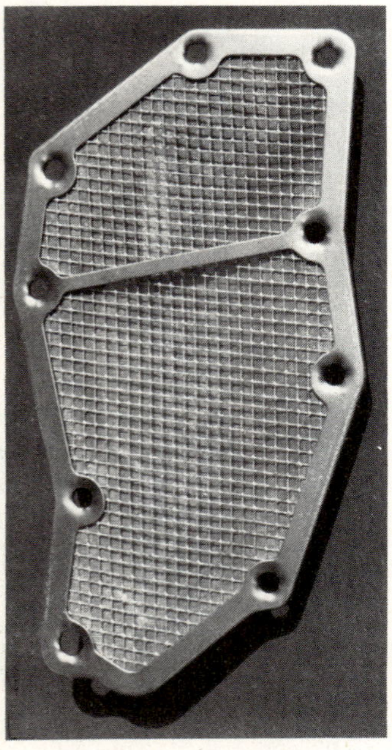

Figure 2-16. A reusable screen is used on some automatic transmissions as a filter.

Screens are removed in the same way as filters. A screen-type filter is shown in Figure 2-16. Use caution when removing screens. Some makes of transmissions—particularly the Ford C4—have small valves and springs between the screen and valve body. Do *not* lose these parts, because they must be reused.

Clean the screen with fresh solvent and a stiff brush. Use low-pressure compressed air to dry the screen.

CAUTION: Do *not* wipe the screen with a cloth. Any lint that may stick to the screen will enter the valve body, resulting in poor transmission operation.

If varnish cannot be removed from the screen during cleaning, the screen must be replaced. Also, if the screen is punctured or bent, replace it.

Install the O-ring and gasket, then push the screen *straight up* into place until it fits snugly in position. Torque the screws or bolts to the proper specifications.

2.8 INSTALLING THE OIL PAN

Remove any traces of the old gasket on the transmission and oil pan by scraping with a gasket scraper or putty knife.

CAUTION: Use care when scraping gasket material off the bottom of the transmission. An aluminum case can be gouged easily by the scraper, resulting in an oil leak. Scrape the transmission pan mating surface only enough to remove old gasket material.

Clean the pan completely with fresh solvent and a stiff brush. Use low-pressure compressed air to dry the oil pan. Do *not* wipe it dry with a cloth. Lint from the cloth will cause the transmission to operate poorly.

Before the oil pan is installed, check the bolt holes. The bolt holes in the pan will be dimpled, or bulged upward where the pan mates with the transmission case. Dimples have sharp edges that will damage the gasket and cause a leak. Dimpling is caused when the bolts are tightened against the gasket. The bolt hole dimples must be removed before the oil pan can be installed.

To remove a dimple, place a flat piece of wood or soft-faced mallet under the bolt hole. Tap gently on the mating side of the oil pan flange with a ball-peen hammer until the dimple is flat. Repeat the procedure for each bolt hole.

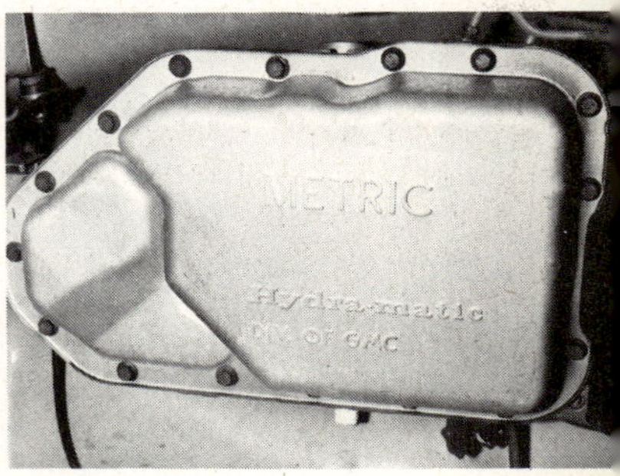

Figure 2-17. Tighten the oil pan bolts in a crisscross pattern to avoid distorting (bending) the pan.

Place a new gasket on the oil pan mounting flange. Do *not* use a gasket sealer or gasket cement. Place the oil pan against the bottom of the transmission case. Be sure the gasket is not pinched and that the bolt holes line up. Install and finger-tighten the attaching bolts. Some oil pan designs use washers with the attaching bolts. Use a torque wrench to tighten the bolts to manufacturer's specifications.

CAUTION: The oil pan must be tightened in a crisscross pattern, as shown in Figure 2-17. Crisscross tightening ensures that the gasket will not be pinched or broken. Otherwise, leaking may develop.

If the oil pan has a drain plug, install the drain plug. Torque the drain plug to manufacturer's specifications.

2.9 REFILLING THE TRANSMISSION

The amount of fluid that is to be added depends on the type of transmission being serviced and whether the torque converter was drained. An automatic transmission will hold between 5 and 8 quarts of ATF. If the torque converter was *not* drained, the transmission may only require 2 to 3 quarts. Check the proper service manual. *Never* overfill the transmission.

Figure 2-18. After servicing the automatic transmission, fill it to the proper level, as described in the text.

Fluid is added through the dipstick tube, using a funnel or special filler tube. See Figure 2-18. Add only the amount of fluid the service manual recommends.

Place the shift lever in park (P) and set the emergency brake. Start the engine and let it idle a few minutes. Hold the brake pedal down with your foot while moving the shift lever into each position. Stop in each gear position to allow the transmission to shift. Place the shift lever in park (P) or neutral (N), depending on the manufacturer's recommendation.

Check the fluid level, as discussed in 2.3. If necessary, add ATF until the fluid level reaches the "add" mark. Do *not* go above the "add" mark.

Warm the engine and transmission to normal operating temperature. Check the fluid level again. The fluid level should be at the "full" mark. If not, add just enough ATF until the dipstick shows a "full" level.

Check the oil pan, drain plug, and other transmission areas for leaks.

Road-test the vehicle for a few miles to make sure the automatic transmission is operating properly.

SAFETY CAUTION: Drive in a safe manner. Jackrabbit starts, hard cornering, or sudden braking should be avoided. Driving conditions other than normal can worsen a problem, or create a new one. Carelessness during the test drive can result in an accident. Remember, you are responsible for someone else's vehicle. Drive only in a manner that is required to check or diagnose proper transmission operation.

NAME ______________________ SECTION ____________ DATE ____________

Job Sheet 2-2

Automatic Transmission Fluid and Filter Change

Before you begin:

Read Topics 2.3 through 2.9.

Vehicle Make ____________________ Model ____________ Year ____________

Time Started ____________________ Time Finished ____________________

Job Completion Time ____________________ Flat Rate Time ____________________

Tools Required

- Hoist
- Safety glasses
- Droplight
- Rag
- Catchpan
- Socket wrenches, extensions, and handle
- Adjustable open-end wrench
- Screwdriver
- Torque wrench
- Soft-faced mallet
- Ball-peen hammer
- Transmission funnel

References

Manufacturer's service manual ____________________________________

Repair manual ____________________________________

Service bulletin ____________________________________

Procedure

1. Check the automatic transmission fluid level. Write in the level (high, low, correct) in the space provided. . . . . ____________________
2. Check the automatic transmission fluid condition, and make a note of the condition in the Job Completion Report.

Instructor Check: ____________________

3. Raise the vehicle on a hoist.
4. Place a catchpan under the transmission pan.
5. Loosen and remove the drainplug, and drain the transmission fluid.*
6. Loosen the transmission pan bolts and drain the transmission fluid.
7. Remove the transmission pan bolts and remove the oil pan.

8. Drain the torque converter.*
9. Remove the transmission fluid filter or screen.*
10. Replace the filter or clean the screen.*

CAUTION: Do *not* wipe a screen with a cloth. Any lint that may stick to the screen will enter the valve body, resulting in poor transmission operation.

11. Inspect the inside of the oil pan for excessive residue. Make a note in the Job Completion Report of the amount and condition of any residue found.
12. Clean the inside and outside of the oil pan, and remove any traces of the old pan gasket.
13. Clean the mounting surface on the bottom of the transmission and remove any traces of the old pan gasket.

CAUTION: Use care when scraping gasket material off the bottom of the transmission. An aluminum case can be gouged easily by the scraper, resulting in an oil leak. Scrape the transmission pan mating surface only enough to remove old gasket material.

14. Place a new pan gasket on the oil pan.
15. Install the oil pan, and finger-tighten the bolts.
16. Torque the pan bolts in a crisscross pattern to the manufacturer's specification.

Write the correct torque in the space provided. ______________________________

CAUTION: The oil pan must be tightened in a crisscross pattern, as shown in Figure 2-17. Crisscross tightening ensures that the gasket will not be pinched or broken. Otherwise, leaking will develop.

17. Install the drain plug, and torque it to the manufacturer's specification.

Write the correct torque in the space provided.* ______________________________

18. Lower the vehicle and refill the transmission with the proper amount and type of transmission fluid.
19. Start the vehicle, and move the shift lever through all gear positions.
20. Check the automatic transmission fluid level, and correct if necessary.
21. Ask your instructor's permission to conduct a road test. If approved, road test the vehicle and check for proper transmission operation in all gear positions.

SAFETY CAUTION: Drive in a safe manner. Jackrabbit starts, hard cornering, or sudden braking should be avoided. Driving conditions other than normal can worsen a problem, or create a new one. Carelessness during the test drive can result in an accident. Remember, you are responsible for someone else's vehicle. Drive only in a manner that is required to check or diagnose proper transmission operation.

Instructor Check: ____________________

22. Check the condition of the transmission fluid after the road test, and make a note of the condition in the Job Completion Report.

*See the text for alternate procedures.

NAME ______________________ SECTION ____________ DATE ____________

Job Completion Report

1. What problems or conditions were indicated by the condition of the transmission fluid before service? ______________________________

   ____

2. Did the transmission serviced have a replaceable filter or a screen? ____________

   ____

3. Was there a drain plug on the torque converter? ____________

4. Describe the residue, if any, found inside the transmission pan when it was removed.

   ____

5. What problems, if any, were indicated by the condition of the fluid after the road test?

   ____

   ____

INSTRUCTOR SIGNOFF ______________________ **DATE** ____________

NAME ______________________ SECTION ____________ DATE ____________

Certification Practice

Test 2

DIRECTIONS:

The following questions are similar to those used on mechanic certification tests. Answer each question by circling the letter of the correct choice.

1. All of the following areas of an automatic transmission should be checked for leaks EXCEPT
   A. the slip joint.
   B. the transmission housing.
   C. around bolt-in or screw-in control devices.
   D. the shift linkage.

2. Mechanic A says that automatic transmissions use cable shift linkage.
   Mechanic B says that automatic transmissions use rod-and-lever shift linkage.
   Who is correct?
   A. A only B. B only C. Both A and B D. Neither A nor B

3. The picture shows
   A. a crack in a transmission housing.
   B. leakage from a slip joint.
   C. oil viscosity.
   D. casting porosity.

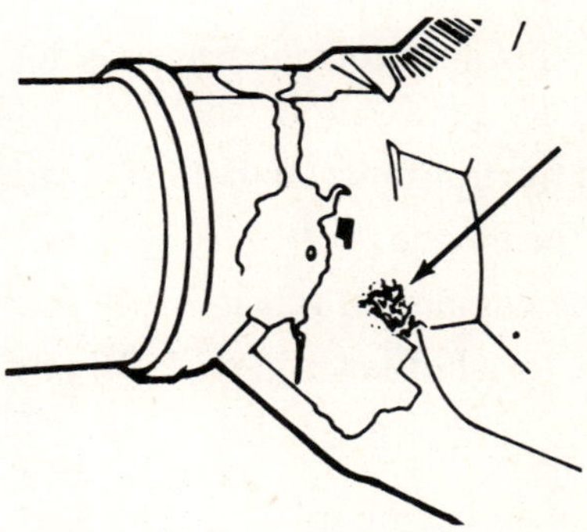

Question 3.
(Ford Motor Company)

4. All of the following conditions should apply when checking A/T fluid level EXCEPT
   A. engine and transmission at normal operating temperature.
   B. engine stopped.
   C. engine running.
   D. shift lever in park (P) or neutral (N).

5. Which of the following indicates engine coolant mixed with A/T fluid?
   A. Dark fluid
   B. Burnt smell
   C. Milky fluid
   D. Metallic particles in fluid

6. Which of the following statements is correct?
   I. All automatic transmissions must be drained by removing the oil pan.
   II. All automatic transmissions have drain plugs.
   A. I only B. II only C. Both I and II D. Neither I nor II

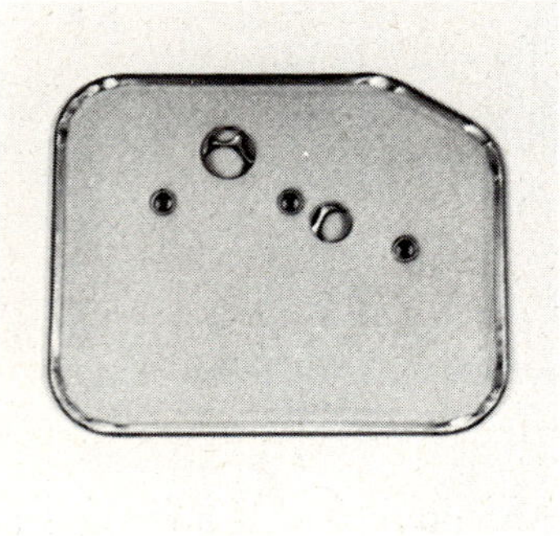

Question 7.

7. Mechanic A says the filtering unit shown in the picture must be replaced.
   Mechanic B says it can be cleaned and reused.
   Who is correct?
   A. A only B. B only C. Both A and B D. Neither A nor B

8. An automatic transmission drain pan is being replaced. How should the bolts be tightened?
   A. From front to back
   B. Left side (driver's side) first, then right side
   C. In any order
   D. In a crisscross pattern

9. After refilling an automatic transmission that has been drained, all of the following should be done EXCEPT
   A. Refill the transmission before starting the engine.
   B. Start the engine.
   C. Let the engine and transmission warm up to normal operating temperature.
   D. Leave the shift lever in park (P) or neutral (N), then check the fluid level.

10. Dimples are
    A. defects in a cast part.
    B. bulges around bolt holes in a stamped part.
    C. bends in a metal rod.
    D. kinks in a metal cable.

3 Shift Linkage and Adjustment

YOUR LEARNING JOB

The shift linkage for the transmission should be checked periodically and adjusted when necessary. The shift linkage is the driver's main control over the automatic transmission. Shift linkage adjustment is important and must be performed whenever necessary, otherwise the transmission may be in gear when it's not supposed to be, or may be in the wrong gear.

JOB COMPETENCY OBJECTIVES

This is a knowledge-building and skill-building unit. When you have completed your assignments and exercises in this unit, you should be able to:

- ☐ Describe how floor-shift and column-shift linkages operate.
- ☐ Explain the function of kickdown linkage.
- ☐ Describe the functions of a neutral start switch.
- ☐ Adjust the linkage systems.

3.1 THE GEARSHIFT SELECTOR

The gearshift selector lever for an automatic transmission may be located on the steering column or on the floor. An indicator near the lever tells whether the transmission is in:

- Park (P)
- Reverse (R)
- Neutral (N)
- Drive (D)
- Second gear (2)
- First gear (marked either 1 or L).

P R N D 2 L

P R N O D L

P R N D 3 2 1

P R N D_4 D_3 L

Figure 3-1. Variations on selector positions in the shift quadrant.

This series of letters, P-R-N-D-2-L, is called the shift quadrant.

When the gearshift selector is placed in P, the transmission output shaft is locked, and the driveline cannot turn. The vehicle cannot, or should not, move.

When the selector is placed in R, the vehicle will move backwards.

In the N position, the transmission is in neutral. Neutral means that no power is transmitted from the engine to the driveline. However, the driveline is free to turn, so the vehicle may roll.

The D position on the quadrant allows the transmission to shift automatically through the forward gear positions. The transmission will shift up or down through first, second, and high gears by itself, depending on speed and load requirements.

If 2 is selected, the transmission will not upshift past second gear. In most automatic transmissions, first or second gear will be selected, depending on speed and load requirements.

When placed in 1 or L, the transmission will remain in first gear.

Not all automatic transmission shift quadrants use the P-R-N-D-2-L designations. Some common variations of the shift quadrant are shown in Figure 3-1.

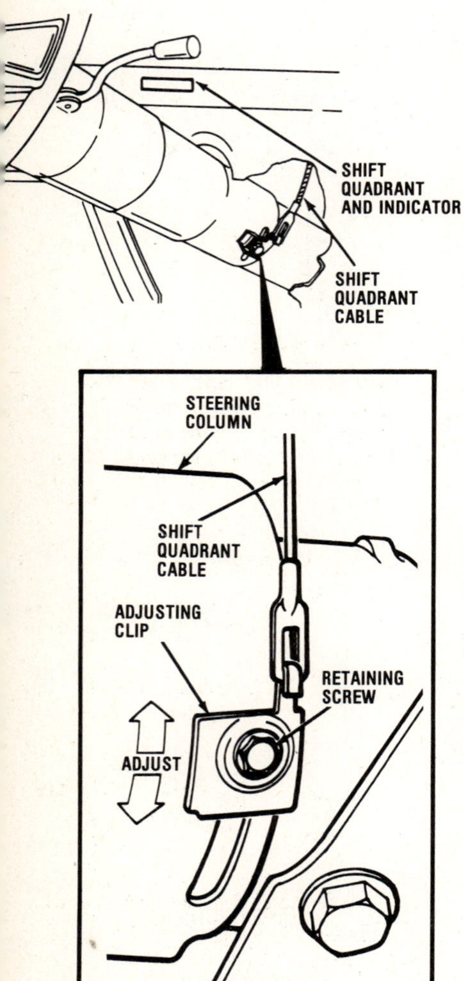

Figure 3-2. One type of shift quadrant linkage. (American Motors Corporation)

3.2 SHIFT LINKAGE SYSTEMS

The shift linkage system for an automatic transmission is more extensive than the linkage system used for manual transmissions. Separate linkage systems—all interrelated—are used to control the automatic transmission. The linkage systems are:

- Quadrant linkage
- Gearshift linkage
- Kickdown linkage.

Though all linkage systems are similar to those discussed below, many variations exist. Always check the proper service manual to determine which linkage systems are used for a specific transmission.

3.3 QUADRANT LINKAGE

The indicator on the quadrant may have a separate linkage-adjusting system. Quadrant linkage is used when the quadrant is located in the dashboard, regardless of where the shift lever is located.

Quadrant linkage consists of an adjusting screw, a retainer plate, and a short cable. The cable is connected to the quadrant indicator. See Figure 3-2 for quadrant adjustment linkage.

3.4 GEARSHIFT LINKAGE

The gearshift lever may be located on the steering column or on the floor of the vehicle. Either a system of rods and levers or a cable may be used to control the shifting process. The gearshift linkage connects directly to a lever arm

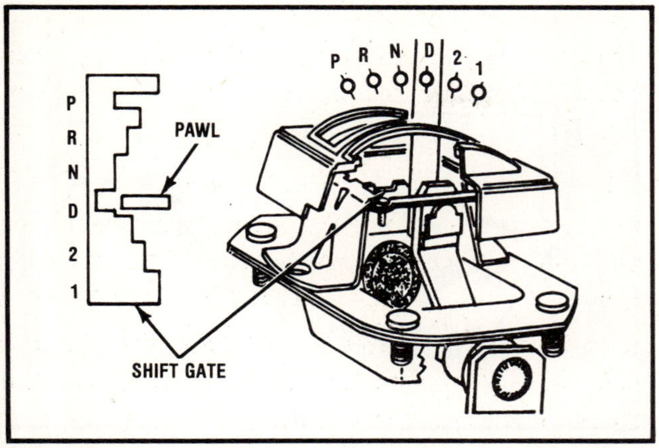

Figure 3-3. The pawl can be moved to various stops in the shift gate. (Ford Motor Company)

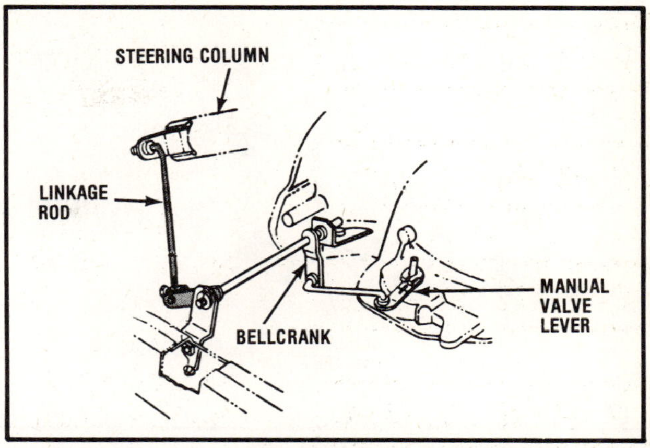

Figure 3-4. Rod-and-lever linkage is used with some column-shift automatic transmissions. (Chrysler Corporation)

on the transmission. The lever arm acts on a manual valve in the valve body. A manual valve is a steel shaft specially designed to direct fluid flow in the valve body according to which quadrant position is selected. As the shift lever is moved, the manual valve slides in and out of a hole, or bore. The manual valve is discussed in more detail in Unit 10.

Another part of the gearshift linkage is in the quadrant. When the gearshift lever is moved, it has to be raised or lowered to engage a position on the quadrant. This rise or drop is caused by a gate and a pawl in the quadrant assembly. See Figure 3-3. A gate is a guide containing a series of indentations, or notches, used to hold or guide another part. A pawl is a pivoted projecting part that engages with the notches in the gate.

The pawl is the first link in the gearshift linkage.

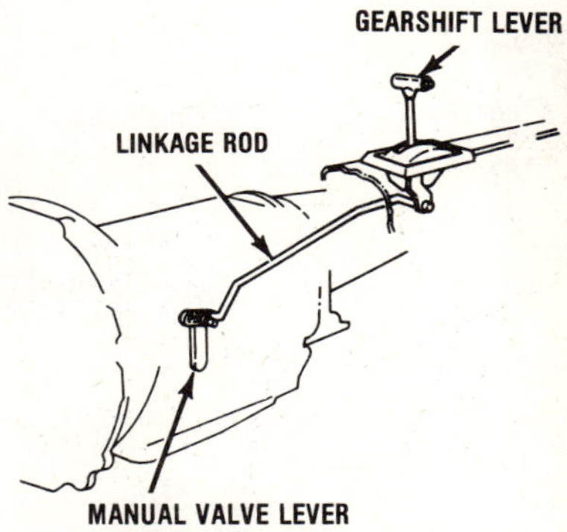

Figure 3-5. Rod-and-lever linkage is also used for some floor-shift automatic transmissions. (Ford Motor Company)

Column-Shift Linkage

Gearshift linkage for steering column shifters is usually operated by a series of rods and levers. However, several models of cars do use a cable linkage.

The pawl in the quadrant connects to a tube that runs through the steering column. An arm welded to the tube is connected to a cable, shown in Figure 3-4. This cable moves the lever arm connected to the manual valve.

Floor-Shift Linkage

Gearshift linkage for floor shifters can be either a rod-and-lever system or a cable system.

The rod-and-lever assembly has fewer parts when the shift lever is mounted to the floor. See Figure 3-5.

Cables are used for shift linkage because they can be turned and bent to get around tight places. Cables are most often found in smaller vehicles. The cable connects directly from the shift lever to the manual-valve arm, or lever, on the transmission case. See Figure 3-6.

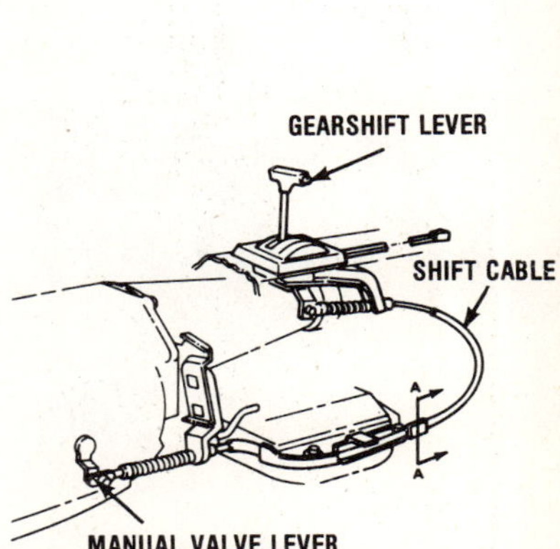

Figure 3-6. Cable linkage can also be used for floor-shift automatic transmissions. (Ford Motor Company)

3.5 KICKDOWN LINKAGE

When the accelerator pedal is depressed hard, the automatic transmission will shift into a lower gear. This is accomplished by using some form of throttle, or kickdown, linkage. The linkage may be operated by a rod-and-lever system,

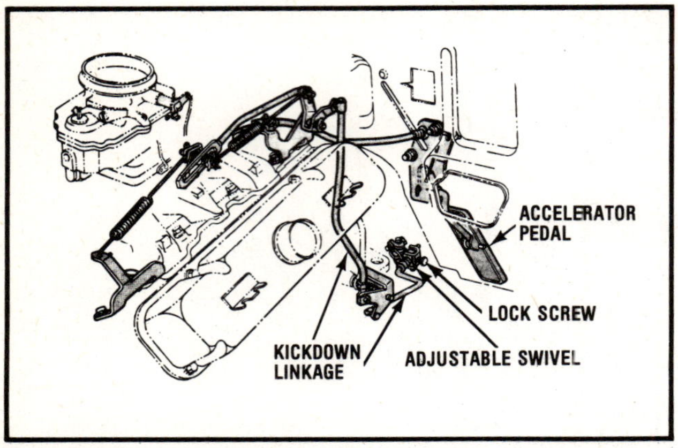

Figure 3-7A. Mechanical kickdown linkage is attached to the accelerator linkage to produce a downshift when the accelerator is pressed down hard. (Chrysler Corporation)

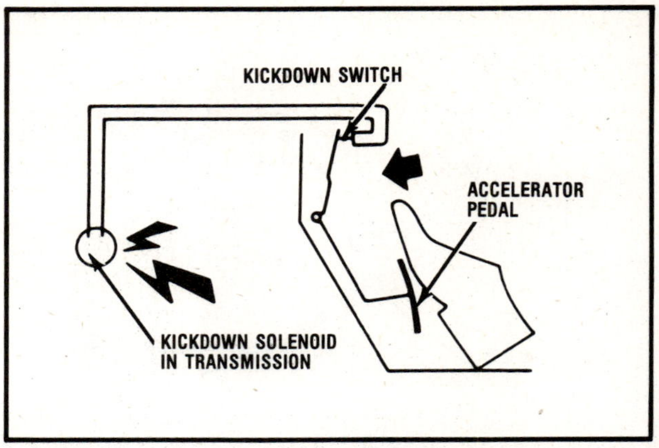

Figure 3-7B. A downshift can also be produced by an electrical solenoid in the automatic transmission, connected to a switch operated by the accelerator pedal. (Nissan Motor Corporation)

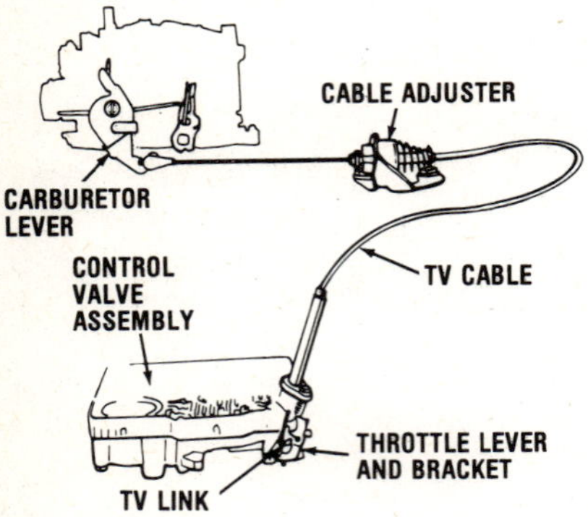

Figure 3-8. Some manufacturers use a throttle valve (TV) cable to produce a downshift. The throttle valve also controls pressure in the automatic transmission hydraulic system. (Chevrolet Motor Division-GMC)

a cable, or electrically. Most kickdown linkages use more than one of these systems to create the downshift. See Figure 3-7. All kickdown systems are activated by pressure on the accelerator, or throttle, pedal.

When the accelerator pedal is depressed, a cable attached to the pedal is pulled. The other end of the cable is attached to linkage on one side of the carburetor. When the throttle is opened to a predetermined setting, the kickdown rod acts on the downshift, or throttle, valve in the transmission. The throttle valve increases fluid pressure in the transmission to create a downshift.

A variation of kickdown linkage, called the TV (throttle valve) cable, is used on newer vehicles. In this system, a cable runs from the carburetor linkage to the transmission. See Figure 3-8. The TV cable transfers carburetor throttle-plate movement to the transmission. In addition to controlling downshifts, the TV cable controls fluid pressure for proper transmission operation.

An electrical kickdown control operates off a switch called a solenoid, or kickdown solenoid. The solenoid may be mounted on the outside of the transmission case or inside, on the valve body.

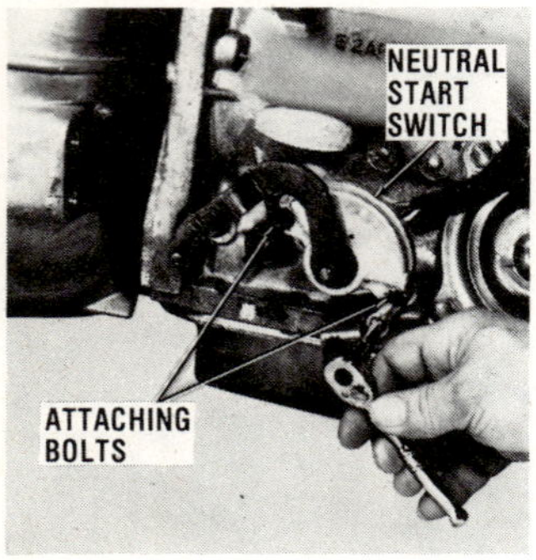

Figure 3-9. The neutral start switch prevents the vehicle from starting if the transmission is in a gear position. (Ford Motor Company)

3.6 NEUTRAL START SWITCH

The neutral start switch is a safety switch. The neutral start switch can be located on the transmission or on the steering column near the floorboard. On vehicles with a floor shift, it may be found at the base of the shift lever. A neutral start switch is shown in Figure 3-9.

For safety, the neutral start switch prevents the engine from starting in any gear position. The switch allows current from the battery to activate the starter only when the shift lever is in P or N. The engine will not start in a gear position if the switch is operating properly and is correctly adjusted.

CHECKPOINT

1. **The gearshift lever for an automatic transmission may be located on the ____________________ or on the ____________________.**
2. **This series of letters, P-R-N-D-2-L, is called the shift ____________________.**
3. **Either a ____________________ and ____________________ system or a ____________________ may be used to control the shifting process.**
4. **The purpose of the ____________________ valve is to direct fluid, according to which quadrant position is selected.**
5. **A ____________________ is a hollow part with a series of indentations, or detents, used to guide another part through a specified motion.**
6. **All kickdown systems are activated by pressure on the ____________________ pedal.**
7. **The neutral start switch prevents the vehicle from starting in any ____________________ ____________________.**

3.7 ADJUSTING LINKAGE SYSTEMS

Linkage that is out of adjustment, or misadjusted, can lead to serious safety problems. The quadrant may indicate that the transmission is in P, when it may actually be in gear or in neutral. This can result in serious personal injury or property damage. Out-of-adjustment linkage can also cause erratic transmission operation.

Adjusting the linkage system is performed only when necessary. Adjustment is not usually necessary unless the linkage has been damaged. All linkage systems must always be adjusted whenever any linkage has been removed, such as during an overhaul.

To determine whether the gearshift linkage is out of adjustment, a few simple checks can be made.

For a vehicle with a column shifter, place the gearshift lever in the drive position. Let go of the lever so that it drops into position against the shift gate. Try moving the shift lever gently toward the next lower gear, without raising the lever. The lever should not move, or should move only slightly. If there is more than 1/2 inch [12.7mm] of movement at the lever, the linkage needs adjusting.

For a vehicle with a floor shifter, place the gearshift lever into drive. Release the button. Try moving the lever gently toward the next lower gear.

More than 1/2 inch [12.7mm] of movement at the shift lever indicates that the linkage needs adjusting. Three linkage adjustments can be made on automatic transmissions:

- Quadrant adjustment
- Gearshift linkage adjustment
- Kickdown linkage adjustment.

3.8 QUADRANT ADJUSTMENT

A separate quadrant adjustment can be made on vehicles with an in-dashboard quadrant. Refer to Figure 3-2, and use the following procedure as a guideline to quadrant adjustment. Always refer to the proper service manual, because quadrant adjustment procedures vary.

CAUTION: The quadrant adjustment is not a gearshift linkage adjustment. This adjustment is used *only* to center the pointer on the quadrant positions. Gearshift linkage adjustment is covered in 3.9.

1. Remove the cover panel on the dashboard that encloses the steering column. The cover is usually held in place by two screws.

2. Unlock the steering column with the ignition key, and place the gearshift lever in N.

3. Loosen the adjusting screw, nut, or bolt until the adjusting clip can be moved easily.

4. Move the clip up or down to center the pointer on N.

5. Tighten the adjusting screw.

6. Move the gearshift through each position on the quadrant to check the pointer at each position. Readjust, if necessary.

7. Replace the steering column cover.

NAME ______________________________ SECTION ______________ DATE ______________

Job Sheet 3-1

Quadrant Linkage Adjustment

Before you begin:

Read Topics 3.1 through 3.8.

Vehicle Make ______________________ Model ____________ Year ____________

Time Started ______________________ Time Finished ______________________

Job Completion Time ________________ Flat Rate Time ______________________

Tools Required

- Screwdrivers
- Open-end or combination wrenches

References

Manufacturer's service manual ____

Repair manual ____

Service bulletin ____

Procedure

CAUTION: The quadrant adjustment is not a gearshift linkage adjustment. This adjustment is used *only* to center the pointer on the quadrant positions.

1. Remove the cover panel on the dashboard that encloses the steering column.
2. Unlock the steering column with the ignition key, and place the gearshift lever in N.
3. Loosen the adjusting screw, nut, or bolt until the adjusting clip can be moved easily.
4. Move the clip up or down to center the pointer on N.
5. Tighten the adjusting screw.
6. Move the gearshift through each position on the quadrant to check the pointer at each position. Readjust, if necessary.

Instructor Check: ____________________

Job Completion Report

1. What difficulties, if any, did you encounter in adjusting the quadrant linkage?

____

INSTRUCTOR SIGNOFF ________________________ **DATE** ______________

3.9 GEARSHIFT LINKAGE ADJUSTMENTS

Rod-and-lever, cable, and electrical gearshift linkage adjustments can be made. Numerous methods are used to adjust the gearshift linkage. However, only general adjustment procedures are discussed here. Refer to the proper service manual for specific procedures.

Rod-and-Lever Adjustment

Slotted rod ends and adjustable swivels are the most commonly used parts to adjust rod-and-lever linkage systems. Use Figures 3-10 (slotted-end) and 3-11 (adjustable swivel) to adjust column shift linkages. Use Figures 3-12 (slotted-end) and 3-13 (adjustable swivel) to adjust floor shift linkages. The following procedure applies to both column shift and floor shift linkages:

NOTE: Plastic grommets may be used between the rods and levers. A grommet is a protective part that fits through an opening to cushion a part against the opening. If the linkage rods and levers are to be disassembled, the grommets must be replaced with new ones. Use a screwdriver to separate the rod from the lever. Cut away the old grommet. Use pliers to snap in a new grommet.

1. Place the gearshift lever in drive or park, depending on the manufacturer's recommendation. If the lever is in drive, be sure the pawl is firmly against the gate stop. An 8-pound [3.6 kg] weight can be hung from a column shift lever to ensure that it remains against the drive stop.

2. Raise the vehicle on a hoist to a comfortable working height.

3. Use safety glasses. Loosen the adjusting nut at the slotted rod end. On an adjustable swivel, loosen the lockscrew.

4. Place the manual-valve lever at the transmission case into drive or park. The position of the manual-valve lever depends on the position in which the gearshift lever was placed. Check the proper service manual to identify the drive or park position detent at the manual-valve lever.

NOTE: Make sure the gearshift lever has not moved from the drive or park position.

5. Tighten the adjusting nut or lockscrew to the manufacturer's torque specifications.

6. Lower the vehicle.

7. Move the gearshift lever through all detent positions to make sure it operates properly.

SAFETY CAUTION: If the linkage is not correctly adjusted, the vehicle may start in a gear position, and jump, or move, forward, causing property damage or personal injury. Before attempting to start the vehicle, make sure no one is in front of or behind the vehicle. Block the vehicle's wheels, apply the emergency brake, and press the brake pedal down while attempting to start.

8. Move the gearshift lever into a gear position, and attempt to start the vehicle.

NOTE: If the vehicle starts in any gear position, the linkage or neutral start switch is not adjusted properly.

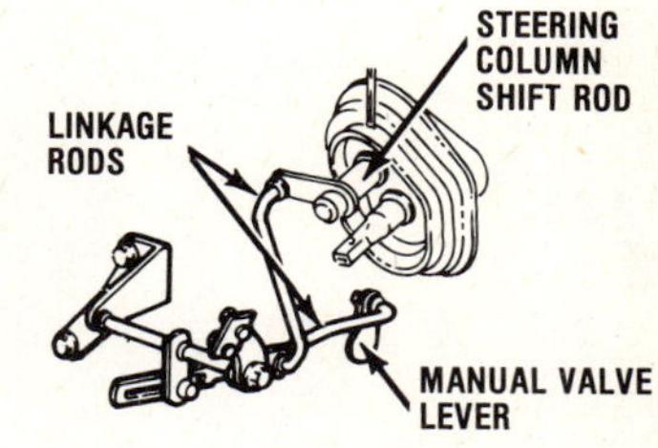

Figure 3-10. Rod-and-lever column-shift linkage. (Ford Motor Company)

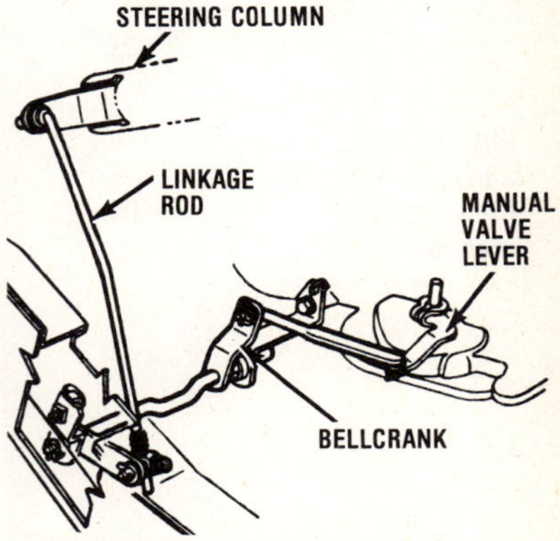

Figure 3-11. Note where adjustments are made on this rod-and-lever column-shift linkage. (Chrysler Corporation)

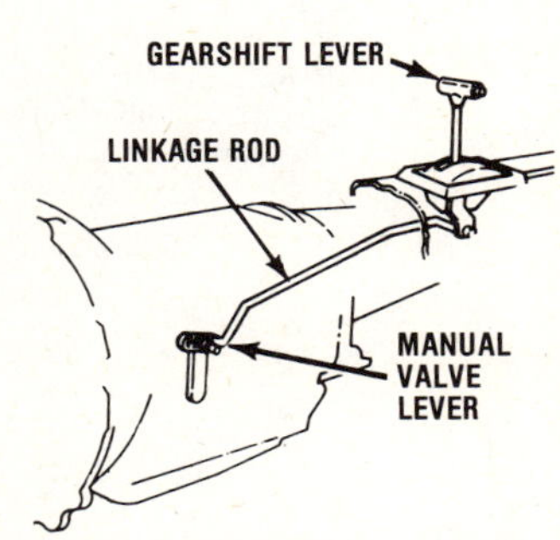

Figure 3-12. Adjustments to this rod-and-lever floor-shift linkage are made by moving a slotted rod end on a fixed stud. (Ford Motor Company)

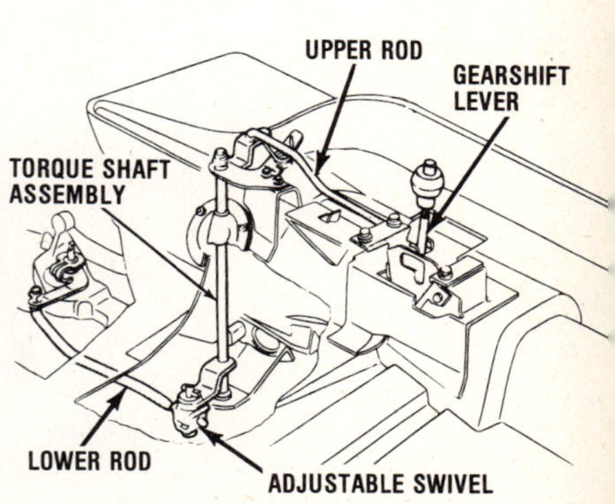

Figure 3-13. This floor-shift rod-and-lever linkage has an adjustable swivel for linkage adjustment. (Chrysler Corporation)

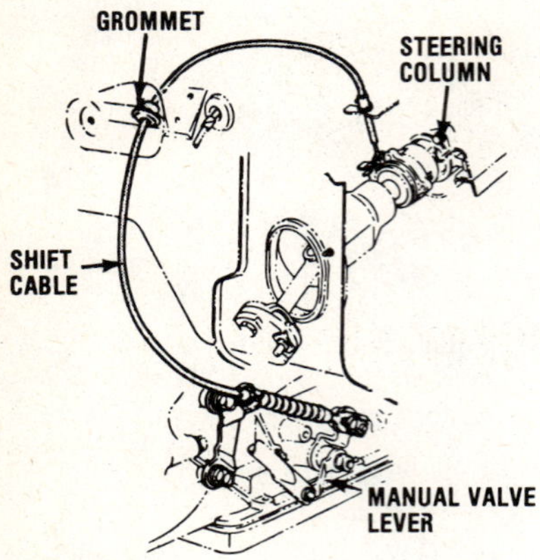

Figure 3-14. Adjustments to this floor-shift cable linkage are made at point A by moving the manual valve lever in the slotted cable end. (Ford Motor Company).

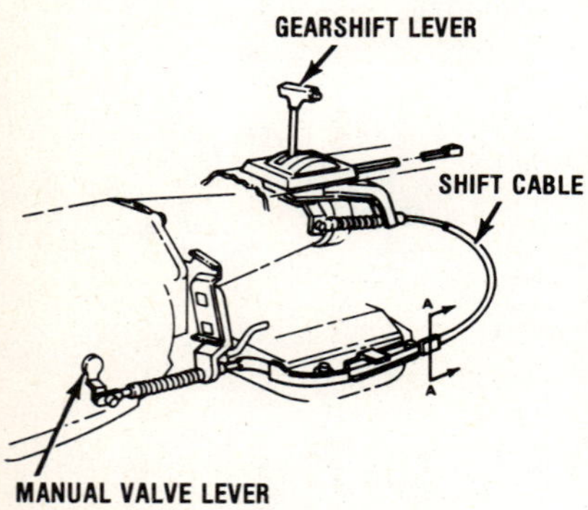

Figure 3-15. This floor-shift cable linkage has a slotted end at the transmission where adjustments are made. (Ford Motor Company)

9. If necessary, readjust the linkage.

10. Move the gearshift lever to the park or neutral position, and check to see that the vehicle starts properly.

Cable Adjustment

Use Figures 3-14 and 3-15 as guides to cable adjustment.

1. Place the gearshift lever in drive, or according to the manufacturer's recommendation. If the lever is in drive, be sure the pawl is firmly against the gate stop.

2. Raise the vehicle on a hoist to a comfortable working height.

3. Use safety glasses. Loosen the retaining nut at the slotted cable end of the manual-valve lever.

4. Place the manual-valve lever at the transmission case into the same detent position as the gearshift lever was placed. Check the proper service manual to locate the proper detent at the manual-valve lever.

NOTE: Make sure the gearshift lever has not moved from its original position.

5. Pull gently on the slotted cable end to remove any slack in the cable. Tighten the retaining nut to the manufacturer's torque specification.

6. Lower the vehicle.

7. Move the gearshift lever through all detent positions to make sure it operates properly.

SAFETY CAUTION: If the linkage is not correctly adjusted, the vehicle may start in a gear position and jump or move forward, causing property damage or personal injury. Before attempting to start the vehicle, make sure no one is in front of, or behind the vehicle. Block the vehicle's wheels, apply the emergency brake, and press the brake pedal down while attempting to start.

8. Move the gearshift lever into a gear position, and attempt to start the vehicle.

NOTE: If the vehicle starts in any gear position, the linkage or neutral start switch is not adjusted properly.

9. If necessary, readjust the linkage and repeat steps 7 and 8.

10. Move the gearshift lever to the park or neutral position, and check to see that the vehicle starts properly.

NAME ______________________ SECTION ____________ DATE ____________

Job Sheet 3-2

Automatic Transmission Gearshift Linkage Adjustment

Before you begin:

Read Topics 3.1 through 3.9.

Vehicle Make ____________________ Model ____________ Year ____________

Time Started ____________________ Time Finished ____________________

Job Completion Time ____________________ Flat Rate Time ____________________

Tools Required

- Hoist
- Safety glasses
- Screwdriver
- Pliers
- Open-end or combination wrenches
- Socket wrenches, handle, and extensions
- Torque wrench
- Wheel chocks

References

Manufacturer's service manual ____________________________

Repair manual ____________________________

Service bulletin ____________________________

Procedure: Check Linkage Adjustment

1. Place the gearshift lever in neutral.
2. Place the gearshift lever in the drive position.
3. Move the lever gently toward the next lower gear, without raising the lever (column shift) or depressing the release button (floor shift).
4. If the lever can be moved 1/2 inch [12.7mm] or more, the linkage needs adjusting.

In the space provided, write down how far the lever moves. . ____________

Procedure: Adjust Rod-and-Lever Gearshift Linkage

Use safety glasses when making this gearshift adjustment.

NOTE: Plastic grommets may be used between the rods and levers. A grommet fits into a hole to protect and cushion parts that extend through the hole. If the linkage rods and levers are to be disassembled, the grommets must be replaced with new ones. Use a screwdriver to separate the rod from the lever. Cut away the old grommet. Use pliers to snap in a new grommet.

1. Place the gearshift lever in drive or park, depending on the manufacturer's recommendation.

Write the correct gearshift position for this vehicle in the space provided.

. . . . ____________________

NOTE: If the lever is in drive, be sure the pawl is firmly against the gate stop. An 8-pound [3.6 kg] weight can be hung from a column shift lever to ensure that it remains against the drive stop.

2. Raise the vehicle on a hoist to a comfortable working height.
3. Use safety glasses. Loosen the adjusting nut at the slotted rod end (slotted rods); lockscrew (adjustable swivels), or slotted cable end (cable).
4. Place the manual-valve lever at the transmission case into drive or park, to correspond with the gearshift lever position in step 1, above.

Write the correct manual valve position for this vehicle in the space provided.

. . . . ____________________

NOTE: Make sure the gearshift lever has not moved from the drive or park position, whichever was set in step 1, above.

5. If adjusting a cable linkage, pull lightly on the cable to remove any slack.
6. Tighten the adjusting nut or lockscrew to the manufacturer's torque specifications.
7. Lower the vehicle.
8. Move the gearshift lever through all detent positions to make sure it operates properly.

SAFETY CAUTION: If the linkage is not correctly adjusted, the vehicle may start in a gear position, and jump, or move, forward, causing property damage or personal injury. Before attempting to start the vehicle, make sure no one is in front of, or behind the vehicle. Block the vehicle's wheels, apply the emergency brake, and press the brake pedal down while attempting to start.

9. Move the gearshift lever into a gear position, and attempt to start the vehicle.

NOTE: If the vehicle starts in any gear position, the linkage or neutral start switch is not adjusted properly.

10. If necessary, readjust the linkage and repeat steps 8 and 9.
11. Move the gearshift lever to the park or neutral position, and check to see that the vehicle starts properly.

Instructor Check: ____________________

Job Completion Report

1. What difficulties, if any, did you encounter in completing these procedures?

____

INSTRUCTOR SIGNOFF ______________________ **DATE** ______________

3.10 ACCELERATOR AND KICKDOWN LINKAGE ADJUSTMENTS

As many as three different types of linkage systems can be controlled by the accelerator pedal. These are: accelerator linkage, kickdown rod linkage, and throttle valve linkage. Many vehicles use more than one of these linkage systems. The Chrysler TorqueFlite transmission combines all three.

Linkage adjustments vary from one type of transmission to another. Always consult the proper service manual for correct procedures.

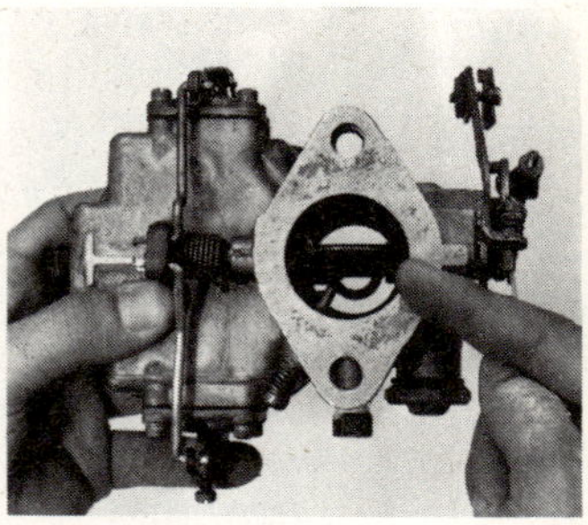

Figure 3-16. When the accelerator is fully depressed, the carburetor throttle valve should be fully open and the linkage should be against the stop.

Accelerator and Throttle Valve Linkage Adjustment

Adjustments for the Chrysler TorqueFlite transmission are discussed here. American Motors Torque-Command transmissions have the same adjustment procedures. Many of the adjustment procedures outlined here are similar to those used by other manufacturers.

Accelerator linkage is checked by depressing the accelerator pedal to the floor. At this point, the carburetor throttle plate should be fully open. Also, the throttle lever should be against the wide-open throttle stop. See Figure 3-16.

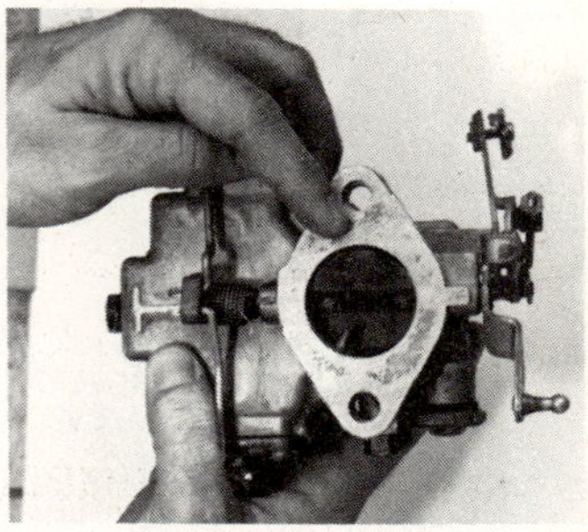

Figure 3-17. When the accelerator is released, the linkage should allow the carburetor throttle valve to return to the idle position.

When the accelerator pedal is released, the linkage must return to the idle position, closing the throttle plate. See Figure 3-17. The action of the linkage should be smooth and immediate.

To perform the following accelerator and throttle valve adjustment, the engine must be at normal operating temperature. If the engine is cold, the linkage can be adjusted by making sure the carburetor is not on the fast idle cam by disconnecting the choke. See Figure 3-18. Use Figure 3-19 as a guide to the following procedure:

Figure 3-18. When the engine is cold, the automatic choke positions the linkage on the fast idle cam for quicker warm-ups.

1. Raise the vehicle on a hoist to a comfortable working height.
2. Use safety glasses. Disconnect the choke linkage if the vehicle is cold and the choke linkage is on the fast idle cam.
3. Loosen the adjusting swivel lockscrew.

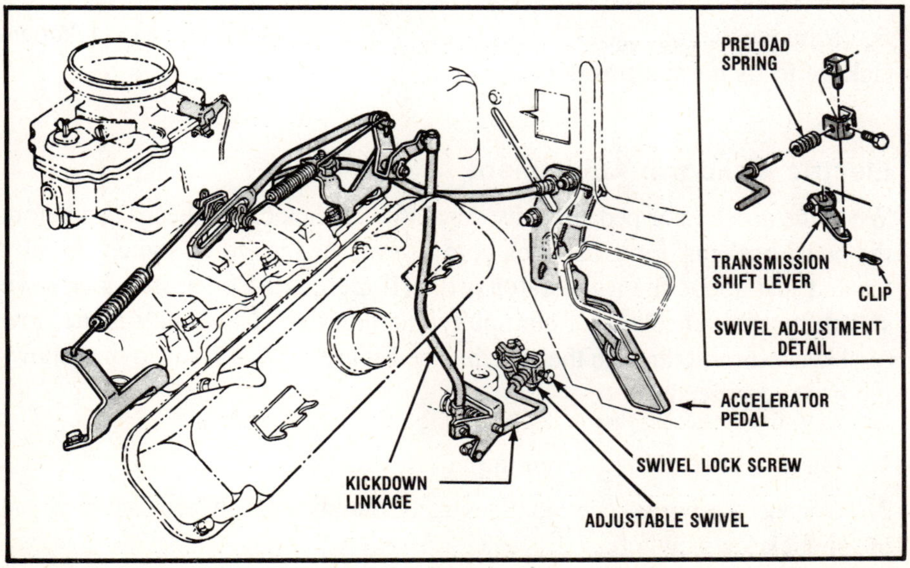

Figure 3-19. This mechanical kickdown linkage has an adjustable swivel. (Chrysler Corporation)

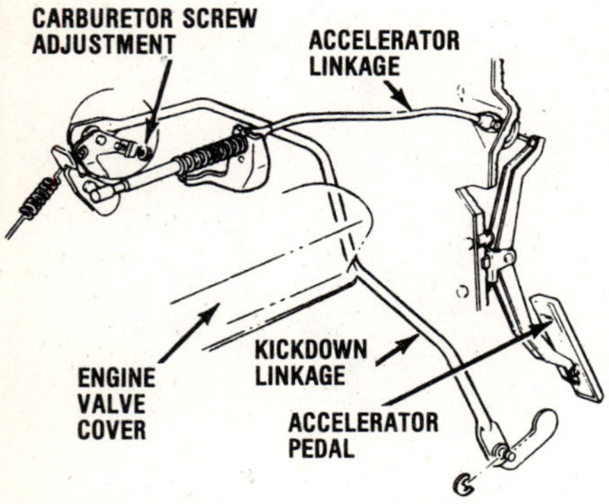

Figure 3-20. This kickdown linkage is adjusted by turning a screw on the linkage near the carburetor. (Ford Motor Company)

NOTE: The swivel must slide freely along the flat end of the throttle rod. If it does not slide freely, the preload spring action will be restricted. If necessary, disassemble and clean or repair any parts that may cause binding.

4. Hold the transmission lever firmly forward against its stop.

5. Tighten the swivel lockscrew with a lb.-in. [Nm] torque wrench to the manufacturer's specification.

6. The adjustment is finished. Linkage backlash, or play, is removed automatically by the preload spring.

7. Lubricate the linkage, if necessary.

8. Lower the vehicle.

9. Reconnect the choke, if necessary.

10. Check linkage operation by moving the throttle rod rearward. Slowly release the throttle rod and make sure it returns to a full forward position. If necessary, readjust the linkage to assure proper linkage operation.

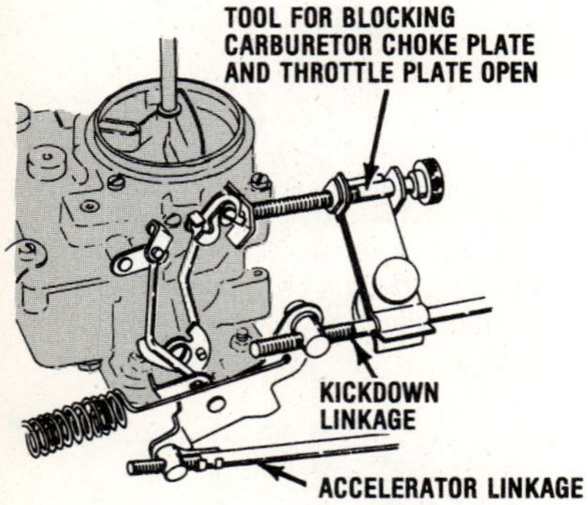

Figure 3-21. Special tools can be used to block open the throttle during kickdown linkage adjustment.

Mechanical Kickdown Adjustment

A separate kickdown rod is often connected from the throttle linkage to a lever on the transmission case. The lever operates a kickdown valve in the valve body to shift the transmission automatically to a lower gear. The kickdown linkage is checked after the accelerator pedal adjustment has been made. Use Figure 3-20 as a guide when using the following kickdown adjustment procedure:

1. Block open the carburetor throttle against its wide-open stop. Use a special tool, shown in Figure 3-21, or a screwdriver.

2. Pull downward on the kickdown rod until it is firmly against its stop on the kickdown valve lever at the transmission case.

3. Loosen the adjusting screw located at the throttle lever on the carburetor.

4. Adjust the screw to the manufacturer's specifications. Adjustment clearance is between the screw head and the throttle lever.

5. Remove the tool used to block open the throttle and return the kickdown linkage to its normal position.

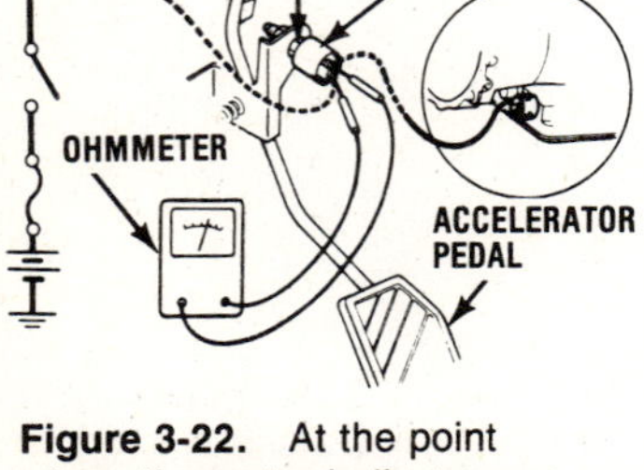

Figure 3-22. At the point where the meter indicates conductivity, the switch allows current to flow to the kickdown solenoid in the transmission. (Toyo Kogyo Co., Ltd.)

Electric Kickdown Adjustment

When the accelerator pedal is fully depressed, a click can be heard just before the pedal bottoms out. This click comes from the kickdown switch, usually located just above the accelerator pedal. If there is no click, the kickdown switch must be checked for continuity and adjusted. Continuity is the flow of electric current through the switch. Use Figure 3-22 as guide to the following general procedure:

1. Disconnect the wires from the kickdown switch.

2. Connect a powered continuity tester, or ohmmeter, between each terminal of the kickdown switch.

3. Depress the accelerator pedal completely and check for continuity. If the light does not light or the ohmmeter does not deflect, there is no electrical flow through the switch. An adjustment is necessary.

4. Loosen the locknut at the back of the kickdown switch.

5. Before making the adjustment, depress the accelerator pedal approximately 7/8 of the way to the floor. Hold the accelerator pedal in that position.

6. Turn the kickdown switch until the voltmeter indicates continuity.

7. Tighten the locknut.

8. Depress the accelerator pedal to make sure a click is heard from the switch when the pedal bottoms out on the floor.

3.11 THROTTLE VALVE CABLE ADJUSTMENT

The throttle valve, or TV, cable connects from the carburetor to the transmission. It is found on General Motors Turbo Hydra-matic transmissions. Use Figure 3-23 to check the cable and make adjustments.

1. Check that the cable is not kinked and is connected at both ends. If the cable is kinked, reroute it.

2. Rotate the snap lock on the TV cable until it unlocks. The cable should be free to slide through the snap lock.

3. Rotate the TV cable lever on the carburetor to the wide-open stop.

4. Hold the cable lever in the wide-open position.

5. Push the snap lock until it is flush with the cable fitting.

6. Ask your instructor's permission to conduct a road test. If approved, road test the vehicle and check for proper transmission operation in all gear positions.

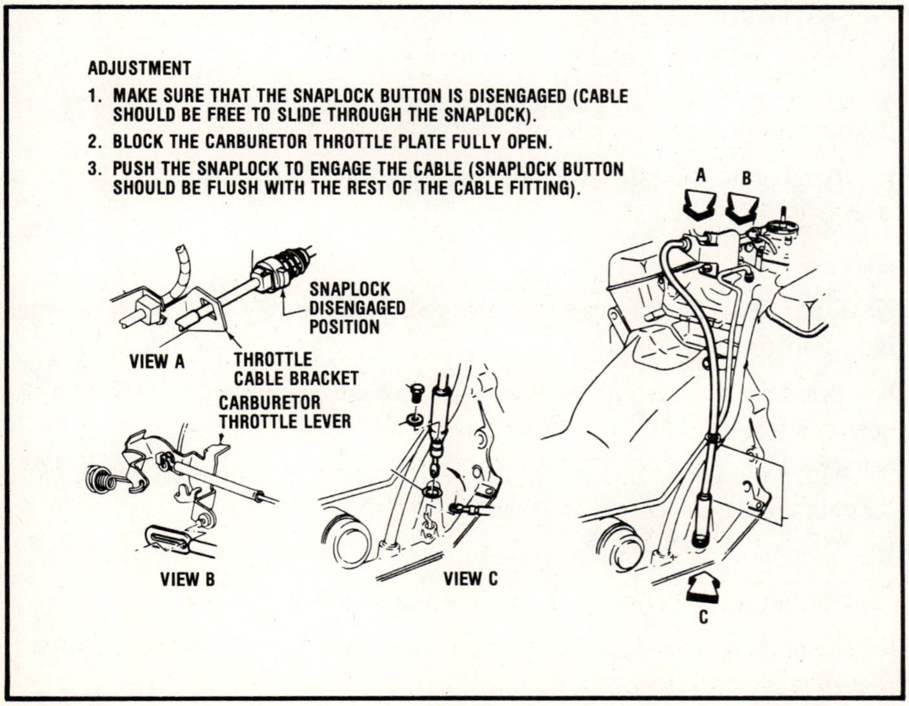

Figure 3-23. The TV cable on General Motors products is adjusted by unlocking the cable "snaplock" and positioning the cable while the carburetor throttle valve is fully opened. (Oldsmobile Division-GMC)

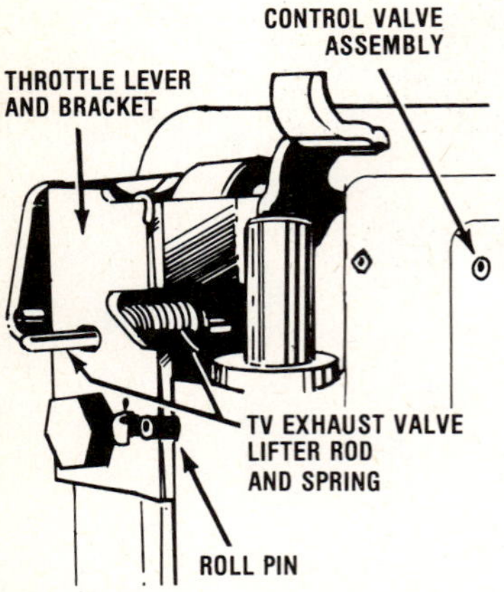

Figure 3-24. The oil pan must be removed to check or adjust the TV exhaust valve assembly inside the transmission. (Oldsmobile Division-GMC)

SAFETY CAUTION: Drive in a safe manner. Jackrabbit starts, hard cornering, or sudden braking should be avoided. Driving conditions other than normal can worsen a problem, or create a new one. Carelessness during the test drive can result in an accident. Remember, you are responsible for someone else's vehicle. Drive only in a manner that is required to check or diagnose proper transmission operation.

6. If there is a delay in the shifting, or if the transmission shifts only at full throttle, continue with the following steps.

7. Raise the vehicle on a hoist, drain the ATF, and remove the oil pan. See Topic 2.5.

8. Inspect the TV exhaust valve lifter rod assembly, shown in Figure 3-24. The rod should not be distorted or binding. The TV exhaust check ball must move up and down as the lifter does. Be sure the lifter spring holds the lifter rod up and against the bottom of the valve body assembly. Make sure the TV plunger is not stuck. Inspect the transmission for the correct throttle lever-to-cable link (see Figure 3-25). Correct any defects.

3.12 NEUTRAL START SWITCH ADJUSTMENT

On many vehicles, the neutral start switch can be adjusted. The first step is to check the continuity between the terminals of the neutral start switch. Always refer to the proper service manual for the correct procedure. If the continuity check indicates the switch is OK, an adjustment, if possible, can be made. The following procedure is only one of several ways to adjust the neutral start switch:

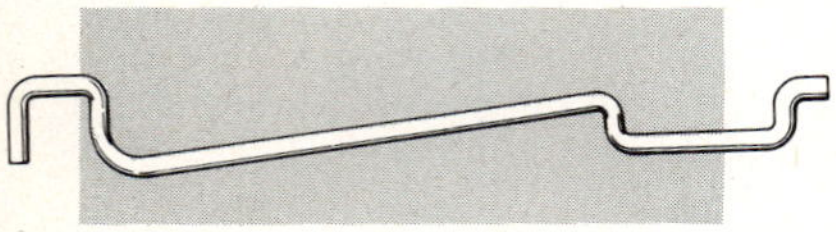

Figure 3-25. General Motors' throttle lever-to-cable link. (Chevrolet Division-GMC)

1. Raise the vehicle on a hoist to a comfortable working height.

2. Use safety glasses. Loosen the neutral start switch bolt, shown in Figure 3-26.

3. Set the gearshift lever in the neutral (N) position.

4. Align the groove and neutral line as shown.

5. Hold the switch in position and torque the nut to the manufacturer's specifications.

6. Disable the vehicle's ignition system so the engine cannot start.

7. Move the gearshift lever into a gear position, and turn the key to the start position.

NOTE: If the engine cranks in any gear position, the neutral start switch is not adjusted properly.

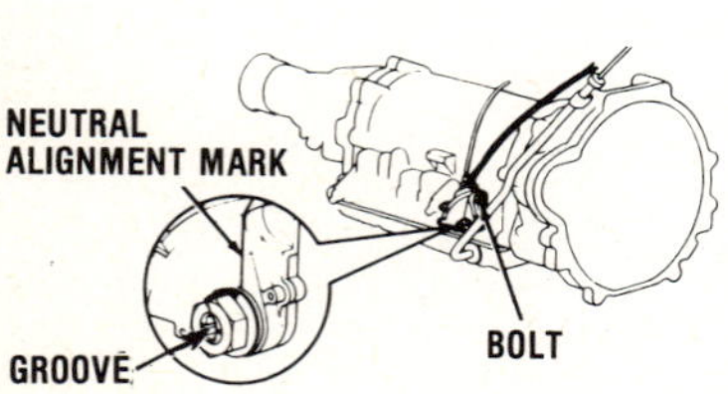

Figure 3-26. This neutral start switch is positioned on the transmission. Adjustment is made by aligning reference lines on the switch and the case. (Toyota Motor Sales U.S.A., Inc.)

8. If necessary, readjust the linkage so that the engine will not crank in any gear position.

9. Move the gearshift lever to the park or neutral positions, and check to see that the engine cranks normally.

10. Reconnect the ignition system.

NAME ______________________ SECTION ______________ DATE ______________

Job Sheet 3-3

Accelerator and Throttle Valve Linkage Adjustment

Before you begin:

Read Topic 3.10.

Vehicle Make ______________________ Model ______________ Year ______________

Time Started ______________________ Time Finished ______________________

Job Completion Time ______________________ Flat Rate Time ______________________

Tools Required

- Hoist
- Safety glasses
- Screwdriver
- Long-nose pliers
- Socket wrench
- Inch-pound torque wrench
- Oil can

References

Manufacturer's service manual ______________________________________

Repair manual ______________________________________

Service bulletin ______________________________________

Procedure

1. Raise the vehicle on a hoist to a comfortable working height.
2. Disconnect the choke linkage if the vehicle is cold and the choke linkage is on the fast idle cam.
3. Loosen the adjusting swivel lockscrew.

NOTE: The swivel must slide freely along the flat end of the throttle rod. If it does not slide freely, the preload spring action will be restricted. If necessary, disassemble and clean or repair any parts that may cause binding.

4. Hold the transmission lever firmly forward against its stop.
5. Tighten the swivel lockscrew with a lb.-in. torque wrench to the manufacturer's specification.

6. Lubricate the linkage, if necessary.
7. Lower the vehicle.
8. Reconnect the choke linkage, if it was disconnected in step 2, above.
9. Check linkage operation by moving the throttle rod rearward. Slowly release the throttle rod and make sure it returns to a full forward position. If necessary, readjust the linkage to assure proper linkage operation.

Instructor Check: ____________________

Job Completion Report

1. Was it necessary to disconnect the choke linkage to perform this adjustment?

____

2. What difficulties, if any, did you encounter in completing this procedure?

____

INSTRUCTOR SIGNOFF ____________________ **DATE** ____________

NAME ______________________ SECTION ____________ DATE ____________

Job Sheet 3-4

Kickdown Linkage Adjustment

Before you begin:

Read Topic 3.11.

Vehicle Make ____________________ Model ___________ Year ___________

Time Started ____________________ Time Finished ____________________

Job Completion Time ______________ Flat Rate Time ____________________

Tools Required

- Safety glasses
- Screwdriver
- Socket wrenches, handle, and extensions
- Special throttle-blocking tool
- Powered continuity tester or ohmmeter
- Open-end or combination wrenches

References

Manufacturer's service manual ____

Repair manual ____

Service bulletin ____

Procedure: Adjust Mechanical Kickdown Linkage

1. Block open the carburetor throttle plate.
2. Pull downward on the kickdown rod until it is firmly against its stop on the kickdown valve lever at the transmission case.
3. Loosen the adjusting screw located at the throttle lever on the carburetor.
4. Adjust the clearance between the screw head and the throttle lever to the manufacturer's specifications.
5. Remove the tool used to block open the throttle and return the kickdown linkage to its normal position.

Instructor Check: ____________________

Procedure: Adjust Electric Kickdown Adjustment

1. Disconnect both wires from the kickdown switch.
2. Connect a powered continuity tester, or ohmmeter, between the terminals of the kickdown switch.

3. Depress the accelerator pedal completely and check for continuity.
4. Loosen the locknut at the back of the kickdown switch.
5. Depress the accelerator pedal approximately 7/8 of the way to the floor. Hold the accelerator pedal in that position.
6. Turn the kickdown switch until the voltmeter indicates continuity.
7. If there is no continuity at any position of the switch, replace the switch and repeat steps 1 through 6.
8. Tighten the locknut.
7. Press the accelerator pedal to make sure a click is heard from the switch as the pedal reaches the bottom of its travel.

Instructor Check: ____________________

Job Completion Report

1. If you adjusted electrical kickdown linkage, was it necessary to replace the kickdown switch? ____

2. What difficulties, if any, did you encounter in adjusting either mechanical or electrical kickdown linkage? ____

____

____

INSTRUCTOR SIGNOFF ______________________ **DATE** ____________

NAME ________________ SECTION ____________ DATE ____________

Job Sheet 3-5

Throttle Valve Cable Adjustment

Before you begin:

Read Topic 3.11.

Vehicle Make ________________ Model ____________ Year ____________

Time Started ________________ Time Finished ________________

Job Completion Time ________________ Flat Rate Time ________________

Tools Required

- Safety glasses
- Pliers
- Hoist
- Oil catchpan
- Socket wrenches, handle, extensions

References

Manufacturer's service manual ________________________________

Repair manual ________________________________

Service bulletin ________________________________

Procedure

1. Check that the TV (throttle valve) cable is not kinked and is connected at both ends. If the cable is kinked, reroute it.
2. Pull out the snap lock on the TV cable until it unlocks the cable.
3. Rotate the TV cable lever on the carburetor to the wide-open stop.
4. Hold the cable lever in the wide-open position.
5. Push the snap lock in until it is flush with the cable fitting. The housing should extend through the snap lock assembly between 1/16 and 5/16 inch [1.6mm and 8.0mm].
6. Return the carburetor lever to the closed position.
7. Ask your instructor's permission to conduct a road test. If approved, road test the vehicle and check for proper transmission operation in all gear positions.

SAFETY CAUTION: Drive in a safe manner. Jackrabbit starts, hard cornering, or sudden braking should be avoided. Driving conditions other than normal can worsen a problem, or create a new one. Carelessness during the test drive can result in an accident. Remember, you are responsible for someone else's vehicle. Drive only in a manner that is required to check or diagnose proper transmission operation.

8. If there is a delay in the shifting, or if the transmission shifts only at full throttle, continue with the following steps.
9. Raise the vehicle on a hoist, drain the ATF, and remove the oil pan. See Topic 2.5.
10. Inspect the TV exhaust valve lifter rod assembly for distortion or binding. Check that the TV exhaust check ball must move up and down as the lifter does. Be sure the lifter spring holds the lifter rod up and against the bottom of the valve body assembly. Make sure the TV plunger is not stuck. Inspect the transmission for the correct throttle lever to cable link. In the Job Completion Report make a note of any defects found.
11. Correct any defects found.
12. Replace the oil pan and refill the transmission with the proper fluid.
13. Ask your instructor's permission to conduct a road test. If approved, road test the vehicle and check for proper transmission operation in all gear positions.

SAFETY CAUTION: Drive in a safe manner. Jackrabbit starts, hard cornering, or sudden braking should be avoided. Driving conditions other than normal can worsen a problem, or create a new one. Carelessness during the test drive can result in an accident. Remember, you are responsible for someone else's vehicle. Drive only in a manner that is required to check or diagnose proper transmission operation.

Instructor Check: ____________________

13. In the Job Completion Report, make a note of any transmission problems noted in the road test.

Instructor Check: ____________________

Job Completion Report

1. What difficulties, if any, did you encounter in adjusting the TV cable? ________

____

____

2. Was it necessary to remove the oil pan to find the cause of a shifting problem?

____

3. What defects, if any, were found and corrected if the oil pan was removed and the TV exhaust valve components were inspected? ____________________

____

____

4. What problems, if any, were noticed during the final road test? ____________

____

INSTRUCTOR SIGNOFF ____________________ **DATE** ____________

NAME ______________________ SECTION ____________ DATE ____________

Job Sheet 3-6

Neutral Start Switch Adjustment

Before you begin:

Read Topic 3.12.

Vehicle Make ____________________ Model ____________ Year ____________

Time Started ____________________ Time Finished ____________________

Job Completion Time ________________ Flat Rate Time ____________________

Tools Required

- Powered continuity tester or ohmmeter
- Hoist
- Open-end or combination wrenches
- Socket wrenches, handle, extensions
- Wheel chocks

References

Manufacturer's service manual ____

Repair manual ____

Service bulletin ____

Procedure

1. Disconnect the wires from the neutral start switch and check for continuity between the switch terminals. If necessary, replace the neutral start switch.
2. Raise the vehicle on a hoist to a comfortable working height.
3. Loosen the neutral start switch bolt.
4. Set the gearshift lever in the neutral (N) position.
5. Adjust the switch to the manufacturer's specifications.
6. Hold the switch in position and torque the nut to the manufacturer's specifications.

Write the correct torque in the space provided. .... ____________________

7. Disable the vehicle's ignition system so that the engine cannot start.
8. Move the gearshift lever into a gear position. Press down on the brake pedal. Turn the ignition key to the start position.

NOTE: If the engine cranks in any gear position, the neutral safety switch is not adjusted properly.

9. If necessary, readjust the neutral start switch and try again to crank the engine with the gearshift selector lever in a gear position.
10. If the engine will not crank in a gear position, move the gearshift lever to the park or neutral positions. If the engine now cranks, the switch is adjusted correctly.

Instructor Check: ____________________

11. Reconnect the vehicle's ignition system.

Job Completion Report

1. What conditions were found in your preliminary check of the neutral start switch for continuity? ____

   ____

2. Was it necessary to replace the neutral start switch? ____________________

3. Did the neutral start switch function correctly after adjustment? ______________

INSTRUCTOR SIGNOFF ________________________ **DATE** ______________

NAME ______________ SECTION ______________ DATE ______________

Certification Practice

Test 3

DIRECTIONS:

The following questions are similar to those used on mechanic certification tests. Answer each question by circling the letter of the correct choice.

1. All of the following about the transmission gearshift positions are true EXCEPT
   A. In P, the transmission output shaft is locked.
   B. In D, the transmission will upshift or downshift through the forward gear positions.
   C. In N, the vehicle can roll.
   D. In 2, the transmission will always be in second gear.

2. Mechanic A says that automatic transmission gearshift linkage uses rods and levers. Mechanic B says that cables are used for automatic transmission gearshift linkage. Who is correct?
   A. A only   B. B only   C. Both A and B   D. Neither A nor B

3. Which of the following is *not* an automatic transmission linkage system?
   A. Quadrant linkage
   B. Clutch linkage
   C. Gearshift linkage
   D. Kickdown linkage

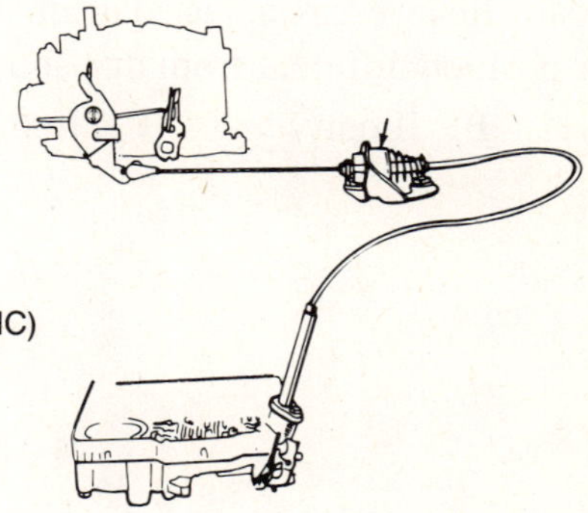

Question 4.
(Chevrolet Motor Division-GMC)

4. Which of the following is true about the linkage shown in the picture?
   I. Its only function is to downshift the transmission into a lower gear.
   II. It controls gear selection and fluid pressure for proper transmission operation.
   A. I only   B. II only   C. Both I and II   D. Neither I nor II

5. A neutral start switch
   A. allows the vehicle to start only in N.
   B. is a theft-prevention device.
   C. allows the vehicle to start only in P.
   D. allows the vehicle to start only in P or N.

6. All of the following are controlled by the accelerator pedal EXCEPT
   A. accelerator linkage.
   B. quadrant linkage.
   C. kickdown linkage.
   D. throttle-valve linkage.

7. Mechanic A says kickdown linkage can only be operated by an electrical switch. Mechanic B says kickdown linkage can only be operated by mechanical linkage. Who is correct?
   A. A only B. B only C. Both A and B D. Neither A nor B

8. In the term, ''automatic transmission TV cable,'' TV stands for
   A. throttle valve.
   B. transmission valve.
   C. transmission variable.
   D. television.

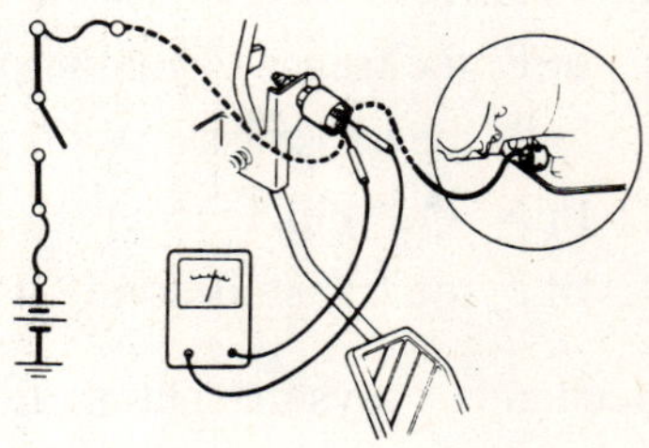

Question 9.
(Toyo Kogyo Co., Ltd.)

9. What is being checked in the picture?
   A. A neutral start switch
   B. A kickdown switch
   C. Battery voltage
   D. RPM or dwell

10. Which of the following statements is true?
    I. If transmission linkages are adjusted incorrectly, the vehicle may start in a gear position.
    II. If transmission linkages are adjusted incorrectly, the transmission may be shifted into a gear position different from that shown in the quadrant.
    A. I only B. II only C. Both I and II D. Neither I nor II

4 Torque Converter Systems

YOUR LEARNING JOB

The torque converter changes the mechanical force of the engine into hydraulic, or fluid, force. This hydraulic force acting on the turbine turns the input shaft of the automatic transmission. The torque converter is made up of different parts. This unit describes the parts that make up the torque converter and how they work together.

JOB COMPETENCY OBJECTIVES

This is a knowledge-building unit. When you have completed your assignments and exercises in this unit, you should be able to:

- ☐ Describe the parts of a torque converter.
- ☐ Explain how a stator operates.
- ☐ Describe the two types of overrunning clutches.
- ☐ Explain the purpose of a lockup torque converter.

4.1 PARTS OF THE TORQUE CONVERTER

A torque converter is a series of parts that forms a fluid coupling and multiplies torque. Torque multiplication is discussed in 1.4. The torque converter transfers engine torque to the transmission gears. A torque converter, shown in Figure 4-1, may consist of six separate parts:

- Torque converter housing
- Impeller
- Turbine
- Stator
- Overrunning clutch
- Lockup clutch assembly.

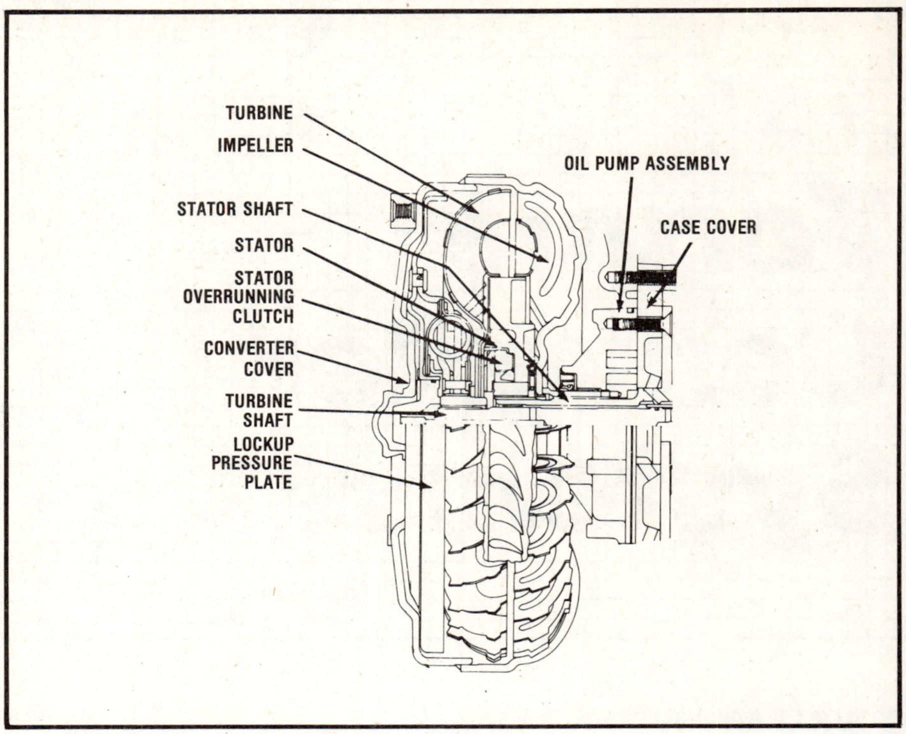

Figure 4-1. A modern torque converter, such as the General Motors 325-4L, consists of the parts shown. (Oldsmobile Division-GMC)

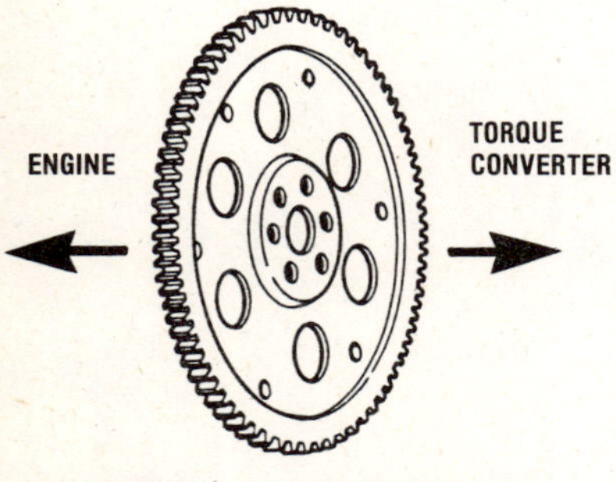

Figure 4-2. The flexplate is attached to the engine crankshaft. The torque converter housing bolts to the flexplate. (Toyota Motor Sales U.S.A., Inc.)

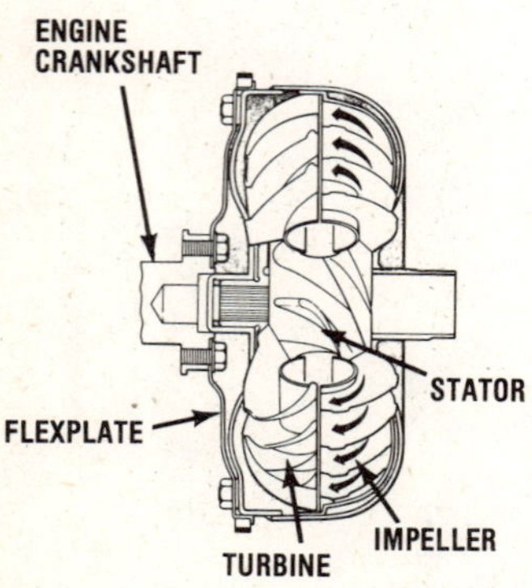

Figure 4-3. The arrows indicate how the fluid is thrown out of the impeller blades. (Ford Motor Company)

4.2 TORQUE CONVERTER HOUSING

All parts of the torque converter are enclosed in a housing, called the torque converter housing. The torque converter housing is sealed and cannot be disassembled without special equipment.

The torque converter housing bolts to the flexplate. With modern automatic transmissions, the torque-converter/flexplate combination replaces the flywheel. The flexplate flexes, or bends, as parts in the transmission rotate and expand or contract because of heat. An external ring gear on the outer edge of the flexplate meshes with the starter drive gear to crank the engine. A flexplate is shown in Figure 4-2.

4.3 IMPELLER

The impeller is part of the *rear* half of the torque converter housing. The impeller is often called the pump, or primary pump, because it pumps fluid to the turbine. The impeller should not be confused with the oil pump, which builds up hydraulic pressure in the automatic transmission. The oil pump is discussed in 10.2.

The impeller turns whenever the torque converter housing turns. The torque converter is bolted to the flexplate, which is bolted to the crankshaft. The impeller pushes, or pumps, fluid inside the torque converter housing and directs the fluid forward toward the turbine. Vanes on the impeller are curved (Figure 4-3) so that fluid is thrown out of the impeller at an angle.

4.4 TURBINE

The turbine is a separate part of the torque converter. It spins inside of the front half of the torque converter housing. Fluid is pumped into the curved vanes on the outside of the turbine by the impeller. The moving fluid that is forced against the turbine blades causes the turbine to rotate.

The turbine is splined at the hub, or center hole, to a shaft, called the input shaft. The rear of the input shaft connects to the planetary gearset to drive the gears inside the transmission.

As the turbine rotates, the fluid that caused it to rotate falls to the center of the turbine. At the center of the turbine, the fluid passes through the stator and back toward the center of the impeller. The turbine vanes are shown in Figure 4-4.

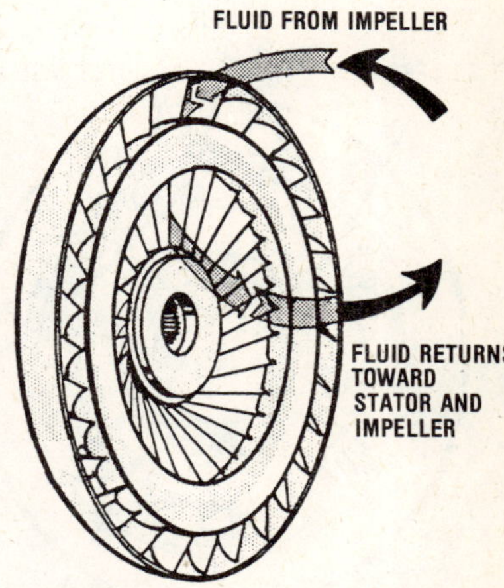

Figure 4-4. The fluid enters the turbine near the outer edge, and returns to the stator and impeller through the center vanes. (American Motors Corporation)

CHECKPOINT

1. **Modern torque converters may consist of the following six separate parts: torque converter ____________, ____________, ____________, ____________, ____________, ____________.**
2. **All parts of the torque converter are enclosed in the torque converter ____________.**
3. **The ____________ turns whenever the torque converter housing turns.**
4. **Vanes in the impeller are ____________ so that fluid is thrown out of the impeller at an angle.**
5. **The ____________ shaft is splined to the turbine hub.**

4.5 STATOR

The stator is located between the impeller and turbine. Its main function is to redirect fluid flow back to the impeller.

As fluid leaves the turbine, it passes into the curved vanes of the stator. See Figure 4-5. The stator reverses the direction of fluid flow and forces it to flow in the same direction as impeller rotation. Instead of fighting impeller motion, the redirected flow now assists the impeller by accelerating the same fluid and developing higher torque multiplication. The principles of hydraulic action are discussed in Unit 11.

For the stator to operate properly, an overrunning clutch is used.

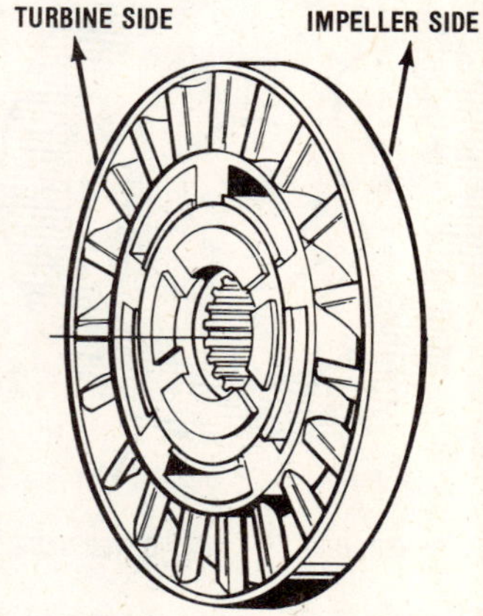

Figure 4-5. The stator blades are curved, and the opening between the blades is narrower on the side next to the impeller. (Hydra-Matic)

4.6 OVERRUNNING CLUTCH

An overrunning, or one-way, clutch allows the stator to rotate only in one direction. This assures that the stator will redirect fluid flow to the impeller in the proper direction during the proper time of operation.

The overrunning clutch is built into the hub of the stator. The inner race is splined to an extension of the oil pump. The oil pump extension is stationary. The outer race of the overrunning clutch is part of the stator hub.

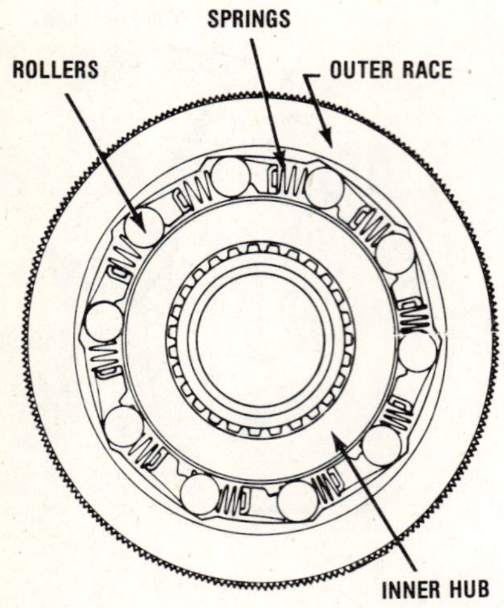

Figure 4-6. The roller clutch in the stator hub will lock when the stator turns in a counterclockwise direction. (Hydra-Matic)

There are two types of overrunning clutches commonly used in modern automatic transmissions. They are the roller clutch and the sprag clutch.

Roller Clutch

The roller clutch is much like a roller bearing. It consists of an inner and an outer race, plus rollers. A race is a smooth surface, or track, on which the rollers can rotate smoothly. In an overrunning roller clutch assembly, a spring is also used, which is not part of a roller bearing assembly.

The inner race of a roller clutch is perfectly round. The outer race, unlike a roller bearing, consists of a series of wide and narrow ramps. A roller fits into each wide spot along with a corrugated, or wavy, spring. See Figure 4-6.

When the stator is rotating in the proper direction, centrifugal, or outward, force pushes the roller against the spring. This allows the clutch to overrun, or freewheel.

When the stator tries to rotate in the wrong direction, the spring pushes the roller to the narrow part of the ramp. The roller is wedged in the narrow spot and will not turn, locking up the clutch assembly.

Sprag Clutch

A sprag clutch is also like a roller bearing, except that each roller is flattened. Between these flattened rollers, or sprags, springs are used. See Figure 4-7.

When the stator is rotating in the proper direction, the sprags are tilted down. This downward tilt allows overrun, or slippage, between the two races. When stator rotation is reversed, spring pressure straightens the sprags to a straight up-and-down position. The sprags now jam against the races to lock them together and stop overrun.

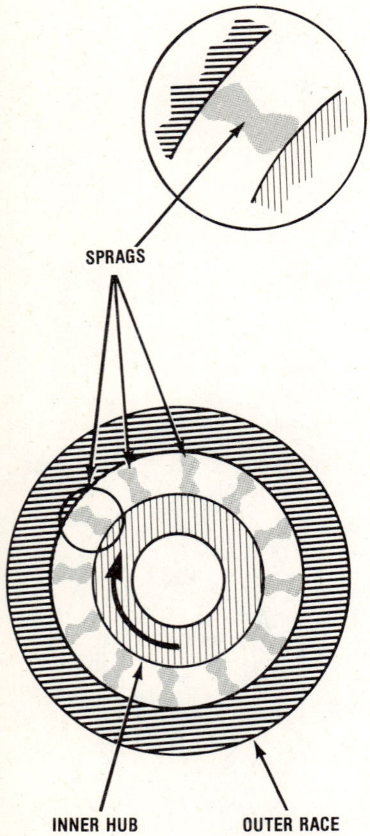

Figure 4-7. A sprag overrunning clutch is used on many automatic transmissions. The sprags will tilt and lock against the hub when the stator turns counterclockwise.

4.7 TORQUE CONVERTER LOCKUP

The torque converter can have up to 10 percent slippage. Slippage results in heat buildup and reduced fuel economy. To eliminate slippage, many new automatic transmissions have a lockup device that locks the turbine and the torque converter housing together. When the lockup is engaged, 100 percent of the engine power goes through the torque converter and to the transmission. The lockup results in a direct drive situation, where the ratio of engine speed to transmission output shaft speed is 1:1.

The parts that lock the turbine and impeller together consist of clutch friction material, a sliding clutch piston, and torsion springs.

The clutch friction material is attached to the front cover of the torque converter housing. The clutch piston is mounted in the turbine. The torsion springs are located on the front, or forward, side of the turbine. The torsion springs absorb shock during lockup and help to dampen, or smooth out, engine firing impulses.

At a predetermined speed—usually around 40 mph—the transmission automatically shifts into direct drive. In three-speed transmissions, lockup will occur only during high-gear operation. In most four-speed transmissions, both third and fourth gears will operate under torque converter lockup.

The position of the gearshift lever on the quadrant and hydraulic and/or solenoid-operated valve action within the valve body activate the lockup.

A typical converter lockup occurs when fluid is channeled through the input shaft and between the clutch piston and turbine. The fluid pressure forces the piston against the clutch friction material to lock the turbine to the impeller. When vehicle speed drops below about 40 mph, fluid pressure is released. The piston releases the clutch friction material. The torque converter will then operate in a conventional manner.

CHECKPOINT

1. **The stator is located between the ____________________ and ____________________.**
2. **An overrunning clutch allows the ____________________ to rotate only in one direction.**
3. **Two types of overrunning clutches are commonly used in modern automatic transmissions: the ________________ clutch and the ________________ clutch.**
4. **A ____________________ clutch is like a roller bearing, except that each roller is flattened.**
5. **Torque converter slippage causes ____________________ buildup and reduced ____________________ economy.**
6. **The parts that lock the turbine and impeller together consist of clutch ____________________ material, a sliding clutch ____________________, and ____________________ springs.**

4.8 TORQUE CONVERTER OPERATION

The engine must be running before the torque converter can operate. As the engine crankshaft turns, it turns the flexplate and the torque converter. The torque converter has a hub that is geared to the transmission oil pump or to the oil pump gears. The torque converter turns the oil pump gears, or vanes. The oil pump assembly is bolted to the transmission case, just behind the converter, and begins pumping fluid as soon as the engine is turning.

When the torque converter housing begins to rotate, the impeller rotates with it in a clockwise direction. The impeller is integral with the torque converter housing. Rotation is determined from the front of the engine. When standing in front of the engine, and rotation is to the right, the rotation is clockwise. If rotation is to the left, it is counterclockwise, or opposite to the rotation of hands on a clock.

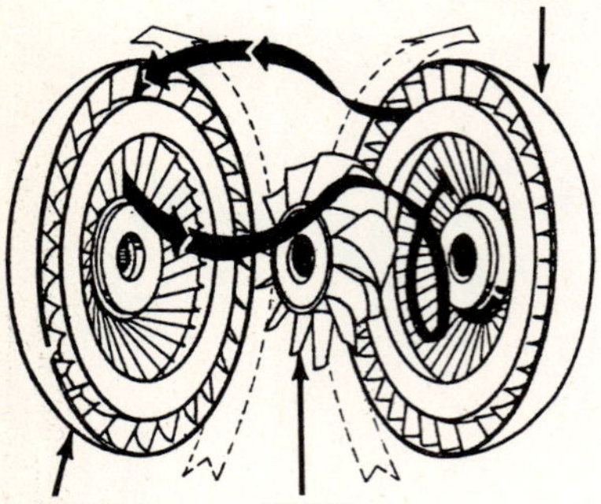

Figure 4-8. The dark arrows show the fluid path through the converter parts. (American Motors Corporation)

The impeller rotates clockwise. It throws fluid to the outer edges of its vanes and into the outer vanes of the turbine. At this point, the turbine is stationary. Much of the fluid from the impeller simply strikes the vanes of the motionless turbine and bounces off. The fluid that bounces off then begins to rotate in a counterclockwise direction. The counterclockwise flow of fluid is fed to the stator through the vanes in the center of the turbine.

The counterclockwise flow from the turbine locks the overrunning clutch so that the stator cannot turn. The stator can only rotate freely in a clockwise direction. The vanes in the stator redirect fluid flow so that it hits the impeller vanes in a clockwise direction. Fluid flow accelerates as it passes through the stator and forces the impeller to turn faster. This increases, or multiplies, the torque transferred from the impeller to the turbine.

As engine speed increases, fluid begins moving the turbine vanes and less fluid bounces off them. As the turbine moves, it begins circulating fluid in a clockwise direction. At this point, the stator begins to turn on the overrunning clutch until it is freewheeling, or rotating, at about the same speed as the turbine and impeller. As the stator freewheels, torque multiplication gradually drops from about 2:1 at breakaway, or startup, to about 1:1 at cruising speeds.

Figure 4-8 shows fluid flow through the torque converter.

Torque multiplication is required anytime the vehicle starts moving from a complete stop, during acceleration, when climbing a hill, or when hauling a load. At no-load cruising speed, the impeller, turbine, and stator are rotating at about the same speed. When the accelerator is depressed, the impeller speeds up. When the impeller speeds up, a counterclockwise fluid flow begins at the turbine, and the stator locks up to multiply torque.

CHECKPOINT

1. **Clockwise rotation is determined from the front of the ____________________.**
2. **Much of the fluid from the impeller simply strikes the vanes of the motionless turbine and ____________________ ____________________.**
3. **The counterclockwise flow from the turbine locks the ____________________ ____________________ so that the stator does not turn.**
4. **Fluid flow ____________________ as it passes through the stator and forces the impeller to turn faster.**
5. **At no-load cruising speed, the ____________________, ____________________, and ____________________ are rotating at about the same speed.**

NAME ______________________ SECTION ____________ DATE ____________

Certification Practice

Test 4

DIRECTIONS:

The following questions are similar to those used on mechanic certification tests. Answer each question by circling the letter of the correct choice.

1. All of the following statements about the torque converter housing are correct EXCEPT
   A. It acts as a reservoir to store ATF.
   B. Its halves are held together by bolts or rivets.
   C. It bolts to the flexplate.
   D. It encloses the impeller, stator, and turbine.

2. Mechanic A says that the turbine is part of the torque converter housing.
   Mechanic B says that the impeller is part of the torque converter housing.
   Who is correct?
   A. A only B. B only C. Both A and B D. Neither A nor B

3. How is the transmission input shaft driven?
   A. The turbine hub is splined to the shaft.
   B. The shaft is splined to the flexplate.
   C. The stator overrunning clutch is bolted to the shaft.
   D. The torque converter housing is bolted to the input shaft.

4. What part is indicated in the picture?
   A. Impeller
   B. Stator
   C. Turbine
   D. Overrunning clutch

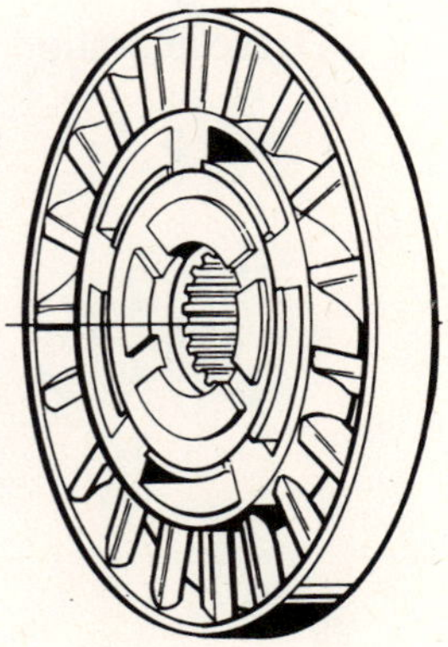

Question 4. (Hydra-Matic)

5. Which of the following statements is correct?
   I. The stator speeds up and redirects fluid from the turbine to the impeller.
   II. The stator speeds up and redirects fluid from the impeller to the turbine.
   A. I only B. II only C. Both I and II D. Neither I nor II

6. All of the following statements about the overrunning clutch are true EXCEPT
   A. It allows rotational movement in two directions.
   B. Its inner race is splined to an extension of the oil pump.
   C. The outer race is part of the stator hub.
   D. It is either a roller clutch or a sprag clutch.

7. The part shown in the picture is
   A. a roller clutch.
   B. a sprag clutch.
   C. a roller bearing.
   D. a sprag bearing.

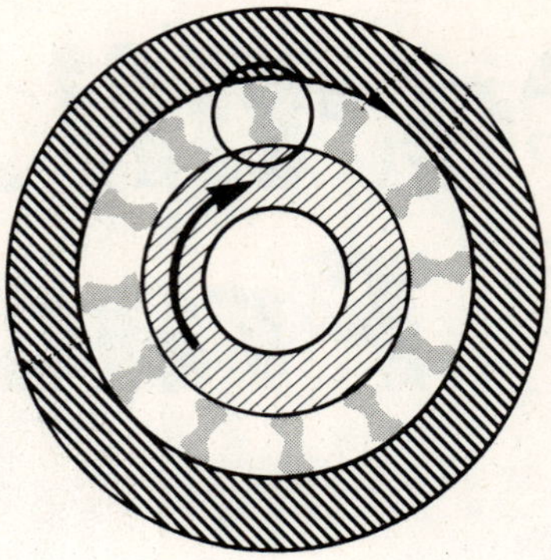

Question 7.

8. Mechanic A says that in a lockup torque converter fluid pressure is used to lock the turbine to the impeller.
   Mechanic B says that in a lockup torque converter friction material is used to lock the turbine to the impeller.
   Who is correct?
   A. A only B. B only C. Both A and B D. Neither A nor B

9. What locks the stator overrunning clutch?
   A. The pressure of fluid bouncing off the turbine vanes
   B. Pressure from the oil pump
   C. The turning of the transmission input shaft
   D. Mechanical pressure from the clutch piston and friction material

10. Torque multiplication is required during all of the following EXCEPT
    A. hill climbing.
    B. highway cruising.
    C. acceleration while moving.
    D. acceleration from a stop.

5 Checking The Torque Converter

YOUR LEARNING JOB

As you have learned, the torque converter controls and couples engine power to the rest of the automatic transmission. A torque converter that does not operate properly can cause many problems in the transmission. This unit discusses a number of checks and tests that can be made to determine if the torque converter is operating properly.

JOB COMPETENCY OBJECTIVES

This is a skill-building unit. When you have completed your assignments and exercises in this unit, you should be able to:

- ☐ Perform a converter stall test.
- ☐ Check torque end play.
- ☐ Check torque converter clearances.
- ☐ Make a torque converter leakage test.

5.1 CHECKING THE TORQUE CONVERTER

Repair shops usually send a faulty torque converter out for rebuilding or install a new one. Highly specialized equipment and experience are necessary to rebuild a torque converter.

Any service that can be performed on torque converters is covered in Unit 2.

A number of tests and checks can be made on the torque converter. Most of these tests and checks require that the torque converter be removed from the transmission. Units 16, 17, and 28 discuss torque converter removal on different types of transmissions.

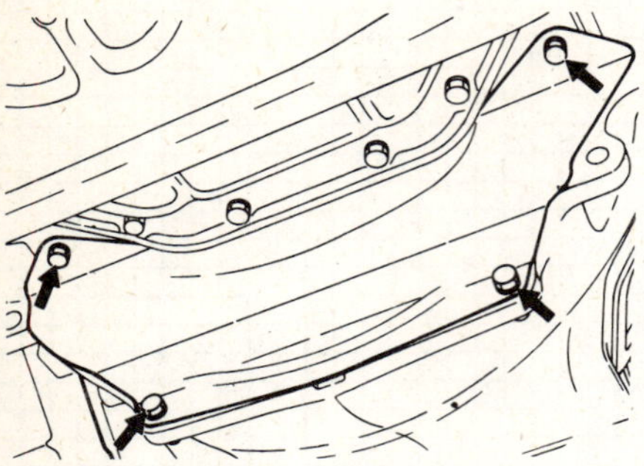

Figure 5-1. Remove the bolts and the access cover to inspect the torque converter. (Chrysler Corporation)

Common indications that a torque converter is not operating properly include:

- Slow acceleration from a stop in all gears
- Normal acceleration, but poor high-speed performance
- Transmission overheating.

Checks for these problems can be made by road-testing the vehicle, as discussed in Unit 14. Other on-car checks of the torque converter are the visual inspection and stall test.

5.2 VISUAL INSPECTION

Before any teardown of the torque converter is done, make a visual inspection. This can tell you a lot about torque converter condition. Whenever making any on-car inspection of the automatic transmission, always check fluid level before anything else. Low fluid levels can cause engine racing and slow acceleration and shifts. Check fluid condition, as discussed in 2.3.

Raise the vehicle on a hoist to a comfortable working height. Use safety glasses. Remove the access cover, shown in Figure 5-1. The access hole is usually a small opening, so it's not always possible to see all of the problem areas. Whenever possible, check for the following:

- Damaged flexplate
- Loosen, misaligned, or missing bolts that connect the flexplate to the converter
- A ballooned, or expanded, torque converter housing
- Cracked torque converter housing
- Heat discoloration on converter housing
- Loose or missing converter balance weights
- Loose or damaged starter ring gear
- Broken or damaged converter pilot.

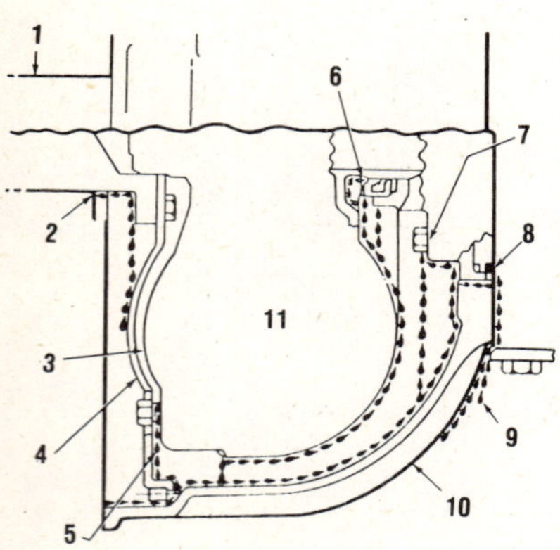

1. ENGINE CRANKSHAFT.
2. CRANKSHAFT SEAL LEAK.
3. CONVERTER ASSEMBLY.
4. FLEXPLATE.
5. CONVERTER DRAIN PLUG LEAK OR CONVERTER-TO-FLYWHEEL-STUD WELD LEAK.
6. FRONT PUMP SEAL LEAK.
7. OIL-PUMP-TO-CASE BOLT LEAK.
8. FRONT PUMP GASKET LEAK.
9. THIS LEAK DOES NOT APPLY TO OVERDRIVE TRANSMISSION.
10. CONVERTER HOUSING.
11. IMPELLER.

NOTE: LEAKAGE FROM THE PUMP GASKET WILL RUN DOWN THE INSIDE OF THE CONVERTER HOUSING ON AUTOMATIC OVERDRIVE TRANSMISSION.

Figure 5-2. Notice which surfaces become coated with ATF. Trace the drops in the picture to locate where leaks start. (Ford Motor Company)

A remote starter switch must be used to turn the flexplate and torque converter while making the visual inspection. The remote starter switch connects to the starter solenoid or starter relay. Check the proper service manual.

SAFETY CAUTION: When using a remote starter switch, the ignition must be off. Do *not* put your fingers near the moving torque converter while the engine is cranking. Serious injury can result.

There should be no fluid leakage around the bellhousing or converter. If leakage is found, it is necessary to trace its source.

5.3 FLUID LEAKAGE IN CONVERTER AREA

The torque converter is sealed from the rest of the transmission, so there should be no oil in the bellhousing. If oil is present, it may come from several different areas. It may be possible to locate the source of the leak before removing the transmission. The paths that fluids use to reach the bottom of the converter housing are shown in Figure 5-2.

NAME ____________________ SECTION ____________ DATE ____________

Job Sheet 5-1

Torque Converter Visual Inspection

Before you begin:

Read Topics 5.1 through 5.3.

Vehicle Make ____________________ Model ____________ Year ____________

Time Started ____________________ Time Finished ____________________

Job Completion Time ________________ Flat Rate Time ____________________

Tools Required

- Absorbent white paper towels
- Remote starter switch
- Hoist
- Droplight

References

Manufacturer's service manual ________________________________

Repair manual ________________________________

Service bulletin ________________________________

Procedure

1. Warm the vehicle to normal operating temperature and check the ATF level and condition. In the Job Completion Report, make a note of the fluid level and condition.
2. Connect a remote starter switch to the vehicle.
3. Raise the vehicle on a hoist to a comfortable working height. Use safety glasses.
4. Remove the access cover or shield.
5. Inspect the converter through the access hole. Use the remote starter to move the converter around.

SAFETY CAUTION: When using a remote starter switch, the ignition must be off. Do *not* put your fingers near the moving torque converter while the engine is cranking. Serious injury can result.

6. In the Job Completion Report, make a note of any visible damage, leaks, or other problems.

Instructor Check: ________________

Job Completion Report

1. Describe the ATF level and condition. ______________________________

____

____

2. What damage, leaks, or other problems, if any, were found during the visual inspection? ______________________________

____

____

INSTRUCTOR SIGNOFF ____________________ **DATE** ____________

5.4 STALL TEST

This test is required by many manufacturers to determine engine speed at full throttle in the drive (D) position. The driving wheels are locked, so that the load is placed on the engine and transmission. Results of this test will tell the mechanic how the stator overrunning clutch, and the clutches and bands inside the transmission, are operating.

Figure 5-3. The tachometer should be placed inside the vehicle where it can be read easily by the mechanic during stall test readings. (Toyota Motor Sales U.S.A., Inc.)

SAFETY CAUTION: Do *not* let anyone stand in front of the vehicle when conducting this test. Be sure the area in front of the vehicle is clear. The vehicle can move forward during this test. The service brakes must be fully applied when making this test, in addition to the emergency brake. However, brake problems can allow the vehicle to move forward, possibly causing property damage or personal injury. Block the wheels.

CAUTION: A stall test can ruin weak clutches and bands inside the transmission. If there is any doubt about the condition of clutches and bands, do *not* perform the stall test.

To perform a stall test, use the following procedure:

1. Connect a tachometer to the engine. Be sure it is properly installed according to the manufacturer's directions.

2. Place the tachometer inside the vehicle where it can be read easily. See Figure 5-3. Mark the dial of the tachometer with a grease pencil at the recommended stall speed in revolutions per minute (rpm). Check the proper service manual for the maximum rpm for this test.

CAUTION: Do *not* exceed the maximum rpm at any time during this test. Severe engine or transmission damage could result.

3. Check engine coolant and transmission fluid levels. If transmission fluid level must be checked when warm, check fluid after Step 5.

4. Block the front wheels and set the emergency brake.

5. Place the transmission in park or neutral and warm the engine and transmission to operating temperatures.

6. Place the gearshift lever into D, or into the gear or gears that the manufacturer recommends.

7. Press the accelerator pedal to the floor. Press down slowly at first, making sure the vehicle does not move. When the tachometer needle stops rising, record the rpm reading.

CAUTION: If the tachometer needle goes above the recommended rpm, *immediately* release your foot from the accelerator pedal. *Never* hold the accelerator pedal down for more than 5 seconds when conducting the stall test. If the rpm rises above the recommended limit, serious transmission damage can result.

8. If another stall test is needed, place the gearshift lever in neutral. Run the engine at about 1,000 rpm for 20 to 30 seconds to cool down the transmission.

5.5 STALL TEST DIAGNOSIS

Compare the rpm readings from the stall test to speed charts in the proper service manual.

Stall Speed Above Recommendation: If the stall speed exceeds the manufacturer's recommendation by more than 200 rpm, it indicates transmission clutch slippage.

Stall Speed Below Recommendation: If stall speeds are 250 to 300 rpm below the manufacturer's recommendation, it is an indication that the stator overrunning clutch is defective. It may also indicate an engine power loss that can be corrected with an engine tune-up.

Stall Speed Within Recommendation: It is always wise to follow up the stall speed test with a road test to doublecheck transmission operation. Road-testing is discussed in detail in Unit 14.

Stall speed noise: A whining or siren-like noise is normal in some transmissions during stall testing. It is caused by fluid flow. However, a loud, metallic noise usually indicates a bad torque converter. To doublecheck the noise, raise the vehicle on a hoist. Be sure the rear wheels will turn freely. Run the vehicle at a light accelerator pedal pressure. Listen for noise from the torque converter or bellhousing with a mechanic's stethoscope or a length of tubing held to your ear. If metallic noise is heard, the torque converter must be replaced.

NAME ______________________ SECTION ____________ DATE ____________

Job Sheet 5-2

Stall Test

Before you begin:

Read Topics 5.4 through 5.5.

Vehicle Make ____________________ Model ____________ Year ____________

Time Started ____________________ Time Finished ____________________

Job Completion Time ____________________ Flat Rate Time ____________________

Tools Required

- Wheel blocks
- Tachometer
- Hoist
- Droplight
- Safety glasses
- Mechanic's stethoscope or length of tubing
- Grease pencil

References

Manufacturer's service manual ______________________________

Repair manual ______________________________

Service bulletin ______________________________

Procedure

SAFETY CAUTION: Do *not* let anyone stand in front of the vehicle when conducting this test. Be sure the area in front of the vehicle is clear. The service brakes must be fully applied when making this test, in addition to the emergency brake. However, brake problems can allow the vehicle to move forward, possibly causing property damage or personal injury. Block the wheels.

CAUTION: A stall test can ruin weak clutches and bands inside the transmission. If there is any doubt about the condition of clutches and bands, do *not* perform the stall test.

1. Connect a tachometer to the engine.
2. Place the tachometer inside the vehicle where it can be read easily. Mark the dial of the tachometer with a grease pencil at the recommended maximum revolutions per minute (rpm) for this test.

In the space provided, write the maximum rpm for this vehicle. ____________

CAUTION: Do *not* exceed the maximum rpm at any time during this test. Severe engine or transmission damage could result.

3. Check engine coolant and transmission fluid levels.*
4. Block the front wheels and set the emergency brake.
5. Place the transmission in park or neutral and warm the engine and transmission to operating temperatures.
6. Place the gearshift lever into the gear that the manufacturer recommends for this test.

In the space provided, write the gear position recommended by the manufacturer.

____

7. Press the accelerator pedal to the floor. Press down slowly at first, making sure the vehicle does not move. When the tachometer needle stops rising, ease off the accelerator pedal.

In the space provided, write the highest rpm reading during the test.

____

CAUTION: If the tachometer needle goes above the recommended rpm, *immediately* release your foot from the accelerator pedal. *Never* hold the accelerator pedal down for more than 5 seconds when conducting the stall test. If the rpm rises above the recommended rpm, serious transmission damage can result.

Instructor Check: ______________________

8. Place the gearshift lever in neutral. Run the engine at about 1,000 rpm for 20 to 30 seconds to cool down the transmission.
9. If noise is heard from the transmission during the test, raise the vehicle on a hoist.
10. Have a helper run the vehicle in drive on the hoist. Make sure the wheels are free to turn. Wear safety glasses. Using the mechanic's stethoscope or length of tubing, listen at the bellhousing for noise from the torque converter.

*See the text for alternate procedures.

Job Completion Report

1. What difficulties, if any, did you encounter in completing this procedure?

____

2. What did the results of the stall test indicate? ______________________

____

3. What noises, if any, were heard from the transmission during the stall test?

____

4. What noises, if any, were heard at the bellhousing? ______________________

____

INSTRUCTOR SIGNOFF ______________________ **DATE** ______________

5.6 BENCH-TESTING THE TORQUE CONVERTER

When a part is removed from the car and tested on a workbench, it is commonly called bench testing. Whenever a torque converter is removed during an overhaul (refer to 14.2) it must be inspected and bench-tested to make sure it is operating properly. A bad torque converter should always be replaced with a new or rebuilt torque converter of the proper rating. The torque converter is replaced as a unit rather than repaired in the shop.

The mechanic is always responsible for the condition of the torque converter during an overhaul. If a faulty torque converter is reused, most shops require that the mechanic who did the original work must redo it without pay. So, careful inspection and testing is important. The inspections and tests performed on a torque converter are:

- Drive stud and drive nut inspection
- Converter hub inspection
- Stator overrunning clutch check
- Converter end play
- Stator-to-impeller interference test
- Stator-to-turbine interference test
- Converter leakage test.

5.7 DRIVE STUD AND DRIVE NUT INSPECTION

The torque converter is held to the flexplate by either drive studs or welded lugs, called drive nuts. See Figure 5-4. Drive studs are threaded shafts welded to a unit to help turn it. The torque converter drive studs extend through the flexplate. Nuts are screwed onto the drive studs to secure the torque converter to the flexplate. Drive nuts are nuts welded to the torque converter housing. Bolts extend through the flex plate and into the drive nuts to secure the torque converter to the flex plate.

Examine the drive studs or drive nuts for stripped threads, bending or twisting, or other damage. If the drive studs or drive nuts are damaged, they must be replaced. Welding and other machine-shop techniques are necessary to repair these parts. If the damage also bent or warped the torque converter housing, the entire torque converter must be replaced.

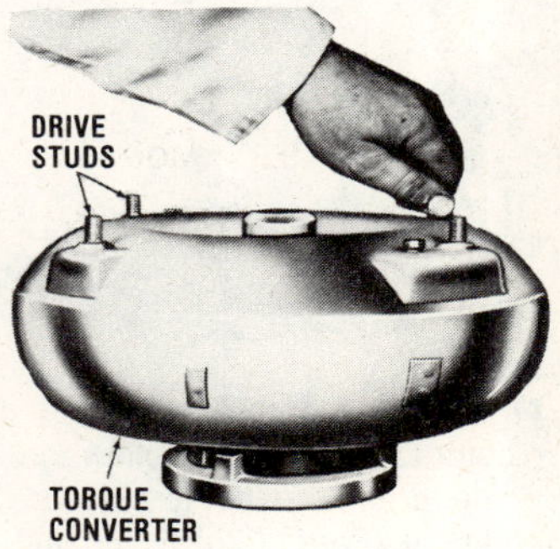

Figure 5-4. Drive studs or drive nuts can be used to attach the converter to the flexplate. (Ford Motor Company)

5.8 CONVERTER HUB INSPECTION

The hub is the part of the torque converter that sticks into the oil pump, as shown in Figure 5-5. The hub should be smooth. Small nicks, burrs, or light scoring can usually be corrected. Scoring is a series of grooves worn into the hub. If the hub is damaged beyond repair, the torque converter must be replaced.

To remove small nicks, burrs, or light scoring, crocus cloth is used. Be sure that the hub opening is covered to prevent dirt and metal particles from entering the torque converter.

TORQUE CONVERTER HUB

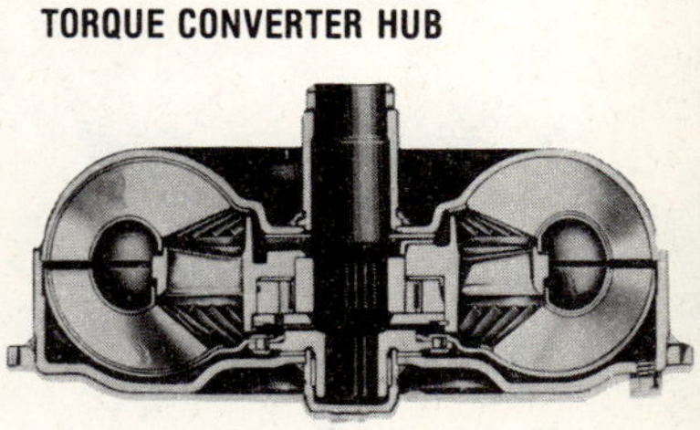

Figure 5-5. The hub of the converter fits into the oil pump. (Chrysler Corporation)

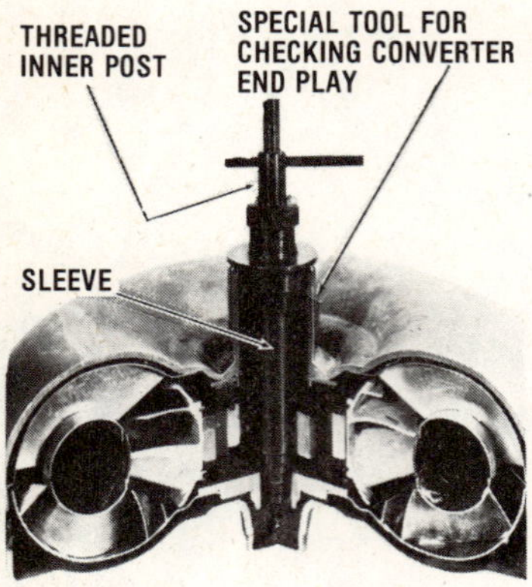

Figure 5-6. Insert the special tool in the converter drive hub until it engages the overrunning clutch splines. (Ford Motor Company)

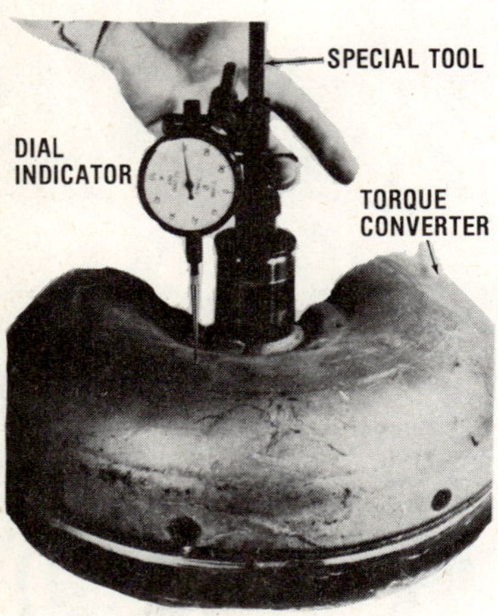

Figure 5-7. Attach a dial indicator to the tool and follow the procedure in the text to measure end play. (Ford Motor Company)

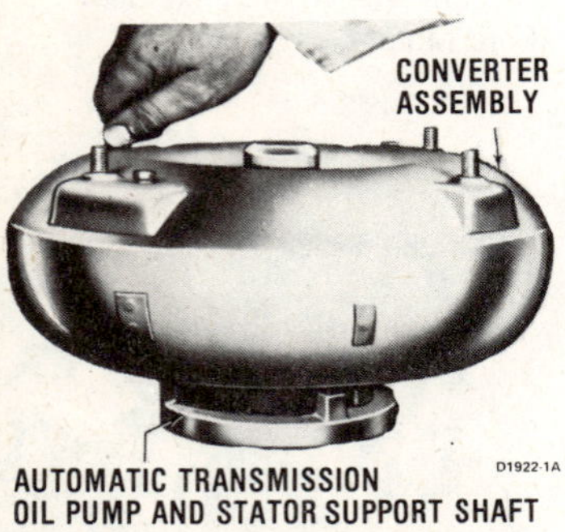

Figure 5-8. Rotate the converter counterclockwise, and check for stator-to-impeller interference. (Ford Motor Company)

5.9 STATOR OVERRUNNING CLUTCH CHECK

The operation of the stator is usually checked by hand. Insert a finger into the rear of the converter and into the splined inner race of the clutch. Try to turn the race in both directions. It should rotate smoothly, with *no* binding, clockwise. It should lock up when turned counterclockwise. If the clutch does not operate properly, the torque converter must be replaced.

Some manufacturers recommend the use of special tools for checking the stator overrunning clutch. Always consult the proper service manual.

5.10 CONVERTER END PLAY

The in and out movement of parts is called end play. This movement must be within the specifications outlined by the manufacturer. To check end play, special tools are required. The following procedure is an example of checking torque converter end play:

1. Insert the proper special tool in the hub until it bottoms, or stops. See Figure 5-6.

2. Tighten the tool until it is locked into the splines in the hub.

3. Place a dial indicator on the tool, as shown in Figure 5-7. The dial indicator plunger should rest on the hub at the rear of the torque converter housing. Set the dial indicator to zero (0).

4. Lift the tool upward as far as it will go. Note the dial indicator reading. This reading is the total end play that the turbine and stator share.

5. If the reading exceeds the manufacturer's specifications, the torque converter must be replaced.

6. Remove the tool.

5.11 STATOR-TO-IMPELLER INTERFERENCE TEST

An interference test is to check for any binding or scraping between the rotating parts of the torque converter. Two interference tests are made on the torque converter: stator-to-impeller interference and stator-to-turbine interference.

The first procedure is for the stator-to-impeller interference test. Figure 5-8 shows how the torque converter assembly is set up.

1. Place the oil pump assembly on a workbench. The splined end of the stator shaft should be pointing upward.

2. Mount the converter on the oil pump assembly. The splines on the inner race of the overrunning clutch must mesh with the splines of the stator support. Be sure the converter engages the drive gear of the oil pump.

3. Hold the oil pump so that it doesn't move.

4. Rotate the converter counterclockwise. The converter should rotate smoothly. There should be no interference, or binding, and no scraping noises.

5. If any interference or scraping noise is heard, the converter must be replaced. This is usually an indication that the impeller vanes and stator vanes are touching.

5.12 STATOR-TO-TURBINE INTERFERENCE TEST

Use Figure 5-9 as a guide to making the stator-to-turbine interference test.

1. Place the converter on a workbench with the front side down.

2. Mount the oil pump assembly on the converter. The splines of the stator support, stator, and oil pump drive gear must mesh.

3. Install the input shaft and mesh the input shaft splines with the splines on the turbine hub.

4. Hold the oil pump so that it doesn't move.

5. Rotate the input shaft to turn the turbine. The turbine should rotate smoothly in both directions. There should be no interference or scraping noise.

6. If any interference or scraping noise is heard,the converter must be replaced. This is usually an indication of a worn stator thrust washer, which causes the stator and turbine to touch. A thrust washer is a part that absorbs the force of one part against another, to protect both parts.

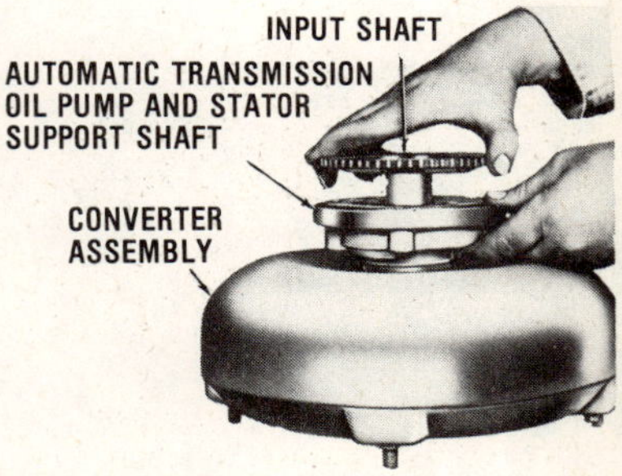

Figure 5-9. Hold the oil pump and rotate the input shaft in both directions to check for stator-to-turbine interference. (Ford Motor Company)

NAME ______________________ SECTION ____________ DATE ____________

Job Sheet 5-3

Torque Converter Bench Test

Before you begin:

Read Topics 5.5 through 5.11.

Vehicle Make ____________________ Model ____________ Year ____________

Time Started ____________________ Time Finished ____________________

Job Completion Time ______________ Flat Rate Time ____________________

Tools Required

- Special torque converter end-play tool
- Dial indicator

References

Manufacturer's service manual ______________________________

Repair manual ______________________________

Service bulletin ______________________________

Procedure: Inspect Converter

1. Examine the drive studs or drive nuts for stripped threads, bending or twisting, or other damage. In the Job Completion Report make a note of any damage found.
2. Inspect the torque converter hub for smoothness. In the Job Completion Report, make a note of the condition of the hub.
3. Cover the hub opening with a clean shop rag. Use crocus cloth to remove small nicks, burrs, or light scoring on the hub.
4. Check the stator overrunning clutch.* In the Job Completion Report make a note of any operating problems.

Procedure: Measure Converter End Play

1. Insert the proper special tool in the hub until it bottoms, or stops.
2. Tighten the tool until it is locked into the splines in the hub.
3. Set up a dial indicator on the tool. Zero the dial indicator.
4. Lift the tool upward as far as it will go.

In the space provided, write the indicator reading. . . . . ____________________

Instructor Check: ____________________

In the space provided, write the manufacturer's specification for maximum end play.

____

5. Remove the tool.

Procedure: Make Stator-to-Impeller Interference Test

1. Place the oil pump assembly on a workbench. The splined end of the stator shaft should be pointing upward.
2. Mount the converter on the oil pump assembly. Be sure the converter engages the drive gear of the oil pump.
3. Hold the oil pump so that it doesn't move.
4. Rotate the converter counterclockwise and check for proper operation. In the Job Completion Report, make a note of any problems found.

Procedure: Make Stator-to-Turbine Interference Test

1. Place the converter on a workbench with the front side down.
2. Mount the oil pump assembly on the converter.
3. Install the input shaft and mesh the input shaft splines with the splines on the turbine hub.
4. Hold the oil pump so that it doesn't move.
5. Rotate the input shaft to turn the turbine. Check if the turbine rotates smoothly in both directions. In the Job Completion Report, make a note of any interference or scraping noises.

Instructor Check: ____________________

*See the text for alternate procedures.

Job Completion Report

1. What problems, if any, were found in the visual inspection of the drive studs or nuts, torque converter hub, and stator overrunning clutch? ____________________

   ____

2. Is the converter end play within the manufacturer's specifications?

   ____

3. What problems, if any, were noticed when checking for stator-to-impeller interference? ____

4. What problems, if any, were noticed when checking for stator-to-turbine interference?

   ____

5. Based on all your inspections and tests, what, if anything, is wrong with the torque converter? ____

INSTRUCTOR SIGNOFF ________________________ **DATE** ______________

5.13 FLUSHING THE CONVERTER

When the transmission fluid becomes badly contaminated, it may be necessary to flush out the torque converter. Contaminated fluid carries sludge, coolant, friction material, or metals that will damage transmission parts.

Flushing is not often done on modern transmissions. Debris left by badly contaminated fluid cannot be completely eliminated from modern torque converters. Contaminants can quickly damage internal transmission parts. So, many service departments simply replace the entire torque converter rather than take a chance on a contaminated converter.

Torque converters with drain plugs can be more easily flushed. Sealed torque converters require drilling a hole carefully into the torque converter housing.

Two methods for flushing torque converters are commonly used: hand flushing and machine flushing. Machine flushing is the preferred method because it usually cleans the converter better.

To flush the torque converter, solvent is poured or forced into the converter housing. The solvent is sloshed around and drained simultaneously, so that particles are poured out while still suspended in the solvent. If the torque converter is to be flushed, always consult the proper service manual.

5.14 TORQUE CONVERTER LEAKAGE TEST

Sometimes leaks in the torque converter are not obvious or may be difficult to locate. In this case, special equipment is necessary. The basic test consists of filling the torque converter with air and submerging it in a tank of water. The torque converter is usually flushed (refer to 5.10) and cleaned with solvent before beginning the following procedure:

SAFETY CAUTION: Always wear protective glasses or goggles when working with compressed air. Dirt, debris, or liquids can be blown into the eyes, causing pain or loss of vision.

1. Install a special leakage test tool, shown in Figure 5-10, according to the manufacturer's directions.

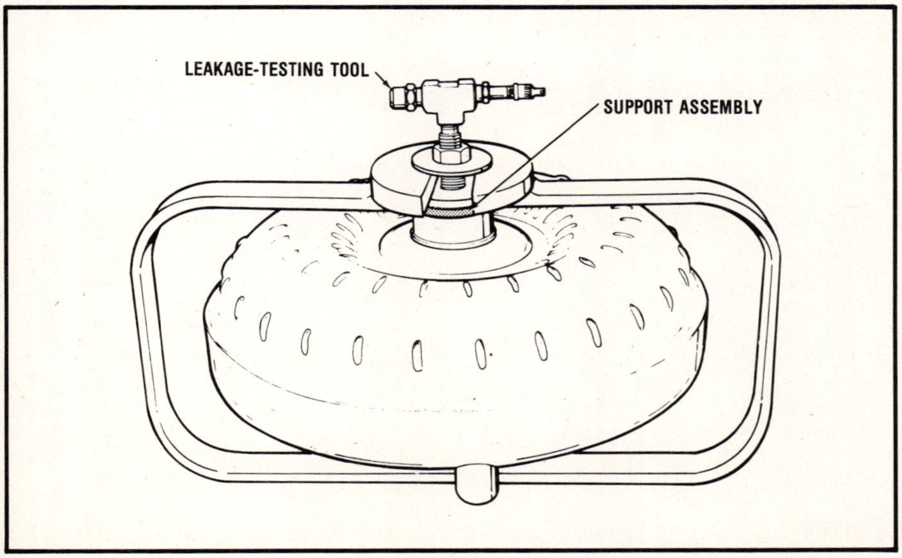

Figure 5-10. Leakage-testing tool and support assembly. Some converters require an additional rubber seal in the hub for sealing. (Chevrolet Motor Division-GMC)

SAFETY CAUTION: Make sure all fasteners are tightened securely. The leakage testing tool must be securely in place before proceeding with the test.

2. Apply about 80 psi [552 kPa] of compressed air, or to the manufacturer's specification.

SAFETY CAUTION: Always refer to the proper service manual for proper air pressure. Too much air pressure can damage the converter or cause it to explode and cause serious personal injury.

3. Place the pressurized converter into a tank of water.

4. Watch for air bubbles, which indicate leakage from the torque converter. Air bubbles may not appear immediately, so allow the converter to remain in the water for at least 10 minutes.

5. If air bubbles appear, replace the torque converter. If *no* air bubbles appear, the torque converter is not leaking.

6. Remove the torque converter from the water.

7. Release the air pressure and remove the test tool.

8. Dry the torque converter thoroughly to avoid rust.

SAFETY CAUTION: Escaping high-pressure air can cause personal injury.

NAME ______________________ SECTION ____________ DATE ____________

Job Sheet 5-4

Torque Converter Leakage Test

Before you begin:

Read Topics 5.10 through 5.11.

Vehicle Make ____________________ Model ____________ Year ____________

Time Started ____________________ Time Finished ____________________

Job Completion Time ______________ Flat Rate Time ____________________

Tools Required

- Safety glasses
- Safety chains or straps
- Converter leakage test tool
- Air compressor
- Water tank

References

Manufacturer's service manual ______________________________

Repair manual ______________________________

Service bulletin ______________________________

Procedure: Make Converter Leakage Test

SAFETY CAUTION: Always wear protective glasses or goggles when working with compressed air. Dirt, debris, or liquids can be blown into the eyes, causing pain or loss of vision.

1. Install a leakage test tool according to the manufacturer's directions.

SAFETY CAUTION: Make sure all fasteners are tightened securely. Safety chains or straps must be in place around the torque converter in case of explosion.

Instructor Check: ____________________

2. Fill the converter with compressed air to the manufacturer's specified pressure. In the space provided, write the correct amount of air pressure.

____________________psi.

SAFETY CAUTION: Always refer to the proper service manual for proper air pressure. Too much air pressure can damage the converter or cause it to explode and cause serious personal injury.

3. Place the pressurized converter into a tank of water.
4. Check the converter for leakage. In the Job Completion Report, make a note of any leakage, where found.
5. Remove the torque converter from the water.
6. Release the air pressure and remove the test tool.
7. Dry the torque converter thoroughly to avoid rust.

SAFETY CAUTION: Escaping high-pressure air can cause personal injury.

Job Completion Report

1. What difficulties, if any, did you encounter in completing this procedure?

____

____

2. What problems were found, if any, with the torque converter? ____________

____

____

____

INSTRUCTOR SIGNOFF ____________________ **DATE** ____________

NAME ____________________ SECTION ____________ DATE ____________

Certification Practice

Test 5

DIRECTIONS:

The following questions are similar to those used on mechanic certification tests. Answer each question by circling the letter of the correct choice.

1. Common indications of torque converter problems include all of the following EXCEPT
   A. slow acceleration from a stop in all gears.
   B. normal acceleration, but poor high-speed performance.
   C. transmission slippage.
   D. transmission overheating.

2. A stall test has been performed. The stall speed is 300 rpm above the recommended speed for the test. What is wrong?
   A. Test improperly performed
   B. Tachometer defective
   C. Transmission clutches slipping
   D. Overrunning clutch defective

3. What is the maximum time the accelerator should be depressed when performing a stall test?
   A. 5 seconds
   B. 0.5 seconds
   C. 20 to 30 seconds to cool off the transmission
   D. One minute

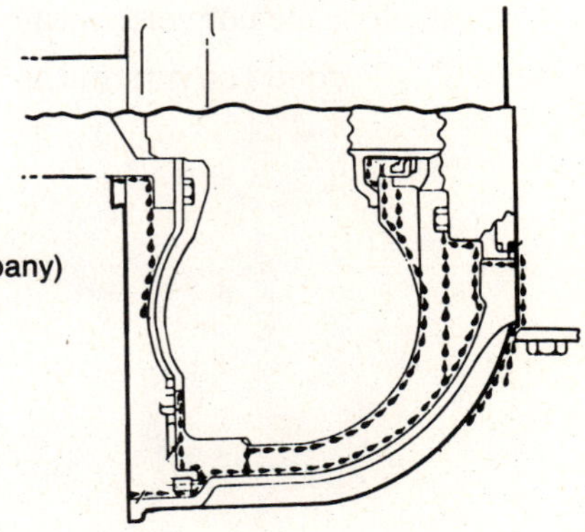

Question 4.
(Ford Motor Company)

4. The picture shows:
   A. fluid flow in the torque converter.
   B. how ATF is added to the transmission.
   C. sources of possible leaks.
   D. fluid flow in the automatic transmission.

5. Mechanic A says that when bench testing a torque converter, the overrunning clutch should lock up when turned clockwise.
   Mechanic B says that it should lock up when turned counterclockwise.
   Who is correct?
   A. A only B. B only C. Both A and B D. Neither A nor B

6. Which of the following statements is correct?

   I. The converter end-play check measures the combined end play of the impeller and the stator.

   II. The converter end-play check measures the combined end play of the turbine and stator.

   A. I only B. II only C. Both I and II D. Neither I nor II

7. A scraping noise heard during a stator-to-turbine interference test indicates a worn

   A. stator.
   B. turbine.
   C. stator thrust washer.
   D. overrunning clutch.

8. The tool in the picture is used to:

   A. flush the torque converter.
   B. make a hole for draining the torque converter.
   C. fill the torque converter with compressed air.
   D. stop leakage in the torque converter.

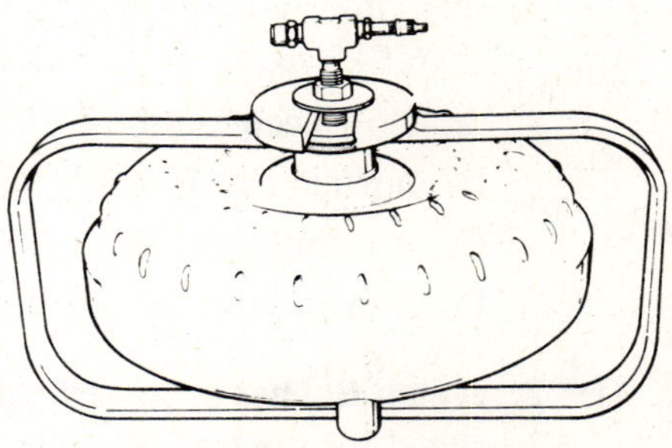

Question 8.
(Chevrolet Motor Division-GMC)

9. Mechanic A says that torque converters can be flushed if they are badly contaminated. Mechanic B says that in general it is better to replace a badly contaminated torque converter.

   Who is correct?

   A. A only B. B only C. Both A and B D. Neither A nor B

10. All of the following statements are correct EXCEPT

    A. The torque converter is replaced as a unit.
    B. Torque converters are sealed.
    C. Torque converters can be flushed if they become contaminated.
    D. Torque converters can be checked for damage or leaks.

6 Transmission Cooling Systems and Repairs

YOUR LEARNING JOB

Transmission fluid must be kept at temperatures that are cool enough to preserve the fluid's lubricating properties. Without proper lubrication, the internal transmission parts are quickly damaged. This unit discusses how the automatic transmission fluid is cooled and how the transmission cooling system can be checked and repaired.

JOB COMPETENCY OBJECTIVES

This is a knowledge-building and skill-building unit. When you have completed your assignments and exercises in this unit, you should be able to:

- ☐ Describe an internal transmission cooling system.
- ☐ Describe an external transmission cooling system.
- ☐ Make a cooler leak test.
- ☐ Flush the transmission cooler system.

6.1 COOLING THE AUTOMATIC TRANSMISSION

The action of the torque converter heats the automatic transmission fluid. In addition, the moving shafts and gears of the transmission rub against each other, creating friction, which creates more heat. In an automatic transmission, fluid temperatures can get very high, especially when a load is applied and during stop-and-go driving. When the fluid is overheated, the internal parts of the transmission overheat quickly. Without properly cooled lubrication, excess heat and friction will cause damage. Some way of cooling the fluid is necessary.

Automatic transmission fluid heat affects the life of the transmission. At about 175 degrees F [89.6 degrees C] the transmission life can be 100,000 miles. However, this is under ideal conditions. Because of varying load con-

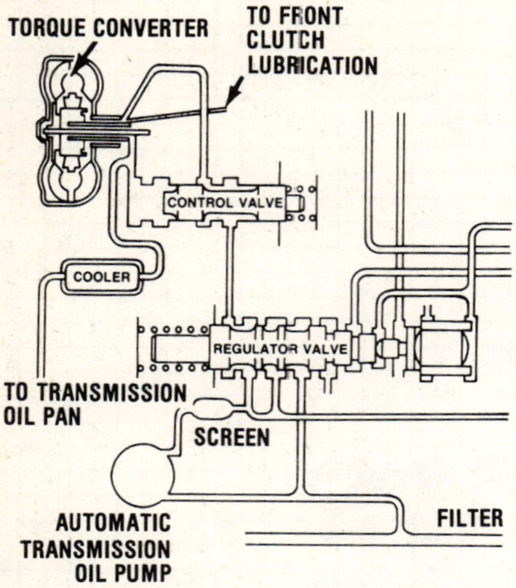

Figure 6-1. Transmission fluid heated by the action of the torque converter passes through a cooler, then returns to the automatic transmission oil pan. (Chrysler Corporation)

ditions, the temperature of the ATF can rise above 175 degrees F. Every 20 degrees F [11 degrees C] of additional fluid temperature causes a drastic decrease in transmission life. For example, at 195 degrees F [91.7 degrees C], transmission life is 50,000 miles. At 215 degrees F [101 degrees C], transmission life is 25,000 miles. At temperatures above 300 degrees F [198 degrees C], the metal parts within the transmission may begin to warp, or bend. A transmission fluid cooler, or transmission cooler, is used to reduce the temperature of the fluid.

Two types or transmission coolers are used: internal cooler and external cooler.

6.2 INTERNAL TRANSMISSION COOLER

Both coolers work in a similar way. Both take hot ATF from the torque converter and direct it to an intercooler, or heat exchanger. The hot ATF is cooled by either liquid, air, or both. The cooled fluid is then returned to the transmission. Fluid flow through a torque converter and cooler valve is shown in Figure 6-1.

Internal coolers are installed at the factory on all modern vehicles equipped with automatic transmissions. Heated transmission fluid travels from the torque converter to the engine radiator. A small internal cooler is located inside of the engine radiator. The cooled transmission fluid is then returned to the transmission lubrication circuit. An internal cooler system is illustrated in Figure 6-2.

An internal cooler is not always sufficient. When heavy loads are placed on a vehicle, the engine heats up. This heat is transferred through the cooling system to the engine coolant. The engine coolant can reach temperatures of up to 280 degrees F [138 degrees C] or more.

Under such conditions the internal transmission cooler will not transfer enough heat to the coolant. In fact, heat may be transferred from the hotter coolant to the ATF, causing *it* to overheat. When high engine temperatures are expected, such as when towing, an external transmission cooler should be used in addition to, or in place of, the internal cooler.

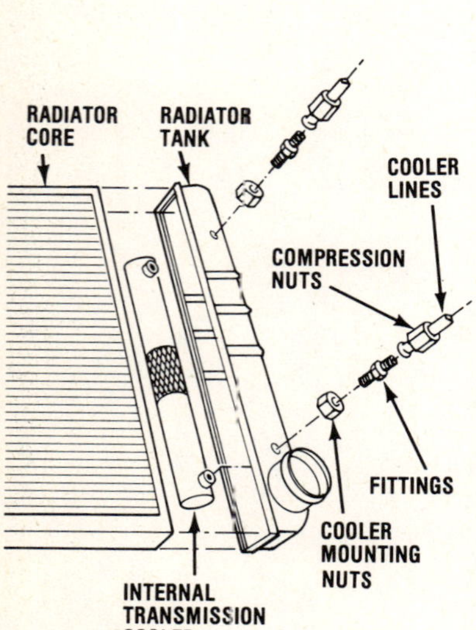

Figure 6-2. An internal transmission cooler is located in the radiator tank that connects to the lower radiator hose.

6.3 EXTERNAL TRANSMISSION COOLER

An external transmission cooler also has two lines to direct fluid to and from the transmission. The external cooler, however, is mounted outside of the engine radiator, usually just in front of it. One line often passes through the internal cooler first to begin cooling the ATF. Instead of returning directly to the automatic transmission, the fluid then passes through the external cooler. When the ATF reaches the external radiator, it is further cooled by air flowing through the front of the vehicle. Then the fluid is returned to the transmission. Figure 6-3 shows an external transmission cooler system.

Many automatic transmission repair shops in hot climates will not guarantee overhauls if an external cooler is not added. With the heat built up in modern engine radiators, internal coolers often do not provide sufficient cooling for the transmission fluid.

CHECKPOINT

1. The action of the torque converter ____________________ the automatic transmission fluid.
2. Automatic transmission fluid deteriorates quickly with ____________________.
3. Two types of transmission coolers are used: ____________ cooler and ________________ cooler.
4. The internal transmission cooler radiator is cooled by the engine radiator ____________________.
5. When the ATF reaches the external radiator, it is further cooled by ____________________ flowing through the front of the vehicle.

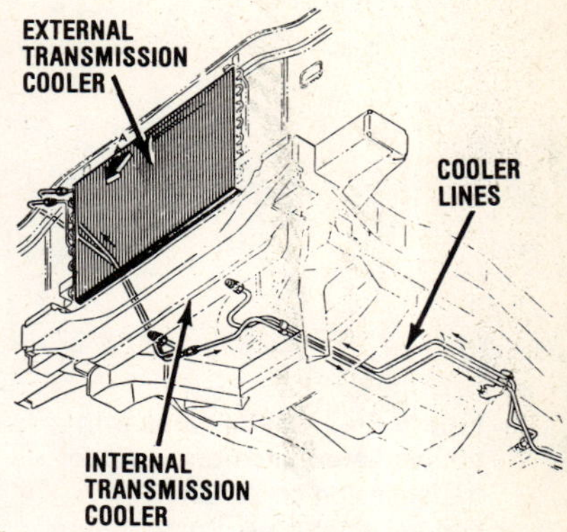

Figure 6-3. Notice how the transmission fluid passes first through the internal cooler, then through the external cooler. The external cooler can also be used alone. (Chrysler Corporation)

6.4 TRANSMISSION COOLER INSPECTION

The engine cooling system is the key to proper transmission cooling when an internal cooler is used. If anything affects engine cooling, it will also affect transmission cooling. The engine cooling system must always be inspected and properly maintained.

Looking for leaks, or potential leaks, is the main concern when visually checking the transmission cooler system.

First, check the transmission dipstick for signs of coolant mixed with transmission fluid. Milky fluid indicates that the transmission internal cooler, or core, is leaking. Engine coolant mixes with the ATF. This produces the milky appearance. Also, ATF leaks into the engine radiator coolant and floats on top of the coolant.

Inspect the two lines that run between the transmission and the transmission cooler. Look for damage, leaks, or looseness. Replace any lines that are damaged or leaking. Check the fittings, and tighten or repair them as necessary.

If it is suspected that the transmission cooler core in the engine radiator is leaking, it can be doublechecked with a leak test.

6.5 TRANSMISSION COOLER LEAK TEST

If a leak is suspected in an internal cooler, a leak test can be made. Use Figure 6-4 as a guide to locating the cooler lines. Leaks in external coolers are usually located by looking for discoloration where fluid has leaked out of the core.

1. Place a catchpan under the fittings that hold the cooler lines to the engine radiator.

2. Disconnect both cooler lines from the radiator. Plug or cap the lines from the transmission. A plug can be almost anything that will fit into an opening to stop fluid or airflow. A cap fits over the opening to stop flow.

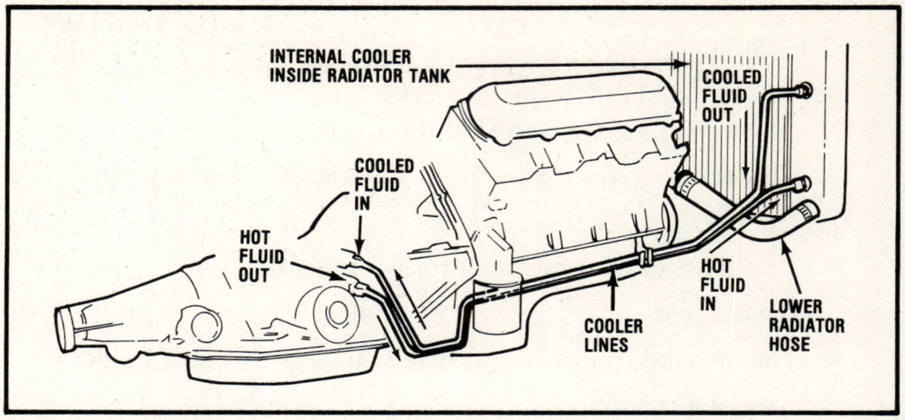

Figure 6-4. Metal tubes with double-flared ends are used for transmission cooler lines on domestic vehicles. Some foreign vehicles use flexible hoses. (Pontiac Motor Division-GMC)

3. Carefully remove the cap from the engine radiator to release pressure from the cooling system.

SAFETY CAUTION: Be sure the engine radiator is cool to the touch. A hot radiator will spray hot coolant up and toward you. This can result in serious eye damage and/or burns. Wearing safety glasses is always a good precaution when working around areas with hot liquids.

4. Plug one fitting in the cooler tightly.

5. Attach an air pressure line to the unplugged cooler fitting. Either compressed air or a hand pump can be used to make this test.

6. Build up air pressure to 50 or 75 pounds per square inch (psi).

SAFETY CAUTION: Wear protective safety glasses or goggles when working with compressed air. Particles, debris, and liquids can be blown into the eye, causing pain or blindness.

SAFETY CAUTION: Be very careful when applying air pressure to any closed container. A weak weld or weak metal can rupture or explode, causing injury.

7. Check the engine coolant through the radiator cap opening to see if bubbles appear. Bubbles indicate escaping air and a leak in the transmission cooler. If bubbles occur, the transmission cooler must be replaced.

NAME ____________________ SECTION ____________ DATE ____________

Job Sheet 6-1

Transmission Cooler Inspection and Intercooler Leak Test

Before you begin:

Read Topics 6.1 through 6.5.

Vehicle Make ____________________ Model ____________ Year ____________

Time Started ____________________ Time Finished ____________________

Job Completion Time ____________________ Flat Rate Time ____________________

Tools Required

- Absorbent white paper towels
- Shop rags
- Safety glasses
- Open-end wrenches or flare-nut wrenches
- Tubing cutter
- Tubing flaring tool
- Double-flare adapters
- Plugs or caps for cooler lines
- Air compressor
- Rubber-tipped air blow gun

References

Manufacturer's service manual ____________________

Repair manual ____________________

Service bulletin ____________________

Procedure: Check for Transmission Cooler Leaks

1. Check the dipstick for signs of coolant mixed with transmission fluid.
2. Open the radiator cap carefully and check for an oily deposit of ATF on the radiator cap and gasket.

SAFETY CAUTION: Be sure the engine radiator is cool to the touch before attempting to open the pressure cap. A hot radiator will spray hot coolant up and toward you. This can result in serious eye damage and/or burns.

3. Check in the radiator for evidence of ATF. In the Job Completion Report, below, make a note of any signs of coolant on the automatic transmission dipstick, and/or ATF in the radiator.
4. Replace the radiator cap.
5. Inspect the transmission cooler metal lines and fittings for leakage or damage.
6. Replace any damaged lines or fittings. In the Job Completion Report, make a note of any repairs done.

Instructor Check: ____________________

Procedure: Make Transmission Intercooler Leak Test

1. Place a catchpan under the fittings that hold the cooler lines to the engine radiator.
2. Disconnect both cooler lines from the radiator. Plug or cap the lines from the transmission.
3. Carefully remove the cap from the engine radiator to release pressure from the cooling system.

SAFETY CAUTION: Be sure the engine radiator is cool to the touch before attempting to remove the pressure cap. A hot radiator will spray hot coolant up and toward you. This can result in serious eye damage and/or burns.

4. Plug one fitting in the cooler tightly.
5. Attach an air pressure line to the unplugged cooler fitting.
6. Build up air pressure to 50 or 75 pounds per square inch (psi).

SAFETY CAUTION: Wear protective safety glasses or goggles when working with compressed air. Particles, debris, and liquids can be blown into the eyes, causing pain or blindness.

SAFETY CAUTION: Be very careful when applying air pressure to any hollow container. A weak weld or weak metal can rupture or explode, causing injury.

7. Check the engine coolant through the radiator cap opening for evidence of escaping air. In the Job Completion Report, make a note of any problems found.

Instructor Check: ____________________

Job Completion Report

1. What problems, if any, were found when checking for coolant on the automatic transmission dipstick or ATF in the radiator? ____________________

____

____

2. What repairs, if any, were made? ____________________

____

____

3. What were the results of the internal cooler leak test? ____________________

____

____

INSTRUCTOR SIGNOFF ____________________ **DATE** ____________

6.6 TRANSMISSION COOLER FLOW CHECK

The flow of fluid inside of the transmission must be constant and uninterrupted. A bad oil pump, a leak in the oil circuit system, or a restricted cooler or cooler line will affect fluid flow. To check fluid flow, use the following procedure:

1. Set the emergency brake and warm up the engine and transmission to operating temperatures.

2. Check fluid level. Also be sure the linkage and fluid control pressure are within the manufacturer's specifications. Fluid control pressure is discussed in 13.11.

3. Remove the dipstick from the filler tube.

4. Place a funnel into the filler tube. The funnel should be 1 quart or larger in size.

5. Raise the vehicle on a hoist.

6. Disconnect the cooler return line where it enters the transmission case.

SAFETY CAUTION: The engine, transmission, and cooler lines will be hot. Be careful when working around hot surfaces and with hot ATF. Serious burns can result.

7. Attach a hose to the disconnected cooler return line. The hose must be long enough to reach from the cooler line to the funnel. Fasten the end of the hose to the funnel to make sure it doesn't fall loose.

8. Lower the vehicle.

9. Place the gearshift lever in neutral (N) and start the engine. Let the engine run at about 1,000 rpm.

10. Watch fluid flow at the funnel. At first, fluid flow may not be constant, which is generally caused by air in the fluid circuit. Air will be bled off in a few minutes, and the flow should be constant and not mixed with air bubbles. An automatic transmission should flow at least 1 quart in 22 seconds. If flow is not constant, the transmission cooler flow may be restricted, or clogged.

11. Proper fluid flow is determined by each manufacturer, so be sure to check the proper service manual. To determine whether poor fluid flow is caused by the transmission cooler, check fluid flow at the converter-out fitting on the transmission case. The converter-out fitting is one of several fittings that can be found on many modern automatic transmissions to check oil pressure. See Unit 13 to locate and identify these oil pressure fittings. A weak flow from the cooler return line and a strong, constant flow from the converter-out fitting indicate that the cooler is restricted.

6.7 FLUSHING THE TRANSMISSION OIL COOLER SYSTEM

Contaminated fluid carries sludge and metal (friction) particles into the cooler system. Sludge is a muddy deposit left by contaminated fluid. After an overhaul, or when overheated fluid is replaced, the cooler system is flushed to remove the contaminants that have settled there.

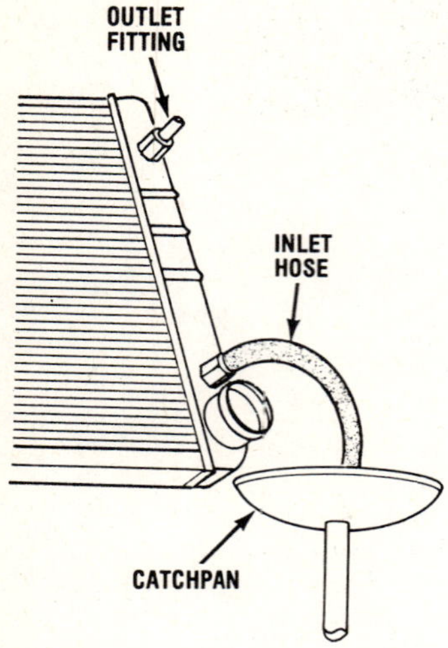

Figure 6-5. The internal transmission can be flushed by hand. Mineral spirits are forced into the outlet fitting and drained out of the inlet filling.

Flushing is important because fluid from the cooler does not return to the reservoir. Instead, fluid from the cooler returns directly to the lubrication circuit of the transmission and is not filtered. This simply contaminates the new fluid and transmission damage will occur.

The transmission oil cooler system can be flushed by hand or by using a machine. The following procedure is for hand flushing (See Figure 6-5). For machine flushing, consult the proper service manual.

1. Raise the vehicle on a hoist.
2. Place a catchpan under the engine radiator or external transmission cooler.
3. Disconnect both cooler lines at the engine radiator or external cooler.
4. Use a small screwdriver to scrape away any foreign material in the cooler inlet fitting.
5. Connect a suitable length of hose to the inlet fitting. Place the other end of the hose into the catchpan.
6. Wear safety glasses, and use mineral spirits and a compressed air gun to clean out the cooler. First, pour some mineral spirits into the outlet fitting of the cooler, then apply short bursts of air to the outlet fitting. This is called reverse flushing, or flushing from a direction opposite to that of normal fluid flow. Alternate between mineral spirits and short bursts of air until the fluid coming out the inlet fitting is clear.
7. Disconnect the cooler lines from the transmission case. Flush each line by pouring mineral spirits through them until the fluid comes out clear.
8. After flushing the cooler core and the lines, run transmission fluid through them to remove the mineral spirits.
9. Remove the hose from the radiator and reconnect the cooler lines.

6.8 TRANSMISSION OIL COOLER REPLACEMENT

Replacing or repairing the internal transmission oil cooler is a job that is usually done in a radiator shop. Special tools and special knowledge of soldering and brazing are required. Many shops do not replace or repair internal coolers. Instead, an external cooler is hooked up to replace the internal cooler.

External coolers are replaced by unbolting the old cooler and bolting in a new cooler.

NAME ______________________ SECTION ______________ DATE ______________

Job Sheet 6-2

Servicing the Transmission Cooler

Before you begin:

Read Topics 6.6 through 6.8.

Vehicle Make ______________________ Model ______________ Year ______________

Time Started ______________________ Time Finished ______________________

Job Completion Time ______________________ Flat Rate Time ______________________

Tools Required

- Absorbent white paper towels
- ATF pressure tester
- Hoist
- Lengths of rubber hose
- ATF funnel
- Open-end or flare-nut wrenches
- Catchpan
- Small funnel
- Safety glasses
- Air compressor
- Rubber-tipped air blow gun
- Mineral spirits

References

Manufacturer's service manual ______________________________________

Repair manual ______________________________________

Service bulletin ______________________________________

Procedure: Transmission Cooler Flow Check

1. Set the emergency brake and warm up the engine and transmission to operating temperatures.
2. Check fluid level and condition.
3. Check the transmission linkage and fluid control pressures. In the Job Completion Report, make a note of any problems found.
4. Remove the dipstick from the filler tube.
5. Place an ATF funnel into the filler tube.
6. Raise the vehicle on a hoist.
7. Disconnect the cooler return line where it enters the transmission case.

SAFETY CAUTION: The engine, transmission, and cooler lines will be hot. Be careful when working around hot surfaces and with hot ATF. Serious burns can result.

8. Attach a hose to the disconnected cooler return line.
9. Lower the vehicle. Place the other end of the hose into the funnel.
10. Place the gearshift lever in neutral (N) and start the engine. Have a helper run the engine at about 1,000 rpm.
11. Check for a constant, strong fluid flow at the funnel. In the Job Completion Report, make a note of the fluid flow.
12. Shut off the engine.
13. Raise the vehicle on a hoist, and remove the hose from the return line. Reconnect the return line and tighten the fitting.
14. Disconnect the converter-out fitting, and attach the hose to the fitting.
15. Lower the vehicle and again check for a constant, strong flow. In the Job Completion Report, make a note of any problems found.

Instructor Check: ____________________

16. Raise the vehicle and reconnect the converter-out fitting. Lower the vehicle, remove the funnel, and replace the dipstick.

Procedure: Hand Flush Transmission Cooler

1. Raise the vehicle on a hoist.
2. Place a catchpan under the engine radiator or external transmission cooler.
3. Disconnect both cooler lines at the engine radiator or external cooler.
4. Clean the opening of the cooler inlet fitting.
5. Connect a hose to the inlet fitting. Place the other end of the hose into the catchpan.
6. Attach a hose and funnel to the cooler outlet fitting.
7. Pour mineral spirits into the funnel. Wear safety glasses and use a rubber-tipped air blowgun to apply short bursts of air to the outlet fitting. Repeat these steps until the mineral spirits coming out of the inlet hose are clean.
8. Disconnect the cooler lines from the transmission case. Flush each line by pouring mineral spirits through them until the fluid comes out clean.

Instructor Check: ____________________

9. Pour transmission fluid through the cooler and the lines to remove the mineral spirits.
10. Remove the hoses from the transmission cooler and reconnect the cooler lines.

NAME ______________________ SECTION ____________ DATE ____________

Job Completion Report

1. What problems, if any, were found in the checks of fluid level, control pressures, and transmission linkage? ____

____

____

2. Describe the fluid flow during the test: ________________________________

____

____

3. What difficulties, if any, did you encounter in flushing the transmission cooler?

____

____

INSTRUCTOR SIGNOFF ______________________ DATE ____________

NAME ______________ SECTION ______________ DATE ______________

Certification Practice

Test 6

DIRECTIONS:

The following questions are similar to those used on mechanic certification tests. Answer each question by circling the letter of the correct choice.

1. Mechanic A says that overheated transmission fluid can melt internal metal transmission parts.
   Mechanic B says that overheated fluid loses its lubricating properties.
   Who is correct?
   A. A only   B. B only   C. Both A and B   D. Neither A nor B

2. ATF that has been heated to 215°F [101°C] has an expected life of
   A. 100,000 miles.
   B. 50,000 miles.
   C. 25,000 miles.
   D. 12,500 miles.

Question 3.
(Chrysler Corporation)

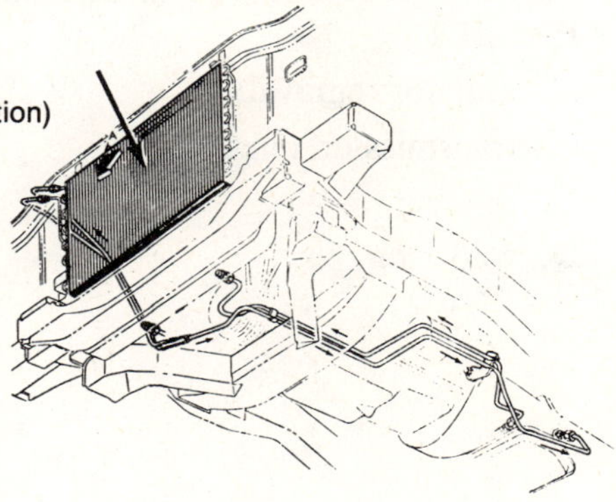

3. The arrow in the picture points to:
   A. an external transmission cooler.
   B. an internal transmission cooler.
   C. an engine radiator.
   D. an engine oil cooler.

4. Which of the following statements is correct?
   I. An internal cooler transfers heat to engine coolant.
   II. An external cooler transfers heat to outside air.
   A. I only   B. II only   C. Both I and II   D. Neither I nor II

5. All of the following may be used on modern vehicles EXCEPT
   A. internal cooler only.
   B. external cooler only.
   C. internal and external coolers connected together.
   D. no cooler.

6. Milky ATF is found on an automatic transmission dipstick. What is the most likely cause?
   A. ATF contaminated with particles
   B. ATF overheated
   C. Transmission cooler leaking
   D. Transmission cooler lines disconnected

7. Mechanic A says that bubbles can be seen in the radiator if an internal cooler leaks during a test with compressed air.

   Mechanic B says bubbles can be seen in the automatic transmission dipstick if an internal cooler leaks during a test with compressed air.

   Who is correct?

   A. A only B. B only C. Both A and B D. Neither A nor B

8. Eye protection should be worn
   A. when safety glasses or goggles are easily available.
   B. at all times, day and night.
   C. whenever a potential eye hazard may be present during work.
   D. only when the instructor is watching.

9. Which of the following statements is correct?
   I. If the flow from the cooler line is weak during testing and the flow from the converter-out fitting is good, the problem is in the cooler.
   II. If the flows from both the cooler return line and the converter-out fitting are weak, the problem is in the transmission.

   A. I only B. II only C. Both I and II D. Neither I nor II

10. Replacing or repairing the internal transmission cooler may require all of the following EXCEPT
    A. radiator removal.
    B. transmission removal.
    C. brazing.
    D. soldering.

7

III
Planetary Gearsets and Apply Devices

Planetary Gearsets and Apply Devices

YOUR LEARNING JOB

The planetary gear system allows the automatic transmission to change gear ratios without shifting, or repositioning, the gears. The planetary gear system is controlled by apply devices. This unit discusses how planetary gearsets and apply devices operate.

JOB COMPETENCY OBJECTIVES

This is a knowledge-building unit. When you have completed your reading assignments and exercises in this unit, you should be able to:

- ☐ Identify the parts of a planetary gearset.
- ☐ Explain input, output, and held members.
- ☐ Describe the differences between a clutch pack and a brake band.
- ☐ Explain how an accumulator and servo operate.

7.1 PLANETARY GEARSET OPERATION

All modern automatic transmissions have either two or three planetary gearsets linked together. These are called compound planetaries. A typical three-speed automatic will have two planetary gearsets. A four-speed overdrive automatic will have three planetary gearsets. Each gearset is controlled by a common or separate sun gear. A simple planetary gearset is discussed in 1.5.

All planetary gearsets are basically the same. The biggest difference is in the type of gears used. Most automatic transmissions use helical-cut gears. The teeth on a helical-cut gear are cut at an angle to the axis of the gear. Straight-cut gears are used on heavy-duty transmissions. Straight-cut gears are stronger and transfer more power, but are very noisy in operation. Helical-cut gears operate more quietly. Figure 7-1 shows the difference between helical-cut and straight-cut gears.

Figure 7-1. Straight-cut, or spur, gears have teeth parallel to the gear axis (left). Helical gear teeth are at an angle to the gear axis. (right)

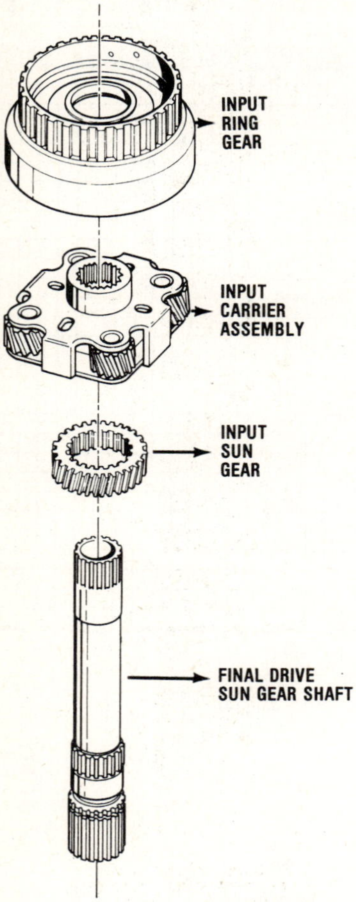

Figure 7-2. The sun gear can be a narrow gear that fits over a hollow shaft, or a long shaft that is splined and hollow to turn the planetary carrier. (Hydra-Matic)

When two planetary gearsets are used, they usually share a common sun gear. The sun gear is actually a long tube that is splined to the front and rear planetary gearsets. See Figure 7-2. The inside of the hollow sun gear rotates freely around the transmission output shaft and rides on bushings.

The planet gears mesh with the sun gear. Three to five planet gears are used around the sun gear in each gearset. The number of planet gears depend on the load the transmission is expected to encounter. Automobile transmissions generally have three planet gears in the planetary gearset. More planet gears are used only for heavy-duty applications.

The planetary carrier holds the planet gears in place. The planet gears fit over pins, or small shafts, on the carrier. Needle bearings between the planet gears and the small shafts allow the gears to rotate smoothly. The planetary carrier is usually splined to the output shaft.

The ring gear meshes with the planet gears. The ring gear can be controlled by a brake band that surrounds the outer surface of the ring gear. Brake band operation is discussed in 7.4. In some three-speed transmissions, the forward friction clutches are splined to the front planetary ring gear. The planetary gear system can operate in five basic ways to provide:

- Gear reduction
- Direct drive
- Reverse
- Neutral
- Overdrive.

A single planetary gearset provides only a limited number of gear ratios. To produce three or four forward speeds in automobile transmissions, two or three planetary gearsets are used. When more than one planetary gearset is used, it is referred to as a compound planetary system. In a two-gearset compound planetary system, the front gearset is called the input unit and the rear gearset is called the reaction unit.

CHECKPOINT

1. **A typical three-speed automatic transmission will have ________________ planetary gearsets.**
2. **The biggest difference in planetary gearsets is in the type of ________________ used.**
3. **When two planetary gearsets are used, they usually share a common ________________ .**
4. **The planetary gear system can operate in five basic ways to provide: ________________ ________________, ________________ ________________, ________________, ________________, and ________________.**
5. **In a compound planetary system, the front gearset is called the ________________ unit and the rear gearset is the ________________ unit.**

7.2 APPLY DEVICES

The planetary gearset is controlled by holding and driving devices, called apply devices. An apply device holds or drives parts of the planetary system. Depending on which apply device is activated, one member of the planetary gearset is driven, and another member is held. There are several kinds of apply devices:

- Multiple-disc clutches
- Transmission bands
- Servos
- Accumulators (hydraulic apply circuit cushions)
- Overrunning clutches.

Apply devices are activated by hydraulic pressure, but apply a mechanical force to the planetary gear system to hold the members. The following sections describe the operation of the different apply devices.

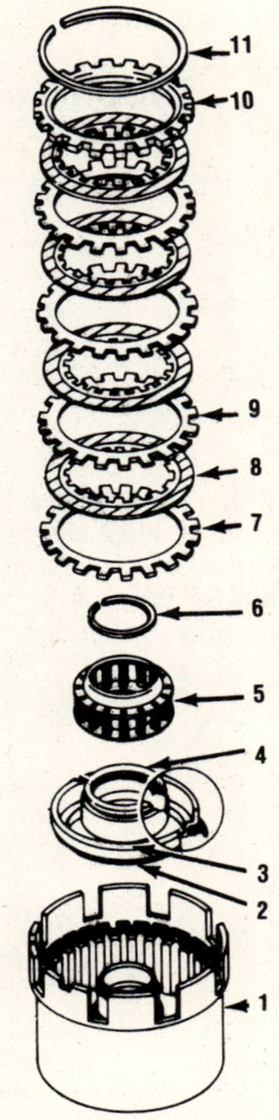

1. REVERSE INPUT HOUSING
2. REVERSE INPUT OUTER SEAL
3. REVERSE INPUT PISTON
4. REVERSE INPUT INNER SEAL
5. REVERSE INPUT SPRING ASSEMBLY
6. REVERSE INPUT SPRING SNAP RING
7. REVERSE INPUT PLATE (WAVED)
8. REVERSE INPUT PLATE (COMPOSITION)
9. REVERSE INPUT PLATE (STEEL)
10. REVERSE INPUT BACKING PLATE
11. REVERSE INPUT SNAP RING

Figure 7-3. Two types of discs help to operate a multiple-disc clutch. The discs are friction discs and steel discs, shown above. (Chevrolet Motor Division-GMC)

7.3 MULTIPLE-DISC CLUTCHES

A multiple-disc clutch consists of a series of friction discs, or plates, sandwiched between steel discs. See Figure 7-3. The number of friction discs and steel discs is determined by how much load the transmission is expected to handle. Heavy-duty transmissions have more discs. Friction discs have rough gripping surfaces. Steel discs have smooth surfaces.

Driving Clutches

The discs are enclosed in a round clutch housing, or drum, attached to one member of a planetary gearset. The friction discs are lugged or splined to the driving member or driven member of the planetary assembly. Lugs are also called splines, teeth, or dogs. The steel discs are lugged to the clutch drum, and the clutch drum is lugged to the input shaft. A piston inside the drum presses the clutch plates firmly together against a pressure plate. A return spring inside the piston retracts the piston to release the clutch plates.

Automatic transmission fluid under pressure operates the multiple-disc clutch assembly. The fluid pressure pushes the piston to compress the return spring and force the clutch plates against the pressure plate. When the friction discs and steel discs are compressed against the pressure plate, the discs, the input shaft, and the drum rotate as one unit.

The clutch pack is released when fluid flow is stopped. The return spring expands to return the piston to the rear of the clutch drum. At this point, the excess fluid is forced to the outside of the spinning drum by centrifugal force. A ball-type check valve at the rear of the drum opens to allow the excess fluid to escape. If the excess fluid did not escape, it would cause bad shifts and, possibly, damage to the clutch assembly. A ball-type check valve consists of a ball and a seat. The seat is where the ball rests. In this instance, the ball rests over a hole, or orifice. Depending on the flow of fluid, the ball will seat, closing up the orifice, or unseat, opening the orifice.

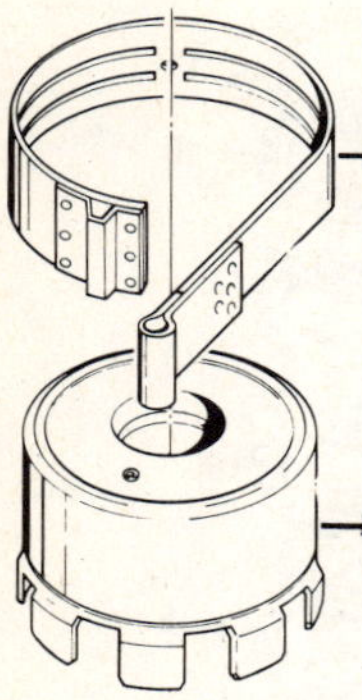

Figure 7-4. The transmission band wraps around a clutch drum. As the band is squeezed closed against the drum surface, the drum stops rotating. (Hydra-Matic)

Holding Clutches

A multiple-disc clutch can also be used as a holding device. In a holding clutch, the outside of the clutch drum is splined to a stationary part, such as the transmission case. The friction discs are splined to the inside of the drum. A shaft connected to one member of a planetary gearset is splined to the steel discs. When the clutch is applied, the shaft locks to the stationary clutch drum, preventing it from turning, and holding the planetary member.

7.4 TRANSMISSION BANDS

A transmission brake band is a flexible piece of metal (Figure 7-4) that wraps around a clutch drum. The inside of a transmission band has a friction surface that grips the clutch drum to stop it. The strength and thickness of a band is determined by the load it is expected to handle.

Transmission bands are holding members. The band tightens to stop a clutch drum and releases to allow the drum to rotate freely. The band is controlled by hydraulic pressure and a servo.

CHECKPOINT

1. **Apply devices are in the form of: ________-________ ________, ________ ________, ________, ________, and ________ ________.**
2. **Friction discs have rough ________ surfaces.**
3. **The steel discs are splined to the ________ ________.**
4. **A piston inside of the drum compresses the clutch plates against a ________ ________.**
5. **The ________ ________ expands to return the piston to the rear of the clutch drum.**
6. **The inside of a ________ ________ has a friction surface that grips the clutch drum to stop it.**

7.5 SERVOS

A transmission band has two lips, or flanges, at the open end of the band. One flange fits into a slot in the transmission case. The other flange is forced inward, toward the stationary flange, by a servo. The servo controls the action of the transmission band.

The servo consists of a piston inside of a cylinder. See Figure 7-5. The piston has a rod on one end that acts against the transmission band to close it. Fluid pressure enters one end of the cylinder and pushes against the piston

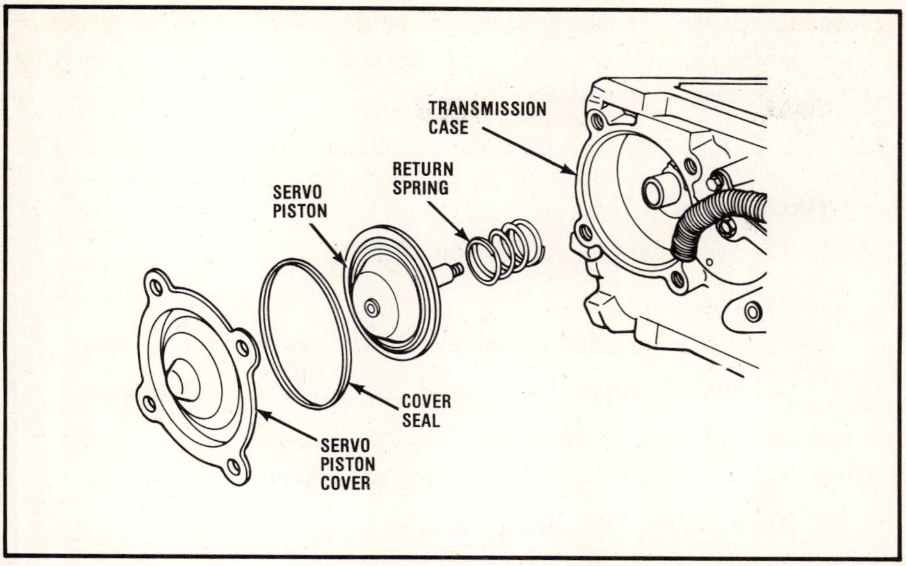

Figure 7-5. A servo controls the action of a transmission band. (Ford Motor Company)

and a return spring. When fluid pressure drops off, the return spring pushes the piston back and the transmission band opens.

The area of the servo piston face is larger than the area of the oil passage that enters the servo. This increases hydraulic force so that a greater amount of pressure can be exerted on the transmission band. Hydraulic principles are covered in Unit 9.

7.6 ACCUMULATORS

An accumulator acts as a shock absorber. It absorbs the shock when a clutch, piston, or band is applied. Essentially, the accumulator accumulates, or saves, excess fluid.

When transmission fluid is forced to stop quickly, or if it changes direction suddenly, it will cause vibration or hard shifts. This will shorten the life of transmission parts. So some sort of cushioning device is necessary, such as an accumulator.

Like a servo, an accumulator consists of a cylinder, a piston, and a return spring. See Figure 7-6. When pressure goes into the servo, it goes into the accumulator at the same time. The accumulator piston then begins to move to accept the incoming fluid. Part of the force of the fluid pressure is used to move the accumulator piston. Pressure builds up gradually in the servo to allow smooth shifts. When the accumulator is full, all of the fluid pressure acts on the servo. When fluid pressure is released, the return spring pushes the accumulator piston back to empty the accumulator cylinder.

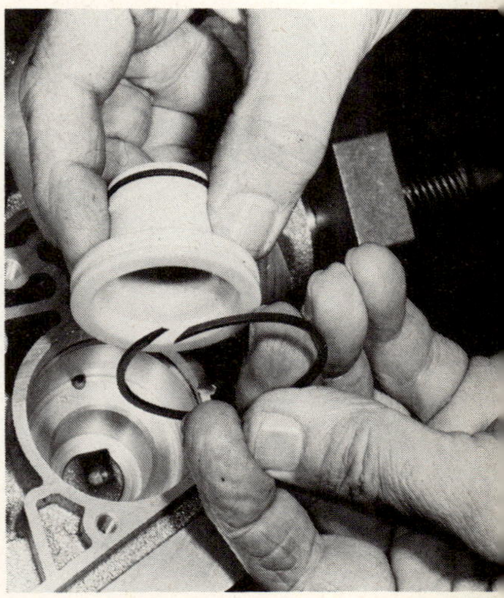

Figure 7-6. The accumulator has parts similar to a servo, except the accumulator is a cushioning device.

7.7 OVERRUNNING CLUTCHES

The one-way, or overrunning, clutch is often used with the planetary gearset system. It is the only apply device that is not operated by hydraulic pressure. In fact, the overrunning clutch is more of a holding device than an apply device. Its function is to prevent counter, or backward, rotation. The overrunning clutch works exactly the same as the overrunning clutch in the torque converter. See 4.6.

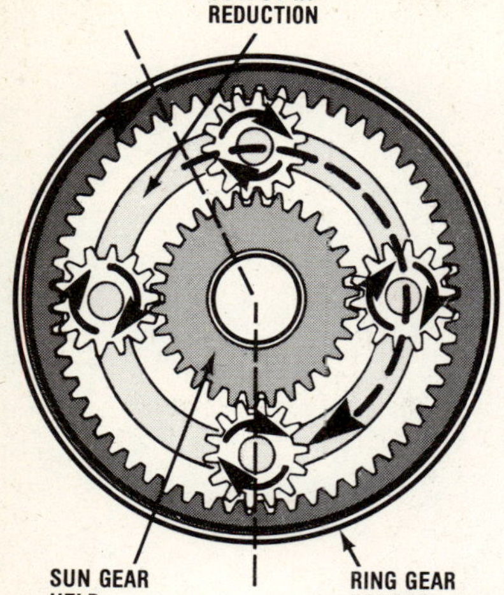

Figure 7-7. A planetary gearset in simple reduction.

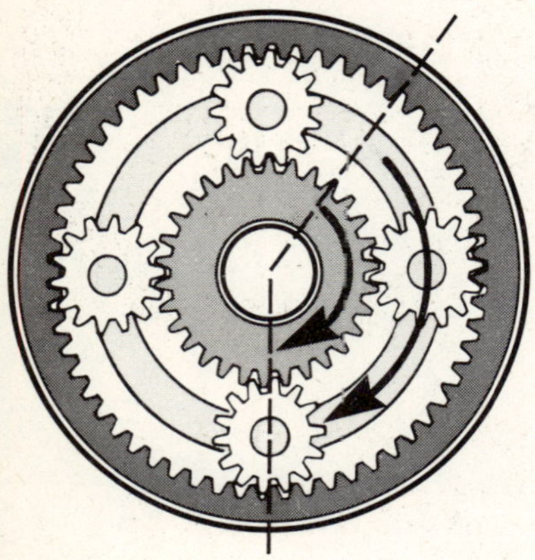

Figure 7-8. A planetary gearset in direct drive.

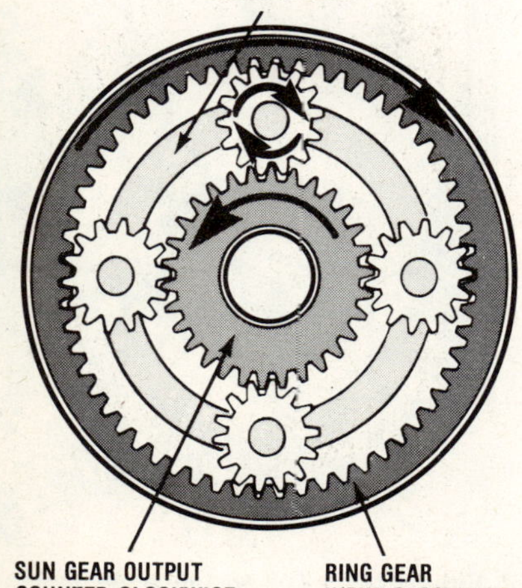

Figure 7-9. A planetary gearset in reverse.

An overrunning clutch can be used to hold one member of the planetary gearset. When the held member turns faster than (overruns) the driving member, the overrunning clutch releases.

CHECKPOINT

1. **The servo controls the action of the ________________ ________________.**
2. **Fluid pressure enters one end of the servo cylinder and pushes against the ________________ and a ________________ ________________.**
3. **An ________________ acts as a shock absorber.**
4. **An overrunning clutch holds a ________________ member in the planetary gearset.**

7.8 PLANETARY GEARSET POWER FLOW

Planetary gears are used to change torque, change speed, reverse the direction of rotation, provide a neutral, and act as a coupling for direct drive.

Multiplying torque is generally known as reduction, because there is always a decrease in speed of the output member. With a constant input speed, the output torque increases as the output speed decreases.

Simple reduction occurs when the sun gear is held and torque is applied to the ring gear in a clockwise direction. See Figure 7-7. In this case, the planetary pinions rotate in a clockwise direction and ''walk'' around the stationary sun gear. This causes the carrier assembly to rotate clockwise, and in reduction.

Direct drive occurs when any two members of the planetary gearset rotate in the same direction at the same speed. See Figure 7-8. This forces the third member to turn at the same speed. In this condition, the pinions do not rotate on their axles. Instead, the pinions lock the entire unit together to form one rotating part.

Reverse occurs whenever the carrier is held and power is applied to either the sun gear or the ring gear. This causes the planet pinions to drive the output member in the opposite direction. See Figure 7-9.

Since all modern automatic transmissions use compound planetary gearsets, the following power-flow descriptions are for two connected planetary systems.

Power Flow in Neutral

In neutral, the band and all clutches are released. No power is transmitted from the forward clutch housing to the planetary gearsets and to the output shaft.

Power Flow in Drive Range First Gear

With the gearshift lever in D, the forward clutch is applied. This delivers torque to the input ring gear in a clockwise direction. The clockwise rotation of the input ring gear causes the input pinions to turn clockwise. This drives the input sun gear clockwise.

The reaction, or output, sun gear drives the reaction pinions clockwise. The reaction pinions drive the reaction ring gear and the output shaft clockwise in reduction. The overrunning clutch is holding the reaction ring carrier. See Figure 7-10.

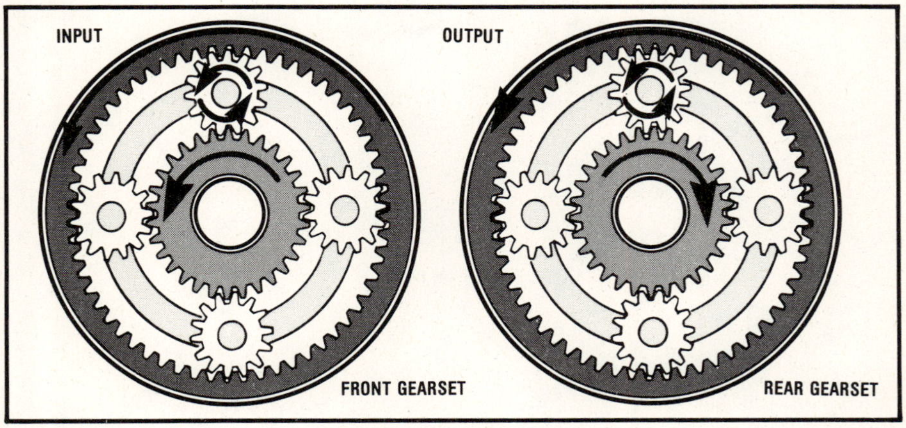

Figure 7-10. When the gearshift lever is in D and first gear begins working, the two planetary gearsets rotate differently.

Power Flow in Drive Range Second Gear

When the transmission shifts automatically to second gear, the intermediate band is applied to hold the sun gear. Converter torque is applied through the forward clutch to the input ring gear in a clockwise rotation. The input ring gear turns the input pinions counterclockwise around the stationary sun gear. This causes the input carrier and output shaft to turn counterclockwise in reduction. See Figure 7-11.

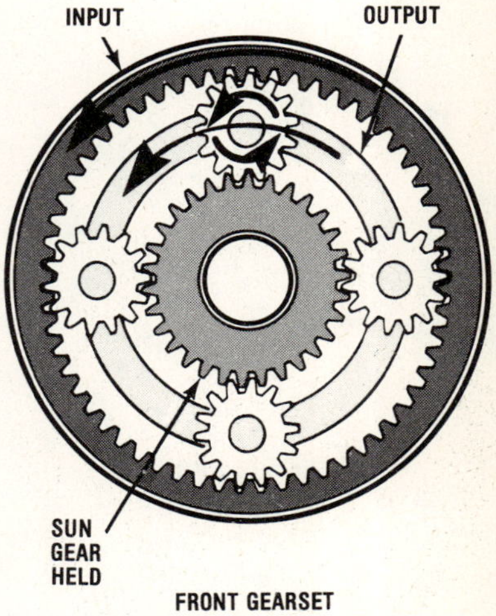

Figure 7-11. When the gearshift lever is in D and second gear is operating, the sun gear is stationary.

Power Flow in Drive Range Third Gear

In third gear, converter torque goes through the forward clutch to the input ring gear in a clockwise rotation. The direct clutch is applied. Clockwise rotation is transmitted through the direct clutch to the input drum and to the sun gear. Both the sun gear and input ring gear are now turning as one unit in direct drive. See Figure 7-12.

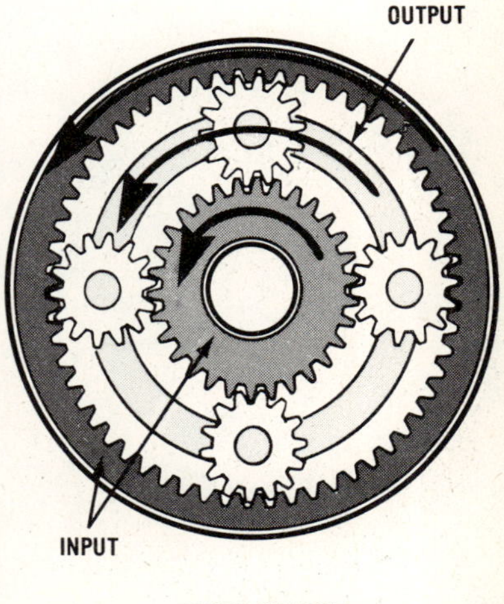

Figure 7-12. When third gear is engaged and the gearshift lever is in D, the sun gear and input ring gear turn as one unit.

CHECKPOINT

1. **Multiplying torque is generally considered as ________________ because there is always a decrease in the speed of the output member.**
2. **Direct drive occurs when any ________________ members of the planetary gearset rotate in the same direction at the same speed.**
3. **Reverse occurs whenever the ________________ is held and power is applied to either the sun gear or ring gear.**
4. **In ________________, the band and all clutches are released.**

NAME ____________________ SECTION ____________ DATE ____________

Certification Practice

Test 7

DIRECTIONS:

The following questions are similar to those used on mechanic certification tests. Answer each question by circling the letter of the correct choice.

1. A typical three-speed automatic transmission will have how many planetary gearsets?
   A. 1
   B. 2
   C. 3
   D. 4

2. Mechanic A says that the clutch drum in a multiple-disc clutch must always be free to turn, so it can drive a planetary gearset member.
   Mechanic B says that the clutch drum must always be splined to a non-moving part of the transmission.
   Who is correct?
   A. A only B. B only C. Both A and B D. Neither A nor B

3. Which of the following statements is correct?
   I. When the accumulator spring is partially compressed, fluid pressure on the servo will be less than maximum.
   II. When the accumulator spring is fully compressed, fluid pressure on the servo will be maximum.
   A. I only B. II only C. Both I and II D. Neither I nor II

4. All of the following are correct EXCEPT
   A. Transmission bands are made of metal with a friction lining.
   B. Transmission band thickness is related to the forces the band must deal with.
   C. Transmission bands are used to hold planetary gearset members.
   D. Transmission bands are used to drive planetary gearset members.

5. Which of the following statements about multiple-disc clutches is correct?
   A. Fluid enters the clutch drum through a hollow output shaft.
   B. The steel discs are splined to the inside of the clutch drum.
   C. The steel discs are splined to the outside of the clutch drum.
   D. The return spring applies pressure to force the clutch disc against the pressure plate.

6. Planetary gearsets are used for all of the following purposes EXCEPT
   A. preventing counter, or back, rotation.
   B. multiplying torque and increasing speed.
   C. reversing the direction of rotation.
   D. acting as a coupling for direct drive.

7. The compound planetary gearset shown is shifted into
   A. drive range first gear.
   B. drive range second gear.
   C. drive range third gear.
   D. reverse.

Question 7.

8. Mechanic A says that when the transmission automatically shifts to second gear, converter torque turns the input ring gear counterclockwise.
   Mechanic B says that when the transmission automatically shifts to second gear, converter torque turns the input ring gear clockwise.
   Who is correct?
   A. A only B. B only C. Both A and B D. Neither A nor B

9. In drive range third gear:
   A. The input ring gear and sun gear turn at the same speed.
   B. Clockwise rotation is transmitted to the input drum and sun gear.
   C. The direct drive clutch is not applied.
   D. The gear ratio is 2:1.

10. Which of the following statements is correct?
    I. In neutral, no power is transmitted to the output shaft.
    II. In direct drive, the input shaft turns at the same speed as the output shaft.
    A. I only B. II only C. Both I and II D. Neither I nor II

8 Repairing Planetary Gearsets and Apply Devices

YOUR LEARNING JOB

The planetary gearsets and apply devices control the power flow of the automatic transmission at the output shaft. If these parts wear badly or break, they can contaminate the transmission fluid and/or ruin the entire transmission assembly. This unit covers the inspection and repair of planetary gearsets and apply devices.

COMPETENCY OBJECTIVES

This is a skill-building unit. When you have completed your assignments and exercises in this unit, you should be able to:

- ☐ Adjust transmission bands.
- ☐ Clean and inspect planetary gearsets and apply devices.
- ☐ Explain the purpose of thrust washers.
- ☐ Adjust multiple-clutch clearance.

8.1 TRANSMISSION BAND ADJUSTMENT

Adjusting the transmission bands may or may not be part of a scheduled maintenance program. Some manufacturers recommend periodic adjustments. Others recommend adjustment only after an overhaul. Some transmissions do not have a band adjustment.

Most band adjustments are performed on the outside of the transmission case, with the vehicle on a hoist. Some band adjustment may require draining the transmission fluid and removing the oil pan.

CAUTION: It is very important that band adjustment procedures be followed *exactly* to the manufacturer's specifications. Serious damage can result if this adjustment is done in a sloppy manner.

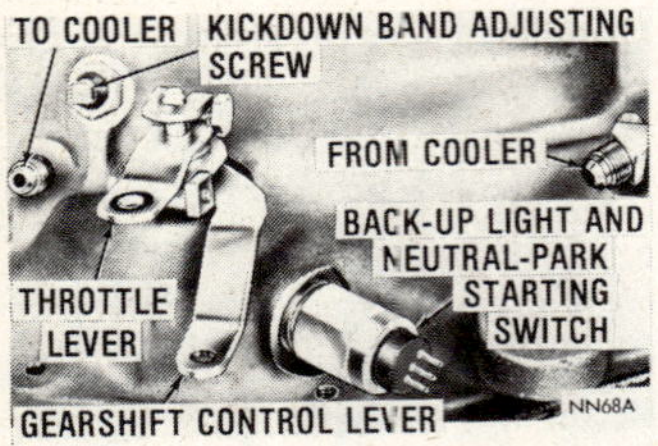

Figure 8-1. Location of kickdown band adjustment screw on TorqueFlite. (Chrysler Corporation)

The following three procedures are typical of the types of band adjustments you can expect to find.

8.2 CHRYSLER TORQUEFLITE BAND ADJUSTMENTS

The Chrysler TorqueFlite and American Motors Torque Command transmissions have two band adjustments. The front intermediate, or kickdown, band is adjusted at the outside of the transmission case. The rear, or low/reverse, band is adjusted from inside of the oil pan. Raise the vehicle on a hoist. Use safety glasses.

Kickdown Band Adjustment

The kickdown band adjusting screw is located on the left side of the transmission. See Figure 8-1. Clean all dirt from the adjusting screw area. Use the following procedure to adjust this band:

1. Loosen the locknut around the adjusting screw approximately five turns. Make sure the adjusting screw is free to turn in the transmission case.

2. Torque the adjusting screw to specifications.

NOTE: Chrysler has two torque specifications, depending on which tool is used. A special Chrysler band-adjusting tool requires a different torque rating than a conventional torque wrench.

3. Back off on the adjusting screw the number of turns that are specified in the proper service manual.

4. Hold the adjusting screw in its set position. Tighten the locknut to specifications.

Low/Reverse Band Adjustment

The low/reverse band is located inside of the transmission case. See Figure 8-2. The fluid must be drained and the oil pan must be removed before any adjustment can be made. Refer to Unit 2. Use the following procedure to adjust the low/reverse band:

1. Loosen the locknut around the adjusting screw approximately five turns. Make sure the adjusting screw is free to turn in the lever.

2. Torque the adjusting screw to specifications.

3. Back off on the adjusting screw the number of turns that are specified in the proper service manual.

4. Hold the adjusting screw in its set position. Tighten the locknut to specifications.

5. Install a new oil pan gasket. Replace the oil pan. Torque the oil pan bolts to specifications.

6. Fill the transmission with the proper fluid.

LOW/REVERSE BAND ADJUSTMENT

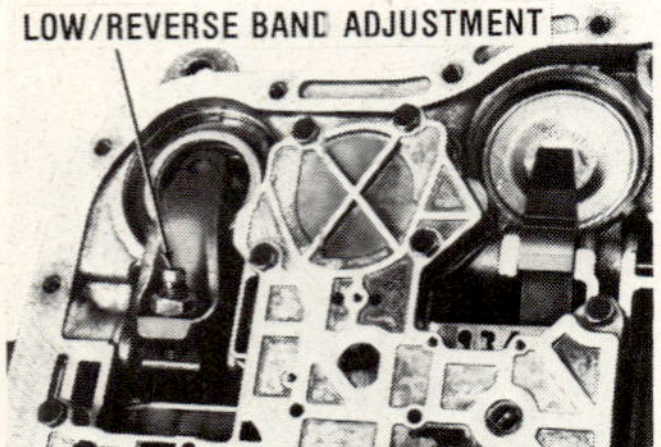

Figure 8-2. The low/reverse band adjustment on a TorqueFlite must be made inside the transmission case. (Chrysler Corporation)

8.3 FORD C-4/C-5 BAND ADJUSTMENTS

Both front and rear band adjustments are made on the outside of the transmission case on these transmissions. Raise the vehicle on a hoist. Wear safety glasses when making the adjustments.

Intermediate Band Adjustment

The front, or intermediate, band adjusting screw is located on the lower left side of the transmission case. See Figure 8-3. Clean all dirt from the adjusting screw area. Use the following procedure to adjust the front band:

1. Remove and discard the locknut. The locknut is a fluid-sealing fastener. It may *not* seal if it is reused.

2. Install a new locknut loosely over the adjusting screw.

3. Torque the adjusting screw to specifications.

4. Back off on the adjusting screw the number of turns that are specified in the proper service manual.

5. Hold the adjusting screw in its set position. Tighten the locknut to specifications.

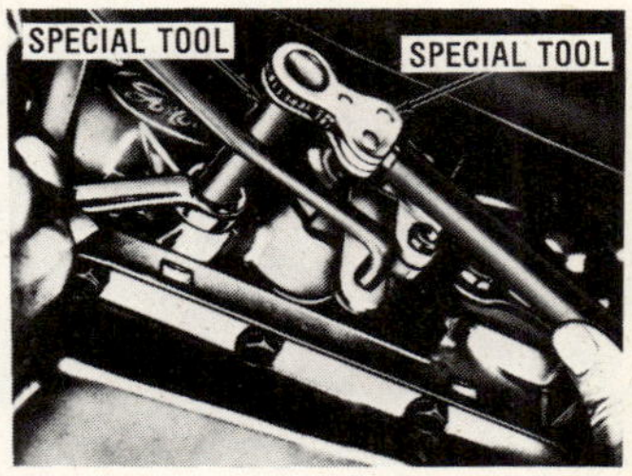

Figure 8-3. Ford intermediate band adjustments are made at the left side of the transmission case. (Ford Motor Company)

Low/Reverse Band Adjustment

The rear, or low/reverse band adjusting screw is located on the lower right side of the transmission case. See Figure 8-4. Clean all dirt from the adjusting screw area. Use the following procedure to adjust the rear band:

1. Remove and discard the locknut. The locknut is a fluid-sealing fastener. It may *not* seal if it is reused.

2. Install a new locknut loosely over the adjusting screw.

3. Torque the adjusting screw to specifications.

4. Back off on the adjusting screw the number of turns that are specified in the proper service manual.

5. Hold the adjusting screw in its set position. Tighten the locknut to specifications.

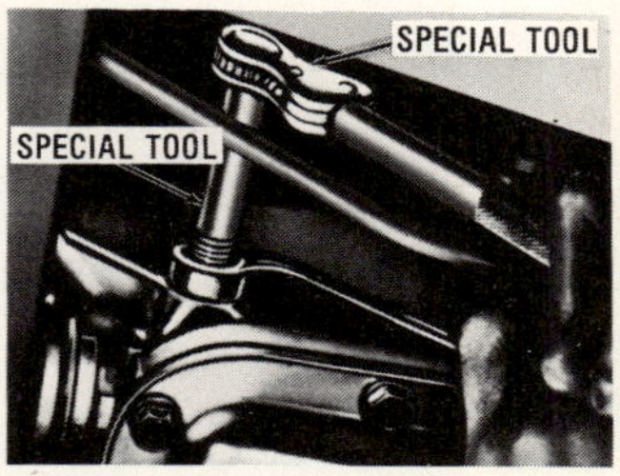

Figure 8-4. Low/reverse band adjustments are made on the right side of Ford transmissions. (Ford Motor Company)

8.4 GM TURBO HYDRA-MATIC 200/200C/200-4R BAND ADJUSTMENT

Only one adjustment is made on this family of transmissions. Adjustment is made at the intermediate servo, which is located at the lower right side of the transmission. Special tools are required for this adjustment. See Figure 8-5. The car must be raised on a hoist to make this adjustment. Wear safety glasses when making the adjustment. Use the following procedure to adjust the intermediate band:

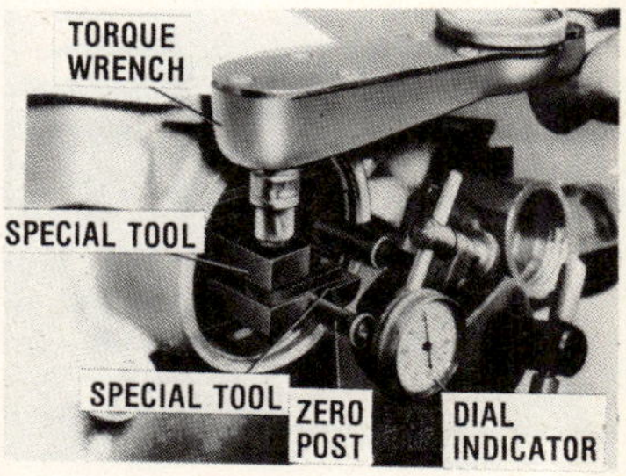

Figure 8-5. Special tools are used to adjust the intermediate band on General Motors vehicles. (Oldsmobile Division-GMC)

1. Remove the intermediate servo assembly by pressing down on the servo cover and removing the snap ring. The servo assembly is shown in Figure 8-6.

SAFETY CAUTION: Wear safety glasses, because the servo assembly is under spring tension. Use extreme care that the cover and servo parts don't fly loose when they are removed. Serious personal injury can result.

2. Inspect all servo parts for wear and damage. Replace any parts that are worn or damaged.

3. Clean the servo bore.

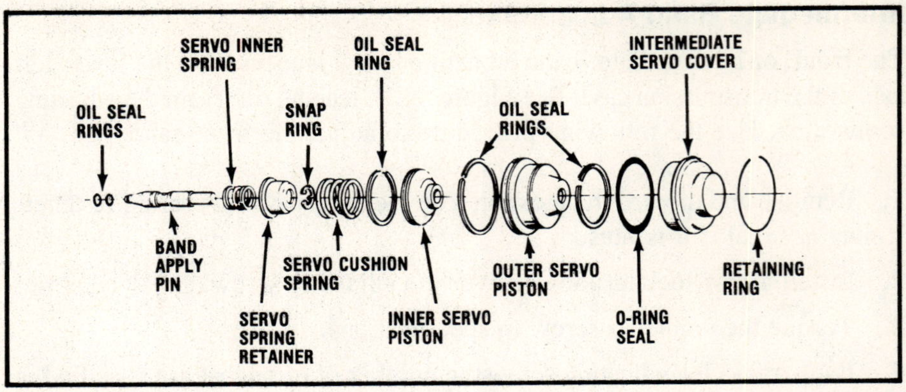

Figure 8-6. A typical General Motors intermediate servo assembly. (Oldsmobile Division-GMC)

4. Install a dial indicator and a special pin gauge into the servo bore, as shown in Figure 8-6. Closely follow the manufacturer's recommendations for installing the parts of the special pin gauge.

5. Zero the dial indicator.

6. Align the stepped side of the band apply pin with the torquing arm of the pin gauge.

7. Torque the hex nut on the side of the pin gauge to specifications.

8. Slide the dial indicator over the pin gauge. Compare the readings to specifications. If the reading does not meet specifications, use a band apply pin of the proper size to correct the readings.

9. Remove the dial indicator and special pin gauge.

10. Install the intermediate servo.

NAME ______________________ SECTION ____________ DATE ____________

Job Sheet 8-1

Adjusting Chrysler Torque Flite/AMC Torque Command Transmission Bands

Before you begin:

Read Topics 8.1 through 8.2.

Vehicle Make ____________________ Model ____________ Year ____________

Time Started ____________________ Time Finished ____________________

Job Completion Time ____________________ Flat Rate Time ____________________

Tools Required

- Hoist
- Safety glasses
- Droplight
- Putty knife
- Shop rags
- Solvent
- Open-end or combination wrenches
- Special band-adjusting torque tool
- Sockets, extensions, handle
- Torque wrench

References

Manufacturer's service manual ________________________________

Repair manual ________________________________

Service bulletin ________________________________

Procedure: Kickdown Band Adjustment

1. Raise the vehicle safely on a hoist. Use safety glasses. Clean all dirt from the adjusting screw area.
2. Loosen the locknut around the adjusting screw approximately five turns. Make sure the adjusting screw is free to turn in the transmission case.
3. Torque the adjusting screw to specifications.

NOTE: Chrysler has two torque specifications, depending on which tool is used. A special Chrysler band-adjusting tool requires a different torque rating than a conventional torque wrench.

In the space provided, write the correct torque. . . . . ____________________

Instructor Check: ____________________

4. Back off on the adjusting screw the number of turns that are specified in the proper service manual.

In the space provided, write the number of turns backed off. ____________

Instructor Check: ____________

5. Hold the adjusting screw in its set position. Tighten the locknut to specifications.

Procedure: Adjust Low/Reverse Band

1. Remove the oil pan and drain the transmission.
2. Loosen the locknut around the adjusting screw approximately five turns. Make sure the adjusting screw is free to turn in the lever.
3. Torque the adjusting screw to specifications.

In the space provided, write the correct torque. . . . . ____________

Instructor Check: ____________

4. Back off on the adjusting screw the number of turns that are specified in the proper service manual.

In the space provided, write the number of turns backed off. ____________

Instructor Check: ____________

5. Hold the adjusting screw in its set position. Tighten the locknut to specifications.
6. Install a new oil pan gasket. Replace the oil pan. Torque the oil pan bolts to specifications.
7. Fill the transmission with the proper fluid.

Job Completion Report

1. What difficulties, if any, did you encounter in completing these procedures?

____

____

____

INSTRUCTOR SIGNOFF ____________________ **DATE** ____________

NAME ______________________ SECTION ____________ DATE ____________

Job Sheet 8-1A

Adjusting Ford C-4/C-5 Transmission Bands

Before you begin:

Read Topics 8.1 and 8.3.

Vehicle Make ____________________ Model ____________ Year ____________

Time Started ____________________ Time Finished ____________________

Job Completion Time ____________________ Flat Rate Time ____________________

Tools Required

- Hoist
- Safety glasses
- Droplight
- Putty knife
- Shop rags
- Solvent
- Open-end or combination wrenches
- Special band-adjusting torque tool

References

Manufacturer's service manual ____________________

Repair manual ____________________

Service bulletin ____________________

Procedure: Adjust Intermediate Band

1. Safely raise the vehicle on a hoist. Use safety glasses. Clean all dirt from the adjusting screw area.
2. Remove and discard the locknut.
3. Install a new locknut loosely over the adjusting screw.
4. Torque the adjusting screw to specifications.

In the space provided, write the correct torque. . . . . ____________________

Instructor Check: ____________________

5. Back off on the adjusting screw the number of turns that are specified in the proper service manual.

In the space provided, write the number of turns backed off. ____________________

Instructor Check: ____________________

6. Hold the adjusting screw in its set position. Tighten the locknut to specifications.

Procedure: Low/Reverse Band Adjustment

1. Use safety glasses. Clean all dirt from the adjusting screw area.
2. Remove and discard the locknut.
3. Install a new locknut loosely over the adjusting screw.
4. Torque the adjusting screw to specifications.

In the space provided, write the correct torque. . . . . ____________________

Instructor Check: ____________________

5. Back off on the adjusting screw the number of turns that are specified in the proper service manual.

In the space provided, write the number of turns backed off. ____________________

Instructor Check: ____________________

6. Hold the adjusting screw in its set position. Tighten the locknut to specifications.

Job Completion Report

1. What difficulties, if any, did you encounter in completing these procedures?

____

____

____

INSTRUCTOR SIGNOFF ____________________ **DATE** ____________

NAME ______________________ SECTION ____________ DATE ____________

Job Sheet 8-1B

Adjusting GM Turbo Hydra-Matic 200/200C/200-4R Transmission Bands

Before you begin:

Read Topics 8.1 and 8.4.

Vehicle Make ______________________ Model ____________ Year ____________

Time Started ______________________ Time Finished ______________________

Job Completion Time ________________ Flat Rate Time ______________________

Tools Required

- Hoist
- Safety glasses
- Dial indicator
- Special pin gauge

References

Manufacturer's service manual ______________________________

Repair manual ______________________________

Service bulletin ______________________________

Procedure

1. Safely raise the vehicle on a hoist. Use safety glasses.
2. Press down on the servo cover and remove the snap ring.

SAFETY CAUTION: The servo assembly is under spring tension. Use extreme care that the cover and servo parts don't fly loose when they are removed. Always wear safety glasses. Serious personal injury can result.

3. Inspect all servo parts for wear and damage. Replace any worn or damaged parts.
4. Clean the servo bore.
5. Install a dial indicator and a special pin gauge into the servo bore.
6. Zero the dial indicator.
7. Align the stepped side of the band apply pin with the torquing arm of the pin gauge.
8. Torque the hex nut on the side of the pin gauge to specifications.

In the space provided, write the correct torque. . . . . ____________________

9. Slide the dial indicator over the pin gauge. Compare the readings to specifications. Replace the band apply pin with one of the proper size, if necessary.

Instructor Check: ____________________

10. Remove the dial indicator and special pin gauge.
11. Install the intermediate servo.

Job Completion Report

1. What difficulties, if any, did you encounter in completing this procedure?

____

____

____

INSTRUCTOR SIGNOFF ____________________ **DATE** ____________

8.5 INSPECTING AND REPLACING APPLY DEVICES AND PLANETARY GEARSETS

The disassembly and reassembly of apply devices, planetary gearsets, and other mechanical parts of the automatic transmission are covered in the overhaul sections. Refer to Units 15, 16, and 26. However, a very important part of the overhaul is knowing how to identify worn and damaged parts. The following sections cover cleaning and inspection of the mechanical parts of the transmission. Use this unit as a guide when it's time to do an overhaul.

8.6 MULTIPLE-DISC CLUTCH INSPECTION

All multiple-disc clutch assemblies are similar. After disassembly, clean all parts—except the friction discs—in fresh solvent. All seals must be replaced with new ones. A multiple-disc clutch is shown in Figure 8-7.

If any parts of the clutch assembly are defective, they must be replaced. The following procedures are used to inspect the parts of the clutch assembly.

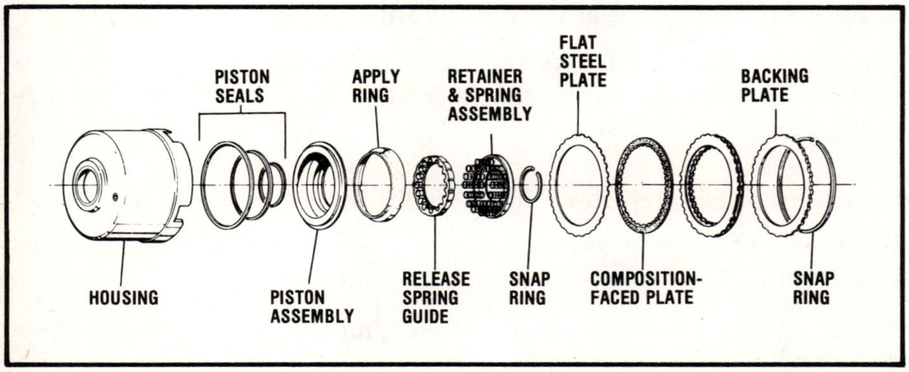

Figure 8-7. Multiple-disc clutch assembly used on the Turbo Hydro-matic 200-4R transmission. (Chevrolet Motor Division-GMC)

1. The steel clutch plates should not be discolored, scored, distorted, or have damaged lugs. These plates can be replaced individually, but most recommendations are that they be replaced only as a set. It's a good idea to rough up the shiny surface of the steel plates with 80-grit sandpaper. The roughness will give the steel plates a better gripping surface.

2. The friction clutch plates are usually automatically replaced, unless the transmission has very low mileage on it. Check the friction plates by looking for any damage, wear, or discoloration. The friction plates should always be soaked in ATF for at least 15 minutes before they are placed in the clutch assembly.

3. The pressure plate is checked the same way as the steel clutch plates. Cracks and discoloration mean that the pressure plate is bad.

4. Piston return springs may be either a set of coils or a Belleville spring. A Belleville spring is a cupped metal disc with fingers pointing toward the open center portion. The coil springs must be round, straight, and not be collapsed. If one spring looks bad, it is usually recommended that the entire set be replaced. A Belleville spring must not show damage or wear, especially at the inner fingers. If there is any evidence of extreme heat, all springs should always be replaced.

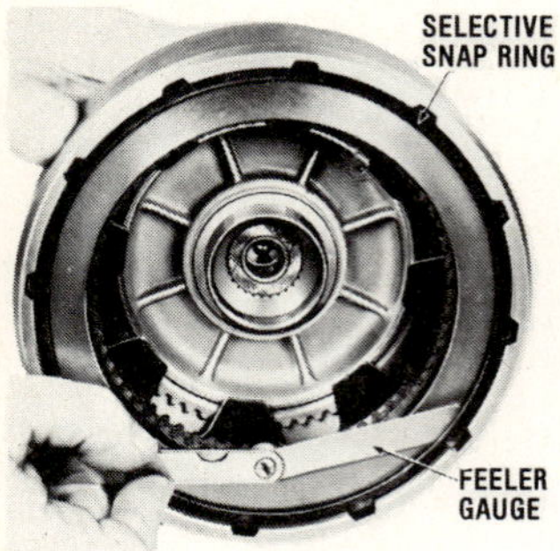

Figure 8-8. Feeler gauge is used to measure clutch plate clearance. (Chrysler Corporation)

5. All drums are checked for damaged lugs, grooves, scoring, and damaged splines. Also check the housing check ball to make sure it operates freely. Minor scoring or burrs on the band-application surfaces can be removed by lightly polishing with crocus cloth.

NOTE: When clutch packs are reassembled, clutch pack clearance must be measured and corrected. Clutch pack clearance is discussed in 8.7.

8.7 ADJUSTING MULTIPLE-CLUTCH CLEARANCE

Multiple-clutch pack clearance affects the shifting of the automatic transmission. Whenever a multiple-clutch system is reassembled, the clearance must be checked. Three places where clearance is checked and adjusted are:

- Snap ring
- Pressure plate
- Spacer.

Snap Ring Clearance and Adjustment

The main job of the clutch-pack snap ring is to hold the clutch assembly parts together. The snap ring may be straight or wavy. Use the following procedure to check snap ring clearance:

1. Have a helper press down firmly on the outer pressure plate.

2. Insert a feeler gauge between the pressure plate and the snap ring. See Figure 8-8.

NOTE: When a waved snap ring is used, measure the clearance where the snap ring is waved *away* from the pressure plate.

3. Check the measurement reading against the manufacturer's specifications.

4. If clearance exceeds the manufacturer's specifications, install a new snap ring. These snap rings are usually selective, so that various thicknesses of snap rings can be tried.

5. Recheck clearance. If not within specifications, continue using thicker snap rings until clearance is proper.

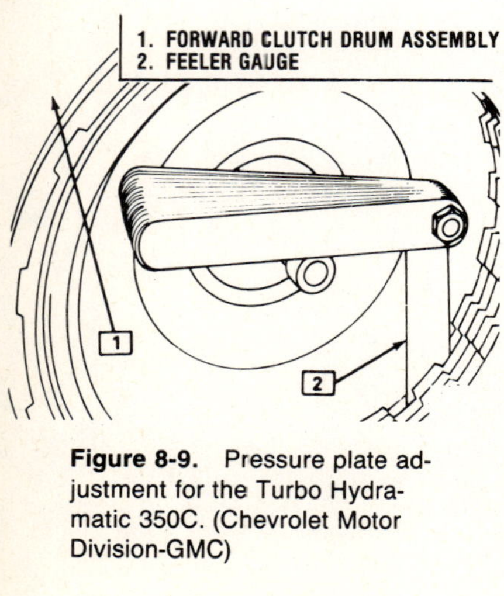

Figure 8-9. Pressure plate adjustment for the Turbo Hydramatic 350C. (Chevrolet Motor Division-GMC)

Pressure Plate Adjustment

With the clutch system fully assembled, use the following procedure to check for pressure plate clearance.

1. Tip the clutch assembly on its side.

2. Insert a feeler gauge between the pressure plate and the adjacent friction clutch plate. See Figure 8-9.

3. Check the measurement reading against the manufacturer's specifications.

4. If not within specifications, use a pressure plate of the proper thickness to give the correct clearance.

5. Recheck clearance.

Spacer Adjustment

A selective spacer is used on the Turbo Hydra-Matic 200/200C/200-4R transmissions. The spacer is located in the low/reverse clutch assembly, shown in Figure 8-10. A special tool is required to check clearance.

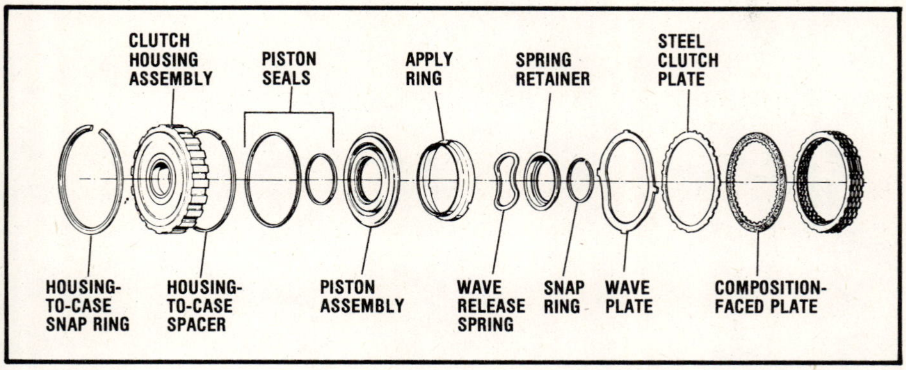

Figure 8-10. Spacer adjustment on the 200-4R transmission is made by varying the thickness of the housing-to-case spacer. (Chevrolet Motor Division-GMC)

Use the following procedure to check for proper clearance:

1. Install the special tool, called the selective spacer gauge, over the clutch housing. The gauge pin must face toward the housing.

2. Tighten the nut on tool to compress the spring until the washer bottoms on the gauge shoulder.

3. Insert a feeler gauge between the gauge pin and the top of the housing.

4. Check the measurement reading against the manufacturer's specifications.

6. If not within specifications, use a spacer of the proper thickness to correct clearance.

7. Recheck clearance.

8. Complete reassembly of the clutch system.

NAME ____________________ SECTION ____________ DATE ____________

Job Sheet 8-2

Overhauling The Multiple-Disc Clutch Assembly

Before you begin:

Read Topics 8.6 through 8.7.

Vehicle Make ____________________ Model ____________ Year ____________

Time Started ____________________ Time Finished ____________________

Job Completion Time ______________ Flat Rate Time ____________________

Tools Required

- Feeler gauges
- Turbo Hydra-Matic 200 selective spacer gauge
- 80-grit emery cloth
- Crocus cloth

References

Manufacturer's service manual ________________________________

Repair manual ________________________________

Service bulletin ________________________________

Procedure: Inspect Multiple-Disc Clutch Parts

1. Check the steel clutch plates for discoloration, scoring, distortion, or damaged lugs. If any of the steels are defective, replace the entire set. Before installation, rough up the shiny surface of the plates with 80-grit emery cloth.
2. Inspect the friction plates for wear, damage, or distortion. Replace the frictions. Soak the new discs in ATF for at least 15 minutes before installation.
3. Check the pressure plate for discoloration, scoring, or distortion.
4. Inspect coil-type piston return springs for roundness, straightness, and heat-caused discoloration. If one spring is bad, replace the entire set.
5. Check Belleville springs for damage or wear, especially at the inner fingers. Check for heat-caused discoloration. Replace the spring if any defects are noted.
6. Check drums for damaged lugs, grooves, scoring, or damaged splines. Make sure the housing check ball operates freely. Lightly polish minor scoring or burrs on the band-application surfaces with crocus cloth.
7. In the Job Completion Report, below, make a note of any defects in the multiple-disc clutch parts.

Procedure: Check and Adjust Snap Ring Clearance

1. Reassemble the multiple-disc clutch with needed new parts and all new seals.
2. Have a helper press down firmly on the outer pressure plate.
3. Insert a feeler gauge between the pressure plate and the snap ring.

NOTE: When a waved snap ring is used, measure the clearance between where the snap ring is waved *away* from the pressure plate.

In the spaces provided, write the correct clearance and the measured clearance.

Manufacturer's specified clearance ....____________________

Measured clearance ....____________________

4. If clearance is not correct, install a thicker or thinner snap ring until the measured clearance is as specified.

Instructor Check: ____________________

Procedure: Check and Adjust Pressure Plate Clearance

1. Tip the clutch assembly on its side.
2. Insert a feeler gauge between the pressure plate and the adjacent friction clutch plate.

In the spaces provided, write the correct clearance and the measured clearance.

Manufacturer's specified clearance ....____________________

Measured clearance ....____________________

3. Install a thicker or thinner pressure plate until the measured clearance is as specified.

Instructor Check ____________________

Procedure: Check and Adjust Turbo Hydra-Matic 200/200C/200-4R Spacer Clearance

1. Install the selective spacer gauge over the clutch housing. The gauge pin must face toward the housing.
2. Tighten the nut on tool to compress the spring until the washer bottoms on the gauge shoulder.
3. Insert a feeler gauge between the gauge pin and the top of the housing.

In the spaces provided, write the correct clearance and the measured clearance.

Manufacturer's specified clearance ....____________________

Measured clearance ....____________________

4. Install a thicker or thinner spacer until the measured clearance is as specified.

Instructor Check: ____________________

Job Completion Report

1. What defects, if any, were found in the multiple-disc clutch parts? ____________

   ____

NAME ______________________ SECTION ____________ DATE ____________

2. What difficulties, if any, were encountered in checking and adjusting multiple-disc clutch clearances? ____

____

____

INSTRUCTOR SIGNOFF ______________________ **DATE** ____________

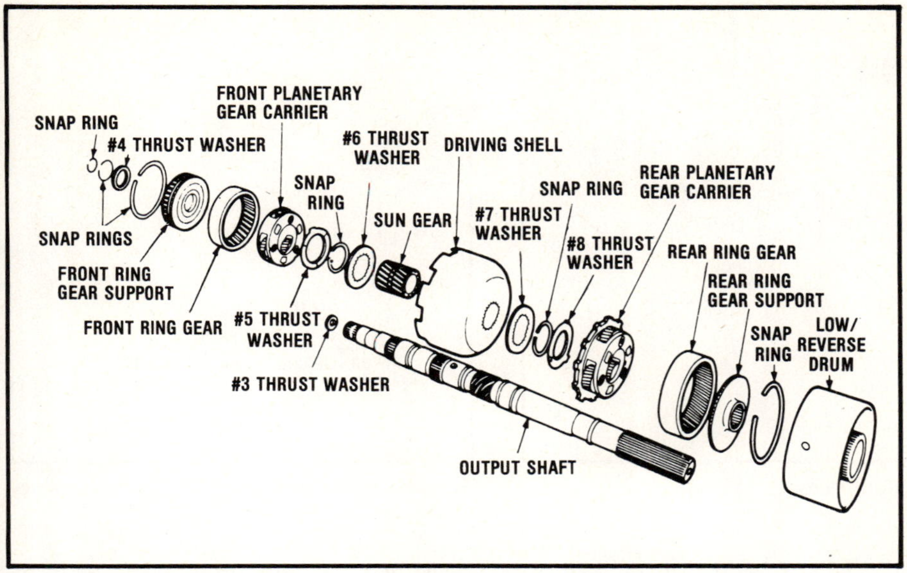

Figure 8-11. Parts of a compound planetary gear system. (Chrysler Corporation)

8.8 PLANETARY GEAR SYSTEM INSPECTION

The entire planetary gearset system must be cleaned in fresh solvent and allowed to air dry. Gearsets can be blown dry with compressed air only if the gears on the planet carrier are *not* allowed to spin.

Figure 8-11 will help you locate the parts of a compound planetary gear system as you inspect for the following defects:

1. Place each planetary gearset on the sun gear. Rotate the gearset slowly. You will feel the gears meshing. However, any binding that tries to stop rotation is cause for replacement.

2. Look for damaged or badly worn gear teeth or splines.

3. Inspect the valley between each gear tooth for foreign material. Remove the material with an awl or pick.

4. Check the carrier assemblies for cracks or breaks.

5. Inspect the planetary gearset for heat-caused discoloration.

6. The output shaft must not be cracked or distorted. Check for wear on the shaft and on the splines.

7. The driving shell should not be worn or damaged.

Individual parts of the planetary gearset system cannot be replaced. The gear system has been meshed together for many miles, so it's necessary to replace the entire assembly. Sometimes the same sun gear can be reused. But, if the sun gear is to be replaced, the entire gear system must be replaced. If the entire gear system is not replaced, it could make a grinding noise.

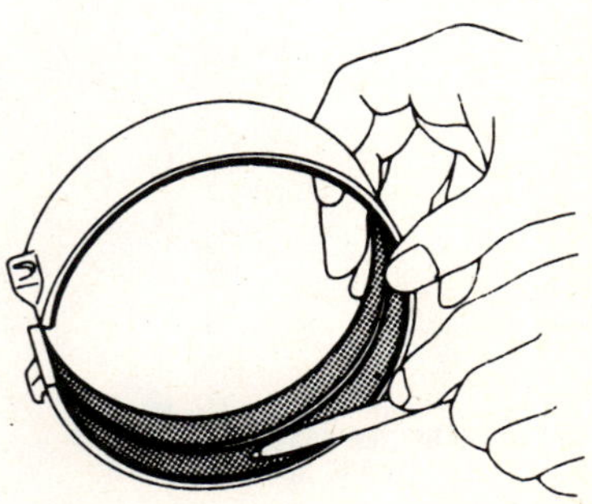

Figure 8-12. Friction material inside the transmission band must be checked for wear. (Nissan Motor Corporation)

8.9 TRANSMISSION BAND INSPECTION

Wipe the transmission bands clean with a dry, lint-free cloth. Unless the transmission has very low mileage, it is usually recommended that the bands be replaced with new ones.

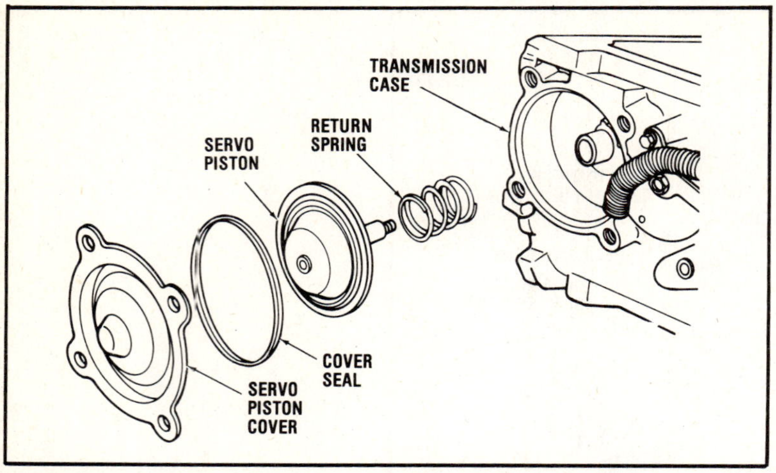

Figure 8-13. Parts in a typical servo. (Ford Motor Company)

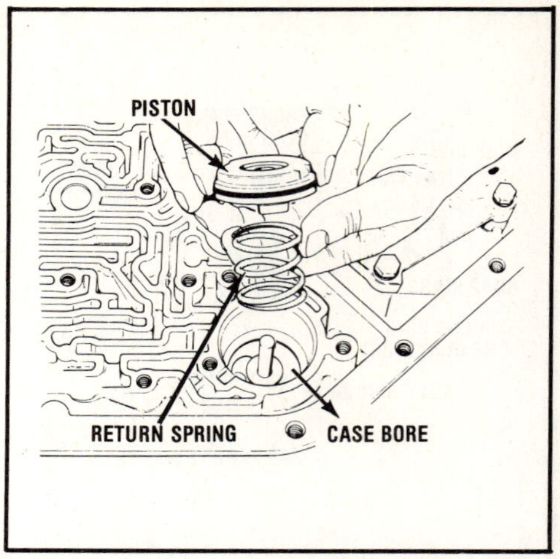

Figure 8-14. A typical accumulator assembly. (Chevrolet Motor Division-GMC)

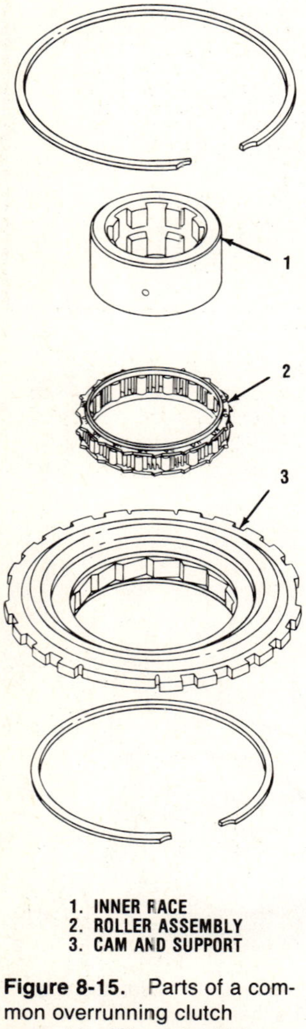

Figure 8-15. Parts of a common overrunning clutch assembly. (Chevrolet Motor Division-GMC)

Check bands for damage, wear, distortion, and any lining imperfections (Figure 8-12).

Band apply struts should not show signs of damage, wear, or distortion.

Always soak transmission bands in ATF for at least 15 minutes before replacing or reinstalling them.

8.10 SERVO AND ACCUMULATOR INSPECTION

Cleaning and inspecting servos and accumulators are similar procedures. A servo system is shown in Figure 8-13. An accumulator system is shown in Figure 8-14. Servo and accumulator parts should be cleaned in fresh solvent and allowed to dry. Remove the seals and replace them with new ones. Some piston and seal assemblies must be replaced as units. Use the following procedures to inspect the parts.

1. Inspect the bore for scoring and damage. If scoring is on the wall of the bore, the transmission case must be replaced.

2. The piston and piston rod should not be nicked, scored, burred, or worn.

3. Spring action must be positive and not weak. Replace any spring that is damaged, collapsed, or distorted.

4. The mating surfaces between the piston and the wall of the bore must be smooth and not restrict movement.

5. Check fluid passages for restrictions. Remove any debris and clean the passages thoroughly.

8.11 OVERRUNNING CLUTCH INSPECTION

Clean each overrunning clutch assembly in fresh solvent and allow to dry. Figure 8-15 shows the parts of an overrunning clutch assembly. Use the following procedures to inspect the parts.

1. Check the roller or sprag assembly first. Look for excessive wear or damage to the rollers or sprags. The springs must not be collapsed, bent, or damaged. The roller or sprag assembly must be replaced as a unit.

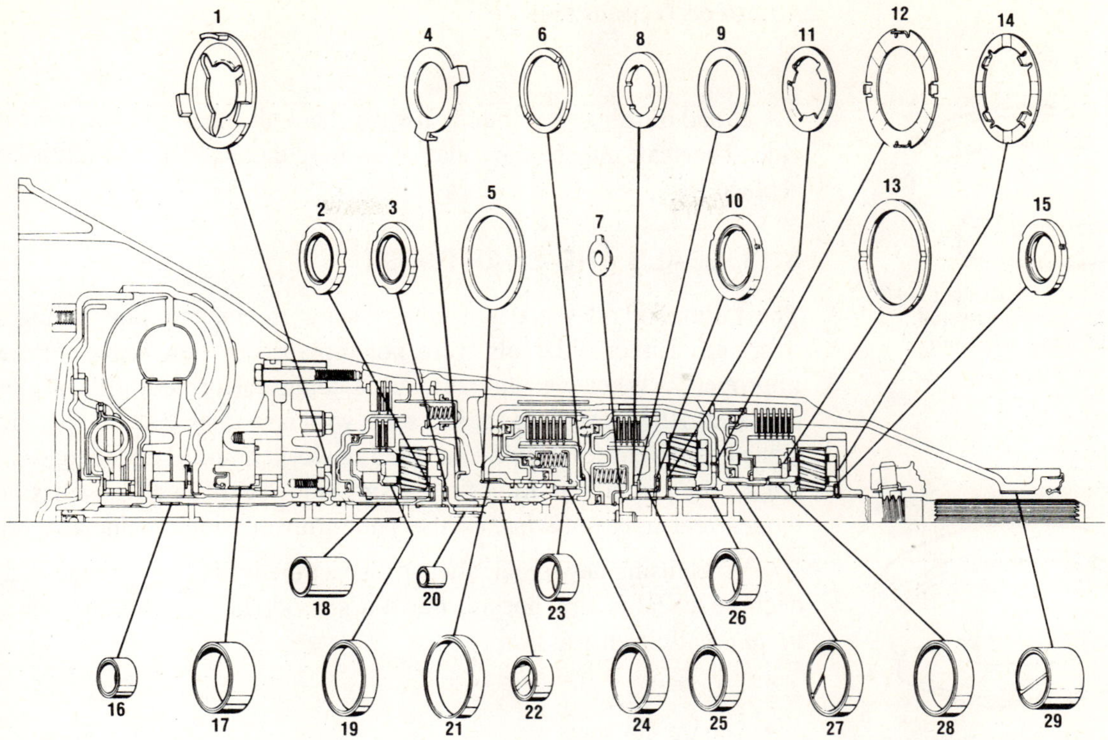

1. OVERRUN CLUTCH TO PUMP THRUST WASHER
2. OVERDRIVE CARRIER TO SUN GEAR THRUST BEARING ASSEMBLY
3. OVERDRIVE CARRIER TO INTERNAL GEAR THRUST BEARING ASSEMBLY
4. OVERDRIVE INTERNAL GEAR TO SUPPORT THRUST WASHER
5. SUPPORT TO DIRECT CLUTCH THRUST WASHER
6. DIRECT TO FORWARD CLUTCH THRUST WASHER
7. TURBINE SHAFT TO OUTPUT SHAFT SEL. THRUST WASHER
8. OUTPUT SHAFT TO FRONT INTERNAL GEAR SEL. THRUST WASHER
9. FRONT INTERNAL GEAR THRUST WASHER
10. FRONT INTERNAL GEAR TO FRONT CARRIER THRUST BEARING ASSEMBLY
11. FRONT CARRIER TO FRONT SUN GEAR THRUST BEARING ASSEMBLY
12. INPUT DRUM TO LOW & REVERSE CLUTCH HOUSING THRUST WASHER
13. REAR CARRIER TO LOW RACE THRUST WASHER
14. REAR CARRIER TO REAR INTERNAL GEAR THRUST WASHER
15. REAR SUN GEAR TO REAR INTERNAL GEAR THRUST BEARING ASSEMBLY
16. STATOR SHAFT BUSHING
17. PUMP BODY BUSHING
18. OVERRUN CLUTCH BUSHING
19. OVERDRIVE CARRIER BUSHING
20. FORWARD CLUTCH SHAFT BUSHING
21. DIRECT CLUTCH BUSHING (FRONT)
22. CENTER SUPPORT BUSHING (FRONT)
23. CENTER SUPPORT BUSHING (REAR)
24. DIRECT CLUTCH BUSHING (REAR)
25. FRONT INTERNAL GEAR BUSHING
26. REAR SUN GEAR BUSHING
27. LOW & REVERSE CLUTCH HOUSING BUSHING
28. REAR CARRIER BUSHING
29. CASE BUSHING

Figure 8-16. Locations of thrust washers, thrust bearings, and bushings in the General Motors 200-4R transmission. (Chevrolet Motor Division-GMC)

2. The inner race and cam surfaces should not be brinneled, scored, or damaged in any way. Replace any parts that show these symptoms.

3. Replace snap rings that are damaged or distorted.

4. Check the lubrication passage in the inner race. If it is restricted, clean it thoroughly.

8.12 WASHER, BEARING, AND BUSHING INSPECTION

Clean all thrust washers, thrust bearings, and bushings in fresh solvent and allow to dry. These parts, shown in Figure 8-16, provide the surface on which other mechanical parts in the transmission ride.

Thrust washers and thrust bearings control the forward and rearward movement of the mechanical parts. Bushings provide lubricated surfaces on which parts can rotate.

If there is the slightest doubt about the quality of one of these parts, it must be replaced. Use the following procedures to inspect these parts.

1. Thrust washers and thrust bearings should be checked for excessive wear, distortion, and damage.

2. Check bushings for scoring and wear.

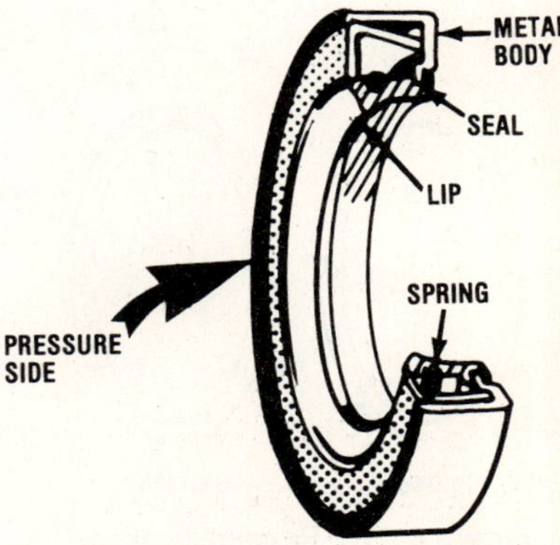

Figure 8-17. Lip seals are enclosed in synthetic coverings.

3. If bad bushings are found, always check the shaft on which the bushings ride. The shaft may be the cause of bushing damage. If the shaft is defective, replace it.

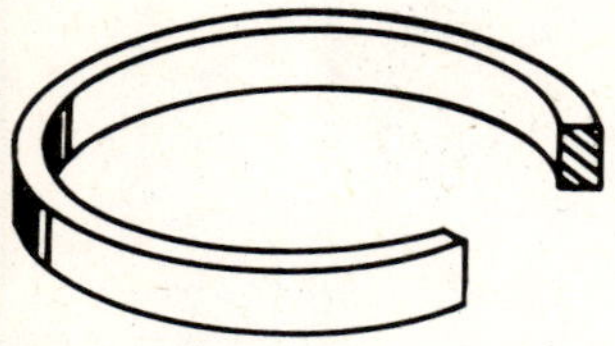

Figure 8-18. Square-cut seals are flexible and allow some movement without leaking.

8.13 SEALS AND SNAP RINGS

The failure of oil seals is a major factor in transmission failure after an overhaul. Either old seals were not replaced or new seals were replaced improperly. Whenever an overhaul is being performed, all seals should be replaced. *Always* replace the old seals with new ones that are specifically designed for the transmission that is being worked on. Many changes are made during the production life of a transmission. Seals on a transmission of one model year may not work on the same transmission produced a year later.

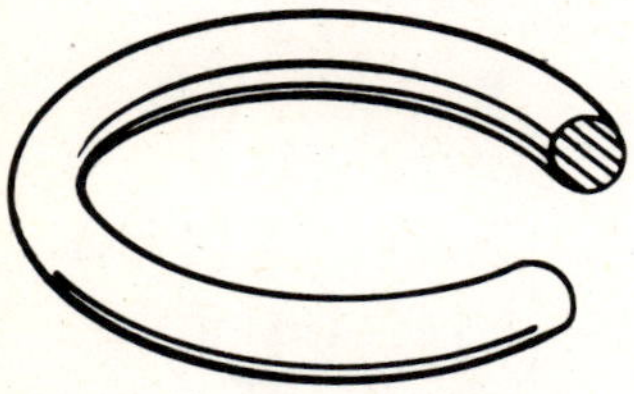

Figure 8-19. O-ring seals are rounded and made of synthetic material.

When using new seals, always inspect them first for damage and fit. Just because a seal is new doesn't mean it's good. Four common seals are used in automatic transmissions:

- Lip seals
- Square-cut seals
- O-rings
- Sealing rings.

Lip Seals

These circular seals are L-shaped and made of a synthetic material, usually enclosed in metal. See Figure 8-17. Lip seals are usually used as the front and rear seals of the transmission. Smaller lip seals may be used in other parts of the transmission.

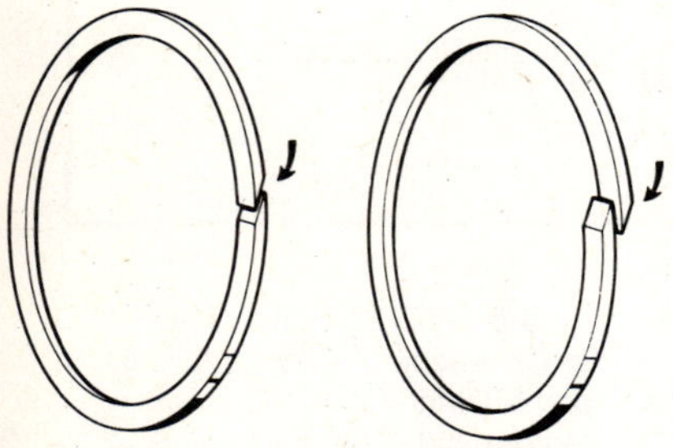

Figure 8-20. Properly fitting Teflon® oil seal rings. (Chevrolet Motor Division-GMC)

Square-Cut Seals

These seals are cut in a square, shown in Figure 8-18. Square-cut seals are made of a synthetic material. Square-cut seals are flexible and allow some movement without leaking. These seals are frequently found on oil pumps and in clutch assemblies.

O-Rings

Seals with rounded surfaces (Figure 8-19) are called O-rings. They are made of synthetic material, and are most frequently found on shafts and pistons. O-rings flatten out with use, and must always be replaced during a teardown of the transmission.

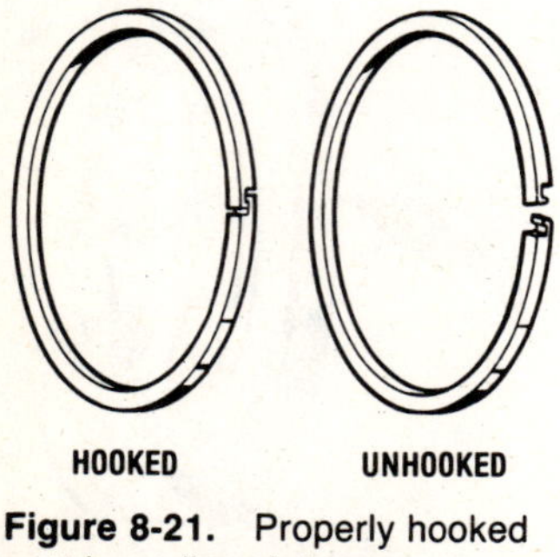

Figure 8-21. Properly hooked cast iron oil seal rings. (Chevrolet Motor Division-GMC)

Sealing Rings

Unlike other seals, sealing rings are made of Teflon® or metal. They do not deteriorate and crack as quickly as synthetic seals. Sealing rings have open ends that butt or snap together. Sealing rings are most frequently found in servos and on transmission shafts. Sealing ring fit is extremely critical. See Figures 8-20 and 8-21 for the proper procedures for closing sealing rings.

NAME ____________________ SECTION ____________ DATE ____________

Job Sheet 8-3

Inspecting Apply Devices

Before you begin:

Read Topics 8.5 through 8.11.

Vehicle Make ____________________ Model ____________ Year ____________

Time Started ____________________ Time Finished ____________________

Job Completion Time ____________________ Flat Rate Time ____________________

Tools Required

- Compressed air
- Shop towels
- Fresh solvent

References

Manufacturer's service manual ______________________________

Repair manual ______________________________

Service bulletin ______________________________

Procedure: Inspect/Replace Planetary Gearset

1. Clean the planetary gearset in fresh solvent and allow to air dry. The gearsets can be blown dry with compressed air, but the pinion bearings *must* be held from moving.
2. Place one planetary gearset on the sun gear at a time. Rotate the gearset slowly and check for binding.
3. Look for damaged or badly worn gear teeth on the sun gear or planetary pinions, or damage to splines.
4. Remove any foreign material between gear teeth.
5. Check the carrier assemblies for cracks or breaks.
6. Inspect the gearset for heat-caused discoloration.
7. Check for wear, cracking or distortion on the output shaft and splines.
8. Check the driving shell and drive lugs for wear and damage.
9. In the Job Completion Report, make a note of any defects found in the planetary gearsets.

Instructor Check: ____________________

10. If the sun gear is undamaged, a damaged planetary carrier and pinions can be replaced. If the sun gear is damaged, replace the entire gearset.

Procedure: Inspect/Replace Bands

1. Wipe the transmission bands clean with a dry, lint-free cloth.
2. Check low-mileage bands for damage, wear, distortion, or any lining imperfections. Replace all other bands.
3. Inspect band apply struts for damage, wear, or distortion.
4. In the Job Completion Report, make a note of any defects found in the bands or apply struts.

Instructor Check: ____________________

5. Soak transmission bands in ATF for at least 15 minutes before installation.

Procedure: Inspect Servos and Accumulators, Replace Parts

1. Clean servo and accumulator parts in fresh solvent and allow to air dry.
2. Inspect the bore for scoring and damage. If scoring is on the wall of the bore, replace the transmission case.
3. Check the piston and piston rod for nicks, scores, burrs, or wear.
4. Replace any spring that is damaged, collapsed, or distorted.
5. Check the mating surfaces between the piston and the wall of the bore for nicks, scores, or damage. Check that the piston moves smoothly in the bore.
6. Check fluid passages for restrictions. Remove any debris and clean the passages thoroughly.
7. In the Job Completion Report, make a note of any defects found in the servos or accumulators.

Instructor Check: ____________________

8. Install new seals and reassemble the servos and accumulators.

Procedure: Inspect/Replace Overrunning Clutch

1. Clean each overrunning clutch assembly in fresh solvent and allow to air dry.
2. Check the rollers or sprags for excessive wear or damage. Check the springs for bending, collapse, or damage.
3. Check the inner race and cam surfaces for scored or damaged surfaces.
4. Replace damaged or distorted snap rings.
5. Thoroughly clean the lubrication passage in the inner race.
6. In the Job Completion Report, make a note of any defects in the overrunning clutches.

Instructor Check: ____________________

7. If defects are found in an overrunning clutch, replace the clutch assembly.

Procedure: Inspect/Replace Washers, Bearings, and Bushings

1. Clean all thrust washers, thrust bearings, and bushings in fresh solvent and allow to air dry.

NAME ____________________ SECTION ____________ DATE ____________

2. Check thrust washers and thrust bearings for excessive wear, distortion, and damage.
3. Check bushings for scoring and wear.
4. Check the shafts on which the bushings ride. If the shaft is defective, replace it.

Instructor Check: ____________________

5. In the Job Completion Report, make a note of any defects found in washers, thrust bearings, and/or bushings.
6. Install new parts where necessary.

Replace Seals

1. Install all new seals in the transmission during reassembly.
2. Fit sealing ring ends properly in the transmission.

Instructor Check: ____________________

Job Completion Report

1. What defects were found during cleaning and inspection of the transmission parts?

____

____

____

____

____

____

____

INSTRUCTOR SIGNOFF ____________________ **DATE** ____________

NAME ______________________ SECTION ____________ DATE ____________

Certification Practice

Test 8

DIRECTIONS:

The following questions are similar to those used on mechanic certification tests. Answer each question by circling the letter of the correct choice.

1. Which of the following statements is correct?
   I. Most automatic transmissions have bands.
   II. Not all transmission bands are adjustable.
   A. I only B. II only C. Both I and II D. Neither I nor II

2. Mechanic A says all band adjustments can be made on the outside of the transmission case.
   Mechanic B says all band adjustments can be made from inside the oil pan area.
   Who is correct?
   A. A only B. B only C. Both A and B D. Neither A nor B

3. The picture shows a tool used for band adjustment on
   A. Chrysler and AMC transmissions.
   B. Ford transmissions.
   C. GM transmissions.
   D. all of the above.

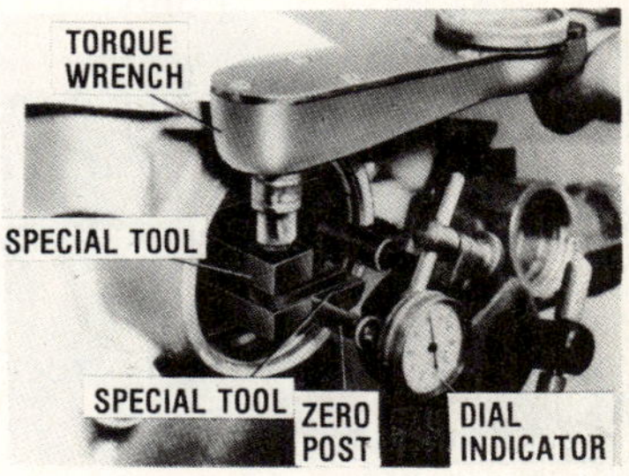

Question 3.
(Oldsmobile Division-GMC)

4. Which of the following should *NOT* be cleaned in solvent?
   A. Friction plates
   B. Steel plates
   C. Planetary gearsets
   D. Thrust washers

5. All of the following are defects in piston return springs EXCEPT
   A. stretched springs.
   B. bent springs.
   C. heat-damaged springs.
   D. belleville springs.

6. Mechanic A says multi-disc clutch clearances are checked at the snap ring and at the pressure plate.
   Mechanic B says multi-disc clutch clearances are checked at the pressure plate and at the spacer.
   Who is correct?
   A. A only B. B only C. Both A and B D. Neither A nor B

7. The picture shows

   A. an accumulator piston.

   B. a servo.

   C. a pressure-regulating valve.

   D. a manual valve.

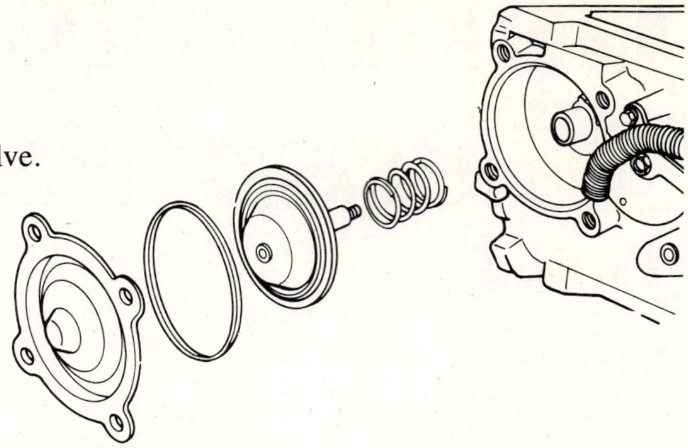

Question 7.
(Ford Motor Company)

8. A year-old car with no reverse is towed into the shop by a novice tow-truck driver. During teardown, the planetary gearsets are found to be discolored by excessive heat. The most likely cause is

   A. engine cooling system clogged.

   B. transmission cooler clogged.

   C. torque converter bad.

   D. car towed improperly.

9. Which of the following statements is true?

   I. Selective-fit snap rings are used to reduce clearance.

   II. Selective-fit spacers are used to increase clearance.

   A. I only B. II only C. Both I and II D. Neither I nor II

10. Mechanic A says that O-ring seal sizes may change slightly on a transmission from model year to model year.

    Mechanic B says that an overhaul O-ring kit for a particular transmission will fit any year.

    Who is correct?

    A. A only B. B only C. Both A and B D. Neither A nor B

IV
Hydraulic Systems

9 Hydraulic Principles

YOUR LEARNING JOB

Liquids, or fluids, can be used to do work. Fluids can transfer motion and force through pressure. The automatic transmission uses fluid pressure to control the action of the planetary gearsets. Fluid pressure is regulated and directed to change gears automatically. This unit covers basic fluid principles and the use of fluid pressure to control the automatic transmission.

JOB COMPETENCY OBJECTIVES

This is a knowledge-building unit. When you have completed your assignments and exercises in this unit, you should be able to:

- ☐ Explain how liquids can act to transmit motion, pressure, and force.
- ☐ Describe the relationship between motion, pressure, and force in a hydraulic system.
- ☐ Name and describe basic hydraulic valves used in the automatic transmission.
- ☐ Explain how different types of hydraulic valves can be used to control and direct fluid pressure.

9.1 HYDRAULICS

The word *hydraulics* comes from a Greek word, *hydros*, that means water, or liquid. Hydraulics is the scientific study of how fluids transmit motion, pressure, and force. A hydraulic system uses a liquid to perform useful work.

More than 300 years ago, a young French scientist named Blaise Pascal figured out the basic principles of all hydraulic systems. His discovery, called Pascal's Law, concerns how liquids act under pressure, which is discussed in 9.2. The metric pressure unit is called a kilopascal (kPa), after this scientist.

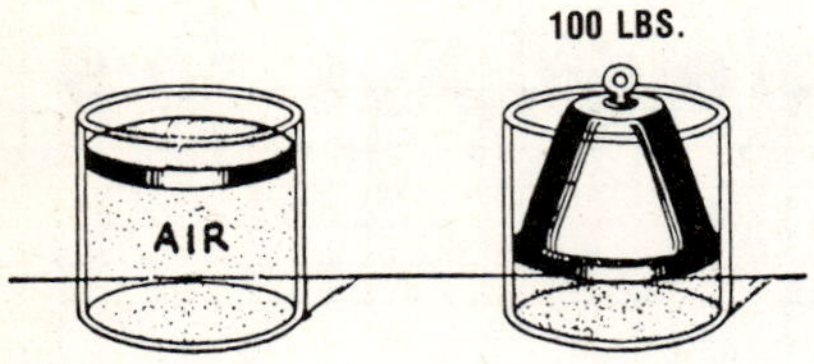

Figure 9-1. Gas can be compressed by adding pressure, or weight. (Pontiac Motor Division-GMC)

9.2 HYDRAULIC FUNDAMENTALS

Gases such as air can be compressed, or squeezed, into a smaller volume. When weight is applied to a plunger tightly sealed in a cylinder filled with air, the air will be compressed, or squeezed, into a smaller volume. See Figure 9-1. However, for all practical purposes, liquids are incompressible—they cannot be compressed. Weight applied to a plunger tightly sealed in a cylinder filled with liquid cannot squeeze the liquid into a smaller volume. See Figure 9-2.

When force is applied to the plunger, it pushes against the liquid, causing pressure to build up. *Pressure* is the amount of force pushing on each unit of the liquid's surface area. Pressure is applied to all the inner surfaces of the liquid's container.

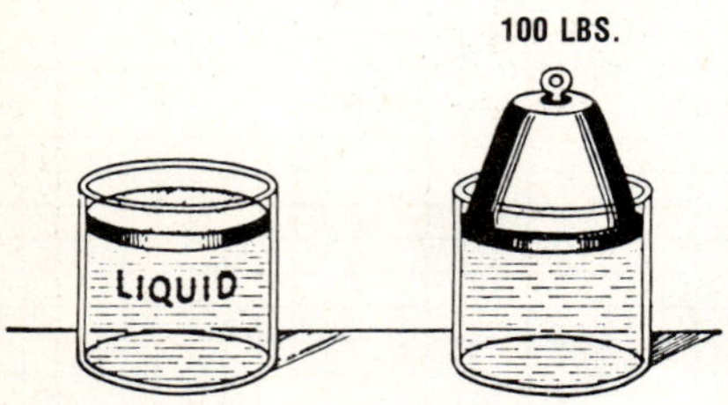

Figure 9-2. Liquid cannot be compressed even when adding the same amount of pressure as was added in Figure 9-1. (Pontiac Motor Division-GMC)

Pressure is measured by dividing the force pushing on the liquid by the amount of surface area to which the force is applied. If the force on the plunger is 100 pounds [45 kilograms], and the surface area of the plunger is 10 square inches [64.5 centimeters], the pressure on the liquid would be: 100 pounds divided by 10 square inches = 10 pounds per square inch [69 kilopascals]. To find the pressure, divide the force by the area of the plunger, or piston, face.

The pressure is applied equally to every surface the liquid touches.

9.3 MOTION TRANSFER

Liquid under pressure can be made to do work. Connecting two piston and cylinder units together will allow liquid pressure to push against the second piston. See Figure 9-3. The second piston will move.

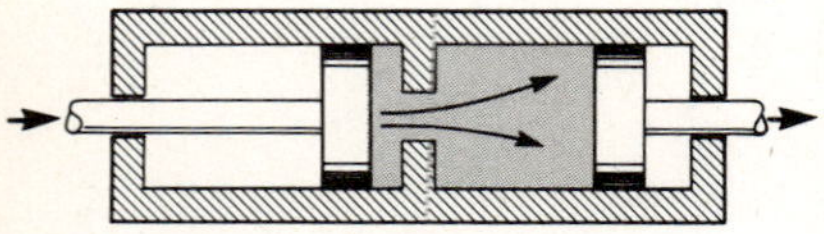

Figure 9-3. Liquid under pressure can be pushed from one cylinder to another by the use of pistons. (Pontiac Motor Division-GMC)

Moving one piston will push a certain volume of liquid into the second cylinder. If both pistons and cylinders are the same size, moving one piston will cause the other piston to move the same distance.

However, if one piston and cylinder have half the area of a connected piston and cylinder, the motion of the second piston will be less. Moving the smaller piston two inches will cause the larger piston to move only one inch. See Figure 9-4.

Moving the smaller piston two inches will only transfer enough fluid volume to move the larger piston one inch. Pushing on the larger piston would transfer more liquid to the smaller cylinder. If the larger piston is pushed 2 inches, the smaller piston will move twice as far, or 4 inches.

It does not matter how far the cylinders are located from each other. As long as the connecting tube and both cylinders are filled with fluid, motion and fluid will be transferred from one cylinder to another.

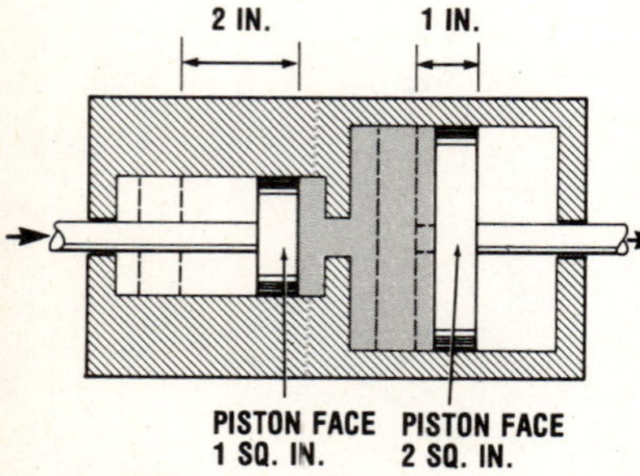

Figure 9-4. If the area between two cylinders and pistons is different, the larger piston will move less distance than the small piston.

9.4 PRESSURE TRANSFER

Pressure is also transferred in a fluid-filled system. The pressure in a closed, sealed system is the same everywhere. If pressure gauges are attached at various points, they will all register the same pressure, as shown in Figure 9-5. The force exerted on the input, or pushing, piston divided by its surface area equals the pressure in the system.

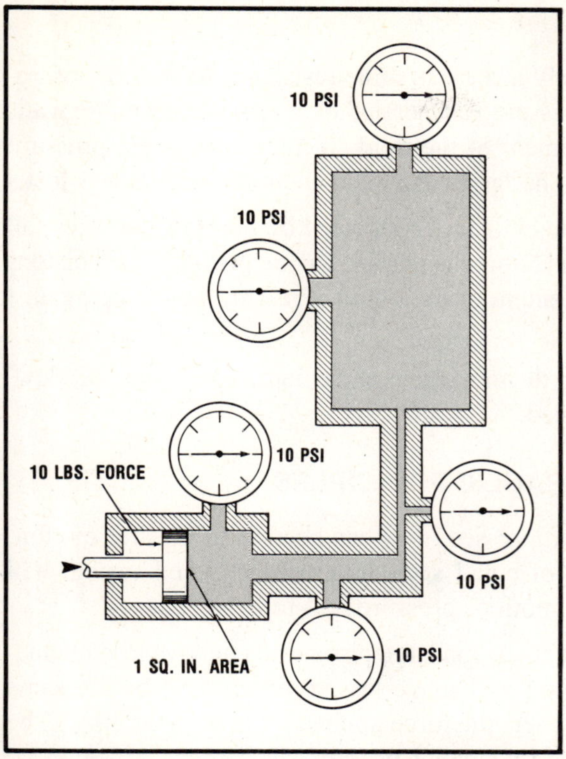

Figure 9-5. Pressure is the same throughout a sealed system.

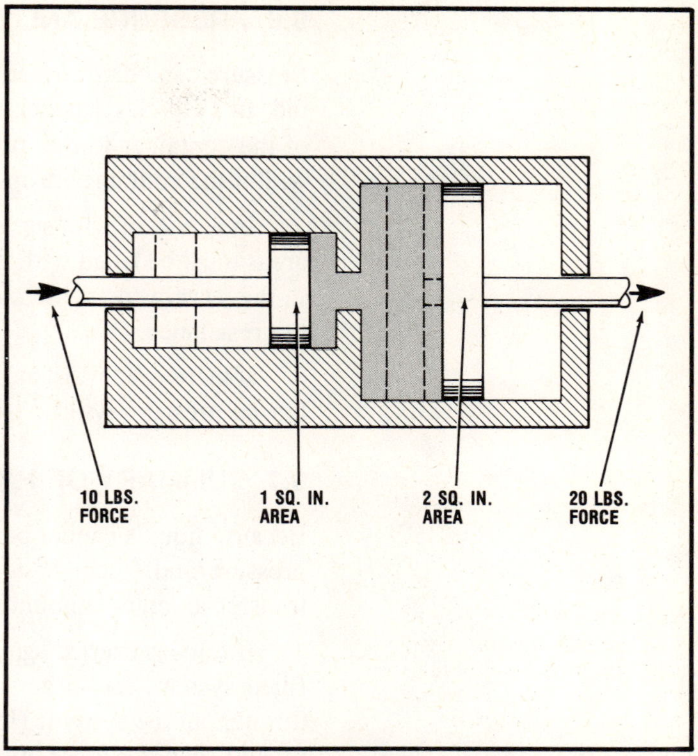

Figure 9-6. Pressure acting against a larger piston will increase the amount of force that is transferred.

9.5 FORCE TRANSFER

The pressure in the connected cylinders is the same. However, If the cylinders in a hydraulic system are of different sizes, different amounts of *force* will be exerted on the pistons.

Pressure is measured in force per unit area, commonly pounds per square inch. When a force of 10 pounds [4.5 kilograms] is exerted against a piston with a surface area of 1 square inch [6.45 square centimeters], the pressure acting on the liquid throughout the system will be 10 pounds divided by 1 square inch = 10 pounds per square inch, or psi [68.95 kPa].

If a piston with a surface area of 2 square inches is connected by a passage or tube, the pressure acting against the larger piston will still be 10 pounds per square inch. However, because the second piston has 2 square inches of area, the total *force* pushing the second piston is 10 pounds × 2 = 20 pounds. See Figure 9-6. By doubling the surface area of the second piston, the amount of force transferred is doubled.

If the larger piston were 3, 5, or 10 times the size of the smaller piston, the force on the larger piston would be 3, 5, or 10 times the force on the smaller. Of course, the larger piston would have less movement than the smaller piston.

When pushing force is applied to a larger piston, the force on the smaller piston would be less. Given equal pressure, the amount of force exerted on a piston is directly proportional to its surface area. The larger the surface area, the greater the force.

9.6 PRESSURE AND FLOW

Pressure can be generated only if there is some resistance to the flow of liquid. In a closed container, pressure is generated by the resistance of the walls of the container to the movement of the fluid. If a leak develops, pressure within the container drops. The larger the leak, the more pressure is lost.

In a system with separate cylinders connected by tubes or passages, all areas must be filled with fluid. Force is applied to one piston. Other pistons in the system resist the movement of the liquid. Pressure builds up against the resistance.

If there is no resistance to the movement of the liquid, the liquid will flow, but no pressure will be exerted.

9.7 SUMMARY OF HYDRAULIC PRINCIPLES

Because liquids cannot be compressed, they can be used to transfer motion, pressure and force. Pistons of equal sizes in a tightly closed system will transfer an equal amount of motion, pressure and force.

If force is exerted against small and large pistons in a tightly closed liquid-filled system, the larger piston will move less. Pressure will be the same throughout the system. However, the force applied by larger pistons will be greater than the force applied by smaller pistons.

By using pistons of different sizes in a hydraulic system, it is possible to increase or decrease motion or force.

There must be resistance to the flow of liquid for pressure to build up.

CHECKPOINT

1. **____________________ is the scientific study of how liquids, or fluids, transmit motion, pressure, and force.**
2. **For all practical purposes, ____________________ are incompressible.**
3. **The pressure in a closed, tightly sealed system is the ____________________ everywhere in the system.**
4. **The amount of ____________________ exerted on a piston is directly proportional to its ____________________ ____________________.**
5. **By using pistons of different sizes in a tightly sealed fluid-filled system, it is possible to ____________________ or ____________________ motion or ____________________.**
6. **Pressure can only be generated if there is ____________________ to the flow of liquid.**

9.8 HYDRAULIC SYSTEMS

To form a complete, working hydraulic system, a number of elements are necessary. All hydraulic systems include:

- Reservoir
- Pressure source
- Control valving
- Output device.

The automatic transmission reservoir is the oil pan. Transmission fluid is drawn from the pan and returned to it. The pressure source for the automatic transmission is the oil pump. The valve control body contains control valving to regulate, or restrict the pressure and flow of fluid within the transmission. Output devices are servos or clutches operated by hydraulic pressure.

A *hydraulic circuit* includes the pressure source, control valving, and one or more output devices used to do work.

9.9 CONTROL VALVES

It is possible to build a system using pistons and valves of varying sizes to control hydraulic pressures to any desired value. Control valves in the automatic transmission are divided into two groups, according to function:

- Pressure-regulating valves
- Switching valves.

Pressure regulating valves control the pressures available to operate or move servos, clutches, and other valves. Switching valves control fluid flow to specific servos, clutches, and other valves.

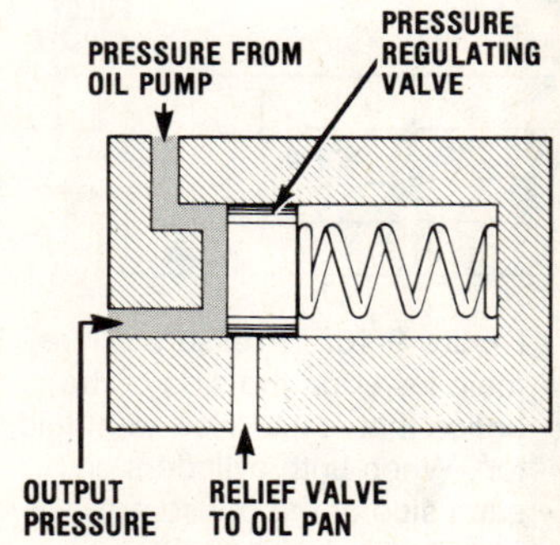

Figure 9-7. Simple pressure-regulating valve.

9.10 PRESSURE-REGULATING VALVES

The faster the engine runs, the more pressure the transmission oil pump generates. Without some way to control the pressure, the pump would create too much pressure at high engine speeds.

A simple pressure-regulating valve consists of a piston in a cylinder with a spring, and a relief, or bleed, opening. See Figure 9-7.

Fluid flows through a passageway from the pump. Hydraulic pressure applies force to the pressure-regulator piston, moving it back into the cylinder. The piston is forced against a compressed spring. When the piston moves back far enough, it uncovers a return passage to the oil pan reservoir.

Pressure drops as the fluid flows through the return passage. The spring pushes the piston outward. At a certain point, the force of the spring is equal to the hydraulic pressure on the piston. The valve stops moving, balanced between the force exerted by the spring and the force exerted by the hydraulic pressure on the piston. The pressures in the input and output lines equalize at a constant pressure determined by the force of the spring.

The pressure-regulating valve can be placed in different locations in relation to the pump pressure line. Most transmissions use an arrangement like that shown in Figure 9-8.

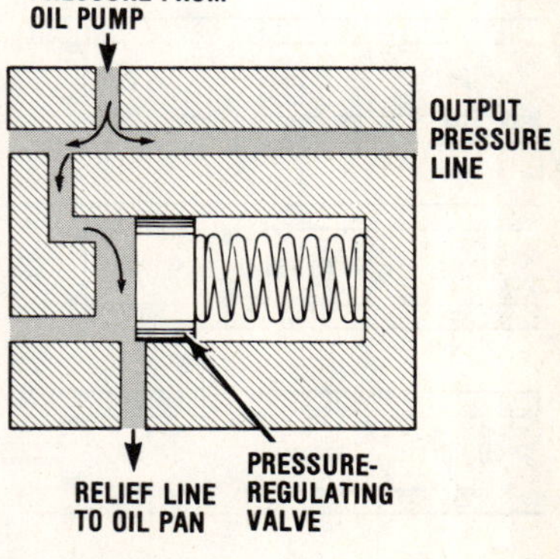

Figure 9-8. The pressure-regulating valve is not always directly connected to the pump pressure line and the output pressure line.

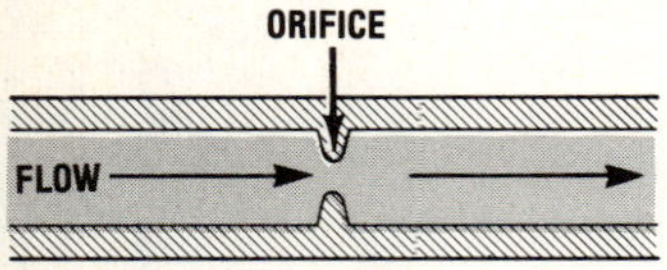

Figure 9-9. An orifice, or narrow passage, is used to regulate pressure.

9.11 ORIFICES

Another way to regulate the hydraulic pressure of liquid flowing through a line or passage is to restrict, or narrow, the size of the passage at some point. Such a restriction is known as an orifice, shown in Figure 9-9.

When the liquid passes through the orifice, the restriction creates resistance to the movement of the fluid. Pressure builds on the supply side of the orifice.

Until the empty space on the other side of the orifice fills, the pressure on that side will be less. When both sides are completely filled, the pressure will equalize. See Figure 9-10. By using an orifice in this way, pressure in the main, or supply, line can be maintained while the empty space fills. As the chamber becomes full, pressure would be applied gradually to a piston. This could be used to operate an apply device gradually, producing a smooth shift.

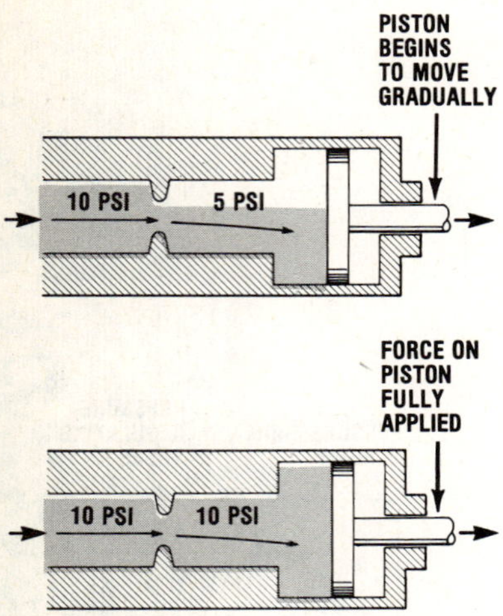

Figure 9-10. Pressure will remain lower on the side of the orifice that is not filled with fluid, top. When both cylinders on each side of the orifice are equally filled with liquid, pressure will be equal in both cylinders.

Several orifices can be placed in a line, or series. As the fluid flows to the empty spaces past each orifice, the pressure is reduced. A longer time is necessary for each extra space behind the orifices to fill. See Figure 9-11. Orifices in series can be used to furnish hydraulic pressure at various rates or at different times.

CHECKPOINT

1. **All hydraulic systems include a ____________________, ____________________ ____________________, ____________________ ____________________, and ____________________ ____________________.**
2. **A simple pressure-regulating valve consists of a ____________________ in a ____________________ with a spring, and a ____________________, or bleed, opening.**
3. **When the pressure-regulating valve piston moves back far enough, it uncovers the ____________________ ____________________ to the oil pan ____________________.**
4. **As liquid attempts to flow through the ____________________, pressure is maintained on the supply side.**
5. **Orifices in series can be used to furnish hydraulic pressure at ____________________ rates or at ____________________ times.**

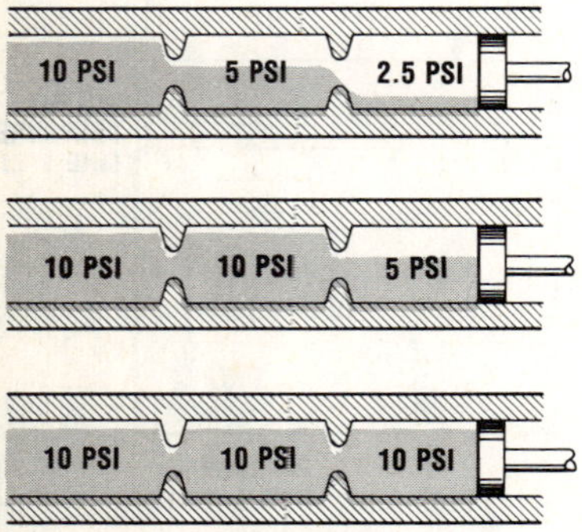

Figure 9-11. More than one orifice may be used in a line. Until liquid is equal in each cylinder, pressures (shown above) will vary.

9.12 SPOOL VALVES

The shape of a spool valve roughly resembles a spool used for sewing thread. The simplest form of spool valve has four piston faces, a valley between the inner faces, and two lands, as shown in Figure 9-12. Spool valves are used throughout the automatic transmission valve body. Many spool valves work both as pressure-regulating valves and as switching valves. Switching valves are discussed in 9.14.

A spool valve may have different sized lands and valve faces at each end. The larger the valve face, the more force will be developed by the hydraulic pressure. Fluid pressure directed to the valley will cause the spool valve to move in the direction of the larger valve face. See Figure 9-13.

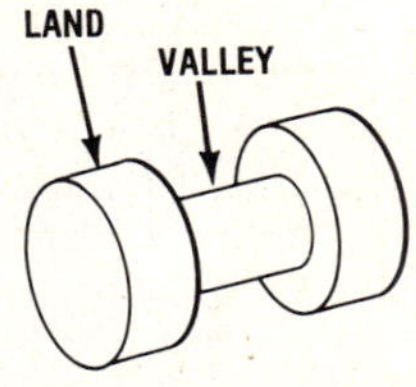

Figure 9-12. Simple spool valve. (Ford Motor Company)

Multiland Spool Valve

A multiland spool valve has several lands and valve faces of different sizes. See Figure 9-14. Fluid pressure acting on the valve faces and valleys will move the multiland spool valve in different directions. The lands can be used to cover and uncover input, output, and relief passages for pressure regulation.

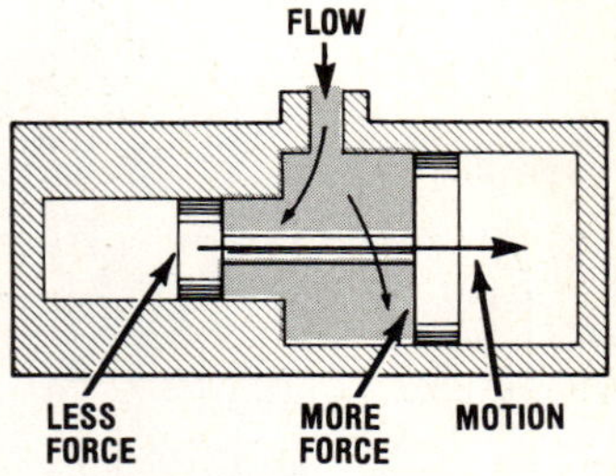

Figure 9-13. Different sizes of lands and valleys will move a spool in different directions.

9.13 BALANCE VALVES

A balance valve is a spool valve used for pressure regulation. The fluid pressure to be regulated pushes against a spring. The spring can be used alone, or can be controlled by lever action or hydraulic pressure.

Balance Valve with Spring

Spring pressure can be used to balance the force of hydraulic pressure, as explained in 9.10, above. A simple spool valve with lands of the same size can be used as a pressure-regulating valve.

Hydraulic pressure is directed to the spool valve valley through an inlet port. A port is an opening. Pressure is equal on both of the inner faces of the spool valve, so the spring pushes the valve to close the outlet port. See Figure 9–15.

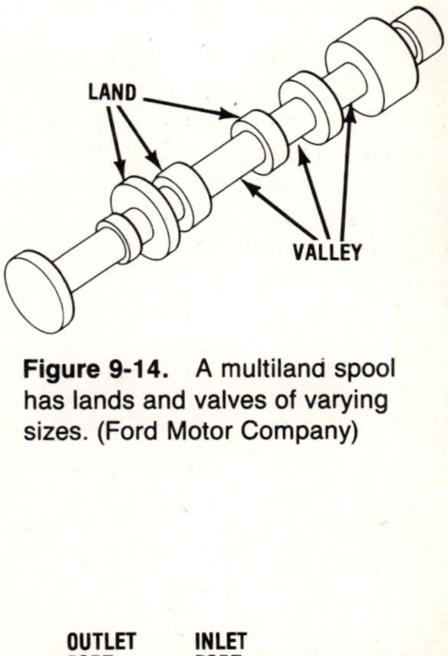

Figure 9-14. A multiland spool has lands and valves of varying sizes. (Ford Motor Company)

The fluid flows through the valley and around to the output chamber. Pressure then pushes against the end face of the valve. The piston moves to the right, opening the outlet port. As the pressure continues to build, it will push the land back far enough to begin to close the inlet port.

The hydraulic pressure pushes the spring back until the hydraulic pressure is equal to the spring pressure. The valve remains balanced between the two opposing forces, partially uncovering the inlet and outlet ports. See Figure 9-16. The pressure in the inlet and outlet lines is regulated to the desired pressure.

The force exerted by the spring determines the hydraulic pressure in the output line. A stronger spring requires more hydraulic pressure to balance the valve. A weaker spring requires less pressure.

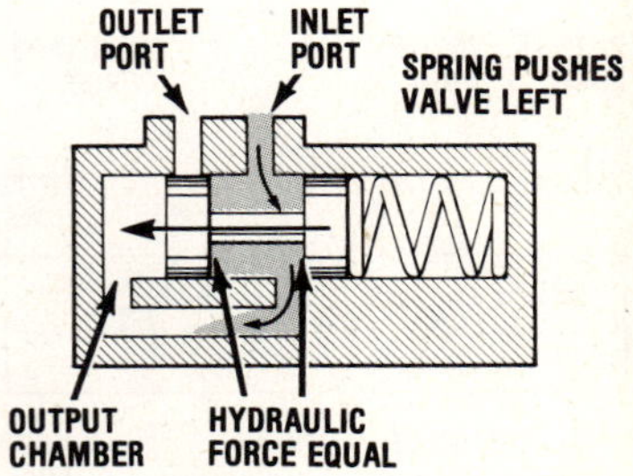

Figure 9-15. When pressure is equal on two inner faces, the spring pushes the valve to close the outlet port.

Balance Valve with Lever-Regulated Spring

A manually operated lever can be used to vary the spring pressure on the balance valve. When the lever is moved to compress the spring, the spring applies greater pressure to the end of the balance valve. As the valve moves to the left, it uncovers more of the inlet port. The pressure in the outlet port increases. The more pressure the spring exerts, the higher the hydraulic pressure becomes. See Figure 9-17.

When the lever is moved in the opposite direction, the spring applies less pressure to the end of the balance valve. The hydraulic pressure pushes the valve back, blocking off more of the inlet port. The pressure in the ports and lines lessens. See Figure 9-18.

Manually operated throttle valves on Chrysler's TorqueFlite and General Motors' Turbo Hydra-Matic 200 transmissions use this principle to control shifting.

Balance Valve with Pressure-Regulated Spring

Fluid pressure can also be used to vary the force on the balance valve. A second spool valve presses against the end of the spring. See Figure 9-19.

The lands on the second valve are of different sizes. As fluid from another circuit enters the second valve valley, the valve shifts toward the end with the larger face. This compresses the spring and raises the pressure in the outlet port line.

Different amounts of force can be exerted to compress the spring by making the second valve faces and lands of different sizes. Varying the pressure operating the second valve will also change the amount the spring is compressed. The more force exerted on the balance valve spring, the higher the pressure in the outlet line.

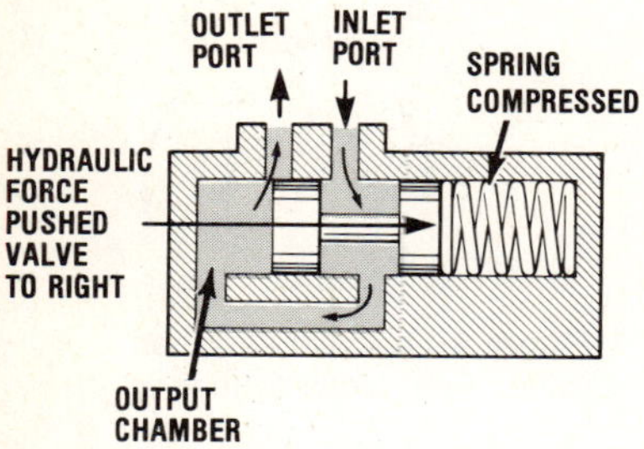

Figure 9-16. When spring pressure is equal to hydraulic pressure, the valve is balanced and holds open both the outlet and inlet ports.

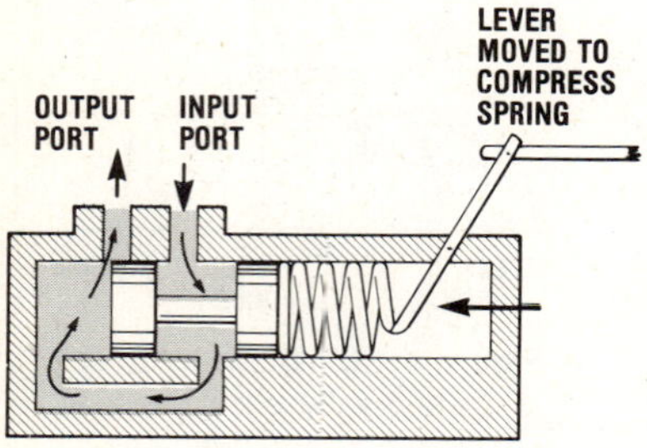

Figure 9-17. Hydraulic pressure increases at the outlet port when spring pressure is increased on the balance valve.

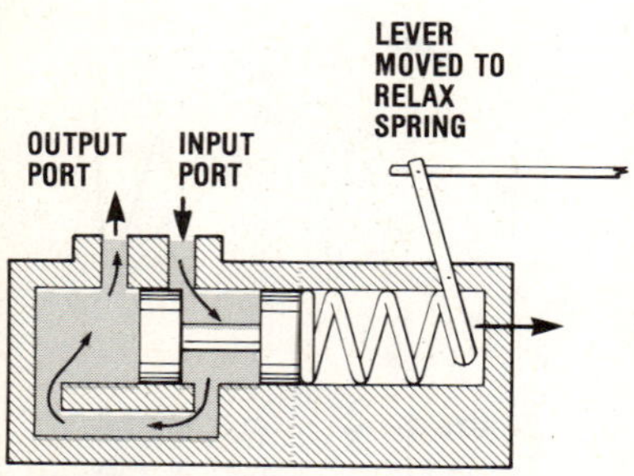

Figure 9-18. Pressure on the balance valve decreases when there is less pressure on the spring.

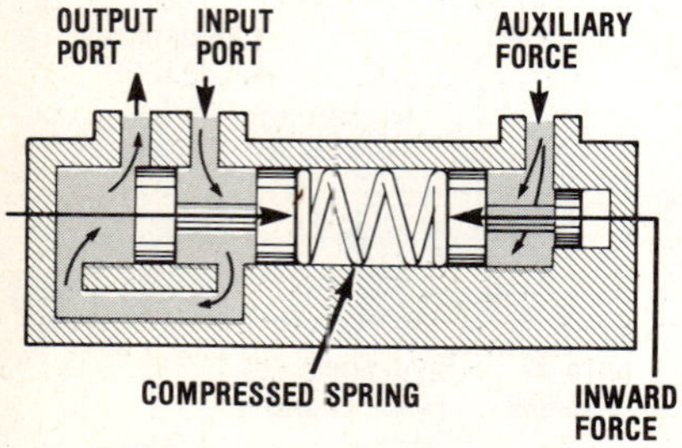

Figure 9-19. A second spool acting against the balance valve spring can add greater pressure to the spring.

CHECKPOINT

1. **Fluid ____________________ directed to the valley between two valve faces will cause the spool valve to move in the direction of the ____________________ valve face.**
2. **A ____________________ spool valve has many lands and valve faces of different sizes.**
3. **The balance valve spring can be used alone, or can be controlled by ________________ action or ________________ pressure.**
4. **The valve remains ____________________ between the two opposing forces, partially ____________________ the inlet and outlet ports.**
5. **The force exerted by the ____________________ determines the hydraulic pressure in the output line.**

9.14 SWITCHING VALVES

Switching valves are used to switch, or direct, the flow of liquid from one passage or hydraulic circuit to another. Types of switching valves include:

- Poppet valve
- One-way ball check valve
- Two-way ball check valve
- Manually operated spool valve
- Hydraulically operated spool valve.

Poppet Valve

A poppet valve is a one-way valve. It allows liquid to flow in one direction only. When spring pressure is greater than fluid pressure, the spring forces the valve against its seat, closing the inlet port. See Figure 9-20.

When fluid pressure builds to exert enough force against the poppet valve face, the spring compresses. Fluid flows through the valve to the outlet port. See Figure 9-21. The force exerted by the spring determines how much pressure is required to open the valve.

Fluid can only flow in through the inlet port. Fluid pressure exerted from the outlet port side will close the poppet valve more tightly.

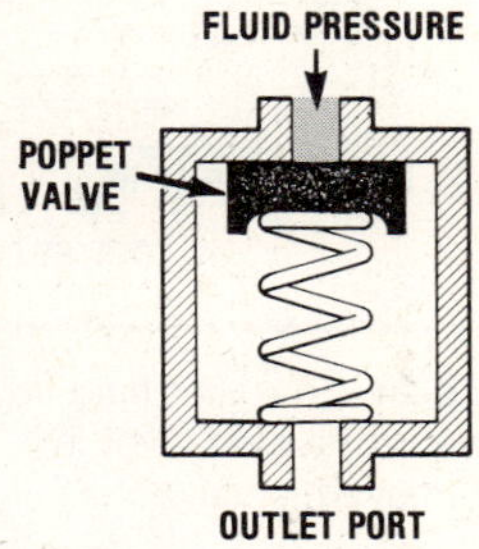

Figure 9-20. A one-way, or poppet, valve stops fluid from flowing when spring pressure is greater than hydraulic pressure.

One-Way Ball Check Valve

A ball and spring can be used in place of the poppet valve and spring to provide the same one-way action. See Figure 9-22.

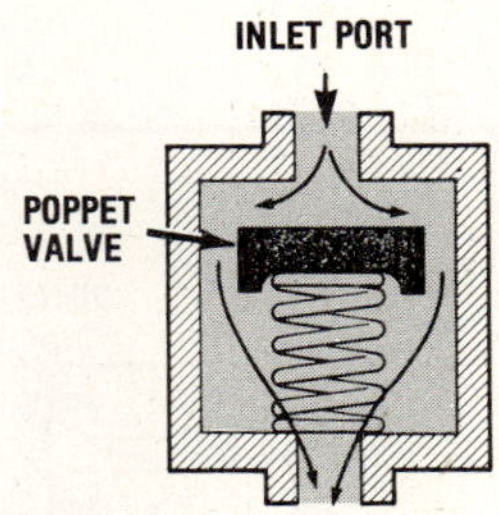

Figure 9-21. When fluid pressure builds to a greater pressure than the spring can exert, the poppet valve is pushed downward.

Two-Way Ball Check Valve

A ball can also be used to make a two-way valve. A two-way valve allows liquid to flow in from either or of two inlet ports. See Figure 9-23.

Two hydraulic circuits can share the same ball check valve. As fluid pressure builds in circuit A, it pushes the ball to close off the passage to circuit B. Fluid then flows through the outlet port.

A two-way ball check valve allows the same apply device to be operated by either of the two circuits. Which circuit operates the apply device depends on driving and load conditions, and the gear range selected.

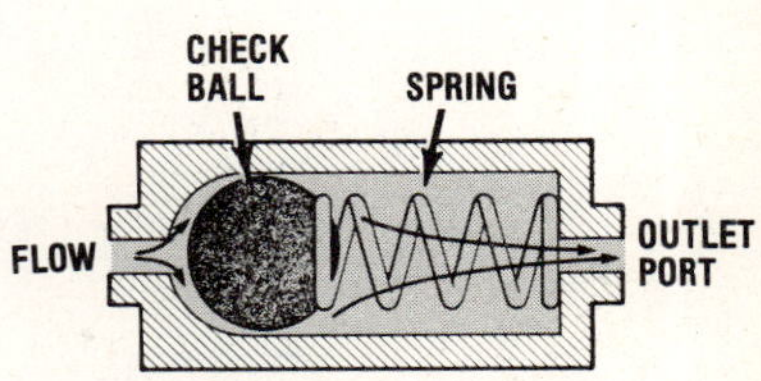

Figure 9-22. A one-way ball check valve, shown, is often used instead of a poppet valve.

Manually Operated Spool Valve

A manually operated spool valve can be used to direct fluid, or to prevent the flow of fluid. Moving a lever pushes the valve lands in position to block the flow of fluid from an inlet port. See Figure 9-24, top.

The lever can be moved to position the spool valve to direct fluid through one or more outlet ports. See Figure 9-24, middle and bottom.

Hydraulically Operated Spool Valve

Hydraulic pressure and a spring can be used to operate a spool switching valve. When pressure from an auxiliary, or alternate, inlet port pushes the spool valve to compress the spring, the valve lands move to uncover inlet and outlet ports. See Figure 9-25.

If the pressure from the auxiliary port lessens, the spring will expand to push the valve and close the inlet port. See Figure 9-26.

A transmission shift valve is similar to a hydraulically operated spool valve. However, pressure can be applied from ports at both ends of the spool valve cylinder. See Figure 9-27.

Hydraulic force from port B is combined with spring pressure to move the valve to the left. When pressure to port A builds to a predetermined level, the force on the left valve face overcomes the combination of spring force and hydraulic force from port B. The valve moves enough to the right to uncover inlet and outlet ports between the lands.

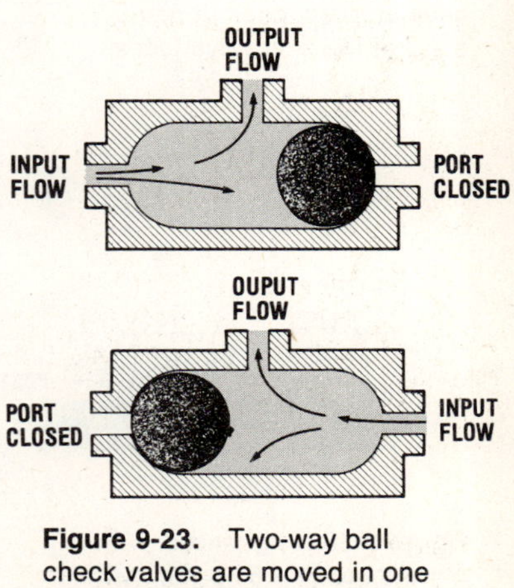

Figure 9-23. Two-way ball check valves are moved in one direction or the other by changes in hydraulic pressure.

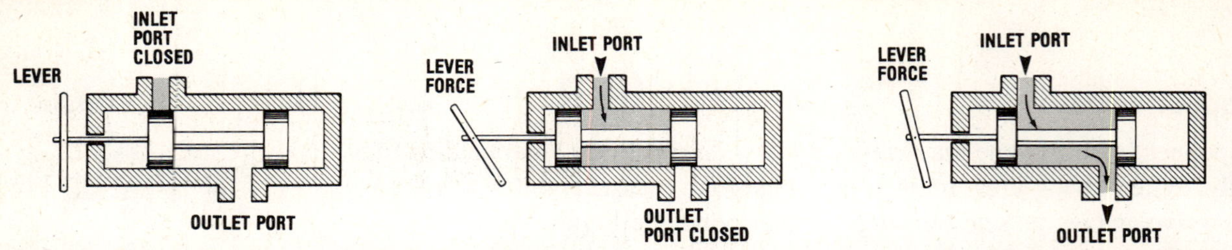

Figure 9-24. To alter fluid flow, a manually operated lever is used to move the valve.

When pressure to port B increases, it combines with the spring force to push the valve to the left and close off the inlet and outlet ports. See Figure 9-28.

Hydraulic force from ports A and B, and spring force determine when the valve moves to direct fluid through the inlet and outlet ports. In operation, the valve shifts quickly, or "snaps" from one position to another. For this reason, this type of valve is often called a "snap" valve.

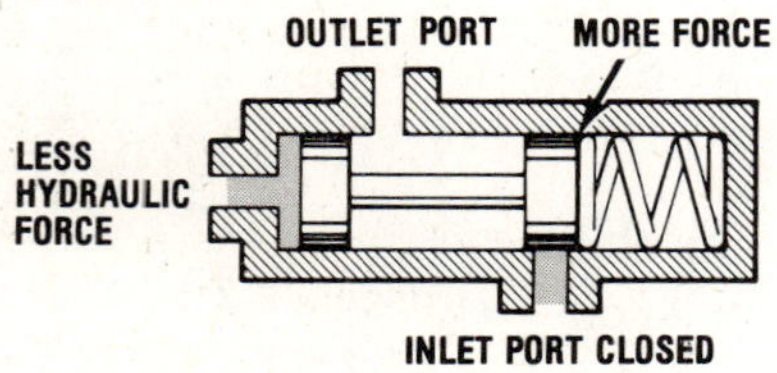

Figure 9-25. Inlet and outlet ports can be opened and closed by a hydraulic port directing fluid against a switching valve face.

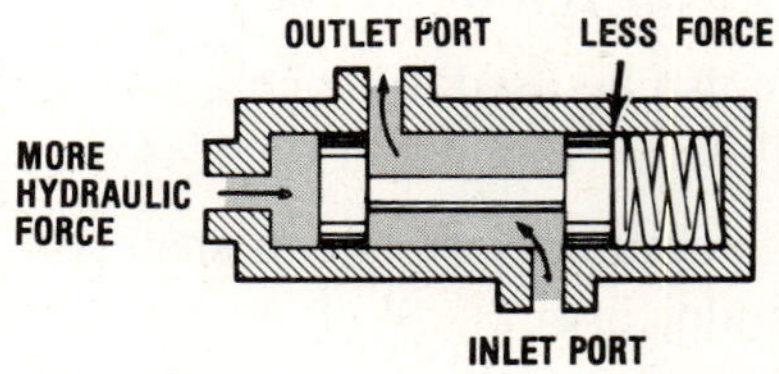

Figure 9-26. The amount of hydraulic pressure from the auxiliary port affects spring tension.

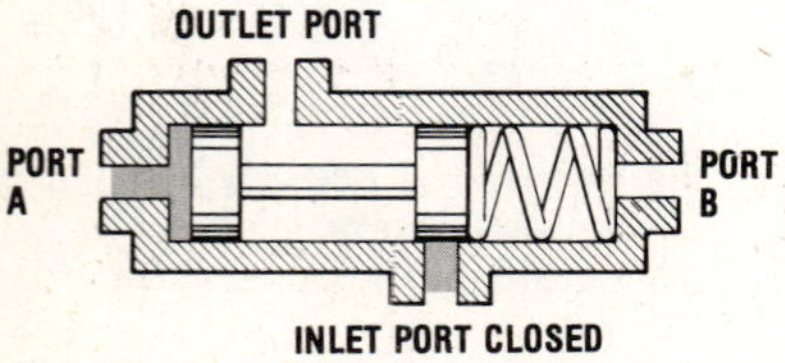

Figure 9-27. Shift valves may have auxiliary ports at both sides of the shift valve faces.

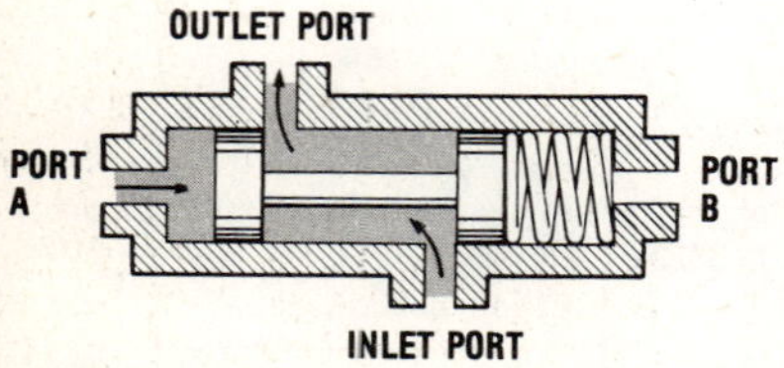

Figure 9-28. Increased pressure to Port B, combined with spring force, pushes the valve to the left.

CHECKPOINT

1. **Types of switching valves include ________________ valve, ________________-way ball check valve, two-way ball ________________ ________________, ________________-operated spool valve, and ________________-operated spool valve.**
2. **A ________________ can be used in place of the flat poppet valve to provide the same ________________-________________ action.**
3. **The lever can be moved to position the ________________ ________________ to direct fluid through one or more outlet ports.**
4. **________________ pressure can be used to overcome the force of a ________________ to operate a spool switching valve.**
5. **A transmission ________________ valve is similar to a ________________ operated spool valve.**

NAME ______________________ SECTION ____________ DATE ____________

Certification Practice

Test 9

DIRECTIONS:

The following questions are similar to those used on mechanic certification tests. Answer each question by circling the letter of the correct choice.

1. The word *hydraulics* comes from a word that means.
   A. oil.
   B. water.
   C. greek.
   D. solid.

2. If a plunger with a surface area of 1.5 square inches pushes against fluid in a sealed container with a force of 15 pounds, the pressure is
   A. 10 psi.
   B. 22.5 psi.
   C. 0.1 psi.
   D. none of the above.

Question 3.
(Pontiac Motor Division-GMC)

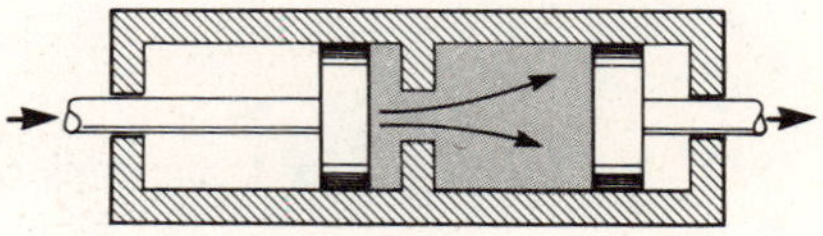

3. Which of the following statements about the picture is correct?
   I. Moving the left piston one inch will cause the right piston to move one inch.
   II. If either piston is moved, an equal volume of fluid will be transferred from one cylinder to another.
   A. I only   B. II only   C. Both I and II   D. Neither I nor II

4. Mechanic A says the pressure by a smaller piston in a closed hydraulic system will be greater than the pressure by a large piston.
   Mechanic B says the pressure by a larger piston will be greater than the pressure by a small piston.
   Who is correct?
   A. A only   B. B only   C. Both A and B   D. Neither A nor B

5. All of the following are correct EXCEPT
   A. Liquids are incompressible.
   B. Pressure within a closed hydraulic system is the same everywhere.
   C. Force on a piston is directly proportional to its surface area.
   D. Pressure can be generated only if there is no resistance to the flow of liquid.

6. The unit shown in the picture is a

   A. simple pressure-regulating valve.
   B. one-way poppet valve.
   C. balance valve.
   D. transmission shift valve.

Question 6.

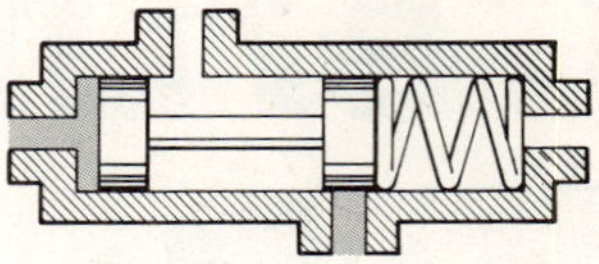

7. A balance valve always has

   A. a spring.
   B. a Multiland spool valve.
   C. one inlet and one outlet port.
   D. a lever to regulate spring pressure.

8. The operation of a simple spool valve with one end larger than the other is being discussed. Which of the following statements is correct?

   I. Pressure directed to the valley will cause the valve to move toward the smaller end.
   II. Less force will be exerted against the larger valve face.

   A. I only B. II only C. Both I and II D. Neither I nor II

9. Mechanic A says a ball and spring can be used to make a one-way ball check valve. Mechanic B says a ball and spring can be used to make a two-way ball check valve. Who is correct?

   A. A only B. B only C. Both A and B D. Neither A nor B

10. All of the following are commonly used to help regulate and control hydraulic pressure in the automatic transmission EXCEPT

    A. orifices, spool valves, multiland spool valves.
    B. springs.
    C. levers.
    D. higher viscosity transmission fluid.

10 Hydraulic Systems

YOUR LEARNING JOB

The hydraulic system controls all of the actions of the automatic transmission. Hydraulic pressure and how it is directed throughout the system is the key to understanding automatic transmission operation. This unit discusses the various parts that direct hydraulic flow and how they operate.

JOB COMPETENCY OBJECTIVES

This is a knowledge-building unit. When you have completed your assignments and exercises in this unit, you should be able to:

- ☐ Describe the common types of automatic transmission fluid.
- ☐ Explain how different types of oil pumps operate.
- ☐ Name and describe the main types of valves in the automatic transmission.
- ☐ Explain how throttle pressure and governor pressure act together to produce shifts at the correct times.

10.1 AUTOMATIC TRANSMISSION FLUIDS

The liquid used to operate the hydraulic system in the automatic transmission must have certain qualities. It must be able to transmit pressure, lubricate, cool, and have the proper viscosity to flow through the narrow orifices. Viscosity is the thickness or thinness of a fluid and its ability to resist flowing.

Proper viscosity is important. If the automatic transmission fluid, or ATF, is too thick, it will not flow properly, especially during cold conditions. A fluid that flows too slowly will result in slower movement of the transmission parts. An ATF that is too thin will permit excessive wear. Compared to engine oil, ATF has a viscosity of approximately SAE 10.

Automatic transmission fluid is a petroleum-based oil, similar to motor oil. Additives are mixed with the fluid to maintain viscosity and to help resist foaming, sludge and varnish buildup, and corrosion. A red dye is added to help in spotting leaks. Three basic types of ATF are widely recommended in modern vehicles:

- DEXRON II
- Type F
- Ford Type CJ.

Each of these fluids has different characteristics. None of them is compatible with all transmissions. The manufacturer's recommendations for ATF must always be followed. Otherwise, poor transmission operation will result.

DEXRON II Fluid

This transmission fluid is recommended for all General Motors cars, some Fords, Chrysler products, and AMC products. Most foreign manufacturers also recommend DEXRON II. DEXRON II replaces earlier GM fluids that were named Type A and DEXRON.

This fluid works well at low temperatures and has a high resistance to oxidation and corrosion. Oxidation is the combination of oxygen with other chemical elements. Oxidation generally damages or destroys the original elements. DEXRON II also has frictional properties that provide for smooth shifting.

Type F Fluid

Type F transmission fluid is recommended for many earlier Ford and foreign vehicles. Like DEXRON II, Type F has a high resistance to oxidation, corrosion, and heat. The biggest difference is that Type F has frictional characteristics suitable for Ford clutches and bands. Type F fluid produces a quicker lockup of these clutch components in Ford transmissions.

Ford Type CJ Fluid

This transmission fluid is used in 1977 and later Ford-produced vehicles with the C-6 and Jatco automatic transmissions. The Type CJ is very similar to DEXRON II. Since 1978, Ford has considered DEXRON II an acceptable substitute for Type CJ. Type CJ has frictional characteristics different from Type F fluid. Like DEXRON II, it is more slippery and produces smoother shifts.

CHECKPOINT

1. **Viscosity is the thickness or thinness of a fluid and its ability to ____________________ flow.**
2. **If the ATF is too thick, it will not ____________________ properly.**
3. **Three basic types of ATF are widely recommended in modern vehicles: __________________________ II, Type ________________, and Ford type ________________.**
4. **The manufacturer's ____________________ for the type of ATF must always be followed.**

10.2 THE OIL PUMP

An automatic transmission will not operate without an oil pump. The oil pump creates the pressure to operate the apply devices and circulate the fluid throughout the system.

One oil pump is used on modern automotive transmissions. It is located just behind the torque converter and is driven by the torque converter. This means the pump will not operate unless the engine is running.

Basically, the oil pump consists of a drive member, a driven member, and two pump body halves. The pump body halves contain passages that direct fluid to various parts of the transmission. Some oil pumps have other controls built into them, but these are discussed later in this unit. Three types of oil pumps are used in modern automatic transmissions:

- Gear and crescent pump
- Rotor pump
- Vane pump.

The pump draws fluid from the reservoir and distributes it throughout the hydraulic system. Regardless of type, all oil pumps have the same basic function.

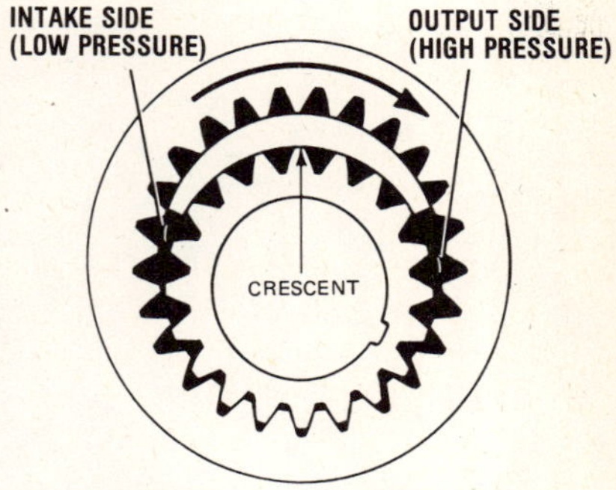

Figure 10-1. In a gear and crescent oil pump, the inner and outer gears rotate. The crescent does not move. (Chevrolet Motor Division-GMC)

Gear and Crescent Pump

A gear and crescent pump has two gears: an inner drive gear and an outer driven gear. See Figure 10-1. The gears are enclosed in a two-piece housing called the pump body. The inner gear is splined or lugged to the torque converter hub.

The inner gear turns at engine speed and drives the outer gear. The two gears mesh only at one side of the pump. A crescent-shaped piece prevents meshing of the gears at the opposite side of the pump.

The crescent plays an important role in the action of this type of pump. As the gears rotate, the gear teeth are separated by the crescent. A low pressure area is formed at the area where the separation occurs. Atmospheric pressure forces the liquid to flow from the reservoir, or oil pan, into the pump. The gears continue to rotate, carrying the fluid with them. At the opposite end of the crescent, the gear teeth begin to mesh, squeezing the fluid between them. The fluid is forced out of the output end under pressure.

Rotor Pump

A rotor pump is similar to a gear and crescent pump in operation. Instead of gear teeth, a rotor pump has inner and outer rotors with lobes, or rounded projections, that mesh. No separating crescent is used. See Figure 10-2. The inner rotor is splined or lugged to the torque converter hub and turns when the engine runs. The inner rotor fits inside the outer rotor off center. Because they are off center, the rotors mesh at one side and separate at the other.

As the rotor lobes separate, a low pressure area is formed. Atmospheric pressure forces the ATF from the oil pan into the separation. As the lobes continue to rotate, they carry the ATF through the pump. At the output side, the lobes begin to mesh tightly, squeezing the fluid between them. The fluid is forced out of the pump under pressure.

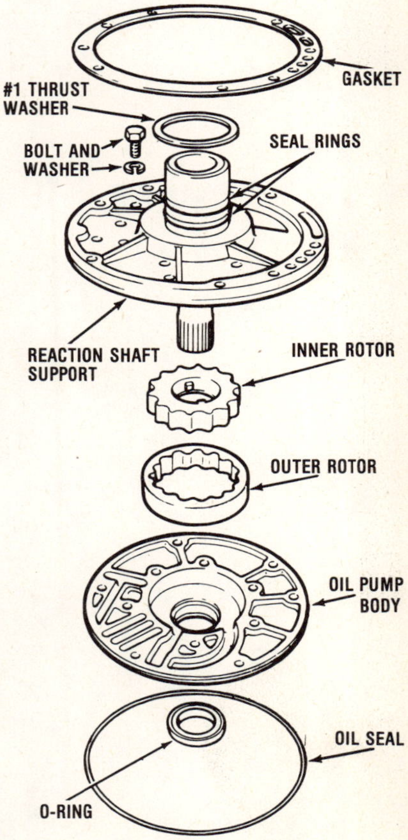

Figure 10-2. A rotor pump operates as lobes on the inner and outer rotors mesh. (American Motors Corporation)

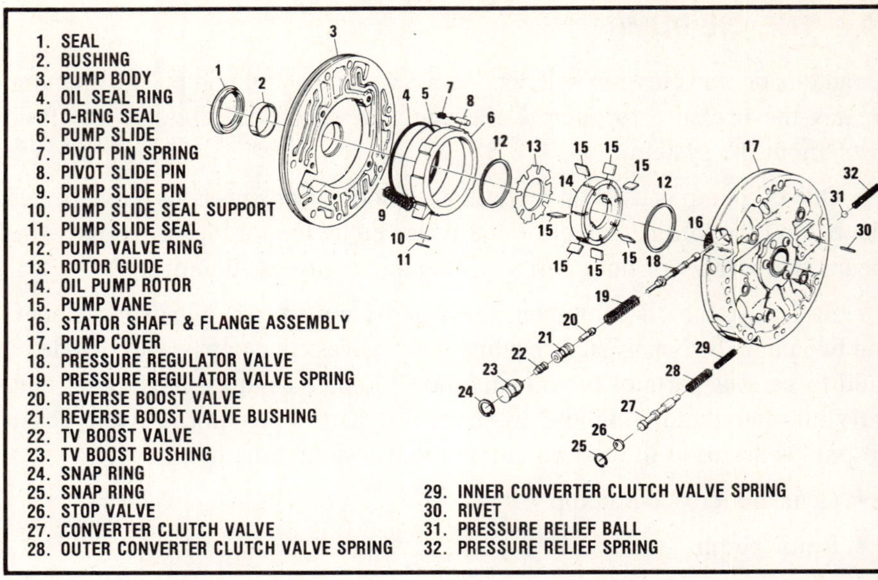

Figure 10-3. The parts of a vane-type oil pump. (Chevrolet Motor Division-GMC)

Vane-type Pump

A vane-type pump operates on the same principles as the gear and crescent and rotor pumps. Vanes, or blades, fit into slots in the vane holder. Springs at the bottom of the slots push the vanes tightly against the bore. The vane holder fits into the bore off-center and turns when the engine rotates. A vane-type pump is shown in Figure 10-3.

As the vane holder rotates, the vanes are forced into the slots, compressing the springs. As the vanes pass the inlet area, they move outward, increasing the area between the blades and creating a low-pressure area. Atmospheric pressure pushes the ATF from the oil pan into the pump. As the vane holder continues to rotate, the ATF is carried between the vanes.

Near the outlet area, the vanes are compressed into the slots, and the area between the vanes becomes smaller. The ATF is squeezed between the vane holder and the bore. The fluid is forced out of the pump under pressure.

CHECKPOINT

1. **The three types of oil pumps used on modern automatic transmissions are the ____________ and ____________, ____________, and ____________-type.**
2. **A ____________-pressure area is formed where the gear teeth separate.**
3. **____________ pressure forces the ATF from the oil pan into the pump.**
4. **As the gear teeth, rotors, or vane holder squeeze the ATF, the fluid is forced out under ____________.**

10.3 HYDRAULIC CIRCUITS

A transmission hydraulic circuit includes a pressure source, lines or passages, control valving, and an output device. The main pressure source for the entire transmission is the oil pump. Control valving consists of pressure-regulating, balancing, and switching valves located inside machined bores in the valve body. Hydraulic pressure is regulated for use in a particular hydraulic circuit. Four main types of fluid pressures are used in automatic transmissions:

- Mainline pressure
- Throttle pressure
- Modulator pressure
- Governor pressure.

Pressure is created by the oil pump. After being regulated, it is known as mainline pressure. All other pressures in the transmission are regulated forms of mainline pressure. Throttle pressure changes with carburetor throttle opening. Governor pressure changes with vehicle speed.

The valve body contains a maze of passages to direct these and other pressures to the proper locations, and to connect hydraulic circuits. A steel separator, or transfer, plate helps to direct and connect fluid pressure. See Figure 10-4. The valve body is located inside the transmission oil pan. A simplified valve body is shown in Figure 10-5.

Fluid pressures travel through hydraulic circuits in the valve body, transmission case, and hollow transmission shafts to the apply devices. The apply device for some circuits may be as simple as one face of a spool valve.

The following topics discuss the main hydraulic circuits and the types of valves used in them.

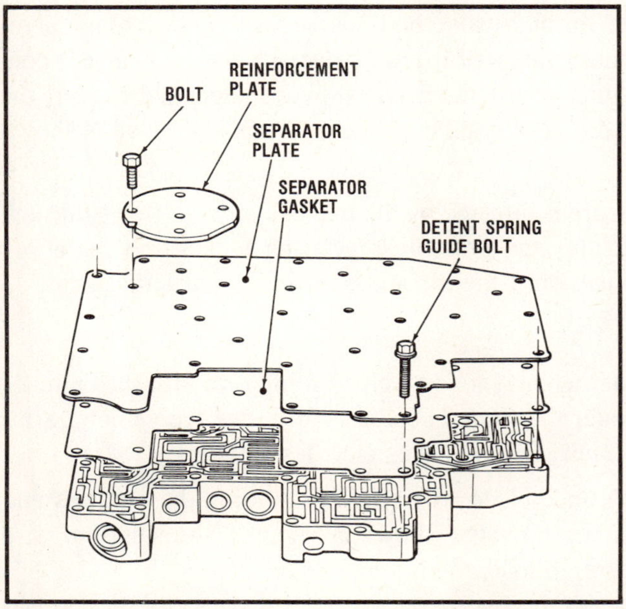

Figure 10-4. A separator plate helps to direct and connect hydraulic pressure in the valve body. (Ford Motor Company)

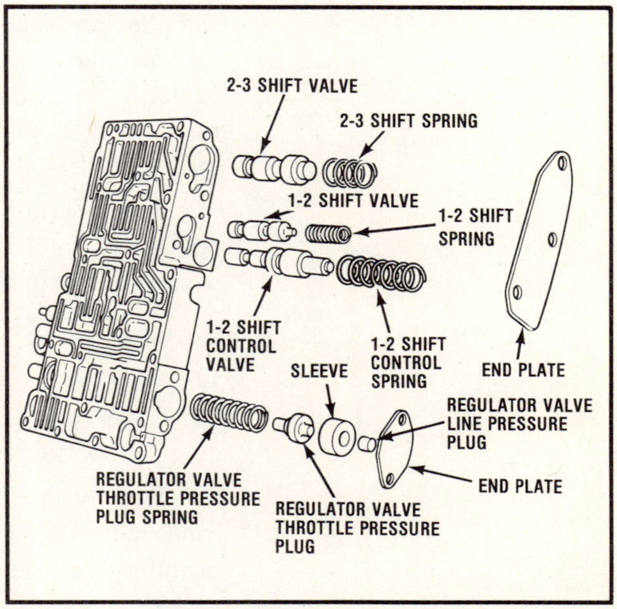

Figure 10-5. Simplified valve body. (American Motors Corporation)

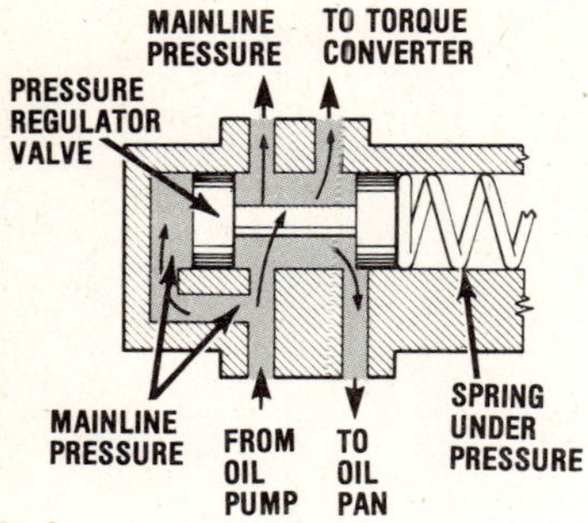

Figure 10-6. The pressure regulator valve controls pressure from the oil pump.

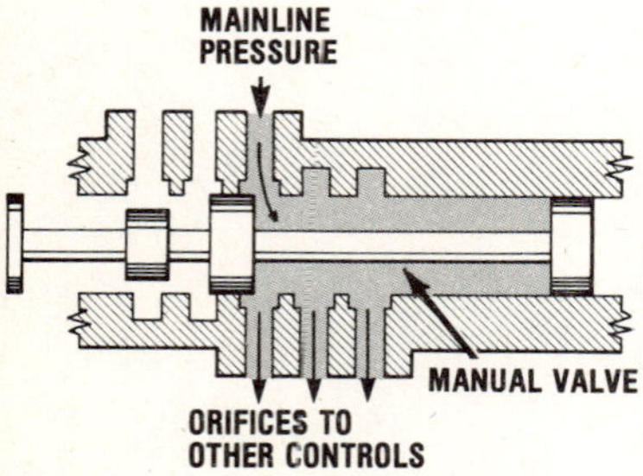

Figure 10-7. The manual valve directs pressure as the gearshift lever is moved

10.4 MAINLINE PRESSURE

As discussed in 9.6, pressure can be generated only if there is some resistance to the flow of liquid. The pump outlet port and connecting passages in the valve body form orifices which help to maintain pressure.

However, a pressure-regulating valve is necessary. The faster the engine turns, the more pressure the oil pump will produce. Without some form of regulation, hydraulic pressure could increase enough to damage parts of the transmission.

Pressure Regulator Valve

A pressure regulator valve, shown in Figure 10-6, is used to control the output pressure from the oil pump.

As the valve balances between the force exerted by the spring and the hydraulic force against the piston face, the oil pump pressure output is regulated. The pressure in the lines from the oil pump and to the transmission will be equal. This is known as mainline, or line, pressure. Some manufacturers refer to mainline pressure as "control pressure" or "drive oil." Typical mainline pressure is between 55 and 150 psi [380 and 1034 kPa] in forward gears and 300 psi [2069 kPa] in reverse.

In a closed hydraulic system, pressure is the same everywhere. Mainline pressure is the same in the pressure regulator input and output lines. This means that the fluid to operate the transmission does not have to pass through the pressure regulator valve. Any sealed connection to the pressure regulator valve input or output passages will have mainline pressure.

The pressure regulating valve is also a switching valve. As the pressure regulator valve moves to regulate mainline pressure, the lands of the spool valve uncover a passage leading to the torque converter. ATF flows into the torque converter.

Manual Control Valve

Mainline pressure flows through valve body passages to a switching valve, known as the manual control valve, or manual valve. The manual valve is connected to the gearshift linkage. As the driver moves the gearshift lever, the manual valve moves to direct hydraulic pressure. A manual valve is shown in Figure 10-7.

After the fluid pressure is directed by the manual valve, it flows through passages to other switching and pressure-regulating valves and the apply devices. The apply devices shift the transmission into a particular gear.

Booster Valve

When the vehicle is under a heavy load, such as going up a hill, the clutches and bands could slip under the torque necessary to move the vehicle. Slipping causes friction and heat that damages clutches and bands.

To prevent slipping under load, the pressure holding the clutches and bands must be higher. A booster valve, shown in Figure 10-8, is used to raise mainline pressure under high load conditions.

A booster valve is a balance valve with a pressure-regulated spring. Under heavy load, or when the accelerator pedal is fully depressed, pressure is

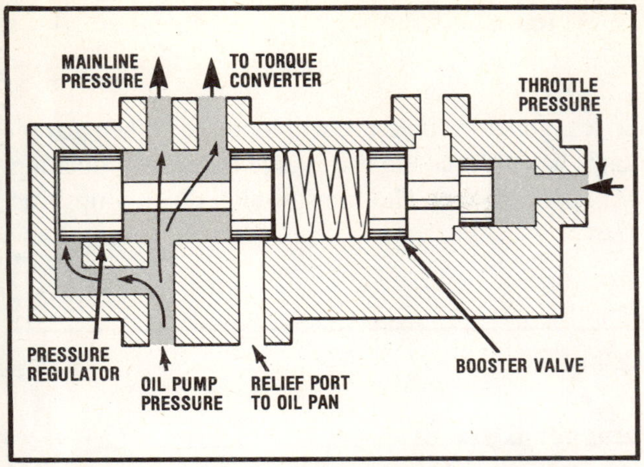

Figure 10-8. A booster valve raises mainline pressure to prevent slippage when the transmission is downshifted or operating under a load.

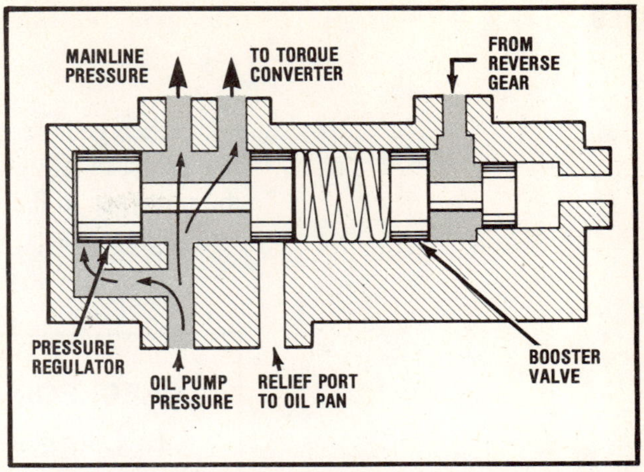

Figure 10-9. The booster valve also doubles mainline pressure to help the reverse apply devices operate without slippage.

directed to the end of the booster valve. This pressure pushes the booster valve, compressing the spring and exerting force against the pressure regulator valve. The pressure regulator valve moves to uncover more of the passages from the pump and to the mainline circuits. Mainline pressure rises.

The accelerator, or throttle, operates the valve that directs pressure to the booster valve. This pressure is called throttle pressure, and is used to help control transmission shifts. Throttle pressure is covered in more detail in 10.5.

The pressure needed for reverse apply devices is usually about double normal mainline pressure, or approximately 100–300 psi [700–2069 kPa]. Pressure from the reverse gear hydraulic circuit is directed into the valley between the lands of the booster valve. See Figure 10-9.

The end of the booster valve closest to the pressure regulator valve is larger than the end operated by throttle pressure. More force is exerted against the larger end, and the booster valve moves to compress the spring. The additional force moves the pressure regulator valve, opening the passages from the pump and to the mainline circuits. Mainline pressure rises to hold the apply devices more firmly.

Shift Valve

To control and time transmission shift points, switching valves are used to direct hydraulic pressure to the apply devices. These switching valves are called shift valves.

Shift valves select the correct gear ratio for road speed and load conditions by operating apply devices at the correct time. The shift valves are operated by a combination of throttle pressure and governor pressure. Throttle pressure increases with throttle opening and engine torque. Governor pressure is controlled by vehicle speed. The faster the vehicle travels, the greater governor pressure becomes. Governor pressure is discussed in more detail in 10.6.

A shift valve is shown in Figure 10-10. The shift valve can be moved to open or close the mainline pressure passage and outlet passage. When the vehicle is under a heavy load but moving slowly, throttle pressure will be high and governor pressure low. Throttle pressure and the force of the spring will combine to push the shift valve to the right. The passage carrying mainline pressure and the passage to the upshift apply device will be closed. The vehicle will not upshift to a higher gear.

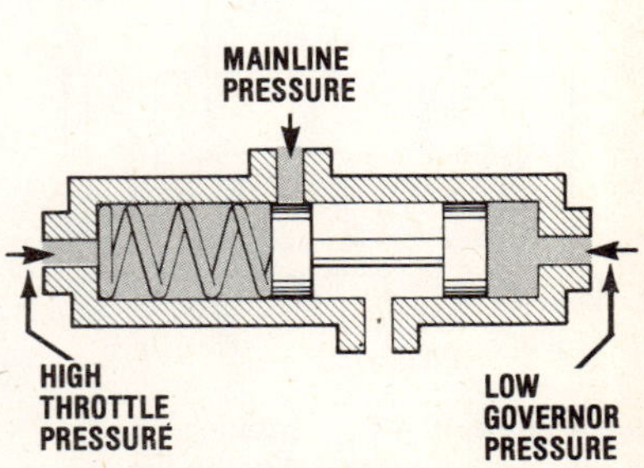

Figure 10-10. The shift valve controls the mainline pressure passages.

When vehicle speed rises, governor pressure will increase. When governor pressure is high enough, it will force the shift valve to the left, opening the passages to the upshift apply device. The transmission quickly upshifts to a higher gear.

CHECKPOINT

1. **Control valving consists of ____________________-regulating, ________________, and __________________ valves located inside machined bores in the valve body.**
2. **As the valve balances between the force exerted by the ____________________ and the ____________________ force against the piston face, the oil pump pressure output is regulated.**
3. **All pressures in the automatic transmission circuits are created from ____________________ pressure.**
4. **As the gearshift lever is moved, the ____________________ ____________________ valve moves to direct hydraulic pressure.**
5. **A ____________________ ____________________ is used to raise mainline pressure under high load conditions.**
6. **____________________ ____________________ select the correct gear ratio for road speed and load conditions by operating apply devices at the correct time.**

10.5 THROTTLE PRESSURE

Throttle pressure is directly related to engine load and throttle, or accelerator, position. Throttle pressure is used to increase mainline pressure and to help control shifting.

When the vehicle is under only a light load, such as when cruising, throttle pressure is low. As vehicle speed increases, governor pressure will rise and move the shift valves to upshift to a higher gear.

If the vehicle begins to climb a hill, the driver pushes harder on the accelerator to maintain speed. Throttle pressure rises to operate the booster valve. Mainline pressure rises to hold the apply devices more firmly. If throttle pressure rises high enough, it will move the shift valve to close off the upshift apply passage. The vehicle will downshift to a lower gear.

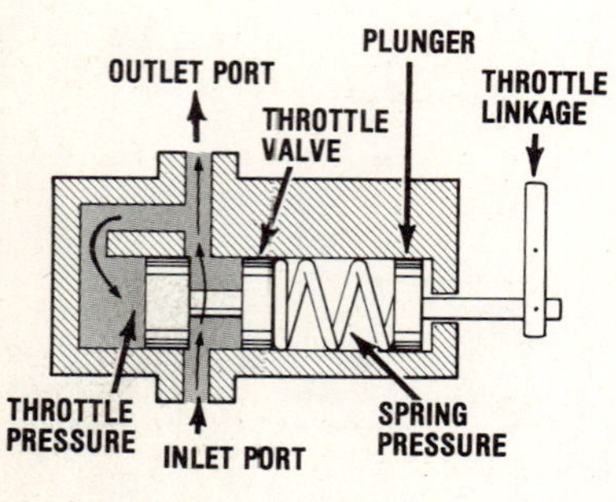

Figure 10-11. Throttle pressure is controlled by the throttle pressure valve. The valve is connected to throttle linkage or operated by vacuum.

Throttle Valve

Throttle pressure is regulated by a throttle valve, shown in Figure 10-11. The throttle valve is a balancing valve with a lever-regulated spring. The throttle valve can be connected mechanically to the throttle linkage, or operated by engine vacuum.

When the engine is under a load, the driver presses harder on the accelerator pedal to maintain speed. On some transmissions, carburetor or accelerator pedal linkage is directly connected to the throttle valve. As the driver presses harder, the throttle-valve spring is compressed. The spring pushes the throttle valve to the left, opening the mainline pressure passage and the throttle pressure outlet port. Mainline pressure flows through the throttle valve, and becomes throttle pressure. When the accelerator is all the way down, the throttle valve opens the mainline and throttle pressure passages fully. Throttle pressure rises to boost mainline pressure. See Figure 10-12.

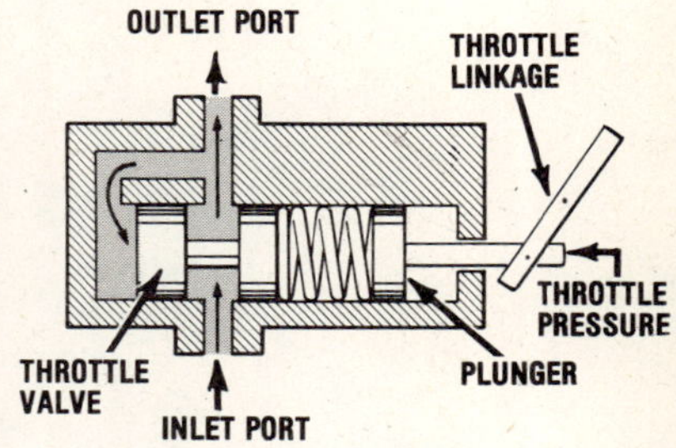

Figure 10-12. When the throttle valve is opened, throttle pressure increases mainline pressure.

10.6 VACUUM MODULATOR PRESSURE

On some transmissions, engine vacuum is used to do the same job. When engine load is high and the carburetor throttle plate is opened wide, intake manifold vacuum drops. A vacuum control diaphragm, or vacuum modulator, is used to exert pressure on the throttle valve spring. See Figure 10-13.

A vacuum control diaphragm consists of a rubber-and-cloth diaphragm and a spring inside a tightly-sealed metal cover. Atmospheric pressure is allowed to act on one side of the diaphragm. Engine intake manifold vacuum on the other side of the diaphragm pulls it in the same direction. The spring exerts force to push the diaphragm in the opposite direction.

When the carburetor throttle plate is opened slightly, such as under light loads, engine intake manifold vacuum is high. The diaphragm is pulled to the right. The throttle valve moves to the right, partially blocking the mainline pressure port. Throttle pressure is low.

When the carburetor throttle plate is pushed open wider, intake manifold vacuum drops. The spring pushes the diaphragm to the left. The throttle valve moves to open the mainline pressure port. Throttle pressure rises.

As the throttle is opened and closed, the vacuum modulator modulates, or changes, throttle pressure. The throttle valve may touch the diaphragm directly, or be connected through a pushrod. Some transmissions use vacuum modulators to change mainline or other pressures or create other transmission control pressures.

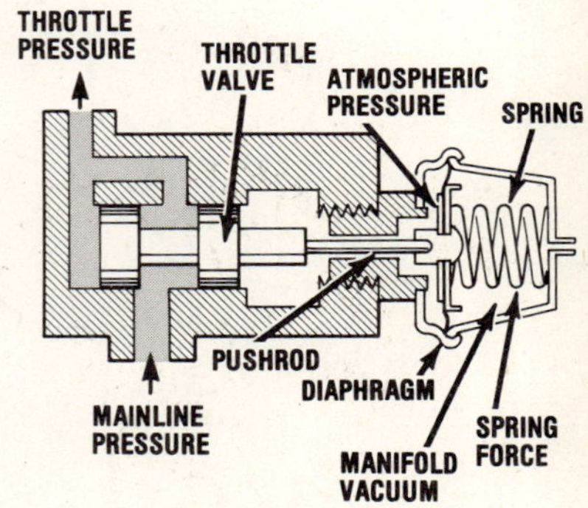

Figure 10-13. A vacuum modulator may be used instead of throttle linkage to exert pressure on the throttle valve spring.

Downshift Valve

Other valves can be used with the throttle valve to produce higher pressures for downshifts. Although known by different names, such as downshift valves, detent valves, or kickdown valves, they all do the same job. Downshift valves can be operated by mechanical linkage, or by electrical solenoids.

A downshift valve applies force to move the throttle valve. When the throttle plate is opened wider, the downshift valve forces the throttle valve to move. The throttle valve uncovers the mainline pressure and throttle pressure ports. Throttle pressure rises.

A downshift valve can also create another control pressure. This pressure is used to overcome governor pressure on a shift valve and force a downshift to a lower gear.

CHECKPOINT

1. **Throttle pressure is directly related to ______________ load and ______________, or accelerator, position.**
2. **Throttle pressure is used to increase ______________ pressure and to help control ______________.**
3. **The throttle valve can be connected ______________ to the throttle linkage, or operated by ______________ ______________ vacuum.**
4. **As the ______________ is opened and closed, the ______________ ______________ modulates, or changes, throttle pressure.**

10.7 GOVERNOR PRESSURE

As discussed in 10.3, governor pressure can be used to help shift the transmission into the proper gear. A governor valve creates governor pressure from mainline pressure, in response to transmission output shaft speed. One type of governor valve, driven at 90 degrees to output shaft, is shown in Figure 10-14.

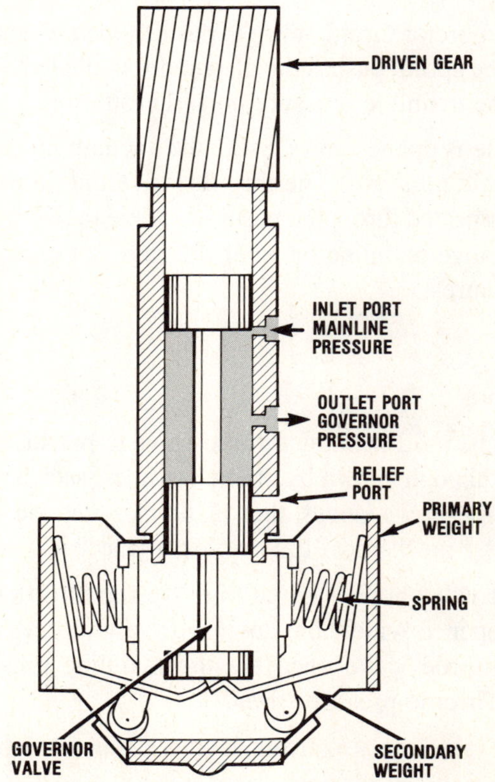

Figure 10-14. A rotating output shaft may turn the governor drive gear. Weights connected to the shaft are thrown outward and act on a spool valve.

As the output shaft rotates, it turns the governor drive gear. Two sets of weights connected to the shaft are thrown outward by centrifugal force. The weights are also connected to a spool valve in the body of the governor.

As the weights move out, they push on the spool valve. The spool valve moves, partially uncovering a mainline pressure port and partially closing an exhaust port. Pressure rises in the governor pressure outlet port.

The faster the vehicle is traveling, the faster the governor spins. Governor pressure rises as vehicle speed rises, generally about 1 psi [7 kPa] for each mile per hour. At a steady speed, the governor pressure will stabilize. When the governor is spinning fast enough, governor pressure may be close to mainline pressure.

Governor pressure pushes against throttle pressure on the opposite end of a shift valve. When governor pressure becomes high enough, it will move a shift valve against throttle pressure to produce an upshift. When vehicle speed drops, governor pressure will drop until throttle pressure pushes the shift valve to close off the upshift apply device. The vehicle will downshift to a lower gear.

Some governor valves are attached directly to the output shaft. No gear drive arrangement is necessary.

10.8 THROTTLE PRESSURE AND GOVERNOR PRESSURE

Throttle pressure and governor pressure work against each other to produce upshifts or downshifts at the correct times.

If the driver accelerates gently, not pressing the accelerator to open the throttle wide, throttle pressure remains low. Governor pressure will upshift the transmission sooner into higher gears for better fuel economy.

If the driver accelerates quickly, opening the throttle wide, throttle pressure will keep the shift valves closed longer. The transmission will remain in each gear longer for better acceleration.

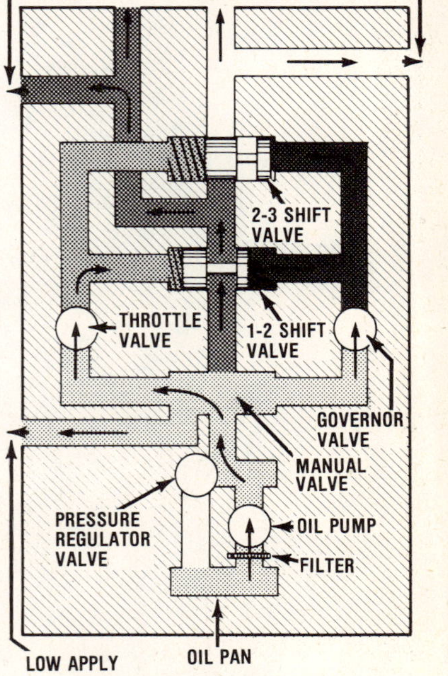

Figure 10-15. Simplified three-speed automatic transmission hydraulic flow chart.

10.9 SIMPLIFIED THREE-SPEED TRANSMISSION

Look at the simplified three-speed transmissions shown in Figure 10-15 as you read the following explanation.

The pump picks up ATF from the oil pan. The pressure regulator valve controls mainline pressure. The gearshift lever has been moved to position the manual valve in D (drive).

The manual valve directs mainline pressure to the first gear apply device circuit, or apply circuit. In addition, the manual valve directs mainline pressure to the throttle and governor valves, and the 1-2 shift valve.

When the output shaft speed is great enough, governor pressure rises above throttle pressure and the spring force at the 1-2 shift valve. The intermediate, or 2, apply device receives mainline pressure. The transmission shifts into intermediate, or second gear. As the 1-2 shift valve moves, it directs mainline pressure to the 2-3 shift valve.

The spring in the 2-3 shift valve is stronger than the spring in the 1-2 shift valve. Governor pressure must be higher to overcome the combined force of throttle pressure and the stronger spring pressure. This prevents the vehicle from shifting into high, or third, gear until the vehicle reaches a greater speed.

When governor pressure is high enough, it moves the 2-3 shift valve to allow mainline pressure to operate the high, or third, gear apply device. A branch of the high gear apply circuit shown at the top of the diagram allows mainline pressure to release the intermediate, or second gear apply device. The transmission is shifted into high gear.

CHECKPOINT

1. **A ____________________ valve creates governor pressure from mainline pressure, in response to transmission ____________________ shaft speed.**
2. **The weights are connected to a ____________________ valve in the body of the governor.**
3. **When the governor is spinning fast enough, governor pressure will become equal to ____________________ pressure.**
4. **If the driver accelerates ____________________, opening the throttle wide, ____________________ pressure will keep the shift valves closed longer.**
5. **The spring in the ____________________ shift valve is stronger than the spring in the ____________________ shift valve.**

NAME ______________________ SECTION ____________ DATE ____________

Certification Practice

Test 10

DIRECTIONS:

The following questions are similar to those used on mechanic certification tests. Answer each question by circling the letter of the correct choice.

1. Mechanic A says a good ATF must be able to transmit pressure, lubricate, cool, and flow through the automatic transmission orifices.
   Mechanic B says a good ATF must resist foaming, high and low temperatures, sludge and varnish buildup, and corrosion.
   Who is correct?
   A. A only B. B only C. Both A and B D. Neither A nor B

2. DEXRON II ATF may be substituted for all of the following fluids EXCEPT
   A. DEXRON.
   B. Type A.
   C. Ford Type F.
   D. Ford Type CJ.

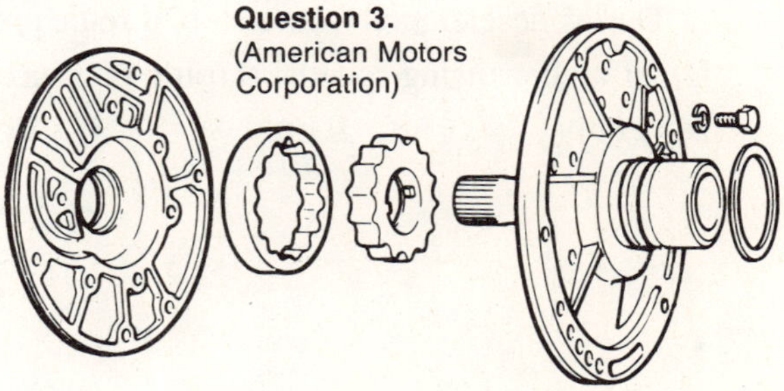

Question 3.
(American Motors Corporation)

3. The unit shown is
   A. a gear and crescent pump.
   B. a rotor pump.
   C. a vane-type pump.
   D. none of the above.

4. Which of the following statements is correct?
   I. As the oil pump turns, a high-pressure area is created at the input side of the pump.
   II. As the oil pump turns, the ATF forces the gears or rotors apart, creating a low-pressure area at the output side of the pump.
   A. I only B. II only C. Both I and II D. Neither I nor II

5. The pressures in the transmission hydraulic circuits can never be greater than
   A. 160 psi [1100 kPa].
   B. mainline pressure.
   C. throttle pressure.
   D. governor pressure.

6. Mechanic A says that the driver always has direct control of the manual valve.
   Mechanic B says that the driver always has direct control of the throttle valve.
   Who is correct?
   A. A only B. B only C. Both A and B D. Neither A nor B

7. The valve shown is a
   A. shift valve.
   B. booster valve.
   C. throttle valve.
   D. governor valve.

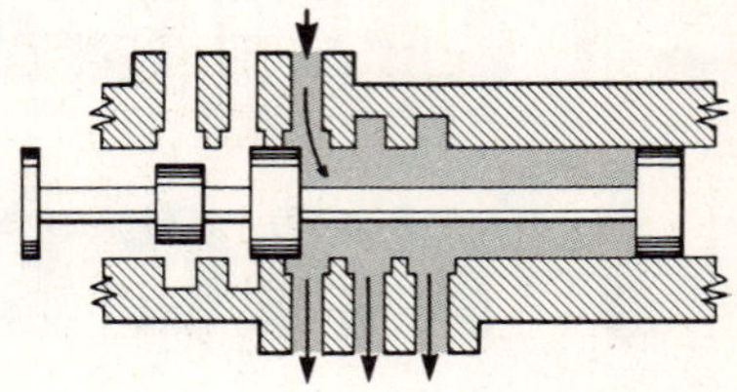

Question 7.

8. The pressure needed for the reverse apply device is usually about
   A. twice mainline pressure.
   B. three times mainline pressure.
   C. mainline pressure.
   D. none of the above.

9. A shift valve includes all of the following EXCEPT
   A. spring.
   B. mainline pressure input port.
   C. governor and throttle pressure input ports.
   D. apply device input port.

10. Which of the following statements is correct?
    I. During acceleration from a stop, throttle pressure is greater than governor pressure.
    II. At high cruising speeds, throttle pressure is greater than governor pressure.
    A. I only B. II only C. Both I and II D. Neither I nor II

11 Reading Flow Charts

YOUR LEARNING JOB

Hydraulic circuits within the transmission regulate and direct pressure to control shifting and operate apply devices. Each gear position requires a different set of hydraulic circuits to operate properly. To help the mechanic understand which circuits are necessary for a particular gear position, diagrams, or flow charts, are printed. Using the manufacturer's flow charts, a mechanic can trace problems to specific valves, servos, clutches and bands. This unit discusses what flow charts show and how to read them.

JOB COMPETENCY OBJECTIVES

This is a knowledge-building unit. When you have completed your assignments and exercises in this unit, you should be able to:

- ☐ Explain what a flow chart is.
- ☐ Explain how different pressures are indicated on a flow chart.
- ☐ Trace the flow of fluid in a typical flow chart.
- ☐ Using a flow chart, correctly describe which valves and apply devices are used for a particular gear position.

11.1 FLOW CHARTS

The flow chart is a map of the automatic transmission. Like a map, the flow chart shows important locations and the paths between them.

Figure 11-1 shows a flow chart for the oil pump and pressure regulating system. The lines show the paths the fluid takes in the hydraulic circuits. Also, there is a *key*, or explanation, that helps you to figure out the flow chart. Different fluid pressures are indicated by different colors or patterns. Refer to Figure 11-1 as you read this explanation.

The key at the bottom of the flow chart tells you what pressures are indicated by the colors or patterns. For example, find the pattern that indicates intake pressure. Then look at the flow chart to find that pressure line.

The intake pressure line runs from the sump at the bottom, through the screen, to the pump. At the pump it tees, or splits, off to the pressure-regulating valve and stops.

Now find the mainline pressure on the key. Find the mainline pressure lines on the flow chart. The mainline pressure begins at the pump. Then it goes through the pressure-regulating valve to the manual valve. For this illustration, mainline pressure stops at the manual valve. The transmission is in park.

Find the converter and lube (lubrication) pressure on the key. The converter and lube pressure starts at the pressure-regulating valve and goes to the torque converter. This line flows from the converter and through the lube line to the oil cooler. After passing through the cooler, it is returned to the lubrication circuit.

The basic flow of fluid will be the same for all transmissions. The oil pump generates the oil pressure used everywhere in the transmission. Pressure from the pump always goes to the pressure-regulating valve. From there, oil under pressure is directed to the manual valve. When the transmission is shifted into a gear, the manual valve moves to direct oil to other valves and apply devices. By tracing the pressures, the mechanic can determine which valves and apply devices should be operating for a particular gear position. Flow charts can be used in this way for troubleshooting.

Flow charts can also be read backwards from the apply devices or valves. By tracing the pressure backwards from a specific valve or apply device, the mechanic can see the origin of the pressure, and which valves must operate to supply pressure.

The principles and techniques necessary for understanding all flow charts are the same. Although different charts may use different colors or patterns to indicate pressures, the charts are read the same way. Always refer to the manufacturer's service manual for specific flow charts.

The following pages show flow charts for the General Motors 325 automatic transmission. Refer to the charts as you read the explanations.

CHECKPOINT

1. **The flow chart is a ____________ of the automatic transmission.**
2. **Different fluid pressures are indicated by different ____________ or ____________.**
3. **The ____________ pressure line runs from the sump at the bottom, through the screen, to the pump.**
4. **The mainline pressure begins at the ____________. Then it goes through the ____________.**

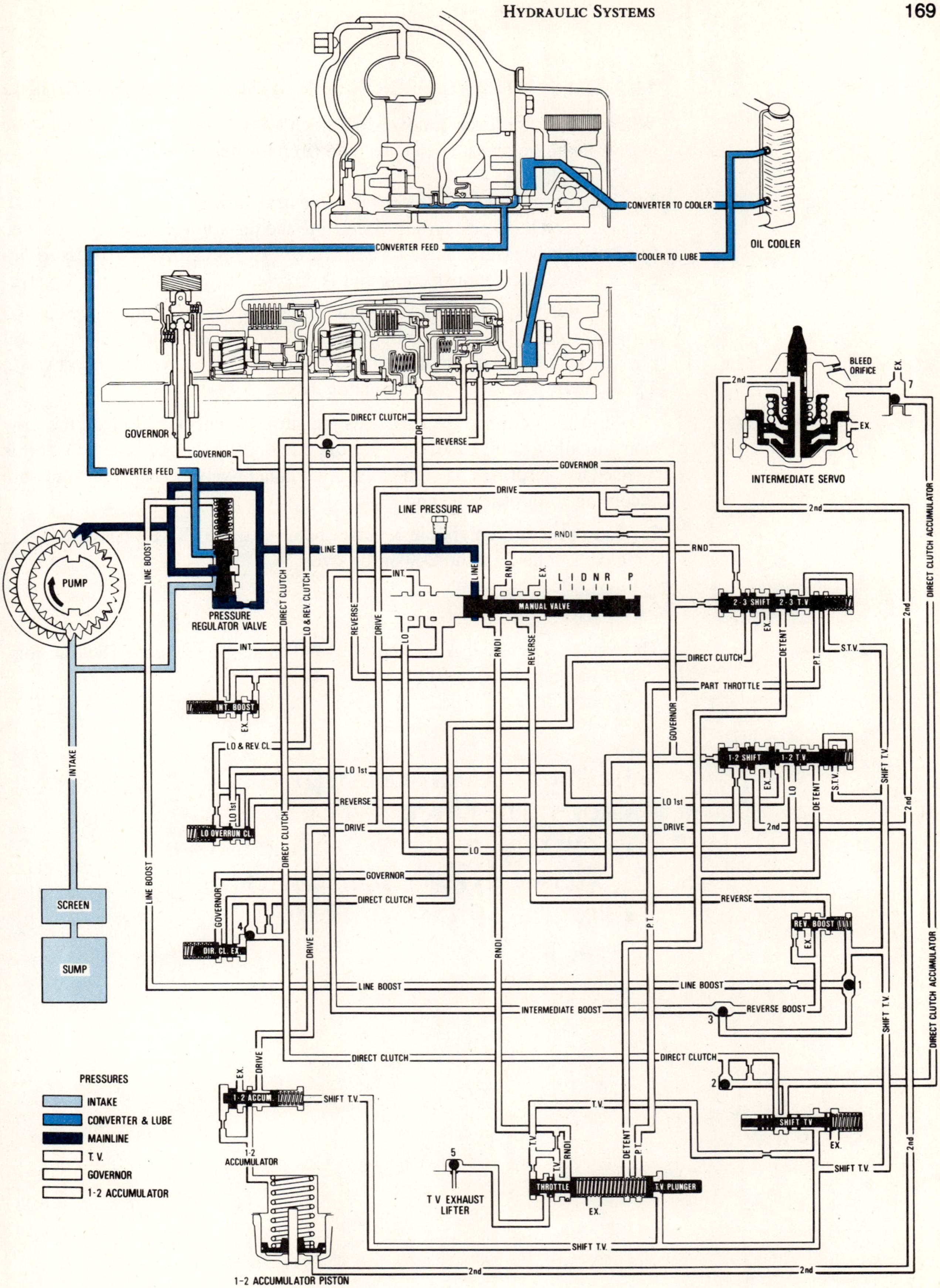

Figure 11-1. Oil pressure and pressure regulating flow chart. (Oldsmobile Division-GMC)

11.2 GM 325 TRANSMISSION IN NEUTRAL—ENGINE RUNNING

When the selector lever is moved to the neutral (N) position, the manual valve allows mainline pressure to enter two (2) passages.

1. Line pressure enters the reverse, neutral, drive, intermediate (RNDI) passage. RNDI oil pressure is directed to the throttle valve and is regulated to a variable pressure called throttle valve (TV) pressure. TV pressure increases with carburetor opening and is directed through the shift TV valve.

The shift TV valve limits shift TV oil pressure from going above 65–105 psi [450 to 725 kPa], depending on the transmission model. Shift TV oil pressure is then directed to the 1-2 and 2-3 throttle valves, reverse boost valve, TV plunger, and the 1-2 accumulator valve.

Shift TV oil pressure also seats the shift TV check ball (1 on the flow chart) to allow shift TV oil to enter the line boost passage. Line boost oil is directed to the spring end of the pressure regulator valve to boost line pressure according to throttle opening.

2. Line pressure enters the reverse, neutral, drive (RND) passage and is directed to a land on the 2-3 shift valve.

SUMMARY

The converter is filled. All the clutches and bands are released. The transmission is in neutral (N).

- **DIRECT CLUTCH—RELEASED**
- **FORWARD CLUTCH—RELEASED**
- **LOW AND REVERSE CLUTCH—RELEASED**
- **INTERMEDIATE BAND—RELEASED**
- **ROLLER CLUTCH—NOT HOLDING**

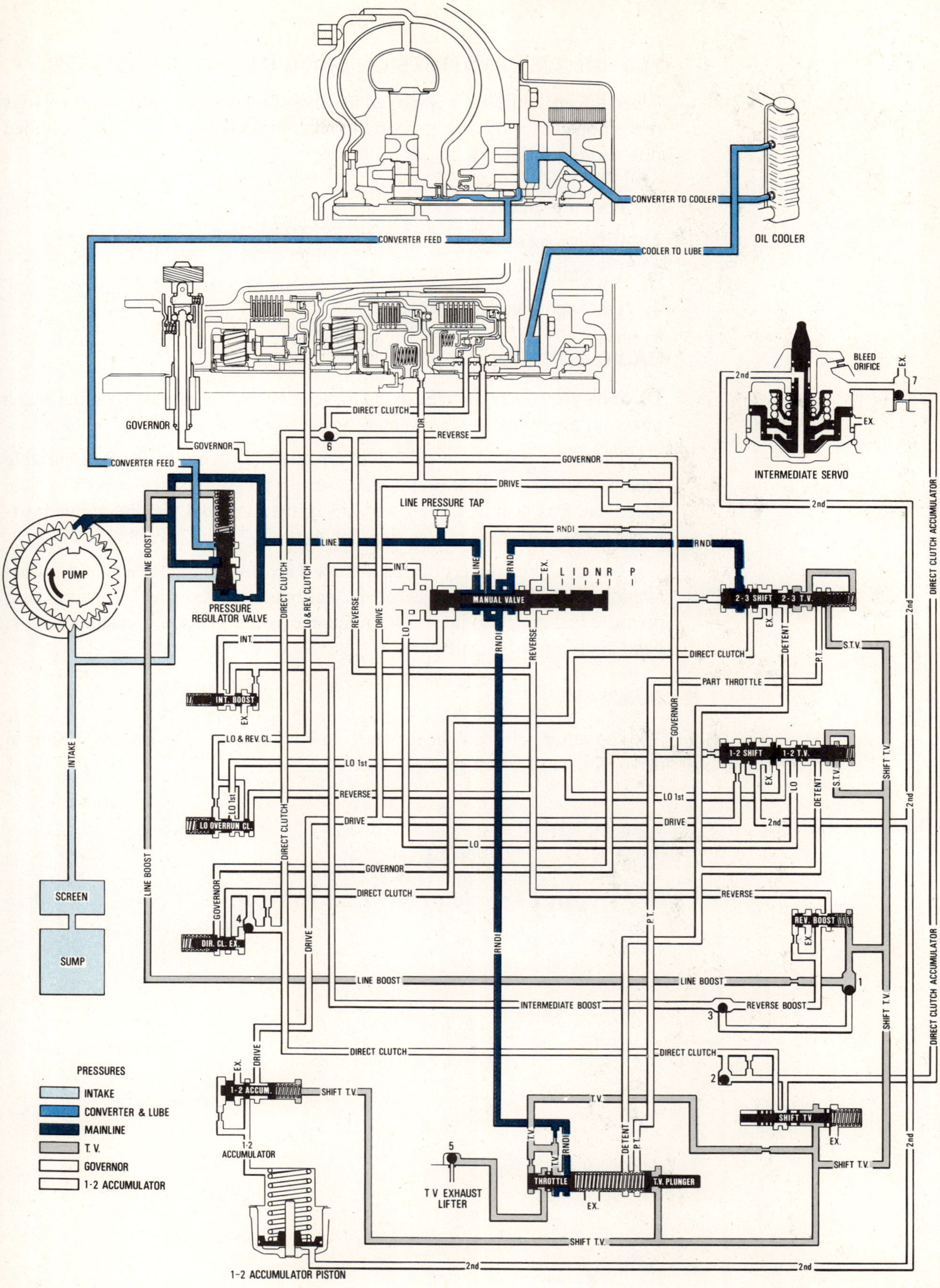

Figure 11-2. Neutral—engine running. (Oldsmobile Division-GMC)

11.3 GM 325 TRANSMISSION DRIVE RANGE—FIRST GEAR

When the gearshift lever is moved to drive (D) position, the manual valve is repositioned to allow line pressure to enter the drive passage. Drive oil then flows to the following:

1. Forward clutch
2. Governor assembly
3. 1-2 Shift valve
4. 1-2 Accumulator valve.

BASIC CONTROL

Drive oil pressure is directed to the forward clutch, where it acts on the clutch piston to apply the forward clutch.

Drive oil pressure at the governor assembly is regulated to a variable pressure called governor pressure. Governor pressure increases with vehicle speed and acts against the 1-2 and 2-3 shift valves and the direct clutch exhaust valve. Drive oil is also directed to the 1-2 shift valve.

Drive oil pressure is directed to the 1-2 accumulator valve and is regulated to another variable pressure called 1-2 accumulator pressure. 1-2 accumulator pressure is directed to the 1-2 accumulator piston to act as a cushion for the band apply on a 1-2 shift.

SUMMARY

The converter is filled. The forward clutch is applied. The transmission is in drive (D) range—first gear.

- **FORWARD CLUTCH—APPLIED**
- **OVERRUNNING CLUTCH—HOLDING**

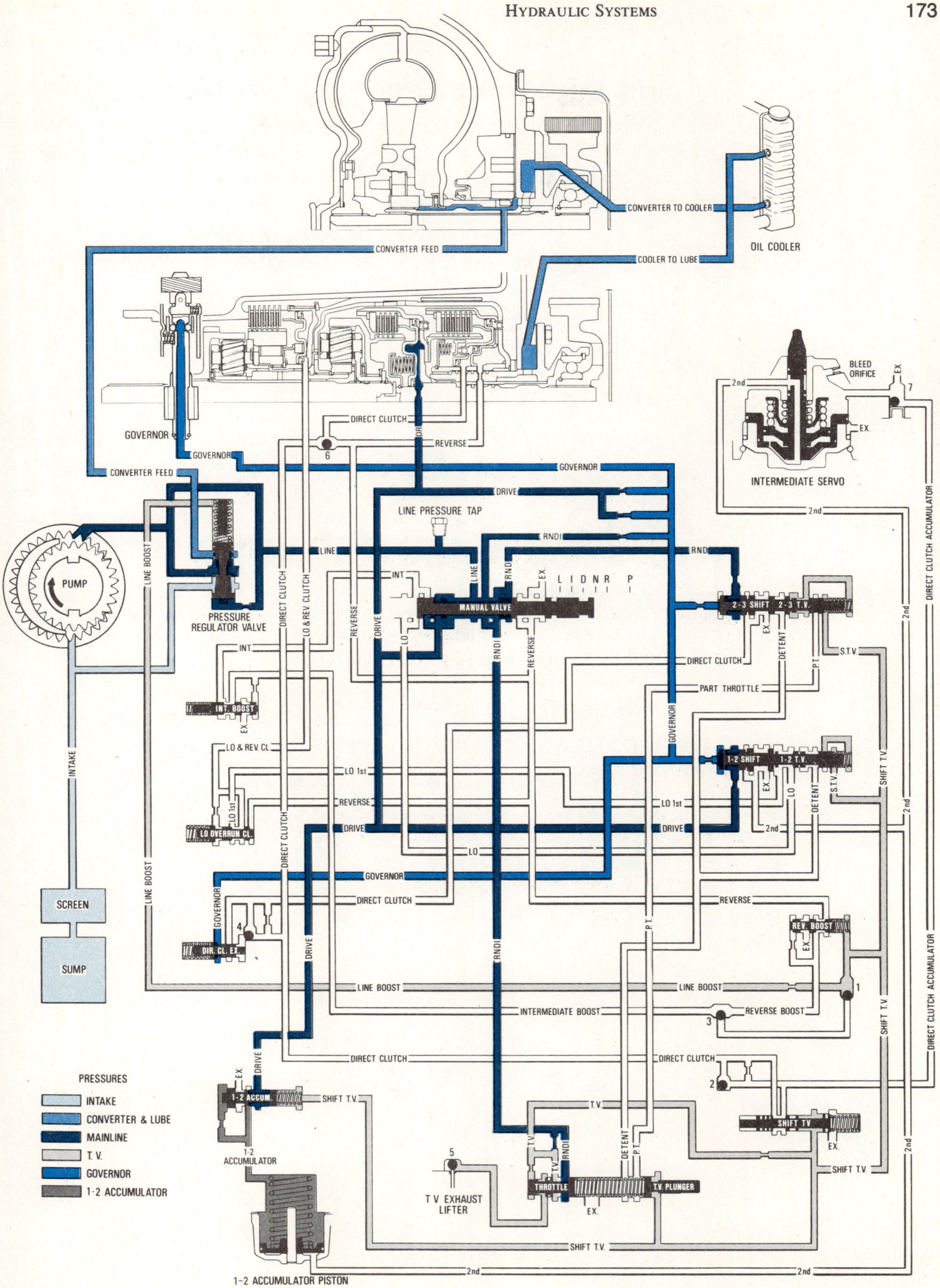

Figure 11-3. Drive range—first gear. (Oldsmobile Division-GMC)

11.4 GM 325 TRANSMISSION DRIVE RANGE—SECOND GEAR

As both vehicle speed and governor pressure increase, the force of governor oil pressure acting on the 1-2 shift valve overcomes the pressure of shift TV oil and 1-2 throttle valve spring force. This allows the 1-2 valve to open, and drive oil enters the second (written as 2nd on the flow chart) oil passage. This oil is called 2nd oil.

2nd oil from the 1-2 shift valve is directed to the following:

1. Intermediate servo
2. 1-2 Accumulator piston.

BASIC CONTROL

Second oil from the 1-2 shift valve is directed to the intermediate servo to apply the intermediate band. At the same time, 2nd oil pressure moves the 1-2 accumulator piston against 1-2 accumulator pressure and the accumulator spring. This action maintains a controlled increase of pressure on the intermediate servo during the 1-2 shift for a smooth band apply.

SUMMARY

The forward clutch and intermediate band are applied. The transmission is in the drive (D) range—second gear.

- **FORWARD CLUTCH—APPLIED**
- **INTERMEDIATE BAND—APPLIED**

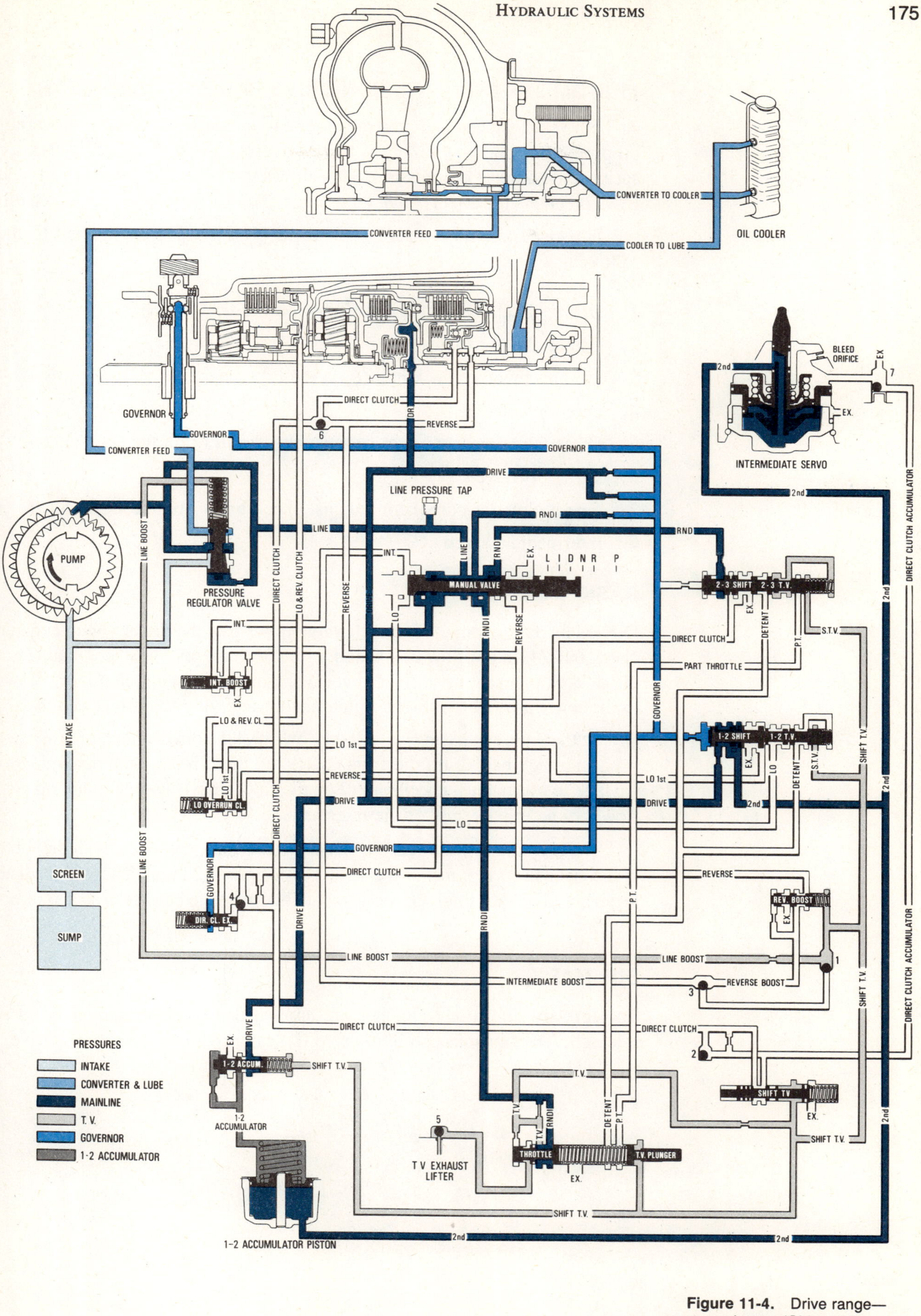

Figure 11-4. Drive range—second gear. (Oldsmobile Division-GMC)

11.5 GM 325 TRANSMISSION DRIVE RANGE—THIRD GEAR

As both vehicle speed and governor pressure increase, the force of governor oil pressure acting on the 2-3 shift valve overcomes the force of the 2-3 TV spring and shift TV oil pressure. This allows the 2-3 shift valve to open, and RND oil enters the direct clutch oil passage. This oil is called direct clutch oil.

Direct clutch oil pressure from the 2-3 shift valve is directed to the following:

1. Direct clutch exhaust check ball (4 on the flow chart)
2. Direct clutch exhaust valve
3. 3rd and reverse check ball (6 on the flow chart)
4. Direct clutch
5. Direct clutch accumulator exhaust check ball (2 on the flow chart)
6. Shift TV valve
7. Direct clutch accumulator exhaust check ball (7 on the flow chart)
8. Intermediate servo.

BASIC CONTROL

Direct clutch oil from the 2-3 shift valve flows past the direct clutch exhaust check ball (4 on the flow chart) and to the 3rd and reverse check ball (6 on the flow chart), seating it in the reverse passage. Direct clutch oil also flows to the inner area of the direct clutch piston, applying the direct clutch.

At the same time, direct clutch oil flows past the direct clutch accumulator exhaust check ball (2 on the flow chart), up the direct clutch accumulator passage to the exhaust check ball (7 on the flow chart). Direct clutch oil also flows to the release side of the intermediate servo.

The pressure of the direct clutch accumulator oil, combined with the force of the cushion spring, moves the servo piston against 2nd oil. The intermediate servo acts as an accumulator for a smooth intermediate band release and direct clutch apply.

SUMMARY

The forward and direct clutches are applied and the intermediate band is released. The transmission is in drive (D) range—third gear (direct drive).

- **FORWARD CLUTCH—APPLIED**
- **DIRECT CLUTCH—APPLIED**

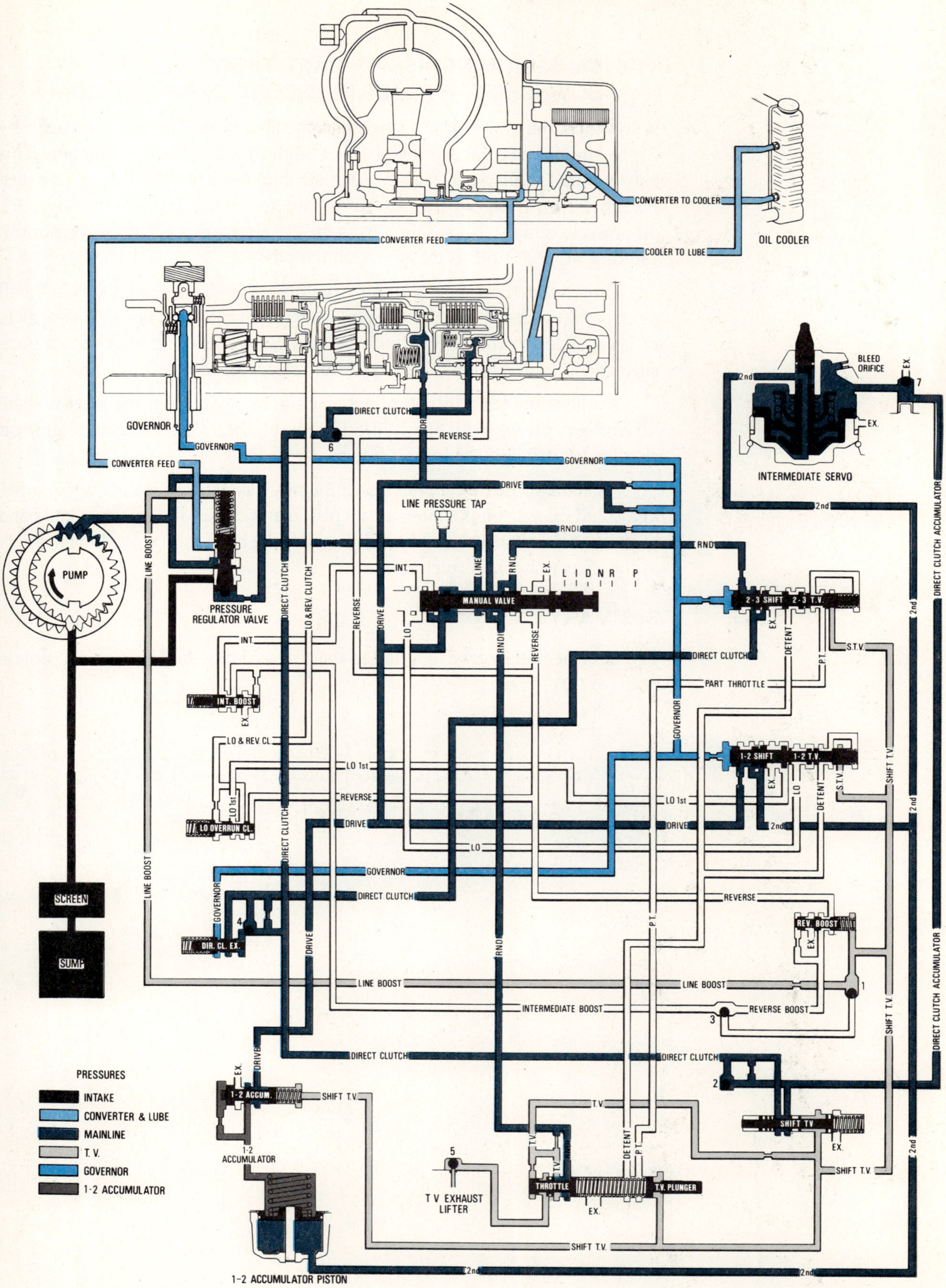

Figure 11-5. Drive range—third gear. (Oldsmobile Division-GMC)

11.6 GM 325 TRANSMISSION PART-THROTTLE 3-2 DOWNSHIFT—VALVES IN SECOND GEAR POSITION

Below 50 mph [80 km/hr] a part throttle 3-2 downshift can be made by depressing the accelerator pedal far enough to move the TV plunger. This allows shift TV oil to enter the PT (part throttle) passage. This oil is then called PT oil. PT oil from the TV plunger is directed to the 2-3 throttle valve. PT oil pressure and the 2-3 spring moves the 2-3 shift valve against governor oil, shutting off RND oil to the direct clutch passage.

Direct clutch oil leaves at the 2-3 shift valve. This seats the check ball (4 on the flow chart). Direct clutch oil is then forced to pass through an orifice, or is regulated by the direct clutch exhaust valve according to governor pressure and vehicle speed. This slows the release of the direct clutch.

At the same time, direct clutch accumulator oil from the intermediate servo flows through the open shift TV valve. Direct clutch accumulator oil also leaves through the same route as direct clutch oil.

As soon as the direct clutch accumulator pressure drops about 6 psi [41 kPa], 2nd oil acts on the inner servo piston and causes the intermediate band to apply. The outer servo piston fully applies the band for a smooth band apply as the direct clutch is released.

SUMMARY

The intermediate band is applied to the forward clutch. The transmission is in drive (D) range.

- **FORWARD CLUTCH—APPLIED**
- **INTERMEDIATE BAND—APPLIED**

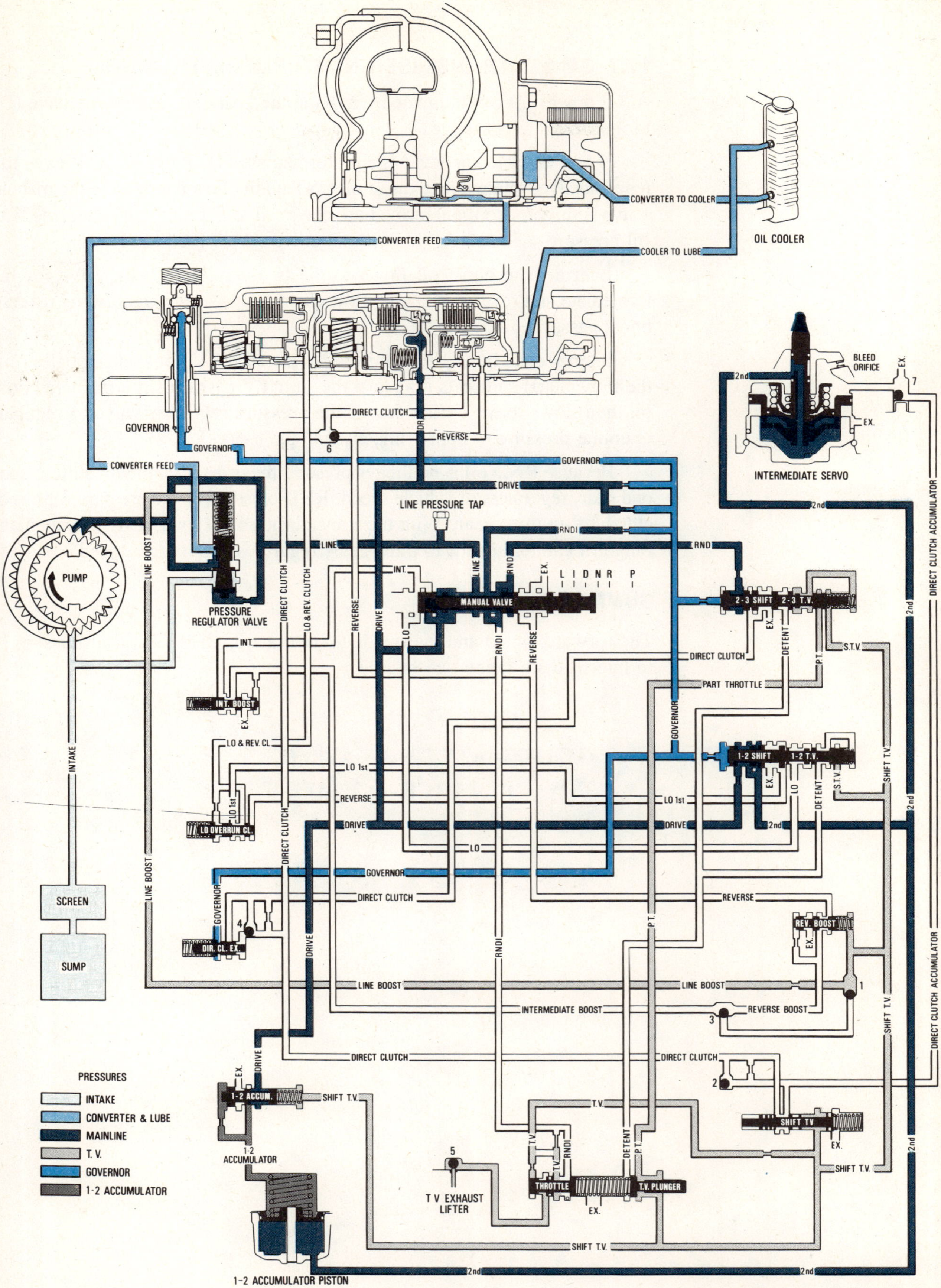

Figure 11-6. Part throttle 3-2 downshift—valves in second-gear position. (Oldsmobile Division-GMC)

11.7 GM 325 TRANSMISSION INTERMEDIATE RANGE

A 3-2 downshift can be made by moving the gearshift lever from drive (D) to intermediate (1) range.

When the selector lever is in intermediate (1), RND oil will leave the manual valve. Intermediate range oil (INT on the flow chart) from the manual valve is directed to the intermediate boost valve. At that point, intermediate oil pressure will stabilize at approximately 90 psi [620 kPa].

Intermediate boost oil flows to the two-way ball check valve for intermediate and reverse (3 on the flow chart). This seats it against the reverse boost passage.

Intermediate boost oil then flows to the shift TV boost check ball (1 on the flow chart) and seats it against the shift TV passage. Intermediate boost oil then flows to the spring end of the pressure regulator valve to regulate mainline pressure to approximately 135 psi [930 kPa].

Because RND oil is no longer present, the transmission will shift to second gear, regardless of vehicle speed. RND oil supplies the pressure for direct clutch oil in third gear. With the transmission in second gear, intermediate range, it cannot upshift to third gear, regardless of vehicle speed.

SUMMARY

The forward clutch and intermediate band are applied. The transmission is in Intermediate range—second gear.

- **INTERMEDIATE BAND—APPLIED**
- **FORWARD CLUTCH—APPLIED**

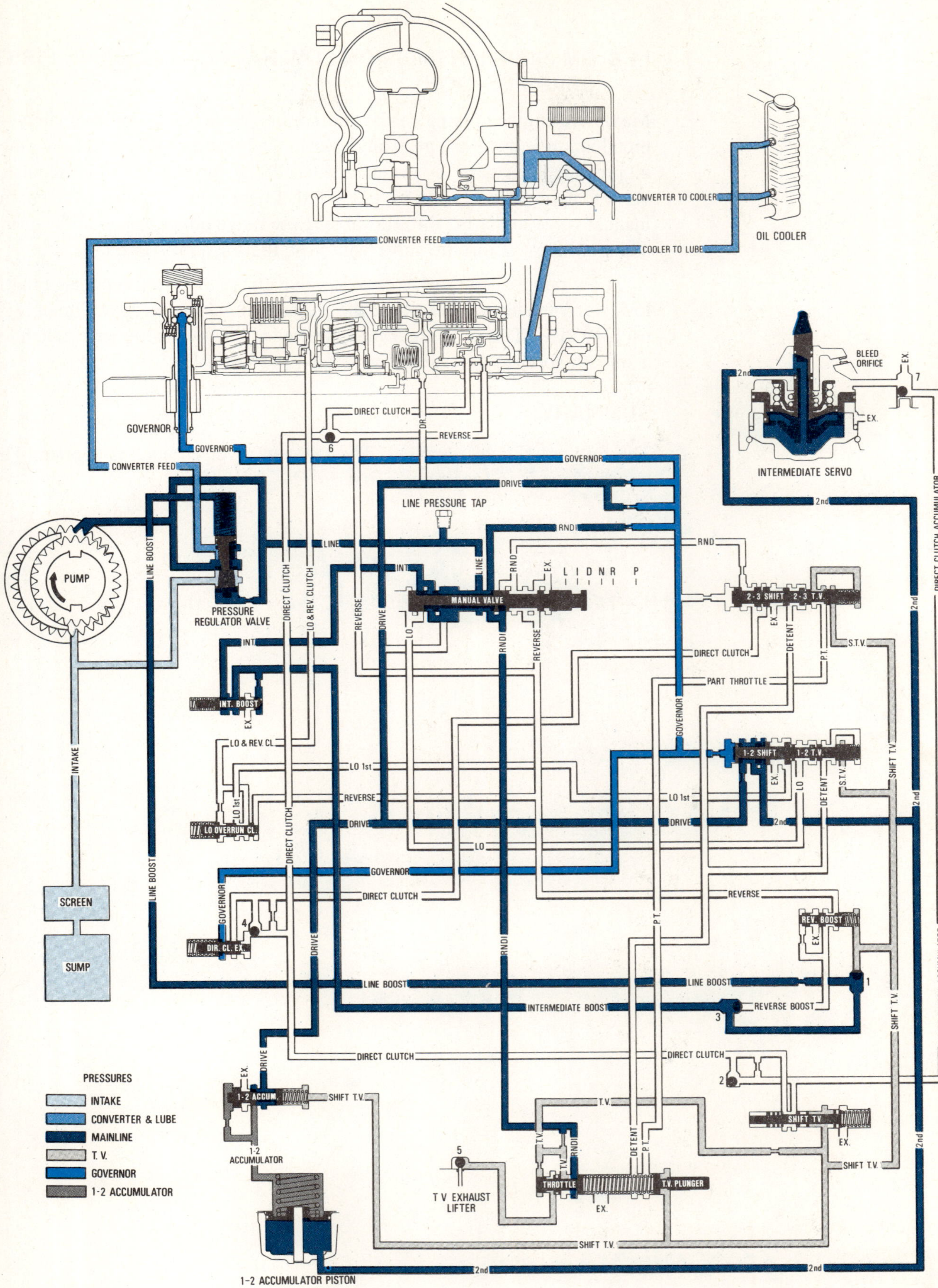

Figure 11-7. Intermediate range. (Oldsmobile Division-GMC)

11.8 GM 325 TRANSMISSION LOW RANGE—VALVES IN FIRST GEAR POSITION

Maximum engine braking can be attained at speeds below approximately 40 mph [64 km/hr] with the gearshift lever in the low range position. Low range oil pressure is regulated by the intermediate boost valve and pressure regulator valve to approximately 135 psi [930 kPa]. Low range oil is directed from the manual valve to the 1-2 throttle valve. Low range oil and the 1-2 TV spring acting on the 1-2 throttle valve will close the 1-2 shift valve.

Low range oil at the 1-2 throttle valve enters the low-1st passage to the low overrrun clutch valve where the pressure is reduced to approximately 35 psi [240 kPa]. This pressure is then directed to the low and reverse clutch for a smooth apply at speeds below 40 mph [64 km/hr].

SUMMARY

The forward clutch is applied. The low and reverse clutch is applied. The transmission is in low range—first gear.

- **FORWARD CLUTCH—APPLIED**
- **LOW AND REVERSE CLUTCH—APPLIED**

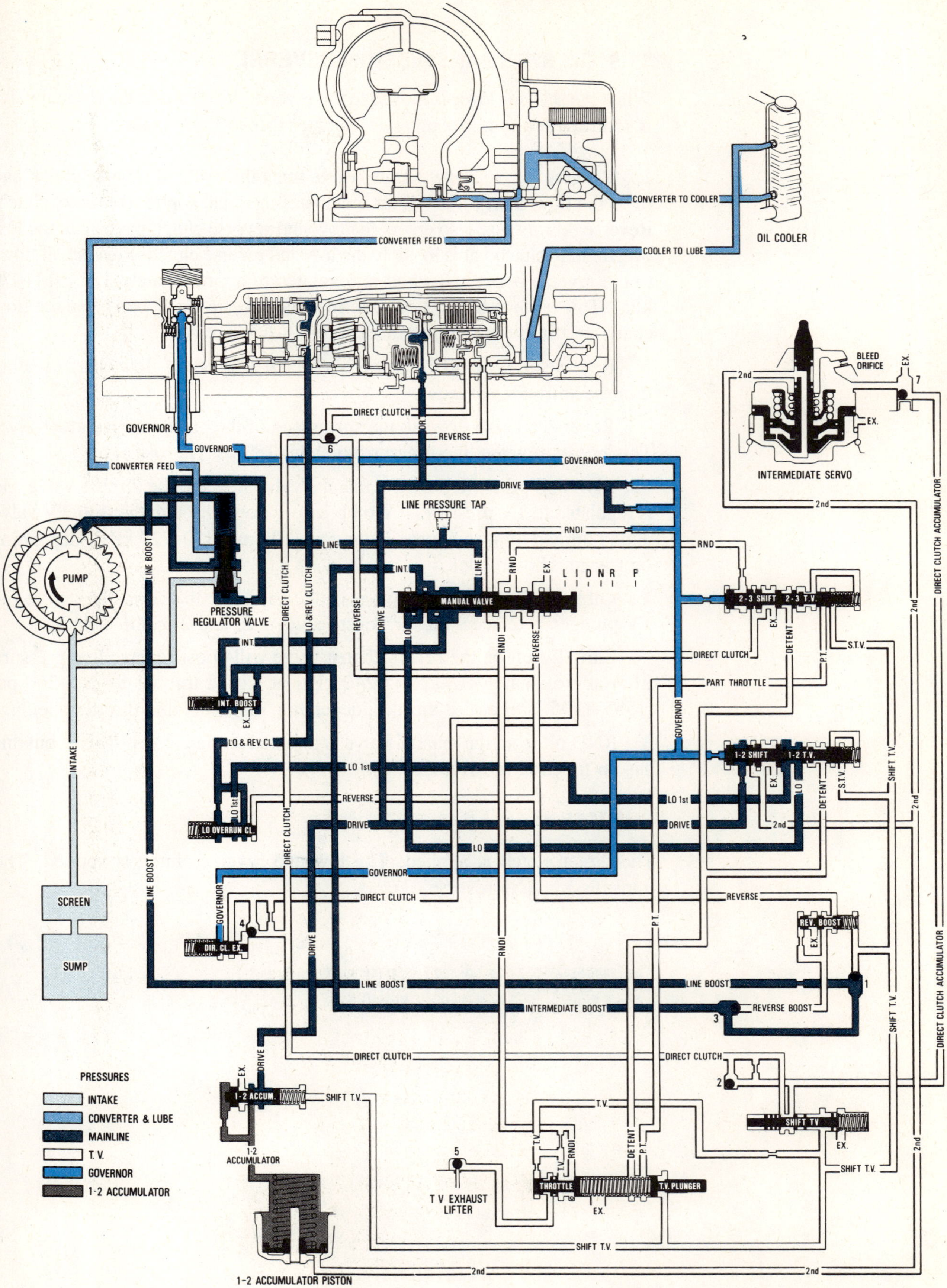

Figure 11-8. Low range—valves in first-gear position. (Oldsmobile Division-GMC)

11.9 GM 325 TRANSMISSION REVERSE

When the selector lever is moved to the reverse (R) position, the manual valve is repositioned to allow pressure to enter three (3) passages:

1. Reverse oil from the manual valve seats the 3rd and reverse check ball (6 on the flow chart) in the direct clutch passage, and applies the direct clutch. Reverse oil pressure also enters the low and reverse clutch passage at the low overrunning clutch valve to apply the low and reverse clutch. Reverse oil flows to the reverse boost valve and is regulated to approximately 155 psi [1070 kPa]. This oil then flows to the intermediate reverse check ball (3 on the flow chart), seating it in the intermediate boost passage.

Reverse oil flows to the shift TV check ball (1 on the flow chart), seating it in the shift TV passage.

Reverse oil also flows to the spring end of the pressure regulator valve. This boosts reverse line pressure to about 180 psi [1240kPa].

2. RNDI oil pressure from the manual valve flows to the throttle valve and is regulated to TV pressure. TV oil pressure flows through the shift TV valve and is limited by it to approximately 65–105 psi [450–720 kPa], depending on transmission model.

Oil from the shift TV valve is called shift TV oil and is directed to the TV plunger, 1-2 accumulator boost valve, 1-2, and 2-3 shift valves.

Shift TV oil on the reverse boost valve will boost reverse line pressure from approximately 180 psi [1240 kPa] at closed throttle to 230–260 psi [1585–1795 kPa] at full throttle, depending on model and throttle opening.

3. RND oil from the manual valve is directed to the 2-3 shift valve, but this has no function in reverse.

SUMMARY

The direct clutch is applied. The low and reverse clutch is applied. The transmission is in reverse (R).

- **DIRECT CLUTCH—APPLIED**
- **LOW AND REVERSE CLUTCH—APPLIED**

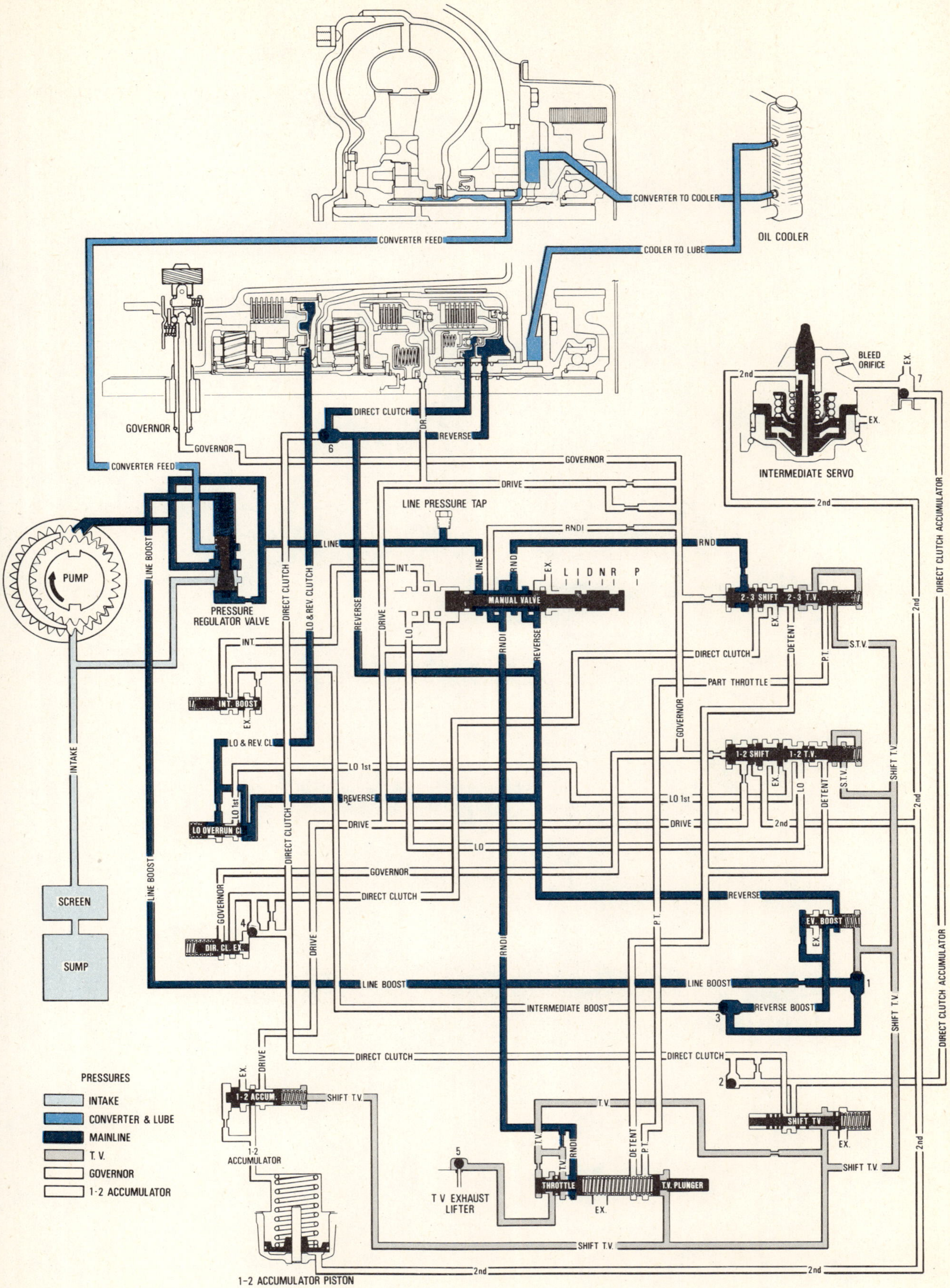

Figure 11-9. Reverse. (Oldsmobile Division-GMC)

NAME ____________ SECTION ____________ DATE ____________

Certification Practice

Test 11

DIRECTIONS:

The following questions are similar to those used on mechanic certification tests. Answer each question by circling the letter of the correct choice.

1. Which of the following statements is correct?

   I. Different fluid pressures can be indicated on a flow chart by different colors.

   II. Different fluid pressures can be indicated on a flow chart by different patterns.

   A. I only B. II only C. Both I and II D. Neither I nor II

2. Mechanic A says that flow charts can be read from the pressure source (the pump) forwards.

   Mechanic B says that flow charts can be read from the apply devices or valves backwards.

   Who is correct?

   A. A only B. B only C. Both A and B D. Neither A nor B

Question 3.
(Oldsmobile Division-GMC)

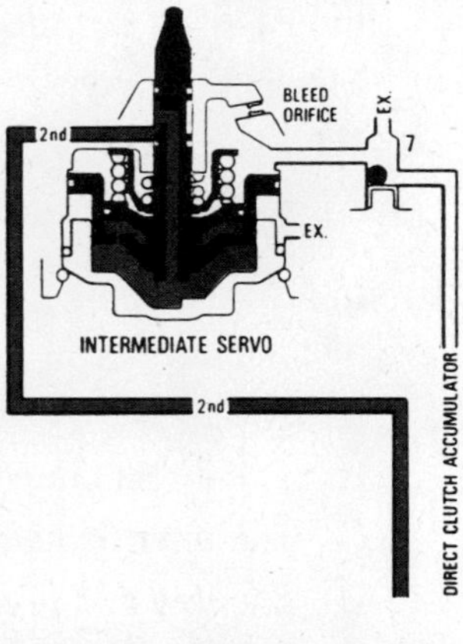

3. The picture shows

   A. hydraulic pressure applied to the intermediate servo.

   B. hydraulic pressure applied to the direct clutch.

   C. both of the above.

   D. neither of the above.

4. When the engine is running and the GM 325 transmission is in drive (D) range, first gear:

   A. The converter is filled. The direct clutch is applied.

   B. The converter is filled. The forward clutch is applied.

   C. The converter is empty. The direct clutch is applied.

   D. The converter is empty. The forward clutch is applied.

5. When the engine is running and the GM 325 transmission is in neutral (N):

   A. The converter is filled. All clutches and bands are released.

   B. The converter is empty. All clutches and bands are applied.

   C. The converter is empty. All clutches and bands are released.

   D. The converter is filled. All clutches and bands are applied.

6. When the engine is running and the GM 325 transmission is in low (L) range, first gear:
   A. The direct clutch is applied. The low and reverse clutch is applied.
   B. The forward clutch is applied. The low and reverse clutch is applied.
   C. Pressure is applied to the intermediate servo.
   D. Pressure is applied to the 1-2 accumulator piston.

Question 7.
(Oldsmobile Division-GMC)

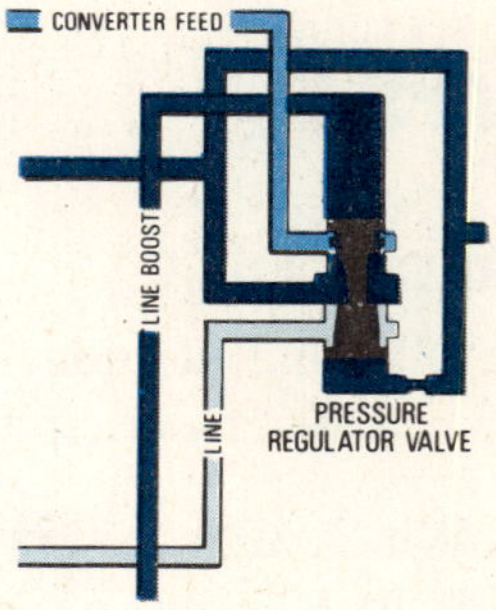

7. The picture shows that
   A. line boost pressure is more than line pressure.
   B. line boost pressure is less than line pressure.
   C. line boost pressure is the same as line pressure.
   D. line boost pressure is 135 psi.

8. Which of the following statements is correct about the GM 325 transmission?
   I. In intermediate range, second gear, the forward clutch and intermediate band are applied.
   II. In intermediate range, second gear, the transmission cannot upshift to third gear.
   A. I only B. II only C. Both I and II D. Neither I nor II

9. Two mechanics are discussing GM 325 transmission operation.
   Mechanic A says that below 50 mph [80 km/hr] a part-throttle downshift can be made by pushing the accelerator far enough to move the TV plunger to allow TV oil to enter the PT (part throttle) passage.
   Mechanic B says that a downshift can only be made by moving the gearshift lever.
   Who is correct?
   A. A only B. B only C. Both A and B D. Neither A nor B

10. In drive (D) range, third gear (direct drive), all of the following statements are correct about the GM 325 transmission EXCEPT
    A. The forward clutch is applied.
    B. The direct clutch is applied.
    C. The intermediate band is released.
    D. The low and reverse clutch is applied.

12 Hydraulic System Repairs

YOUR LEARNING JOB

Problems in the valve body can cause the transmission to operate poorly, or not at all. Apply devices can slip or operate improperly if hydraulic pressure is low or not applied at the proper time. Slipping clutches or bands, or shifts that occur at improper speeds can cause transmission and/or engine damage. Inspecting and servicing the valve body is an important part of overhaul procedure. This unit discusses valve body checks and repairs.

JOB COMPETENCY OBJECTIVES

This is a skill-building unit. When you have completed your assignments and exercises in this unit, you should be able to:

- ☐ Remove the valve body.
- ☐ Clean and inspect the valve body and internal parts.
- ☐ Repair or replace defective valves and/or valve body parts.
- ☐ Correctly reassemble and install the valve body.

12.1 VALVE BODY AND INTERNAL PARTS

The valve body is a delicate and precise assembly. Hydraulic pressures are directed and regulated to cause the transmission to shift at the proper moment for best performance and economy. A valve body and internal parts are shown in Figure 12-1.

Transmission problems can be caused by dirt or metal contaminants in the valve body passages. Spool valves and their bores can become worn. The transmission cannot shift properly if the valves, passages, springs and other parts are dirty or damaged. Cleaning, inspection, and repair or replacement of worn valve body parts is done during transmission overhaul.

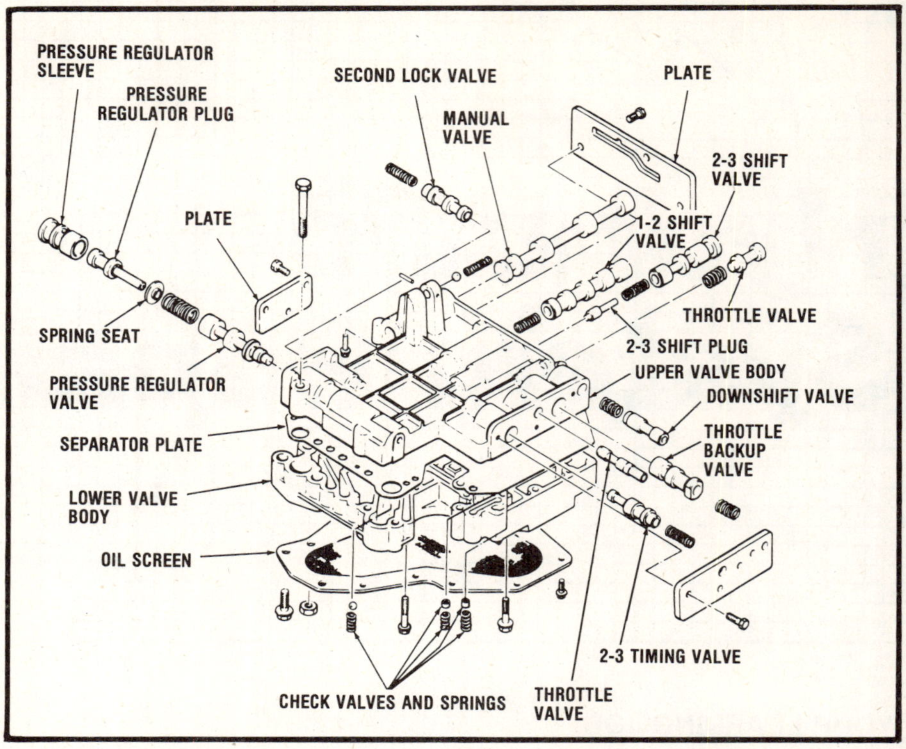

Figure 12-1. Complete valve body and all internal parts. (Ford Motor Company)

Manufacturers often make changes in valve bodies and internal parts for different vehicles and different needs. Before beginning work, check the manufacturer's service manual for drawings or photographs of valve body parts and locations.

12.2 VALVE BODY REMOVAL

The valve body is attached to the lower part of the transmission, inside the oil pan. Always refer to the manufacturer's service manual for specific valve body removal instructions. The following procedure can be used as a guide to removing valve bodies:

1. Safely raise the vehicle on a hoist. Use safety glasses.

2. Drain the oil and remove the oil pan.

3. Remove the transmission filter or screen.

4. Remove gearshift, kickdown, and/or throttle linkage.

5. Loosen the valve body attaching screws, starting at the edges and working toward the center.

6. If necessary, rotate the driveline to align the parking gear and internal linkage for clearance. Remove the valve body and place it in a clean washing container. In some cases, the vacuum modulator may have to be loosened or removed.

12.3 VALVE BODY DISASSEMBLY AND CLEANING

The valves inside the valve body bores operate with only a few thousandths of an inch [hundredths of a millimeter] clearance. The smallest speck of dirt or contamination can cause a valve to bind, or stick, and scratch the bore.

During valve body servicing, everything must be kept spotlessly clean. Your tools, work area, and hands must be clean.

Do not use a vise to hold the valve body or transfer plate during disassembly or reassembly. The slightest distortion of the aluminum valve body or transfer plate will result in sticking valves and/or leakage. When removing or installing valves or plugs, carefully slide them in or out. Do not use force. Use the following general procedure to disassemble the valve body:

1. Remove the end plates and covers one at a time. Hold the end plates and covers while removing the retaining screws to prevent the springs and other parts from flying out.

2. Make notes or drawings as each part is removed to show where and how all parts, including retaining screws, should be replaced. Turning a valve end for end, or placing a spring between the wrong two parts will prevent the transmission from operating properly. Lay the parts out in assembly order on clean shop towels. See Figure 12-2.

NOTE: The check balls are *not* all the same size. Measure each check ball as it is removed, and keep check balls of different sizes separate.

3. Remove all gaskets, and save to compare with the new gaskets.

4. Wash all parts in clean carburetor cleaner to remove all dirt, grease, gum, and varnish. Rinse the parts in hot running water. Place all parts in a clean container and cover with clean mineral spirits to remove water from the passages. Use filtered, moisture-free, compressed air to blow the passages dry.

SAFETY CAUTION: Compressed air can blow liquids or metal chips into the eyes, causing serious physical injury or blindness. Always wear safety glasses or goggles when working with compressed air.

5. Do not drop the valves or place parts or tools on top of them. Lay out all parts neatly in assembly order on a clean work surface.

Figure 12-2. Valve body parts should be laid out in assembly order so that reassembly is correct. (Ford Motor Company)

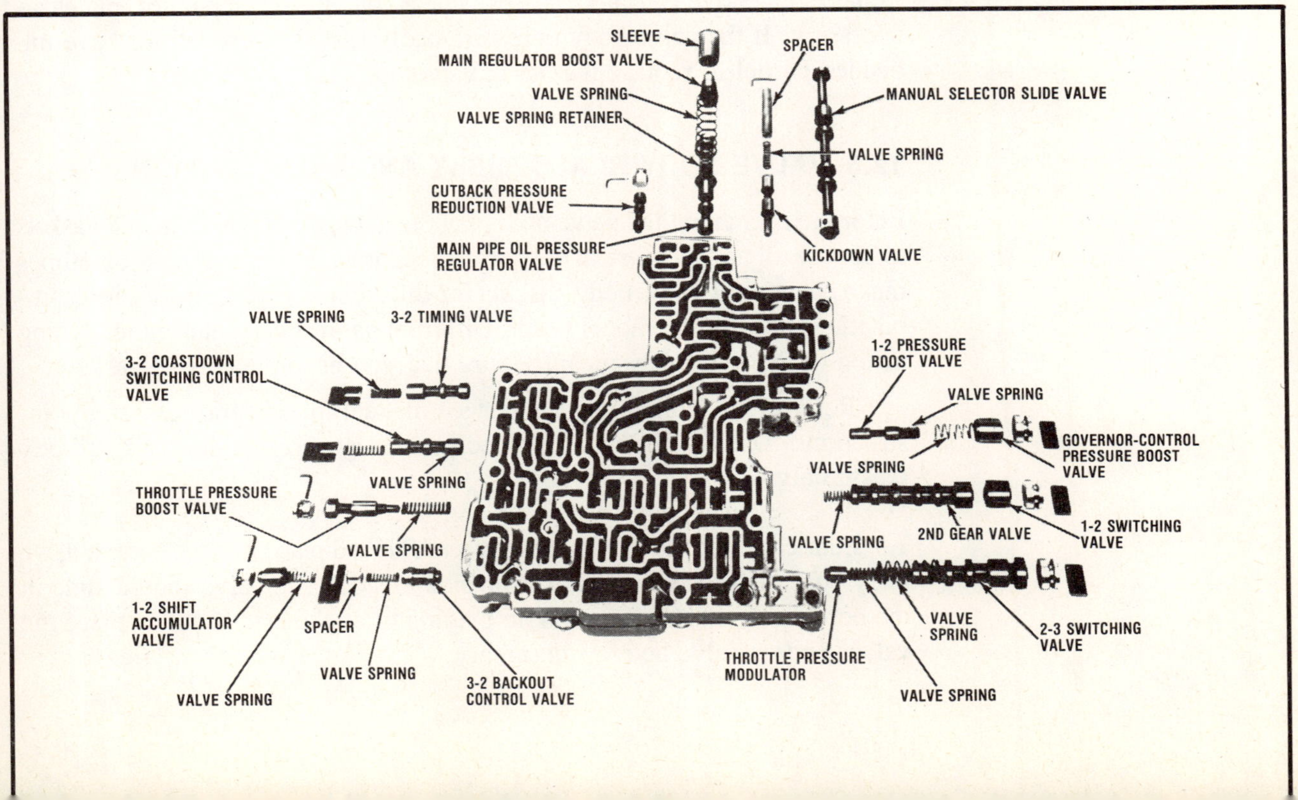

12.4 VALVE BODY AND INTERNAL PARTS INSPECTION

After cleaning and drying, inspect the valve body for defects. Use an inspection flashlight to see defects in valve bores. Look for the following defects:

- Scored, rusted, or corroded bores in the valve body
- Imbedded metal particles in the bores
- Fluid passages filled or partially blocked with foreign material
- Cracks, nicks, or warped mating surfaces
- Stuck check valves.

Inspect the matching plugs and valves if defects are found in the valve body bores. Parallel lines on valve lands indicate particles imbedded in the valve bores. Roll the manual valve on a clean, flat surface to check for bending. Check the valves and related parts for the following defects:

- Scoring, cracks or burrs on valves and plugs
- Bent, broken, or worn springs
- Bent manual valve
- Rusted or scored check balls.

Shiny areas inside the bores and on the valve lands show where the valve lands have been rubbing. If the valve is not badly worn, polish the shiny areas with crocus cloth and clean the valve again.

Check the separator plate for the following defects:

- Bent or warped areas
- Rust or corrosion
- Enlarged or worn check-ball holes.

Replace defective springs, valves, check balls, and separator plates as necessary. If the valve body bores are badly rusted, corroded, or have imbedded particles, replace the entire valve body and all valves.

12.5 VALVE BODY REASSEMBLY AND INSTALLATION

Before reassembling the valve body, lay the new gaskets over the old gaskets to be sure all the holes are in the correct locations. Manufacturers sometimes make changes in valve body gaskets for different engine/transmission combinations or different model years. Different gaskets can change the shifting action of the transmission. *Make sure* that you have the correct gaskets.

Refer to the manufacturer's service manual for specific reassembly instructions. Use the following procedure as a general guide to valve body reassembly:

1. Lubricate each valve with the correct ATF and place it in the correct valve body bore. Tilt the valve body from side to side. The valve should slide in the bore freely and easily, with no binding or stickiness. If not, replace the valves and/or valve body as necessary.

2. Reassemble the valves and springs correctly into the bores one at a time. Refer to notes or drawings made during disassembly. Replace the end plates or covers. Install the retaining screws finger-tight. Torque the screws to the manufacturer's specification.

3. Install all check balls in the correct locations.

4. Check the new gaskets against the old gaskets to make sure they are the correct parts. Place the new gasket on the valve body. Align spring-loaded check balls and other parts as necessary. Position the transfer plate. Place the new gasket on top of the transfer plate. Align the gasket holes with the transfer plate and valve body holes. Discard the old gaskets.

5. Position the valve body on the transmission. Align the parking linkage and other internal linkage as necessary.

6. Install the retaining screws finger-tight. Starting at the center and working outward, torque the screws to the manufacturer's specification.

CAUTION: A torque wrench reading in pound-inches [Nm] *must* be used during reassembly. See Figure 12-3. Improper tightening can warp and ruin the valve body and/or transfer plate.

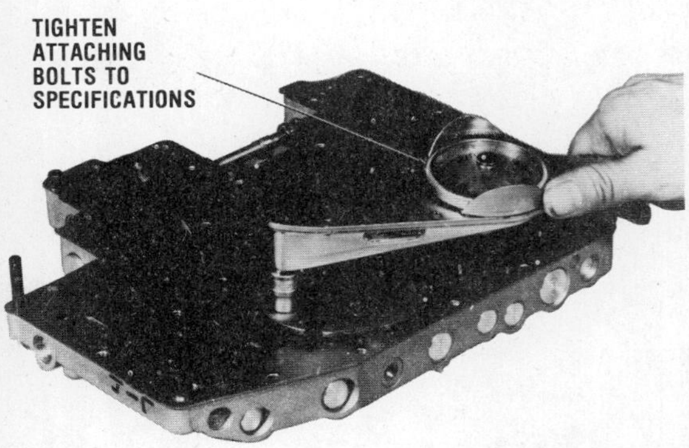

Figure 12-3. A torque wrench is necessary when tightening valve-body attaching bolts. (Ford Motor Company)

NAME ______________________ SECTION ______________ DATE ______________

Job Sheet 12-1

Servicing the Valve Body

Before you begin:

Read Topics 12.1 through 12.5.

Vehicle Make ______________________ Model ______________ Year ______________

Time Started ______________________ Time Finished ______________________

Job Completion Time ______________ Flat Rate Time ______________________

Tools Required

- Hoist
- Safety glasses
- Droplight
- Sockets, extensions, handle
- Open-end or combination wrenches
- Screwdrivers
- Pliers
- Inspection flashlight
- Measuring calipers
- Air blow gun
- Inch-pound [Nm] torque wrench

References

Manufacturer's service manual ______________________________________

Repair manual ______________________________________

Service bulletin ______________________________________

Procedure: Remove Valve Body

1. Safely raise the vehicle on a hoist. Wear safety glasses.
2. Drain the oil and remove the oil pan.
3. Remove the transmission filter or screen.
4. Remove gearshift, kickdown, and/or throttle linkage.
5. Loosen the valve body attaching screws, starting at the edges and working toward the center.
6. If necessary, rotate the driveline to align the parking gear and sprag to permit linkage to clear. Remove the valve body and place it in a clean washing container.

Procedure: Clean and Inspect Valve Body and Internal Parts

NOTE: During valve body servicing, everything must be kept spotlessly clean. Your tools, hands, and work area must be clean.

CAUTION: Do not use a vise to hold the valve body or transfer plate during disassembly or reassembly. The slightest distortion of the aluminum valve body or

transfer plate will result in sticking valves and/or leakage. When removing or installing valves or plugs, carefully slide them in or out. Do not use force.

1. Remove the end plates and covers one at a time. Apply pressure to the end plates and covers while removing the retaining screws.
2. Make notes or drawings as each part is removed to show where and how all parts, including retaining screws, should be replaced. Lay the parts out in assembly order on clean, lint-free shop rags.

NOTE: The check balls are *not* all the same size. Measure each check ball as it is removed, and keep check balls of different sizes separate.

Instructor Check: ____________________

3. Remove all gaskets, and save to compare with the new gaskets.
4. Properly clean the valve body and internal parts. Use compressed air to blow the parts dry.

SAFETY CAUTION: Compressed air can blow liquids or metal chips into the eyes, causing serious physical injury or blindness. Always wear safety glasses or goggles when working with compressed air.

5. Lay out all parts neatly in assembly order on a clean work surface.
6. Inspect the valve body and internal parts for defects. In the Job Completion Report, below, make notes of any defects found.

Instructor Check: ____________________

7. Replace parts as necessary.

Procedure: Reassemble and Install Valve Body

1. Check all valves in the bores for binding or sticking. Replace the valves and/or valve body as necessary.
2. Reassemble the valves and springs correctly into the bores one at a time. Refer to notes or drawings made during disassembly. Replace the end plates or covers. Install the retaining screws finger tight. Torque the screws to the manufacturer's specification. In the space provided, write the correct torque specification.

Manufacturer's Specified Torque .... ____________________

Instructor Check: ____________________

3. Install all check balls in the correct locations.
4. Check the new gaskets against the old gaskets to make sure they are the correct parts. Place the gasket on the valve body. Align spring-loaded check balls and other parts as necessary. Position the transfer plate. Place the gasket on top of the transfer plate. Align the gasket holes with the transfer plate and valve body holes. Discard the old gaskets.

NAME ______________________ SECTION ____________ DATE ____________

5. Position the valve body on the transmission. Align parking linkage and other internal linkage as necessary.
6. Install the retaining screws finger-tight. Starting at the center and working outward, torque the screws to the manufacturer's specification.

CAUTION: A torque wrench reading in pound-inches [Nm] *must* be used during reassembly. See Figure 12-3. Improper tightening can warp and ruin the valve body and/or transfer plate.

In the space provided, write the correct torque specification.

Manufacturer's Specified Torque . . . . ____________________

Instructor Check: ____________________

Job Completion Report

1. What defects, if any, were found in the valve body and internal parts? ________

____

____

2. What parts, if any, were replaced? ____________________________________

____

____

3. What difficulties, if any, were encountered in these procedures? __________

____

____

INSTRUCTOR SIGNOFF ______________________ **DATE** ____________

NAME ______________________ SECTION ____________ DATE ____________

Certification Practice

Test 12

DIRECTIONS:

The following questions are similar to those used on mechanic certification tests. Answer each question by circling the letter of the correct choice.

1. Transmission problems can be caused by all of the following EXCEPT
   A. dirt, contaminants, or corrosion in the valve body bores.
   B. damaged spool valves.
   C. warped or distorted valve body or transfer plate.
   D. small shiny areas on spool valve lands.

2. Mechanic A says that the valve body attaching screws can be loosened and tightened in any manner.
   Mechanic B says that the screws should be loosened from the edges toward the center, and tightened from the center outward.
   Who is correct?
   A. A only B. B only C. Both A and B D. Neither A nor B

3. What is shown in the picture?
   A. Valve body and internal parts in assembly order
   B. Correct locations of parts inside the valve body
   C. Both A and B
   D. Neither A nor B

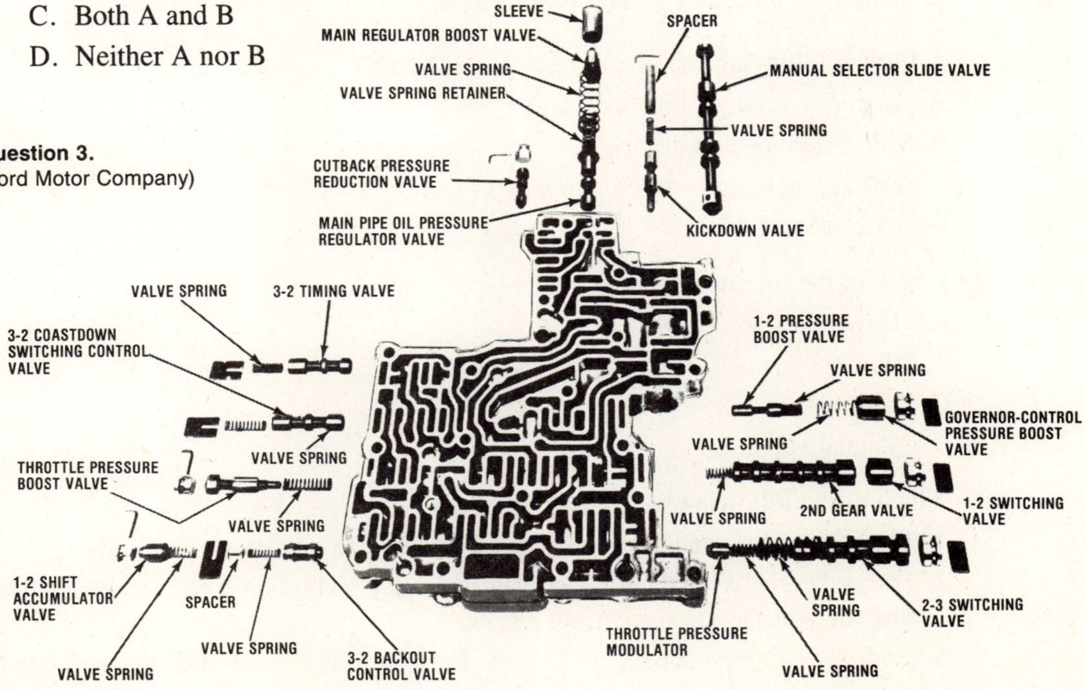

Question 3.
(Ford Motor Company)

4. Which of the following statements is correct?
   I. During valve body service, everything must be kept spotlessly clean.
   II. During valve body service, the valve body should be securely mounted in a vise.
   A. I only B. II only C. Both I and II D. Neither I nor II
5. All of the following should be done during valve body disassembly EXCEPT
   A. Securely hold end plates and covers when removing retaining screws.
   B. Make reassembly notes or drawings as each part is removed.
   C. Wash and save all check balls in the same container.
   D. Keep valves from being dropped or damaged.
6. Mechanic A says valve body parts should be blown dry.
   Mechanic B says valve body parts should be allowed to air dry.
   Who is correct?
   A. A only B. B only C. Both A and B D. Neither A nor B

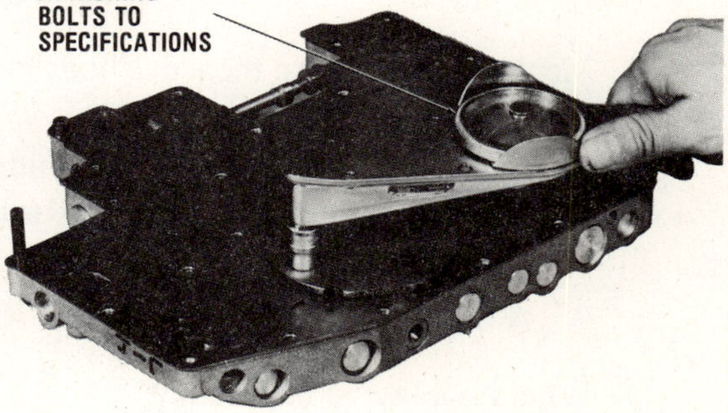

Question 7.
(Ford Motor Company)

7. The tool shown should always be used
   A. during disassembly.
   B. during reassembly.
   C. if the mechanic is inexperienced.
   D. unless the mechanic feels it is not necessary.
8. All of the following are valve body defects EXCEPT
   A. scored, rusted, or corroded bores.
   B. imbedded metal particles.
   C. cracks, nicks, or warped mating surfaces.
   D. loose check valves.
9. Which of the following is most likely to be bent?
   A. Throttle valve
   B. Governor valve
   C. Shift valve
   D. Manual valve
10. Which of the following statements is correct?
    I. The gaskets between the transfer plate and valve body should be replaced.
    II. The gaskets between the transfer plate and valve body should be checked to make sure the holes are in the correct places.
    A. I only B. II only C. Both I and II D. Neither I nor II

13 Troubleshooting and Diagnosis

YOUR LEARNING JOB

Before repairs on the automatic transmission are attempted, the mechanic must find the cause of a problem. Worn parts in the transmission may be the *symptom*, not the *cause*, of a problem. If the underlying cause of a problem is not found and corrected, the problem will occur again. The procedures for finding the cause of a problem are known as troubleshooting and diagnosis. This unit covers the tests and procedures used in troubleshooting and diagnosing the automatic transmission.

COMPETENCY OBJECTIVES

This is a knowledge-building and skill-building unit. When you have completed your assignments and exercises in this unit, you should be able to:

- ☐ Describe the basic steps in diagnosing automatic transmission problems.
- ☐ Perform troubleshooting checks and inspections.
- ☐ Conduct a road test to determine transmission problems.
- ☐ Perform a pressure test.
- ☐ Use troubleshooting charts to find the causes and corrections of transmission problems.

13.1 WHAT IS TROUBLESHOOTING AND DIAGNOSIS?

The most important responsibility a mechanic has is finding problems. To do this, the mechanic must know *exactly* what is wrong. Many service complaints arise when the wrong part is repaired or replaced. These complaints are created when an incorrect diagnosis is made.

Troubleshooting is the method mechanics use to diagnose problems. Troubleshooting is simply a process of elimination. Checks are made to eliminate certain parts or functions as the cause of problems. Troubleshooting progresses from simple checks and inspections to more involved test procedures.

Most service manuals furnish the mechanic with procedures and guides for troubleshooting various systems on the vehicle.

13.2 DETERMINING THE PROBLEM

Begin transmission troubleshooting by speaking to the driver. A test ride with the driver along can be helpful in determining the problem. Using the driver's information, an experienced mechanic may be able to find the cause of the problem more quickly. When asking questions about the complaint, concentrate on four areas:

- **What** is wrong? (For example: harsh shifting, no reverse, delay when gearshift lever is put into gear.)
- **When** does it occur? (For example: when the car is cold, when fully warmed up, when making turns.)
- **Where** does the problem occur? (For example: up or down hills; on flat roads; on highways or expressways.)
- **How long** has the problem been noticeable? (For example: gradually; over a period of months; for a week; just began today.)

However, do not rely on the driver's information alone. Make your own evaluation. Although information from the driver can be helpful, it is the mechanic who must correctly diagnose and fix the problem.

13.3 ELIMINATING ENGINE-RELATED TRANSMISSION PROBLEMS

Transmission problems can be caused by engine malfunctions. For example, a poorly tuned or worn-out engine that does not produce enough vacuum can cause shifting problems in the transmission. A clogged engine cooling system can cause the transmission fluid to overheat. Worn-out engine bearings can cause downward pressure on the input shaft, damaging or destroying internal transmission parts. Any problem that affects engine power can affect transmission operation.

Before assuming that the problem is in the transmission, check the following:

- Engine compression
- Engine oil pressure
- Engine vacuum
- Engine vacuum lines and components
- Tune-up specifications
- Emission control systems
- Air and fuel filters
- Automatic choke operation.

Engine problems must be corrected before proceeding with further transmission checks and/or service.

CHECKPOINT

1. **If the underlying __________________ of a transmission problem is not corrected, the symptoms will occur again.**
2. **Begin transmission troubleshooting by talking to the __________________.**
3. **When asking questions about the complaint, concentrate on 4 areas: ______________, ______________, ______________, and ______________ ______________**
4. **A ______________ ______________ with the driver along can be helpful in determining the problem.**
5. **______________ problems must be corrected before proceeding with further transmission checks and/or service.**

13.4. BASIC TRANSMISSION PROBLEMS

If basic problems are found, corrections must be made before proceeding with further tests. Many transmission problems are identified during preliminary troubleshooting procedures.

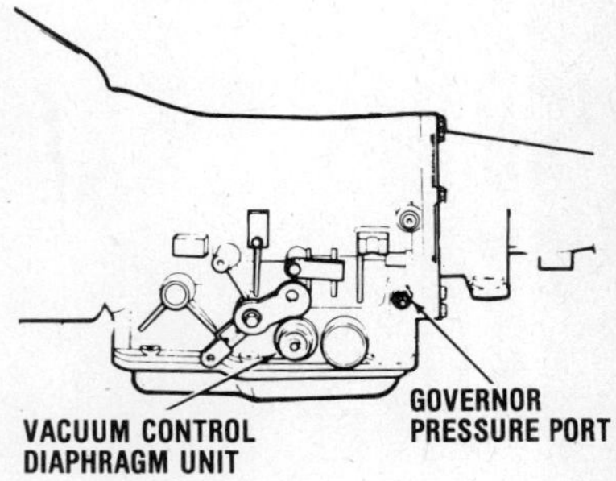

Figure 13-1. Control test ports on a Ford transmission.

Fluid Level, Condition, Leakage

The transmission must have adequate fluid in order to operate properly. Insufficient fluid will lead to low pressures in the hydraulic system, and can cause shifting problems.

The fluid must be clean and uncontaminated. Metal particles, friction materials, or engine coolant in the fluid indicate basic problems.

A transmission leak can be caused by a faulty gasket or seal, or simply by loose bolts on the oil pan.

Vacuum Lines and Control Units

On most transmissions, engine vacuum is used to help control and shift the transmission. A metal tube with rubber vacuum hoses at the ends connects the intake manifold to a control unit on the transmission. The control unit is called a vacuum modulator, or vacuum control diaphragm. See Figure 13-1. Ford Motor Company transmissions sometimes use a dual-diaphragm vacuum control, shown in Figure 13-2.

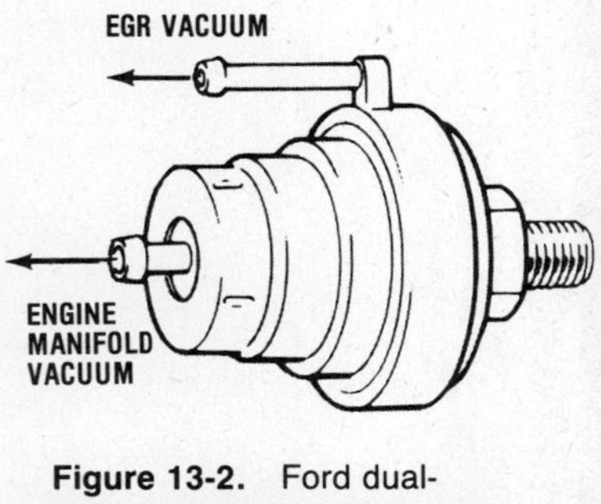

Figure 13-2. Ford dual-diaphragm with two vacuum lines. (Ford Motor Company)

The synthetic rubber and cloth diaphragm inside the vacuum modulator reacts to engine vacuum. As the diaphragm moves, it controls the movement of a pushrod or a valve inside the transmission. The pushrod or valve controls, or modulates, throttle pressure. Some units have an adjustment to vary the vacuum necessary to move the diaphragm.

Broken, leaking, or disconnected vacuum lines will prevent the vacuum modulator from operating correctly. In addition, the engine will have a vacuum leak. An engine vacuum leak may cause the engine to run roughly, and can lead to serious engine damage.

A leaking diaphragm will allow transmission fluid to be sucked into the intake manifold and burned by the engine. This can produce clouds of dense bluish smoke from the exhaust and lower the fluid level in the transmission rapidly. However, the driver may never notice the engine smoking until it is too late for the transmission.

Linkage

Misadjusted linkage can cause serious damage to the automatic transmission. If the gearshift linkage is misadjusted, the manual valve may be held between the correct positions. Fluid pressures to the clutch assemblies and servos may be incorrect, leading to clutch and band damage. Kickdown and TV cable linkage problems can also cause erratic shifting. Linkage adjustment is often overlooked as a cause of transmission shifting problems.

Electrical Problems

Loose, disconnected, or corroded electrical connections to solenoids within the transmission can cause shifting problems. Connections must be tight and clean for proper shifting operation. Electrical wiring that is frayed or shorted to ground can prevent solenoid operation.

CHECKPOINT

1. **Check fluid ________________ and ________________, ____________________ lines and control units, and ____________________ adjustments.**
2. **Many transmission problems are solved at a ____________________ level of troubleshooting.**
3. **A transmission leak can be caused by a faulty gasket or seal, or simply by a ________________________________ ____________________ on the oil pan.**
4. **Broken, leaking, or disconnected vacuum lines will prevent the ____________________ ____________________ from operating correctly.**
5. **____________________ adjustment is often overlooked as a cause of transmission shifting problems.**
6. **Loose, disconnected, or corroded ____________________ ____________________ to solenoids within the transmission can cause shifting problems.**

13.5 STALL TEST

Stall testing can indicate operational problems in the engine, stator overrunning clutch, and transmission clutches and bands. However, stall tests can damage weak clutches and bands in the automatic transmission, or cause engine damage. For this reason, not all manufacturers recommend stall testing. Stall testing is discussed in 5.2.

13.6 ROAD TEST

Road testing involves driving the vehicle to check transmission operation. A properly conducted road test can indicate to the mechanic how the apply devices are operating. Apply devices are multiple-disc clutches, bands, and overrunning clutches. A road test checks for slippage, harsh or delayed shifting, or shifts at the incorrect road speed. Slippage means that the transmission's connection to engine power is erratic. The engine speeds up, but vehicle speed does not keep up with engine speed. Engine power is used up as friction within the transmission, creating heat and damaging clutches and bands. Road test procedures are discussed in 13.9.

13.7 PRESSURE TEST

A road test enables the mechanic to determine which apply devices might not be working properly. For example, a slipping band might be detected. The next question to ask is, what caused the band to slip?

The problem might be mechanical—improper band adjustment or a worn friction lining. However, low hydraulic pressure could have caused the band to slip and wear out. A valve in the valve body could be stuck or leaking. Just replacing the worn-out band would *not* solve the problem. If the sticking valve is not found and replaced, the new band would soon slip and wear out.

A pressure test checks hydraulic pressures in the transmission by using gauges which are attached to the transmission. The transmission is shifted into a specific gear, and the engine is run at a recommended speed while pressure readings are recorded. Pressure readings will reveal possible problems in the oil pump, governor, and throttle circuits.

Pressure test openings, or ports, are located on the transmission case. See Figure 13-3. After removing a plug, a fitting can be screwed into the port. The pressure gauge connects to the fitting. See Figure 13-4. In addition, a tachometer and vacuum gauge may be required. The use of a combination tester, shown in Figure 13-5, will make the procedure easier. Consult the manufacturer's service manual for the recommended gauge or tester.

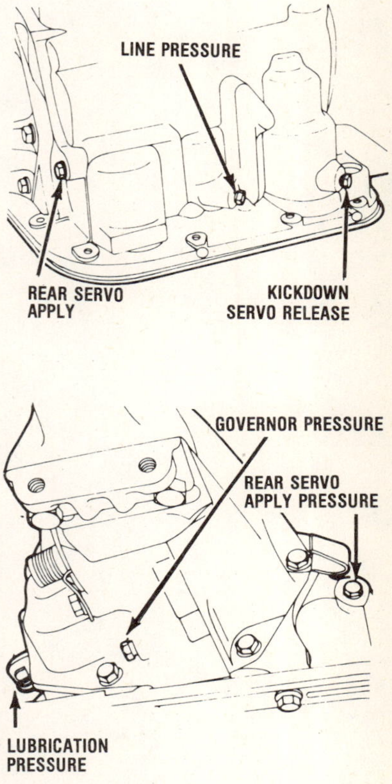

Figure 13-3. Pressure test locations on the right side (top) and left side (bottom) of a Chrysler transmission case. (Chrysler Corporation)

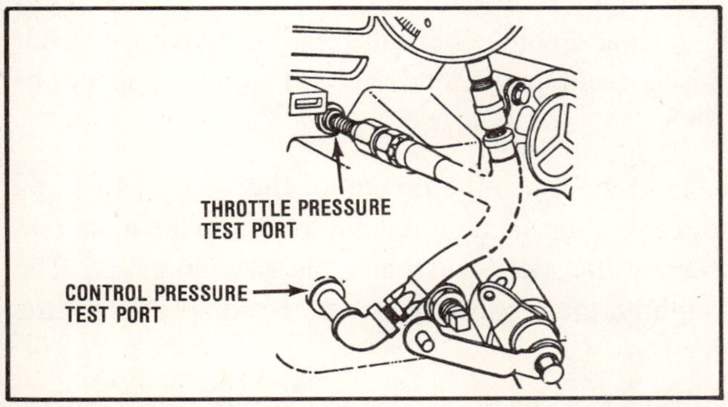

Figure 13-4. A pressure gauge connects to fittings placed in the test ports to take pressure readings.

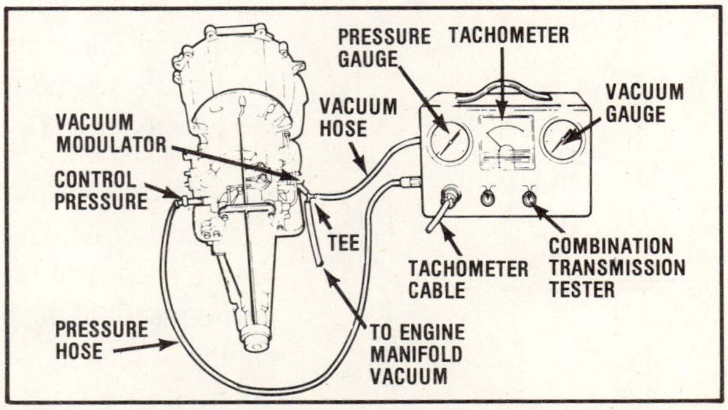

Figure 13-5. A combination transmission tester is often recommended when taking different readings. (Ford Motor Company)

CHECKPOINT

1. **Stall testing can indicate operating problems in the engine, ________________ overrunning clutch, and transmission ________________ and ________________.**
2. **Apply devices are ________________-________________ clutches, ________________, and ________________ clutches.**
3. **A road test checks for ________________, ________________ or delayed shifting, or shifts at the incorrect ________________ speed.**
4. **The manufacturer's service manual contains ________________ that list the correct shift ________________ for vehicles with various engine, transmission, differential and tire combinations.**
5. **A ________________ ________________ checks hydraulic pressures in the transmission.**

13.8 CHECKING THE TRANSMISSION

Begin by checking the transmission fluid level and condition. Checking the transmission fluid is discussed in 2.3. and 2.4.

Raise the vehicle on a hoist. Check for transmission fluid leaks. Checking for fluid leaks is described in 2.1

Checking Vacuum Lines and Control Devices

Use the following procedure to check the vacuum lines and vacuum modulator, or vacuum control diaphragm:

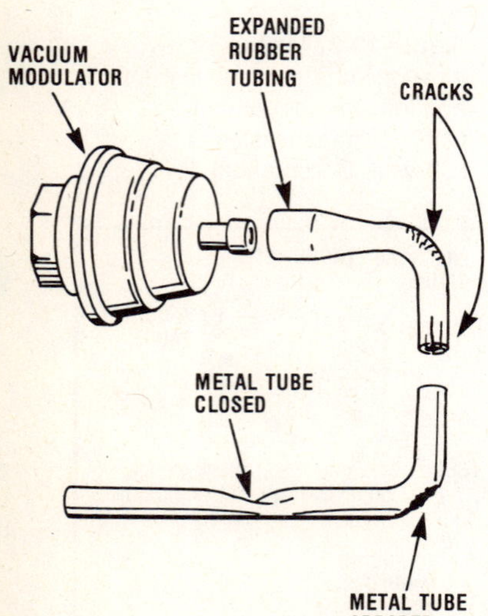

Figure 13-6. Inspect the vacuum lines and connections for damage.

1. Inspect the transmission vacuum lines and connections. The rubber connectors at the ends of the metal tubing should fit tightly and have no cracks or breaks. The metal tubing should not be worn through or smashed closed. Check at both the transmission and engine ends and all along the length of the tubing. See Figure 13-6.

2. Attach a vacuum gauge to the transmission end of the vacuum line and start the engine. Check to see that the vacuum reading is within the manufacturer's specifications. Increase idle speed and watch the vacuum gauge. The vacuum should drop off slightly, then return to a steady reading. Shut off the engine.

3. Pull the rubber hose off the vacuum connector on the vacuum modulator or vacuum control diaphragm unit. If fluid is evident at the connection or in the rubber hose, the modulator diaphragm is leaking and must be replaced.

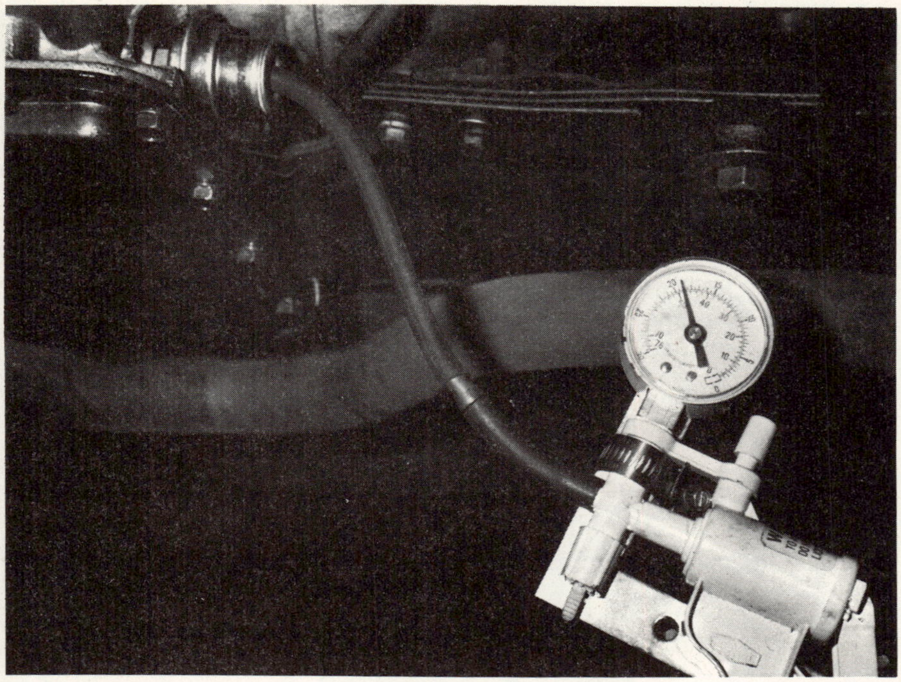

Figure 13-7. Use a vacuum pump to check vacuum modulator condition.

4. If no fluid is visible, attach a vacuum pump to the unit and draw 17 to 20 inches [432 mm to 508 mm] of vacuum. See Figure 13-7. The modulator diaphragm should hold the vacuum for at least 30 seconds without dropping. Ford vacuum control diaphragm units may have two vacuum connections. Follow the manufacturer's recommended service procedures in the appropriate service manual to check such units.

5. If the diaphragm is leaking transmission fluid, or fails the vacuum-holding test, it must be replaced. Some modulator units are held in the transmission case with a retainer bracket and a bolt. Other modulators simply screw directly into the case. Follow the manufacturer's recommendations in the appropriate service manual to replace and adjust the modulator or vacuum control diaphragm.

Before proceeding with transmission tests, check the transmission linkage. Disconnect the gearshift linkage at the transmission and check that the adjustment is correct. Check kickdown linkage and/or TV cable adjustment. Readjust the linkage if necessary. Linkage adjustments are discussed in 3.7.

Checking Electrical Connections

Poor or nonconductive electrical connections will prevent solenoids within the transmission from functioning properly. To check electrical connections, use the following procedure:

SAFETY CAUTION: Remove rings or jewelry before working on any part of the electrical system. Avoid touching grounded tools to powered electrical leads or connections, which can result in severe burns or dangerous sparks. Sparks can cause an explosion if gasoline or volatile solvents are present.

1. Refer to the manufacturer's service manual for the location of transmission electrical connectors.

2. Remove electrical connectors and check for oxidation and corrosion on the terminals. Clean any corrosion with sandpaper or a small file. Spray the connections with a light lubricant spray, such as dielectric grease.

3. Connect a voltmeter or continuity tester to the electrical wires according to the manufacturer's instructions. Operate kickdown switches or linkage to connect electrical power to the leads. Check for power at the wires.

4. If no power is present, check for power at the kickdown switch or linkage switch. If no power is present at the switch, check the wiring to the switch. Replace the wiring if necessary. Check the switch for continuity. Continuity testing is covered in topic 3.10. Replace the switch if defective.

5. If the switch is OK, but no power is present at the transmission end of the electrical wires, the problem is in the wires. Replace the wires.

6. After replacing a switch or wire, repeat Step 3.

7. If power is present at the wires when the kickdown or linkage switch is operated, reconnect the electrical wire to the solenoid connection on the transmission.

After basic checks and/or service of the transmission, road test the vehicle. Road test procedures are discussed in 13.10.

NAME ______________________ SECTION ____________ DATE ____________

Job Sheet 13-1

Checking the Transmission

Before you begin:

Read Topics 13.1 through 13.7.

Vehicle Make ____________________ Model ____________ Year ____________

Time Started ____________________ Time Finished ____________________

Job Completion Time ______________ Flat Rate Time ____________________

Tools Required

- Absorbent white paper towels
- Hoist
- Vacuum gauge
- Vacuum pump
- Socket wrenches, extensions, handle
- Open-end or combination wrenches
- Special tool for Ford vacuum control diaphragms
- Screwdrivers
- Pliers
- Voltmeter or continuity tester
- Sandpaper
- Small file
- Light lubricant spray

References

Manufacturer's service manual ______________________________

Repair manual ______________________________

Service bulletin ______________________________

Procedure: Check Transmission Fluid Level, Condition, Leakage

1. Check transmission fluid level and condition. In the Job Completion Report, below, make a note of any problems.
2. Safely raise the vehicle on a hoist. Check for transmission fluid leaks. In the Job Completion Report, below, make a note of the source of any leakage.

Procedure: Check Vacuum Lines and Control Devices

1. Inspect the transmission vacuum lines and connections. Check at both the transmission and engine ends and all along the length of the tubing. In the Job Completion Report, below, make a note of any problems found.

2. Attach a vacuum gauge to the transmission end of the vacuum line and start the engine. In the space provided, write the manufacturer's vacuum specification.

____

Increase idle speed to 1,500 rpm and watch the vacuum gauge. In the Job Completion Report, make a note of the readings of the vacuum gauge as idle speed is increased. Shut off the engine.

3. Pull the rubber hose off the vacuum connector on the vacuum modulator or vacuum control diaphragm unit. In the Job Completion Report, make a note of any leakage found.
4. If no fluid is visible, attach a vacuum pump to the vacuum modulator or vacuum control diaphragm and draw 17–20 inches [432–508 mm] of vacuum. In the Job Completion Report, make a note of the condition of the diaphragm. Follow the manufacturer's recommended service procedures in the appropriate service manual to check Ford Motor Company dual-diaphragm units.

Instructor Check: ____________________

5. Replace the vacuum modulator, or vacuum control diaphragm, if defective.

Procedure: Check Transmission Linkage

1. Check the gearshift linkage and adjust if necessary.
2. Check the kickdown linkage and adjust if necessary.
3. Check the TV cable linkage and adjust if necessary. In the Job Completion Report, make a note of any problems found or adjustments made in the linkage.

Instructor Check: ____________________

Procedure: Check Electrical Connections

SAFETY CAUTION: Remove rings or jewelry before working on any part of the electrical system. Avoid touching grounded tools to powered electrical leads or connections, which can cause severe burns or dangerous sparks. Sparks can cause an explosion if gasoline or volatile solvents are present.

1. Remove and clean transmission electrical connectors. Spray the connections with a light lubricant spray, or dielectric grease.
2. Connect a voltmeter or continuity tester to the electrical wires. Operate kickdown switches or linkage to connect electrical power to the wires. Check for power at the wires.
3. If no power is present, check for power at the kickdown switch or linkage switch. Check the wiring to the switch. Check the switch for continuity. Replace the wiring to the switch, or the switch, if defective.

Instructor Check: ____________________

4. If necessary, replace the electrical wires from the kickdown switch to the transmission solenoid connection.
5. After switch or wire replacement, repeat step 2.
6. If power is present at the leads when the kickdown or linkage switch is operated, reconnect the lead to the solenoid connection on the transmission. In the Job Completion Report, make a note of any electrical repairs or replacements made.

NAME ____________________ SECTION ____________ DATE ____________

Job Completion Report

1. What were the level and condition of the transmission fluid? ____________

____

2. Was any leakage found during inspection? ____________

If so, what was the source of the leakage? ____________

____

3. How did the vacuum gauge react as engine idle speed was increased? ____________

____

4. What were the results of checking the vacuum modulator or vacuum control diaphragm? ____________

____

5. Was it necessary to replace the vacuum control modulator or vacuum control diaphragm? ____________

____

6. What problems, if any, were found and corrected in the transmission linkage?

____

____

7. What problems, if any, were found and corrected in the transmission electrical connections? ____________

____

INSTRUCTOR SIGNOFF ____________________ **DATE** ____________

13.9 STALL TESTING

Refer to the appropriate manufacturer's service manual to determine if stall testing is recommended. If so, conduct a stall test according to the manufacturer's instructions. Stall test procedures are covered in 5.2.

13.10 ROAD TESTING

To find out what symptoms or problems are present, the mechanic must know the correct shift speeds for the transmission being tested. The manufacturer's service manual contains charts that list correct shift speeds for vehicles with various engine, transmission, differential, and tire combinations. See Figure 13-8. Read the appropriate chart for the vehicle to be tested.

AUTOMATIC SHIFT SPEEDS AND GOVERNOR PRESSURE CHART (APPROXIMATE MILES PER HOUR)

Carline	VL	VL	RW	SX	PDC	PDC	Y
Engine Cu. In.	225	318	360-4 Hi. Perf.	360-2	400-2 400-4	400-4 &440 Hi. Perf.	440
Axle Ratio	2.76	2.45	3.21	2.45	2.71	2.71	2.71
Tire Size	6.95x14	E78x14	H78x14	GR78x15	HR78x15	JR78x15	LR78x15
Throttle Minimum							
1-2 Upshift	9-16	8-16	8-15	9-16	9-16	8-15	8-16
2-3 Upshift	15-25	15-25	15-23	17-25	15-25	15-23	16-25
3-1 Downshift	8-13	9-14	8-13	9-14	8-13	8-13	9-14
Throttle Wide Open							
1-2 Upshift	31-43	39-54	43-56	41-57	37-52	38-53	39-54
2-3 Upshift	63-76	79-95	78-93	83-100	77-92	77-93	79-95
Kickdown Limit							
3-2 WOT Downshift	60-73	76-92	75-90	79-96	73-89	74-90	76-92
3-2 Part Throttle Downshift	46-61	30-56	34-57	31-58	30-56	29-54	30-56
3-1 WOT Downshift	28-35	30-44	34-47	31-46	29-43	29-43	30-44
Governor Pressure*							
15 psi	20-22	21-23	20-22	22-24	21-23	21-23	21-23
40 psi	47-53	60-66	60-66	62-69	59-66	58-65	59-66
60 psi	66-72	83-90	81-88	87-94	82-90	81-88	82-90

*Governor pressure should be from zero to 1.5 psi at stand still or downshift may not occur.

NOTE: Figures given are typical for other models. Changes in tire size or axle ratio will cause shift points to occur at corresponding higher or lower vehicle speeds.

Figure 13-8. Shift speeds and governor pressures for a variety of engine/transmission/differential/tire combinations are listed in this chart. (Chrysler Corporation)

The idea of a road test is to drive the vehicle so that any transmission problems will show up. If the driver indicated the problem occurred under certain conditions, such as driving up or down hills, the road test should be conducted under the same conditions.

See the manufacturer's service manual for specific road test procedures. The following procedure is intended as a general guide to properly conducting a road test.

1. Ask your instructor's permission to conduct a road test. If approved, conduct a safe road test.

SAFETY CAUTION: Drive in a safe manner. Jackrabbit starts, hard cornering, or sudden braking should be avoided. Driving conditions other than normal can worsen a problem or create a new one. Carelessness during the test drive can result in an accident. Remember, you are responsible for someone else's vehicle. Drive only in a manner that is required to check or diagnose proper transmission operation. Observe all speed limits, and obey all traffic laws. Conduct road tests only on isolated roads with little traffic.

2. Check the D (drive) range. Place the gearshift lever in D. Accelerate moderately. Check for proper 1-2 and 2-3 shift, according to the manufacturer's directions. At approximately 30 mph, after the transmission has shifted into high, depress the accelerator part way to produce a 3-2 downshift. Ease off the accelerator. Accelerate gently to 45 mph. Push the accelerator down slightly, enough to produce a 3-2 downshift. Notice how the transmission downshifts. Ease off the accelerator.

3. At about 30 mph, push the accelerator down to the floor to produce a 3-1 downshift. Ease off the accelerator as soon as the shift is made. Coast to a stop in high, and try to notice if the transmission downshifts from high directly to first at about 10–15 mph. Bring the vehicle to a stop.

NOTE: Transmission shift points vary from vehicle to vehicle.

4. Check the 2 (second or intermediate) range. Place the gearshift lever in 2. During light throttle, check at what road speed the 1-2 upshift occurs. The transmission should not shift past 2. Engine braking should be noticeable when the accelerator is released.

NOTE: Ford Motor Company vehicles should start in 2, not 1.

Move the lever to D (drive). Accelerate to 30 mph. Move the selector back to 2. Check how the vehicle downshifts. Check for proper engine braking. Slow down. At approximately 10–15 mph, the transmission should downshift to 1. Bring the vehicle to a stop.

NOTE: Ford Motor Company vehicles will remain in 2 (second or intermediate) when coasting down and stopping.

5. Check the L or 1 (low or first) range in a three-speed or four-speed transmission. Place the gearshift lever in L or 1. Accelerate from a stop. Check whether the transmission upshifts to 2 (second or intermediate) at moderate speeds.

NOTE: Some transmissions will not upshift past the low or first gear. Other transmissions will upshift to 2 (second or intermediate).

Move the gearshift lever into D (drive). Using light throttle, accelerate to 30 mph. Move the gearshift lever back to L or 1. Check whether the transmission downshifts all the way to low or first. Check how much engine braking occurs. Bring the vehicle to a stop.

NOTE: Some transmissions will downshift to 2 (second or intermediate) when the gearshift lever is moved into L or 1. Then, as the car slows, the transmission will downshift again to low or first.

6. Drive to a straight, level road, with smooth shoulders at the side of the road. When there is no traffic coming, stop, place the gearshift lever in R (reverse).

SAFETY CAUTION: Be extremely careful when backing up. The vehicle will turn sharply in reverse, and it is easy to lose control. Conduct this test only on a flat, level road with no ditches or dropoffs to the side. Do not conduct this test near a curve where oncoming traffic can not be seen.

Press the accelerator gently and check transmission operation. Steer carefully, and press the accelerator to the floor briefly. Check for gear noise, unusual shifts, slippage, or other improper transmission operation in reverse.

Analyzing Road Test Information

Road testing provides information about the transmission. After the road test has been completed, this information must be analyzed. Manufacturer's charts similar to that shown in Figure 13-9 indicate what apply devices are used in each gearshift lever position for a particular transmission. By using road test information and the chart, transmission apply device problems can be isolated.

For example, suppose you noticed transmission slippage when the gearshift lever was in L or 1, but not when downshifted in D (drive) to low or first. The chart in Figure 13-9 indicates that with the gearshift lever in 1 (low or first), the rear clutch and the low-reverse band are applied. With the lever in D (drive), and the transmission downshifted to 1 (low or first), the rear clutch and the overrunning clutch are applied. Since the rear clutch is applied in both applications, and the transmission did not slip in D, the problem would not seem to be in the rear clutch. It is reasonable to assume that if the rear clutch was slipping, it would slip in D as well as in 1.

With the lever in 1, the low/reverse band is applied along with the rear clutch to produce 1. Since the rear clutch seems to work in D, the trouble is probably in the low/reverse band. By using a process of elimination, you can figure out which apply device doesn't seem to be working.

Remember that each transmission will have a different chart.

ELEMENTS IN USE AT EACH POSITION OF THE SELECTOR LEVER

Lever Position	Gear Ratio	Start Safety	Parking Sprag	Clutches			Bands	
				Front	Rear	Over Running	(Kickdown) Front	(Low-Rev.) Rear
P—PARK		X	X					
R—REVERSE	2.21			X				X
N—NEUTRAL		X						
D—DRIVE								
First	2.45				X	X		
Second	1.45				X		X	
Direct	1.00			X	X			
2—SECOND								
First	2.45				X	X		
Second	1.45				X		X	
1—LOW (First)	2.45				X			X

Figure 13-9. Chart shows which elements of a transmission are in use at each position of the shift selector lever. (Chrysler Corporation)

NAME ______________________ SECTION ____________ DATE ____________

Job Sheet 13-2

Testing the Automatic Transmission

Before you begin:

Read Topics 13.1 through 13.9.

Vehicle Make ____________________ Model ___________ Year ___________

Time Started ____________________ Time Finished ____________________

Job Completion Time ______________ Flat Rate Time ____________________

Tools Required

- Shift-speed chart
- Apply device chart

References

Manufacturer's service manual ________________________________

Repair manual ________________________________

Service bulletin ________________________________

Procedure: Stall Test

1. If recommended by the manufacturer, safely conduct a stall test.

Instructor Check: ____________________

2. In the Job Completion Report, make a note of the results of the test.

Procedure: Road Test

1. Ask your instructor's permission to conduct a road test. Before making a road test, check the transmission fluid level and fluid condition. Also check for transmission leaks or damage. If approved, conduct a safe road test.

SAFETY CAUTION: Drive in a safe manner. Jackrabbit starts, hard cornering, or sudden braking should be avoided. Driving conditions other than normal can worsen a problem, or create a new one. Carelessness during the test drive can result in an accident. Remember, you are responsible for someone else's vehicle. Drive only in a manner that is required to check or diagnose proper transmission operation. Observe all speed limits, and obey all traffic laws. Conduct road tests only on isolated roads with little traffic.

2. Check the D (drive) range. Check for 1–2 and 2–3 shift conditions according to the manufacturer's recommendations. Press the accelerator to produce a 3–2 downshift. Accelerate gently to 45 mph. Push the accelerator down slightly, enough to produce a 3–2 downshift. Notice how the transmission downshifts. Ease off the accelerator.
3. At about 30–40 mph, push the accelerator down to the floor to produce a 3–1 downshift. Ease off the accelerator as soon as the shift is made. Coast to a stop in high, and try to notice if the transmission downshifts from high directly to first at about 10–15 mph. Bring the vehicle to a stop.
4. In the Job Completion Report, make a note of any transmission problems noticed in the D (drive) range.

Instructor Check: ____________________

5. Check the 2 (second or intermediate) range. Check at what road speed the 1–2 upshift occurs. Check whether the transmission shifts past 2. Check for engine braking when the accelerator is released.

NOTE: Ford Motor Company vehicles should start in 2, not 1.

Move the lever to D (drive). Accelerate to 30 mph. Move the selector back to 2. Check how the vehicle downshifts. Check for proper engine braking. Slow down. At approximately 10–15 mph, the transmission should downshift to 1. Bring the vehicle to a stop.

NOTE: Ford Motor Company vehicles will remain in 2 (second or intermediate) when coasting down and stopping.

6. In the Job Completion Report, make a note of any transmission problems noticed in the 2 (second or intermediate) range.

Instructor Check: ____________________

7. Check the L or 1 (low or first) range. Accelerate from a stop. Check whether the transmission upshifts to 2 (second or intermediate) at moderate speeds.

NOTE: Some transmissions will not upshift past the low or first gear. Other transmissions will upshift to 2 (second or intermediate).

Move the gearshift lever into D (drive). Accelerate to 35 mph. Move the gearshift lever back to L or 1. Check whether the transmission downshifts all the way to low or first. Check how much engine braking occurs. Bring the vehicle to a stop.

NOTE: Some transmissions will downshift to 2 (second or intermediate) when the gearshift lever is moved into L or 1. Then, as the car slows, the transmission will downshift again to low or first.

8. In the Job Completion Report, make a note of any transmission problems noticed in the L or 1 (low or first) range.

Instructor Check: ____________________

NAME ______________________ SECTION ____________ DATE ____________

9. Drive to a straight, level road, with smooth shoulders at the side of the road. When there is no traffic coming, place the gearshift lever in R (reverse).

SAFETY CAUTION: Be extremely careful when backing up. The vehicle will turn sharply in reverse, and it is easy to lose control. Conduct this test only on a flat, level road with no ditches or dropoffs to the side. Do not conduct this test near a curve where oncoming traffic cannot be seen.

Press the accelerator gently and check transmission operation. Steer carefully, and press the accelerator to the floor briefly. Check for proper transmission operation in reverse.

Instructor Check: ____________________

10. Refer to the manufacturer's shift speed and apply device charts to determine transmission problems. In the Job Completion Report, below, make notes of any transmission problems found. Make a note of your diagnosis of the problems.

Job Completion Report

1. What problems, if any, were noticed when driving in the D (drive) range?

____

____

2. What problems, if any, were noticed when driving in the 2 (second or intermediate) range? ____

____

____

3. What problems, if any, were noticed when driving in the L or 1 (low or first) range?

____

____

4. What problems, if any, were noticed when driving in the R (reverse) range?

____

____

5. Based on the manufacturer's charts, what is your diagnosis of any problems found?

____

____

INSTRUCTOR SIGNOFF ____________________ **DATE** ____________

13.11 PRESSURE TESTING

Procedures for pressure testing vary greatly between manufacturers. Some manufacturers' procedures call for reading more than one pressure gauge at the same time, in addition to a vacuum gauge. To perform a pressure test correctly, refer to the proper manufacturer's service manual.

When conducting pressure tests on different transmissions, all of the following steps may be necessary. However, some manufacturers may omit one or more of the steps. Other vehicle procedures may replace some steps with alternate, or different, steps. Additional steps may be required. Use the following procedure as a general guide to pressure testing.

1. Start the engine and warm it up to normal operating temperature. Shut the engine off.

2. Connect a tachometer to the engine according to the manufacturer's instructions. Be sure all gauges and instruments are connected before the engine is started.

3. Have a helper inside the vehicle to start the engine and operate the transmission, accelerator, and brake controls.

4. Safely raise the vehicle on a hoist to a comfortable working height.

5. Connect pressure gauges to the specified test ports.

6. Use a T-fitting and vacuum tubing to connect a vacuum gauge with a shut-off or bleed-down valve to the vacuum modulator line. Refer to Figure 13-5.

SAFETY CAUTION: If the pressure test procedures or equipment require you to be under the vehicle while the test is performed, be extremely careful. The driveline and rear wheels will be turning for most tests. Contact with the spinning driveline or wheels can cause serious physical injury.

7. Start the engine. Move the selector lever to the first test position. Run the engine at the specified rpm. Take readings from the gauges, and write them down. Move the transmission controls as specified, and note the readings or change in readings. Make notes of the pressures or pressure changes.

8. Slow the engine and press gently on the service brakes to stop the spinning wheels and driveline. Shut off the engine.

9. Attach the pressure gauges and/or vacuum gauge to the proper pressure ports for the next test.

10 Start the engine. Move the gearshift lever to the next test position. Run the engine at the specified rpm. Take pressure readings as specified. Slow the engine, and press gently on the service brakes to stop the spinning wheels and driveline.

11. Repeat the steps for each test procedure. Make notes of all pressure readings and vacuum readings obtained for each test.

Analyzing Pressure Test Readings

Manufacturers' shop manual charts list correct specifications for the pressure tests, as well as possible causes of incorrect readings. See Figure 13-10. Refer to the proper manufacturer's service manual to determine the causes of pressure readings other than as specified.

Control Pressure Test Results:	
Control Pressure Condition	**Possible Causes**
Low in P	Valve body bolts loose, low reverse servo leakage.
Low in R	Reverse clutch leakage, low reverse servo leakage.
Low in N	Valve body bolts loose.
Low in D	Forward clutch leakage, overdrive servo leakage, valve body bolts loose.
Low in 3	Forward clutch leakage.
Low in 1	Forward clutch leakage, low reverse servo leakage.
Low at idle in all ranges.	Low fluid level, restricted oil filter, loose valve body bolts, pump leakage, case leakage, valve body leakage, excessively low engine idle, fluid too hot, main oil regulator valve sticking.
High at idle in all ranges.	T.V. linkage, valve body (throttle valve or main oil regulator valve sticking).
Okay at idle but low at W.O.T.	Internal leakage, pump leakage, restricted inlet screen, T.V. linkage, valve body (T.V. or T.V. limit valve sticking), main oil regulator valve sticking.

Control Pressure Test Results Chart—AOD

T.V. Pressure	Line Pressure	Range	Possible Cause
High at WOT	High at WOT	All	TV limit valve not regulating pressure — replace valve body.
Low at WOT	Low at WOT	All	Throttle valve stuck — replace valve body. TV linkage out of adjustment damaged binding — adjust and/or repair as necessary.

Figure 13-10. Shop manuals contain charts showing possible causes of incorrect pressure test readings. (Ford Motor Company)

NAME ______________________ SECTION ____________ DATE ____________

Job Sheet 13-3

Pressure Testing the Automatic Transmission

Before you begin:

Read Topics 13.1 through 13.10.

Vehicle Make ____________________ Model ____________ Year ____________

Time Started ____________________ Time Finished ____________________

Job Completion Time ________________ Flat Rate Time ____________________

Tools Required

- Tachometer
- Pressure gauges
- Vacuum gauge
- T-fitting and vacuum tubing
- Hoist
- Safety glasses
- Droplight
- Manufacturer's pressure test chart
- Manufacturer's pressure test troubleshooting chart

References

Manufacturer's service manual ____

Repair manual ____

Service bulletin ____

Procedure: Make Transmission Pressure Test

1. Start the engine and warm it up to normal operating temperature.
2. Connect a tachometer to the engine according to the manufacturer's instructions.
3. Have a helper inside the vehicle to start the engine and operate the transmission, accelerator, and brake controls.
4. Raise the vehicle on a hoist to a comfortable working height.
5. Connect pressure gauges to the specified test ports.
6. Use a T-fitting and vacuum tubing to connect the specified vacuum gauge to the vacuum modulator or vacuum control diaphragm line.

SAFETY CAUTION: If the pressure test procedures or equipment requires you to be under the vehicle while the test is performed, use extreme caution. The driveline and rear wheels will be turning for most tests. Contact with the spinning driveline or wheels can cause serious injury.

7. Start the engine. Move the selector lever to the first test position. Run the engine at the specified rpm. Move transmission controls as specified, and note the readings, or change in readings. In the Job Completion Report, make a note of the gearshift lever position, rpm, vacuum gauge reading, and pressure reading, or pressure change.
8. Slow the engine and press gently on the brake pedal to stop the spinning wheels and driveline. Shut off the engine.
9. Attach the pressure gauges to the proper pressure ports for the next test.
10. Move the gearshift lever to the next test position. Run the engine at the specified rpm. Take pressure readings as specified. Slow the engine, and press gently on the brake pedal to stop the spinning wheels and driveline.
11. Repeat the steps for each test procedure as specified in the manufacturer's service manual. In the Job Completion Report, make notes of all tests and their results.

Instructor Check: ____________________

Job Completion Report

1. What difficulties, if any, did you encounter in completing these procedures?

____

____

2. Describe all tests performed and the results obtained. ____________________

____

____

____

____

____

____

____

____

____

____

INSTRUCTOR SIGNOFF ________________________ **DATE** ______________

13.12 AIR PRESSURE TESTS

The source of a transmission hydraulic problem can be difficult to locate. If road testing and pressure testing (discussed in 13.6 and 13.9) have indicated a slipping clutch or band, air pressure tests can be made. Air Pressure tests can determine whether the problem is in the apply device itself, or in the valve body that supplies hydraulic pressure.

After the valve body has been removed, air pressure is applied to the case holes and passages leading to the apply devices. Correct operation of servos and clutches can be heard and/or felt. Air pressure tests are done during overhaul disassembly to find leaking seals. Air pressure tests are also done during reassembly to check for proper clutch and servo operation.

Figure 13-11. Compressed air is applied to the transmission case fluid passages to check for proper clutch and servo operation. (Ford Motor Company)

Filtered, moisture-free compressed air of 40 psi [276 kPa] is usually used for the test. The air is blown into the transmission case fluid passages, shown in Figure 13-11. The proper operation of a servo can be heard and/or felt as a dull thud or thunk. Air pressure of up to 90 psi [620 kPa] can be used if it is difficult to hear the clutches apply.

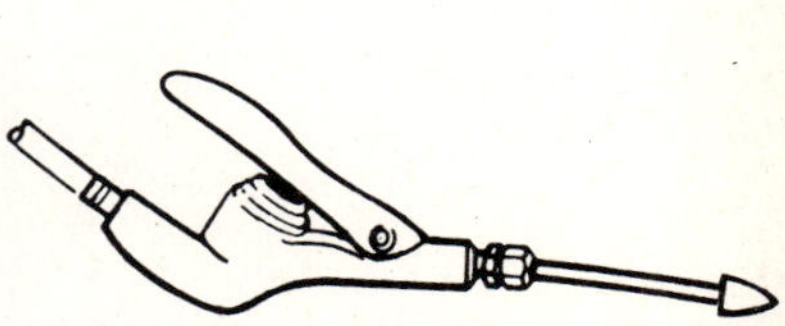

Figure 13-12. An air blow gun with conical rubber nozzle is used to apply the compressed air.

An air blow gun with a conical rubber nozzle is used to apply the air pressure. See Figure 13-12.

Some manufacturers recommend the use of special metal plates that bolt to the transmission to seal the case apply passages. See Figure 13-13. The metal plate and gasket are bolted to the case. Identification of the holes is stamped into the plate. Air is applied through the plate holes. Refer to the manufacturer's service manual for specific instructions.

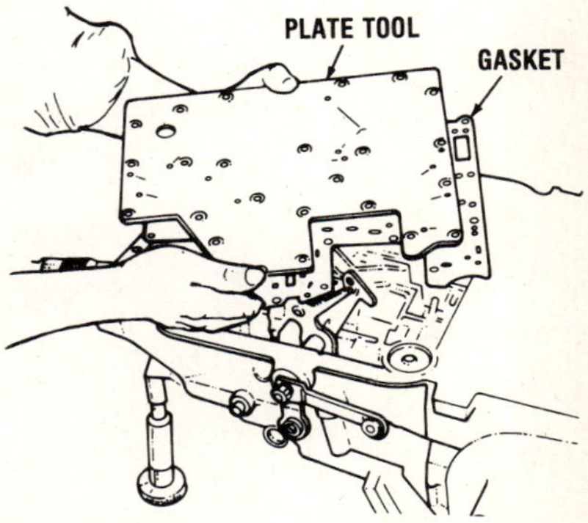

Figure 13-13. Installing the air-pressure-plate service tool. (Ford Motor Company)

The following procedure can be used as a general guide to conducting air pressure tests.

1. Safely raise the vehicle on a hoist.
2. Drain the transmission oil and remove the oil pan.
3. Remove the filter.
4. Remove the valve body. Valve body removal is discussed in 12.2.

5. Refer to the manufacturer's service manual to identify the fluid passages. Attach special plates and gaskets as necessary. Torque the plate attaching screws to the manufacturer's torque specification.

6. Apply air pressure to the servo and clutch apply passages and listen for 5 to 10 seconds. Listen for hissing. Hissing indicates air leakage from faulty seals.

SAFETY CAUTION: Compressed air can blow liquids or metal chips into the eyes, causing serious physical injury or blindness. Always wear safety glasses or goggles when working with compressed air.

7. Release the air pressure and reapply. Listen for thuds or thunks. If you cannot hear a servo operate, watch to see if the band tightens and loosens properly as air pressure is applied and released. Repair or replace apply devices during overhaul as necessary.

NAME ______________ SECTION ______________ DATE ______________

Job Sheet 13-4

Air-Pressure Testing

Before you begin:

Read Topic 13.11.

Vehicle Make ______________ Model ______________ Year ______________

Time Started ______________ Time Finished ______________

Job Completion Time ______________ Flat Rate Time ______________

Tools Required

- Special air-pressure testing plate (if necessary)
- Pound-inch [Nm] torque wrench
- Rubber-tipped blow gun

References

Manufacturer's service manual ______________

Repair manual ______________

Service bulletin ______________

Procedure

1. Safely raise the vehicle on a hoist.
2. Drain the transmission oil and remove the oil pan.
3. Remove the filter.
4. Remove the valve body.
5. Refer to the manufacturer's service manual to identify the fluid passages. Attach special plates and gaskets as necessary. Torque the special plates to the manufacturer's specification.

In the space provided, write the correct torque specification.

Manufacturer's Specified Torque .... ______________

Instructor Check: ______________

6. Apply air pressure to the servo and clutch passages and listen for 5 to 10 seconds. Listen for hissing. In the Job Completion Report, make a note of any defects found. Replace faulty seals as necessary during overhaul.

SAFETY CAUTION: Compressed air can blow liquids or metal chips into the eyes, causing serious physical injury or blindness. Always wear safety glasses or goggles when working with compressed air.

7. Release the air pressure and reapply. Listen for thuds or thunks. If you cannot hear a servo operate, watch to see if the band tightens and loosens as air pressure is applied and released. In the Job Completion Report, make a note of any defects found.
8. Remove any special plates and gaskets. Repair or replace apply devices as necessary during overhaul.

Job Completion Report

1. What defects, if any, were found during air pressure testing? ____________

   ____

   ____

2. What apply devices, if any, were repaired or replaced? ____________

   ____

   ____

3. What difficulties, if any, were encountered during this procedure? ____________

   ____

   ____

INSTRUCTOR SIGNOFF ____________________ **DATE** ____________

13.13 TRANSMISSION TROUBLESHOOTING CHARTS

Finding the cause of transmission problems can be complicated. One transmission problem may have 15 or more possible causes. To aid in troubleshooting and diagnosis of the automatic transmission, manufacturers furnish troubleshooting charts in service manuals. The charts have columns for conditions, or problems, possible causes, and the resolutions, or fixes. The accompanying chart refers to Ford Motor Company automatic transmissions. Each manufacturer's charts will be different.

DIAGNOSIS — AUTOMATIC TRANSMISSION

CONDITION	POSSIBLE CAUSE	RESOLUTION
Slow initial engagement.	1. Improper fluid level.	1. Perform fluid level check.
	2. Damaged or improperly adjusted linkage.	2. Service or adjust linkage.
	3. Contaminated fluid.	3. Perform fluid condition check.
	4. Improper cluch and band application, or low main control pressure.	4. Perform control pressure test.
	5. Damaged 2-3 accumulator valve (AOD).	5. Service as required.
Rough initial engagement in either forward or reverse. Rough initial engagement — AOD.	1. Improper fluid level.	1. Perform fluid level check.
	2. High engine idle.	2. Adjust idle to specifications.
	3. Automatic choke on (warm temp.).	3. Disengage choke.
	4. Looseness in the driveshaft, U-joints or engine mounts.	4. Service as required.
	5. Improper clutch or band application, or oil control pressure.	5. Perform control pressure test.
	6. Damaged 2-3 accumulator valve (AOD).	6. Service as required.
	7. Sticking or dirty valve body.	7. Clean, service or replace valve body.
Harsh engagements — (warm engine).	1. Improper fluid level.	1. Perform fluid level check.
	2. Throttle linkage misadjusted long/disconnected sticking/damaged/return spring disconnected.	2. Adjust linkage.
	3. Engine curb idle too high.	3. Check engine curb idle.
	4. Valve body bolts — loose/too tight.	4. Tighten to specification.
	5. Valve body dirty/sticking valves.	5. Determine source of contamination. Service as required.
No/delayed forward engagement (reverse OK).	1. Improper fluid level.	1. Perform fluid level check.
	2. Manual linkage — misadjusted/damaged.	2. Check and adjust or service as required.
	3. Low main control pressure.	3. Control pressure test, note results.
	4. AOD ONLY — 2-3 accumulator piston seal (small) cut/worn — piston bore in case damaged.	4. Replace seals/determine cause of condition. Service as required.
	5. Forward clutch assembly burnt/damaged.	5. Peform air pressure test.
	6. Valve body bolts — loose/too tight.	6. Tighten to specification.
	7. Valve body dirty/sticking valves.	7. Determine source of contamination. Service as required.
No/delayed reverse engagement (forward OK).	1. Improper fluid level.	1. Perform fluid level check.
	2. Manual linkage misadjusted/damaged.	2. Check and adjust or service as required.
	3. Low main control pressure in reverse.	3. Control pressure test.
	4. Reverse clutch assembly burnt/worn.	4. Perform air pressure test.
	5. Valve body bolts loose/too tight.	5. Tighten to specification.
	6. Valve body dirty/sticking valves.	6. Determine source of contamination. Service as required.

DIAGNOSIS — AUTOMATIC TRANSMISSION

CONDITION	POSSIBLE CAUSE	RESOLUTION
No/delayed reverse engagement and/or no engine breaking in manual low (1).	1. Improper fluid level.	1. Perform fluid level check.
	2. Linkage out of adjustment.	2. Service or adjust linkage.
	3. Low reverse band/servo piston burnt/worn.	3. Perform air pressure test.
	4. Bands out of adjustment — C5 only.	4. Adjust reverse band — C5 only.
	5. Polished, glazed band or drum — C5 only.	5. Service or replace as required.
No engine braking in manual second gear.	1. Improper fluid level.	1. Perform fluid level check.
	2. Linkage out of adjustment.	2. Service or adjust linkage.
	3. Intermediate band out of adjustment.	3. Adjust intermediate band.
	4. Improper band or clutch application, or oil pressure control system.	4. Perform control pressure test.
	5. Intermediate servo leaking.	5. Perform air pressure test of intermediate servo for leakage. Service as required.
NOTE: No manual second gear AOD.	6. Polished or glazed band or drum.	6. Service or replace as required.
Forward engagement slips/shudders/chatters.	1. Improper fluid level.	1. Perform fluid level check.
	2. AOD ONLY — Low throttle pressure/throttle valve rod misadjusted (short).	2. Adjust throttle valve rod.
	3. Manual linkage misadjusted/damaged.	3. Check and adjust or service as required.
	4. Low main control pressure.	4. Control pressure test.
	5. Valve body bolts — loose/too tight.	5. Tighten to specification.
	6. Valve body dirty/sticking valves.	6. Determine source of contamination. Service as required.
	7. Forward clutch piston ball check not seating.	7. Replace forward clutch cylinder. Service transmission as required.
	8. Forward clutch piston seal cut/worn.	8. Replace seal and service clutch as required.
	9. Contamination blocking forward clutch feed hole.	9. Determine source of contamination. Service as required.
	10. Low one way clutch (planetary) damaged.	10. Determine cause of condition. Service as required.
Reverse shudder/chatters/slips.	1. Improper fluid level.	1. Perform fluid level check.
	2. Low main control pressure in reverse.	2. Control pressure test.
	3. Reverse servo/servo bore damaged.	3. Determine cause of condition. Service as required.
	4. Low (planetary) one-way clutch damaged.	4. Determine cause of condition. Service as required.
	5. Reverse clutch drum bushing damaged.	5. Determine cause of condition. Service as required.
	6. Reverse clutch stator support seal rings/ring grooves worn/damaged.	6. Determine cause of condition. Service as required.
	7. Reverse clutch piston seal cut/worn.	7. Determine cause of condition. Service as required.
	8. Intermediate servo seals damaged (C5).	8. Service as required.
	9. Reverse band out of adjustment or damaged.	9. For C5 — Adjust reverse band. AOD has no adjustment. Service as required.
	10. Looseness in the driveshaft, U-joints or engine mounts.	10. Service as required.

DIAGNOSIS — AUTOMATIC TRANSMISSION

CONDITION	POSSIBLE CAUSE	RESOLUTION
No drive, slips or chatters in first gear in D. All other gears normal. First gear in D or 3 Automatic Overdrive transmission.	1. Damaged or worn one-way clutch.	1. Service or replace one-way clutch.
No drive, slips or chatters in second gear.	1. Improper fluid level. 2. Damaged or improperly adjusted linkage. 3. Intermediate band out of adjustment. (Except AOD). 4. Intermediate friction clutch or one-way clutch (for Automatic Overdrive transmission only). 5. Improper band or clutch application, or oil pressure control. 6. Damaged or worn servo and or internal leaks. 7. Dirty or sticking valve body. 8. Polished, glazed intermediate band or drum. (Except AOD).	1. Perform fluid level check. 2. Service or adjust linkage. 3. Adjust intermediate band. 4. Service as required. 5. Perform control pressure test. 6. Perform air pressure test. 7. Clean, service or replace valve body. 8. Replace or service as required.
Starts up in 2nd or 3rd.	1. Improper fluid level. 2. Damaged or improperly adjusted linkage. 3. Improper band and or clutch application, or oil pressure control system. 4. Damaged or worn governor. Sticking governor. 5. Valve body loose. 6. Dirty or sticking valve body. 7. Cross leaks between valve body and case mating surface.	1. Perform fluid level check. 2. Service or adjust linkage. 3. Perform control pressure test. 4. Perform governor check. Replace or service governor, clean screen. 5. Tighten to specification. 6. Clean, service or replace valve body. 7. Service or replace valve body and/or case as required.
Shift points incorrect.	1. Improper fluid level. 2. Vacuum line damaged, clogged or leaks (Except AOD). 3. Improper operation of EGR system. (Except AOD). 4. Throttle linkage out of adjustment (AOD). 5. Improper speedometer gear installed. 6. Improper clutch or band application, or oil pressure control system. 7. Damaged or worn governor. 8. Dirty or sticking valve body.	1. Perform fluid level check. 2. Perform vacuum supply test. 3. Service or replace as required. 4. Service or adjust linkage. 5. Replace gear. 6. Perform shift test and control pressure test. 7. Service or replace governor — clean screen. 8. Clean, service or replace valve body.
All upshifts harsh delayed or no upshifts.	1. Improper fluid level. 2. AOD Only — Throttle linkage — misadjusted (long) disconnected sticking damaged return spring disconnected. 3. Manual linkage — misadjusted damaged. 4. Governor sticking. 5. Main control pressure too high. 6. Valve body bolts — loose too tight. 7. Valve body dirty sticking valves. 8. Vacuum leak to diaphragm unit (C5 only).	1. Perform fluid level check. 2. Adjust throttle linkage. Service as required. 3. Check and adjust or service as required. 4. Perform governor test. Service as required. 5. Control pressure test. Service as required. 6. Tighten to specification. 7. Determine source of contamination. Service as required. 8. Check vacuum lines to diaphragm unit. Service as necessary. Check diaphragm unit. Service as necessary.

DIAGNOSIS — AUTOMATIC TRANSMISSION

CONDITION	POSSIBLE CAUSE	RESOLUTION
Mushy/early all upshifts/pile up upshifts	1. Improper fluid level.	1. Perform fluid level check.
	2. AOD ONLY — Throttle linkage misadjusted (short)/sticking/damaged.	2. Adjust throttle linkage. Service as required.
	3. Low main control pressure.	3. Control pressure test. Note results.
	4. Valve body bolts loose/too tight.	4. Tighten to specification.
	5. Valve body valve sticking.	5. Determine source of contamination. Service as required.
	6. Governor valve sticking.	6. Perform governor test. Repair as required.
No 1-2 upshift.	1. Improper fluid level.	1. Perform fluid level check.
	2. AOD ONLY — Throttle linkage — misadjusted (long)/disconnected/sticking.	2. Adjust throttle linkage.
	3. C5 — Kickdown linkage misadjusted.	3. Adjust linkage.
	4. Manual linkage — misadjusted/damaged.	4. Check and adjust or service as required.
	5. AOD ONLY — Low main control pressure to intermediate friction clutch.	5. Control pressure test. Note results.
	6. Governor valve sticking. Intermediate band out of adjustment (C6, C5 only).	6. Perform governor test. Service as required. Adjust intermediate band.
	7. Valve body bolts — loose/too tight.	7. Tighten to specification.
	8. Valve body dirty/sticking valves.	8. Determine source of contamination. Service as required.
	9. Intermediate clutch/band and or servo assembly burnt.	9. Perform air pressure test.
Rough/harsh/delayed 1-2 upshift.	1. Improper fluid level.	1. Perform fluid level check.
	2. Poor engine performance.	2. Tune engine.
	3. AOD ONLY — Throttle linkage — misadjusted (long)/damaged.	3. Adjust linkage. Service as required.
	4. Intermediate band out of adjustment (C5 only).	4. Adjust intermediate band.
	5. Main control pressure too high.	5. Control pressure test. Note results.
	6. Governor valve sticking.	6. Perform governor test. Service as required.
	7. Damaged intermediate servo (C6, C5 only).	7. Air pressure check intermediate servo.
	8. Engine vacuum leak (C6, C5 only).	8. Check engine vacuum lines. Service as necessary. Check vacuum diaphragm unit. Service as necessary.
	9. Valve body bolts — loose/too tight.	9. Tighten to specifications.
	10. Valve body dirty/sticking valves.	10. Determine source of contamination. Service as required.

DIAGNOSIS — AUTOMATIC TRANSMISSION

CONDITION	POSSIBLE CAUSE	RESOLUTION
Mushy/early/soft/slipping 1-2 upshift.	1. Improper fluid level.	1. Perform fluid level check.
	2. Incorrect engine performance.	2. Tune adjust engine idle as required.
	3. AOD ONLY — Throttle/linkage misadjusted (short)/sticking/damaged.	3. Adjust throttle linkage. Service as required.
	4. Intermediate band out of adjustment (C5 only).	4. Adjust intermediate band.
	5. Low main control pressure.	5. Control pressure test. Note results.
	6. Valve body bolts loose/too tight.	6. Tighten to specification.
	7. Valve body dirty/sticking valves.	7. Determine source of contamination. Service as required.
	8. AOD ONLY — Intermediate friction clutch burnt/worn.	8. Determine cause of condition. Service as required.
	9. Governor valve sticking.	9. Perform governor test. Service as required.
	10. Damaged intermediate servo or band.	10. Perform air pressure test. Service as required.
	11. Polished, glazed band or drum (C5 only).	11. Service or replace as required.
No 2-3 upshift.	1. Low fluid level.	1. Perform fluid level check.
	2. AOD ONLY — Throttle linkage misadjusted (long)/sticking/damaged.	2. Adjust linkage. Service as required.
	3. Low main control pressure to direct clutch.	3. Control pressure test. Note results.
	4. Valve body bolts — loose/too tight.	4. Tighten to specification.
	5. Valve body dirty/sticking valves.	5. Determine source of contamination, then service as required.
	6. AOD ONLY — Direct clutch assembly burnt/worn.	6. Stall test. Determine cause of condition. Service as required.
	7. AOD ONLY — Converter damper hub weld broken.	7. Perform converter damper hub weld check. Replace torque converter if required.
Harsh/delayed 2-3 upshift.	1. Low fluid level.	1. Perform fluid level check.
	2. Incorrect engine performance.	2. Check engine tune-up.
	3. Engine vacuum leak (C5, C6 only).	3. Check engine vacuum lines. Service as necessary. Check vacuum diaphragm unit. Service as necessary.
	4. AOD ONLY — Throttle linkage misadjusted (long)/sticking damaged.	4. Adjust linkage. Service as required.
	5. AOD ONLY — 2-3 accumulator piston drain hole plugged/omitted.	5. Remove 2-3 accumulator piston and visually inspect for plugging condition or omission.
	6. AOD ONLY — 2-3 accumulator piston seals cut/worn.	6. Replace seals, determine cause of condition. Service as required.
	7. AOD ONLY — Damaged 2-3 accumulator.	7. Service as required.
	8. Damaged or worn intermediate servo release and high clutch piston check ball. (C6, C5 only).	8. Air pressure test the intermediate servo apply and release the high clutch piston check ball. Service as required.
	9. Valve body bolts — loose/too tight.	9. Tighten to specification.
	10. Valve body dirty/sticking valves.	10. Determine source of condition. Service as required.

DIAGNOSIS — AUTOMATIC TRANSMISSION

CONDITION	POSSIBLE CAUSE	RESOLUTION
Soft/early/mushy 2-3 upshift.	1. Improper fluid level.	1. Perform fluid level check.
	2. AOD Only — Throttle linkage — misadjusted short bent sticking.	2. Adjust linkage. Repair as required.
	3. Valve body bolts loose too tight.	3. Tighten to specification.
	4. Valve body dirty sticking valves.	4. Determine source of contamination. Service as required.
	5. AOD ONLY — Direct clutch assembly burnt worn.	5. Stall test. Determine cause of condition. Service as required.
AOD Only — No 3-4 upshift.	1. Low fluid level.	1. Perform fluid level check.
	2. Throttle linkage — misadjusted long bent sticking.	2. Adjust linkage. Service as required.
	3. Low pressure to overdrive band servo.	3. Line pressure test. Note results.
	4. Valve body bolts — loose too tight.	4. Tighten to specification.
	5. Valve body dirty sticking valves.	5. Determine source of contamination. Service as required.
	6. Overdrive band assembly burnt worn.	6. Determine cause of condition. Service as required.
AOD Only — No 3-4 upshift.	1. 3-4 accumulator piston seals worn cut.	1. Determine cause of condition. Service as required.
	2. 3-4 accumulator piston drain passage blocked.	2. Determine cause of condition. Service as required.
AOD Only — Harsh/delayed 3-4 upshift.	1. Low fluid level.	1. Perform fluid level check.
	2. Throttle linkage — misadjusted long bent sticking.	2. Adjust linkage. Service as required.
	3. Valve body bolts — loose too tight.	3. Tighten to specification.
	4. Valve body dirty sticking valves.	4. Determine source of contamination. Service as required.
	5. Incorrect engine performance.	5. Tune. adjust engine idle as required.
AOD Only — Slips/shudder soft/early 3-4 upshift.	1. Low fluid level.	1. Perform fluid level check.
	2. Throttle linkage — misadjusted short bent sticking.	2. Adjust linkage. Service as required.
	3. Low main control pressure to overdrive band servo.	3. Control pressure test. Note results.
	4. Valve body bolts — loose too tight.	4. Tighten to specification.
	5. Valve body dirty sticking valves.	5. Determine source of contamination. Service as required.
	6. Overdrive band assembly burnt worn.	6. Determine cause of condition. Service as required.
	7. Damaged or glazed reverse clutch drum or overdrive band.	7. Service as required.

DIAGNOSIS — AUTOMATIC TRANSMISSION

CONDITION	POSSIBLE CAUSE	RESOLUTION
Erratic shifts	1. Improper fluid level. 2. Poor engine performance. 3. Throttle linkage — binding/sticking. 4. Valve body bolts — loose/too tight. 5. Valve body dirty/sticking valves. 6. Governor valve stuck. 7. Output shaft collector body seal rings damaged.	1. Perform fluid level check. 2. Check engine tune-up. 3. Inspect throttle linkage. Service as required. 4. Tighten to specification. 5. Line pressure test, note results. Determine source of contamination. Service as required. 6. Perform governor test. Service as required. 7. Service as required.
Shifts 1-3 in D. (D or 3 range for AOD)	1. Improper fluid level. 2. Intermediate band out of adjustment (C6,C5 only). 3. Intermediate friction (AOD Only) clutch burnt/damaged. 4. AOD Only — Intermediate one-way clutch damaged. 5. Damaged intermediate servo and/or internal leaks (C6, C5 only). 6. Improper band or clutch application, or oil pressure control system. 7. Polished, glazed band or drum (C5 only). 8. Dirty or sticking valve body.	1. Perform fluid level check. 2. Adjust band. 3. Determine cause of condition. Service as required. 4. Determine cause of condition. Replace/service as required. 5. Perform air pressure test. Service front servo and/or internal leaks. 6. Perform control pressure test. 7. Service or replace band or drum. 8. Clean, service or replace valve body.
Engine over-speeds on 2-3 shift.	1. Improper fluid level. 2. Linkage out of adjustment. 3. Throttle linkage — AOD. 4. Improper band or clutch application, or oil pressure control system. 5. Damaged or worn high clutch and/or intermediate servo. 6. Dirty or sticking valve body.	1. Perform fluid level check. 2. Service or adjust linkage. 3. Service or adjust. 4. Perform control pressure test. 5. Perform air pressure test. Service as required. 6. Clean, service or replace valve body.
AOD Only — Shift hunting 3-4, 4-3.	1. Poor engine performance — EGR solenoid defective. 2. Throttle linkage misadjusted. 3. Manual linkage misadjusted.	1. Tune-up engine — Replace solenoid. 2. Inspect and adjust throttle linkage as required. 3. Check and adjust or service as required.
Rough/shudder 3-1 shift at closed throttle in D.	1. Improper fluid level. 2. Incorrect engine idle or performance. 3. Improper linkage adjustment. 4. Improper clutch or band application or oil pressure control system. 5. Improper governor operation. 6. Dirty or sticking valve body.	1. Perform fluid level check. 2. Tune, and adjust engine idle. 3. Service or adjust linkage. 4. Perform control pressure test. 5. Perform governor test. Service as required. 6. Clean, service or replace valve body.
AOD Only — Rough or mushy 4-2 or 3-1 shift.	1. Improper fluid level. 2. Incorrect engine performance. 3. Improper throttle or manual linkage adjustment. 4. Improper application of intermediate friction and one way clutch. 5. Dirty or sticking valve body.	1. Perform fluid level check. 2. Tune, adjust engine idle as required. 3. Service or adjust linkage. 4. Service as required. 5. Clean, service or replace valve body.

DIAGNOSIS — AUTOMATIC TRANSMISSION

CONDITION	POSSIBLE CAUSE	RESOLUTION
No forced downshifts.	1. Improper fluid level. 2. Kickdown linkage out of adjustment (C6, C5 only). 3. Damaged internal kickdown linkage (C6, C5 only). 4. Damaged, misadjusted (short) throttle linkage (AOD). 5. Improper clutch or band application, or oil pressure control system. 6. Dirty or sticking governor. 7. Dirty or sticking valve body.	1. Perform fluid level check. 2. Service or adjust linkage. 3. Service internal kickdown linkage. 4. Inspect and adjust throttle linkage. 5. Perform control pressure test. 6. Service or replace governor, clean screen. 7. Clean, service, or replace valve body.
Engine over-speeds on 3-2 downshift (C6, C5 only).	1. Improper fluid level. 2. Linkage out of adjustment. 3. Intermediate band out of adjustment. 4. Improper band or clutch application, and one way clutch, or oil pressure control system. 5. Damaged or worn intermediate servo. 6. Polished, glazed band or drum. 7. Dirty or sticking valve body.	1. Perform fluid level check. 2. Service or adjust linkage. 3. Adjust intermediate band. 4. Perform control pressure test service clutch. 5. Air pressure test check the intermediate servo. Service servo and/or seals. 6. Service or replace as required. 7. Clean, service or replace valve body.
Shift efforts high.	1. Manual shift linkage damaged/misadjusted. 2. Inner manual lever nut loose. 3. Manual lever retainer pin damaged.	1. Check and adjust or service as required. 2. Tighten nut to specification. 3. Adjust linkage and install new pin.
Transmission overheats.	1. Improper fluid level. 2. Incorrect engine idle, or performance. 3. Improper clutch or band application, or oil pressure control system. 4. Restriction in cooler or lines. 5. Seized converter one-way clutch. 6. Dirty or sticking valve body.	1. Perform fluid level check. 2. Tune, or adjust engine idle. 3. Perform control pressure test. 4. Service restriction. 5. Replace one-way clutch. 6. Clean, service or replace valve body.
Transmission leaks.	1. Case breather vent. 2. Leakage at gasket, seals, etc.	1. Check the vent for free breathing. Repair as required. 2. Remove all traces of lube on exposed surfaces of transmission. Check the vent for free breathing. Operate transmission at normal temperatures and perform fluid leakage check. Service as required.
Poor vehicle acceleration.	1. Poor engine performance. 2. Torque converter one-way clutch locked up.	1. Check engine tune up. 2. Replace torque converter.
Transmission noisy — valve resonance. NOTE: Gauges may aggravate any hydraulic resonance. Remove gauge and check for resonance level.	1. Improper fluid level. 2. Linkage out of adjustment. 3. Improper band or clutch application, or oil pressure control system. 4. Cooler lines grounding. 5. Dirty or sticking valve body. 6. Internal leakage or pump cavitation.	1. Perform fluid level check. 2. Service or adjust linkage. 3. Perform control pressure test. 4. Free up cooler lines. 5. Clean, service or replace valve body. 6. Service as required.

NAME ______________________ SECTION ______________ DATE ______________

Certification Practice

Test 13

DIRECTIONS:

The following questions are similar to those used on mechanic certification tests. Answer each question by circling the letter of the correct choice.

1. All of the following apply to troubleshooting EXCEPT
   A. The mechanic should ask the driver questions about the problem.
   B. A test ride with the driver can be helpful in determining transmission problems.
   C. The mechanic should always rely on the driver to correctly diagnose a problem.
   D. The mechanic is responsible for correctly diagnosing and fixing the problem.

2. Mechanic A says that transmission problems can be caused by engine malfunctions. Mechanic B says that transmission operation is not affected by engine problems. Who is correct?
   A. A only B. B only C. Both A and B D. Neither A nor B

3. The transmission of a vehicle with a worn-out engine is being disassembled. When inspected, major damage to the torque converter, oil pump, and planetary gears is found. The most likely cause is
   A. wrong type of ATF used.
   B. vehicle driven in stop-and-go traffic.
   C. engine main bearings worn out.
   D. transmission fluid changed too often.

4. The unit shown in the picture is
   A. a vacuum gauge.
   B. a vacuum pump.
   C. a vacuum control diaphragm unit.
   D. all of the above.

Question 4.
(Ford Motor Company)

5. A customer drives into an auto repair shop. The exhaust is pouring out clouds of dense, bluish smoke. The driver says the problem just began today.
   The most likely cause is
   A. worn-out engine rings.
   B. worn-out engine valve guides.
   C. leaking vacuum modulator diaphragm.
   D. all of the above.

6. Which of the following statements is correct?

I. Misadjusted linkage can cause major transmission damage.

II. Misadjusted linkage can cause erratic shifting.

A. I only B. II only C. Both I and II D. Neither I nor II

7. Slippage refers to

A. lack of traction on wet or icy roads.

B. lack of traction on dry roads.

C. the engine-to-transmission power connection.

D. engine power output.

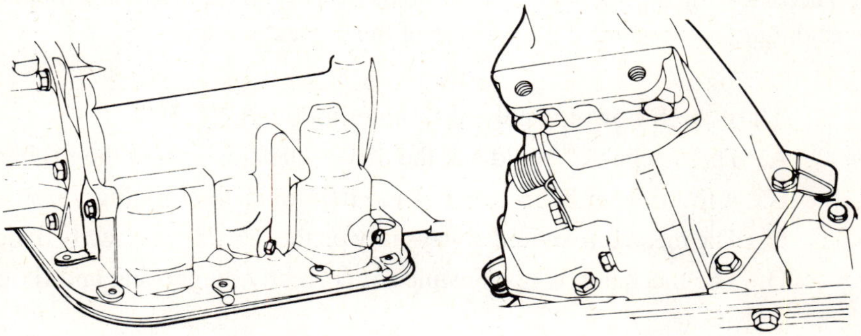

Question 7
(Chrysler Corporation)

8. The items indicated in the picture are

A. pressure test gauges.

B. pressure test ports.

C. core plugs.

D. vacuum ports.

9. An air-pressure test checks

A. apply device operation.

B. linkage adjustment.

C. vacuum modulator operation.

D. transmission hydraulic pressures.

10. Mechanic A says a pressure test will always pinpoint the exact cause of any transmission problem.

Mechanic B says troubleshooting charts may be necessary to help find the cause of transmission problems.

Who is correct?

A. A only B. B only C. Both A and B D. Neither A nor B

14

VI Automatic Transmission Overhaul

Removing the Transmission

YOUR LEARNING JOB

This unit discusses the removal of an automatic transmission for overhaul. Safety precautions are emphasized. The weight of the transmission and the necessity to provide support for the engine are two major factors. Correct procedures are vital in this phase of transmission work.

JOB COMPETENCY OBJECTIVES

This is a skill-building unit. When you have completed your assignments and exercises in this unit, you should be able to:

- ☐ Remove the shift linkage.
- ☐ Support the engine and remove the transmission support crossmember.
- ☐ Remove the transmission.
- ☐ Remove the torque converter.

14.1 BEFORE YOU BEGIN

There are as many differences in transmission removal methods as there are automatic transmissions and the vehicles they're used in. Crossmembers, linkage systems, exhaust systems, wiring, and several other parts must be disconnected before the transmission can be removed.

Allow the exhaust system parts to cool down before beginning the transmission removal process. Hot parts can cause severe burns.

The automatic transmission is extremely heavy. Always have a helper assist you when removing the transmission. The engine must be supported to keep it from dropping down when the transmission is removed. A transmission jack must always be placed under the transmission to support it (Figure 14-1). Failure to support the engine or transmission can lead to vehicle damage and serious personal injury.

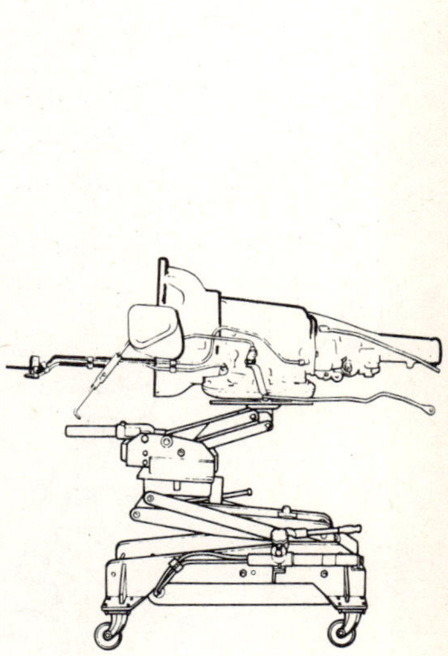

Figure 14-1. A transmission jack must be used to support the transmission during removal.

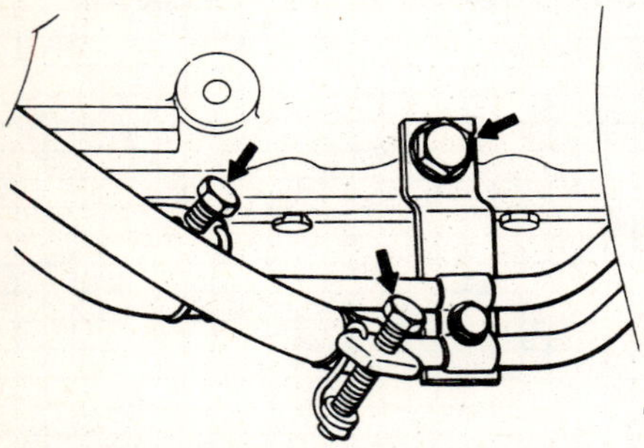

Figure 14-2. Disconnecting oil cooler return tube and supply tube. Arrows indicate bolts to be loosened. (Chrysler Corporation)

Always refer to the proper service manual before removing any automatic transmission. Go over the removal procedures carefully. Make a visual check of the job that is required before any part is removed.

The removal procedure discussed in this unit is for the Chrysler TorqueFlite. TorqueFlite overhaul is covered in Unit 15.

14.2 REMOVING THE TORQUEFLITE TRANSMISSION

The TorqueFlite transmission and torque converter must be removed as an assembly. If not, the converter flexplate, pump bushing, or oil seal may be damaged.

1. Place the gearshift lever in park. Set the emergency brake. Use safety glasses.

2. Disconnect the grounded cable from the battery. This will prevent a short circuit when disconnecting the starter motor or other electrical devices.

3. Raise the vehicle on a hoist.

4. Remove whatever parts of the exhaust system are necessary for the transmission to clear during removal.

5. Remove any struts that connect the transmission to the engine.

6. Disconnect and remove the transmission cooler lines at the transmission. See Figure 14-2.

7. Disconnect the wires to the starter. Unbolt (Figure 14-3) and remove the starter motor and the bracket that holds the cooler line.

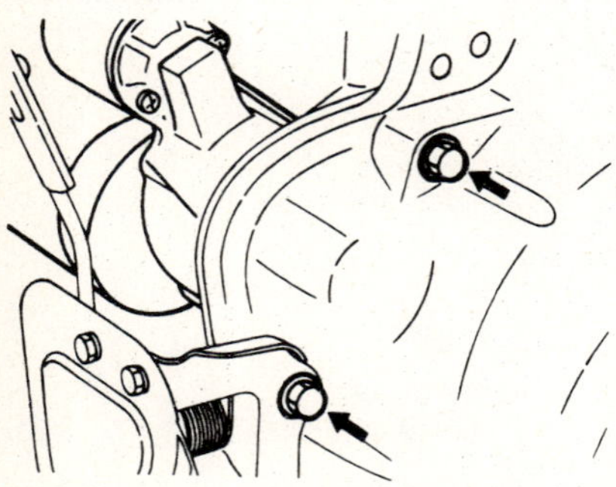

Figure 14-3. Remove starter bolts shown with arrows. (Chrysler Corporation)

8. Remove the converter inspection cover. To expose the converter drain plug, rotate the engine clockwise. Use a socket wrench on the vibration dampener bolt at the front of the crankshaft to rotate the engine.

9. Drain the converter.

NOTE: Check the torque converter for evidence of contamination or defect. If any defect is apparent, it is recommended that torque converters without drain plugs be replaced with new or rebuilt torque converters.

10. Loosen the oil pan bolts and drain the transmission.

11. Loosely install at least two of the pan bolts to hold the oil pan in place.

12. Mark the converter and flexplate to help in reassembly.

NOTE: The crankshaft flange bolt circle, the inner and outer circle of holes in the flexplate, and the four tapped holes in the front face of the converter all have one hole that is offset. These offset holes are aligned so that these parts can be installed only in their original position. This maintains balance beween the engine and the converter.

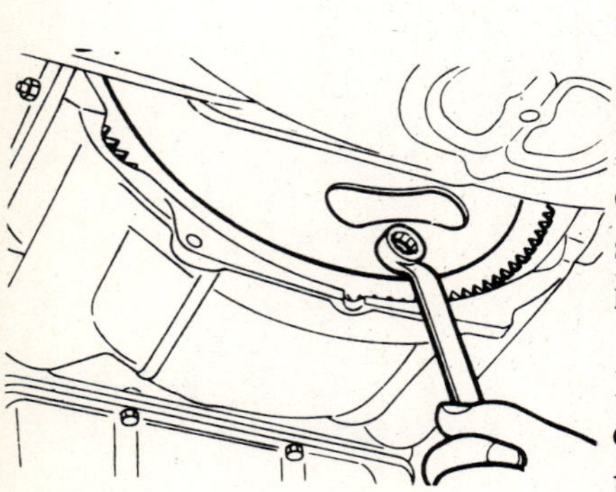

Figure 14-4. Remove bolts holding flexplate to torque converter. (Chrysler Corporation)

13. Rotate the engine clockwise with a socket wrench on the vibration dampener bolt. This exposes the bolts attaching the torque converter to the flexplate. Remove the bolts, as shown in Figure 14-4.

CAUTION: Do *not* rotate the converter or flexplate by prying with a screwdriver or similar tool. The flexplate can be distorted.

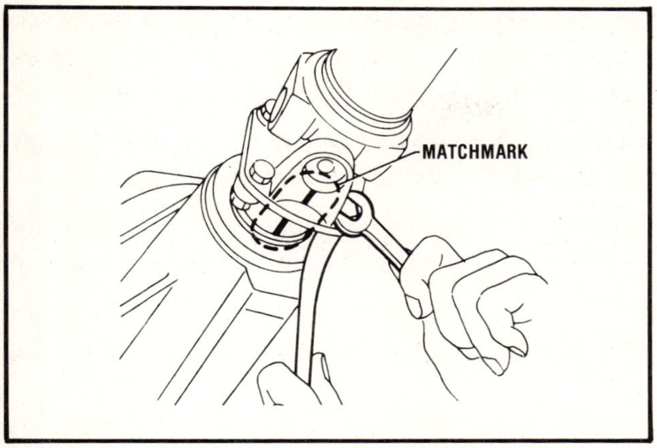

Figure 14-5. Apply matchmarks on companion flange and driveshaft for proper reassembly. (Nissan Motor Corporation)

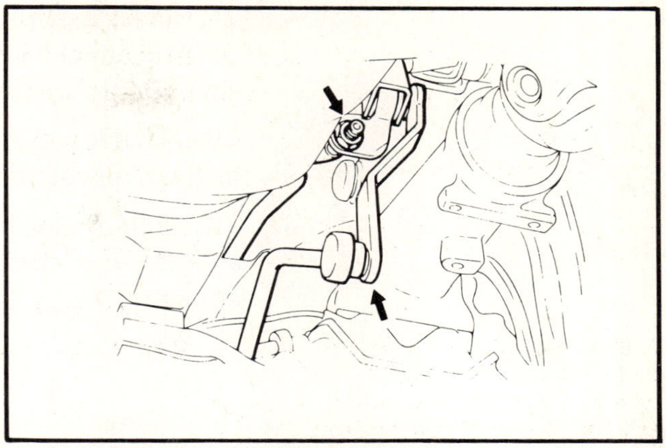

Figure 14-6. Disconnect the transmission control rod from the lever and from the support. (Chrysler Corporation)

14. Paint matchmarks on the rear universal joints on the driveshaft for disassembly, as shown in Figure 14-5. Disconnect the driveline at the rear universal joint. Carefully pull the driveline out of the extension housing.

15. Disconnect the wire connector from the backup light and neutral safety switch.

16. Disconnect the gearshift rod and torque shaft assembly from the transmission.

NOTE: If disconnecting linkage rods from levers that use plastic grommets, the grommets must be replaced with new ones. Use a prying tool to force out the grommets. Cut away the old grommets. Use pliers to snap new grommets into place.

17. Disconnect the throttle rod from the lever at the left side of the transmission. See Figure 14-6. If so equipped, remove the linkage bellcrank from the transmission.

18. Remove the oil filler tube.

19. Disconnect and remove the speedometer cable, pinion, and sleeve. Figure 14-7 shows the location.

20. Place an engine support under the rear of the engine. See Figure 14-8.

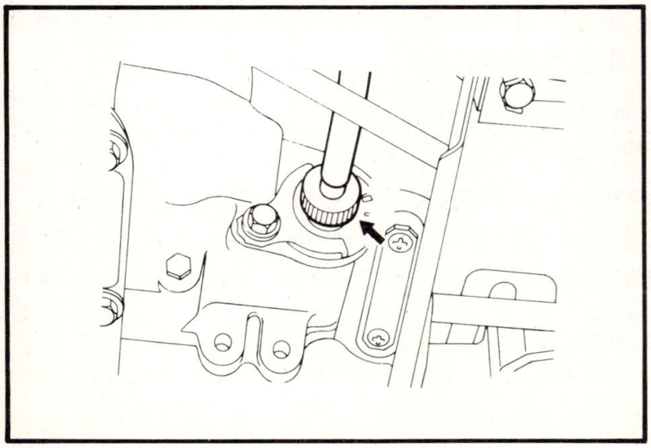

Figure 14-7. Disconnecting speedometer cable, arrow. (Chrysler Corporation)

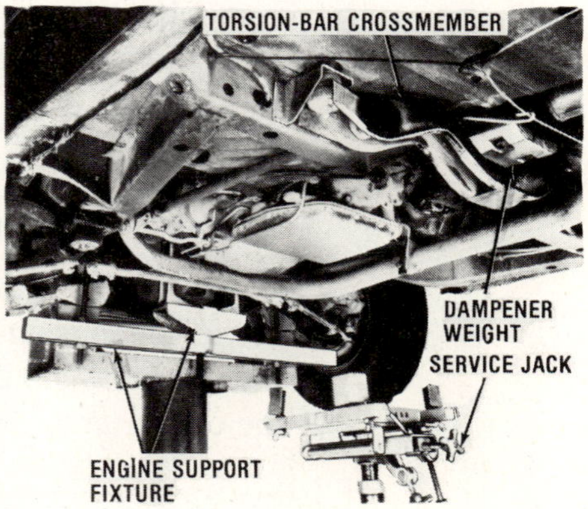

Figure 14-8. Engine support fixture in place. (Chrysler Corporation)

21. Place a transmission jack under the transmission. Position a safety strap or chain around the transmission to keep it from falling. Raise the transmission slightly with the transmission jack to lift the load off the supports.

CAUTION: The flexplate will not support a load. None of the weight of the transmission must rest on the flexplate during removal.

22. Remove the bolts holding the transmission mount to the crossmember and the crossmember to the frame. See Figure 14-9. Remove the crossmember.

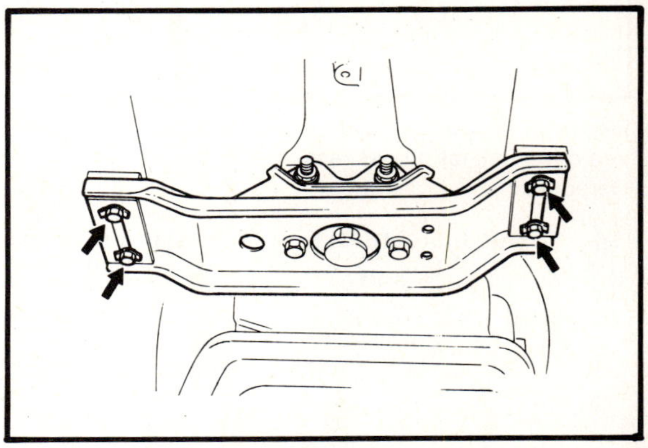

Figure 14-9. Remove bolts (arrows) holding rear engine support bracket. (Chrysler Corporation)

NOTE: Some models have a torsion bar anchor crossmember, which must remain in place. This requires a careful downward tilt of the front of the transmission as it is being removed. If these models have a vibration-dampening weight bolted to the rear of the extension housing, it must be removed to provide additional clearance. Some models have access holes through the crossmembers for the three bolts that hold the weight-mounting bracket.

23. Remove all bellhousing bolts.

24. Carefully work the transmission and torque converter assembly rearward and off the engine block dowels. At the same time, disengage the converter hub from the end of the crankshaft. Attach a small C-clamp to the edge of the bellhousing to hold the converter in place during transmission removal.

CAUTION: Do *not* leave the converter attached to the engine while removing the transmission. This could cause damage to the input shaft, the converter, and/or the oil pump.

25. Lower the transmission and remove the assembly from under the vehicle. Carefully move the transmission jack and the transmission rearward.

26. Place the transmission in a transmission support or on a workbench.

27. To remove the converter assembly, remove the C-clamp from the edge of the bellhousing. Carefully slide the assembly straight forward and out of the transmission. Be sure to hold it tightly and use adequate support because the converter is heavy.

NAME ______________________ SECTION ____________ DATE ____________

Job Sheet 14-1

Removing The Transmission

Before you begin:

Read Topics 14.1 through 14.2.

Vehicle Make ____________________ Model ____________ Year ____________

Time Started ____________________ Time Finished ____________________

Job Completion Time ____________________ Flat Rate Time ____________________

Tools Required

- Hoist
- Safety glasses
- Droplight
- Sockets, extensions, handle
- Open-end or combination wrenches
- Pliers
- Screwdrivers
- Paint and small brush
- Engine support stand
- Transmission jack and safety straps

References

Manufacturer's service manual ________________________________

Repair manual ________________________________

Service bulletin ________________________________

Procedure

1. Place the gearshift lever in park. Set the emergency brake.
2. Disconnect the grounded cable from the battery.
3. Safely raise the vehicle on a hoist.
4. Remove exhaust system parts as necessary.

SAFETY CAUTION: Allow the exhaust system parts to cool down before working. Hot parts can cause severe burns.

5. Remove any braces or struts that connect the transmission to the engine.
6. Disconnect the transmission cooler lines at the transmission.
7. Disconnect the wires to the starter. Unbolt and remove the starter motor and the bracket that holds the cooler line.
8. Remove the access plate to the torque converter. Rotate the engine clockwise to locate a converter drain plug.
9. Remove the drain plug, if fitted, and drain the converter.

NOTE: Check the torque converter for contamination or defect. If anything is found wrong, it is recommended that torque converters without drain plugs be replaced with new or rebuilt torque converters during overhaul.

10. Loosen the oil pan bolts and drain the transmission.
11. Loosely install at least two of the pan bolts at opposite corners of the pan to hold it in place.
12. Make matchmarks on the converter housing and flexplate to help in reassembly.

NOTE: Some crankshafts and flexplates have one offset bolt hole. These offset holes are aligned so that the flexplate can be installed only in its original position.

13. Rotate the engine clockwise to expose the bolts attaching the torque converter to the flexplate. Remove the bolts.

CAUTION: Do *not* rotate the converter or flexplate by prying with a screwdriver or similar tool. The flexplate can become distorted.

14. Make matchmarks on the rear universal joints and driveshaft for reassembly. Disconnect the driveshaft at the rear universal joint. Carefully pull the driveline out of the transmission extension housing.
15. Disconnect any wire connectors.
16. Disconnect the gearshift linkage from the transmission.

NOTE: If disconnecting linkage rods from levers that use plastic grommets, the grommets must be replaced with new ones. Use a prying tool to force out the grommets. Cut away the old grommet. Use pliers to snap new grommets into place.

17. Disconnect the throttle rod from the lever at the left side of the transmission. Disconnect all other linkage from the transmission.
18. Remove the oil filler tube.
19. Disconnect and remove the speedometer cable, pinion, and sleeve.
20. Place an engine support under the rear of the engine.
21. Place a transmission jack under the transmission. Position a safety strap or chain around the transmission to keep it from falling. Raise the transmission slightly with the transmission jack to lift the load off the transmission mounts.

CAUTION: The flexplate will not support a load. The weight of the transmission must not rest on the flexplate during removal.

Instructor Check: ____________________

22. Disconnect and remove any emergency brake linkage connected to the crossmember. Remove the bolts holding the transmission mounts to the crossmember and the crossmember to the frame. Remove the crossmember.

NOTE: Some vehicles have a torsion bar anchor crossmember that must remain in place. This requires a careful downward tilt of the front of the transmission as it

is being removed. These vehicles may have a vibration-dampening weight bolted to the rear of the transmission extension housing. Remove the weight to provide additional clearance.

23. Remove all bellhousing bolts or nuts.

SAFETY CAUTION: The transmission is extremely heavy. If not properly supported and secured on the transmission jack, the transmission can fall, causing damage or serious injury. Always work with a helper to assist in removing the transmission from the vehicle.

24. Carefully move the transmission and torque converter assembly rearward and off the engine block dowels. Disengage the converter hub from the end of the crankshaft. Attach a small C-clamp to the edge of the bellhousing to hold the converter in place during transmission removal.

CAUTION: Do *not* leave the converter attached to the engine while removing the transmission. This could cause damage to the input shaft, the converter, and/or the oil pump.

25. Lower the transmission and remove it from under the vehicle. Carefully move the transmission jack and the transmission rearward.
26. Place the transmission in a transmission support or on a workbench.
27. To remove the converter assembly, remove the C-clamp from the edge of the bellhousing. Carefully slide the converter forward and straight out of the transmission. Use extreme care when removing the converter because it is heavy.

Job Completion Report

1. What difficulties, if any, were encountered during transmission removal?

____

____

____

INSTRUCTOR SIGNOFF ____________________ **DATE** __________

NAME ________________ SECTION ________ DATE ________

Certification Practice

Test 14

DIRECTIONS:

The following questions are similar to those used on mechanic certification tests. Answer each question by circling the letter of the correct choice.

1. All of the following may need to be disconnected and removed before removing the transmission EXCEPT
   A. linkage.
   B. crossmember.
   C. driveline.
   D. differential.

2. Mechanic A says that the engine must be supported during transmission removal. Mechanic B says that the transmission must be supported during transmission removal. Who is correct?
   A. A only B. B only C. Both A and B D. Neither A nor B

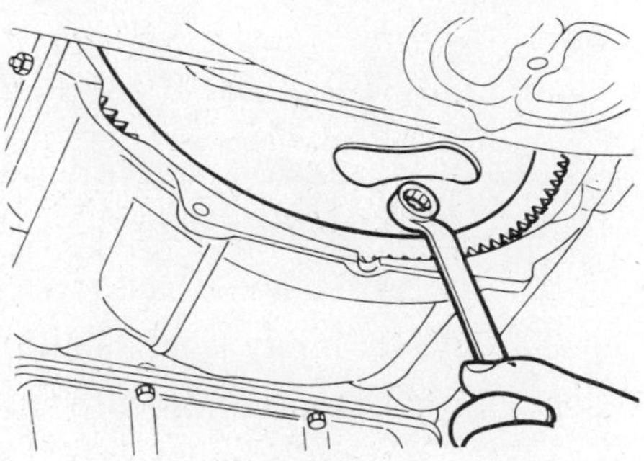

Question 3.
(Chrysler Corporation)

3. The picture shows
   A. crankshaft removal.
   B. flexplate removal.
   C. the flexplate being disconnected from the torque converter.
   D. the crankshaft being disconnected from the torque converter.

4. Which of the following statements is correct?
   I. Torque converters without drains should be replaced during overhaul if contaminated or defective.
   II. All torque converters must be replaced during overhaul.
   A. I only B. II only C. Both I and II D. Neither I nor II

5. Which of the following must be done before removing the transmission?
   A. Drain the radiator.
   B. Disconnect the battery.
   C. Drain the engine oil.
   D. Disable the ignition system.

6. Which must be done before the driveline is removed?
   A. The driveline must be marked for reassembly.
   B. The crossmember must be removed.
   C. The transmission linkage must be disconnected.
   D. The transmission must be removed.

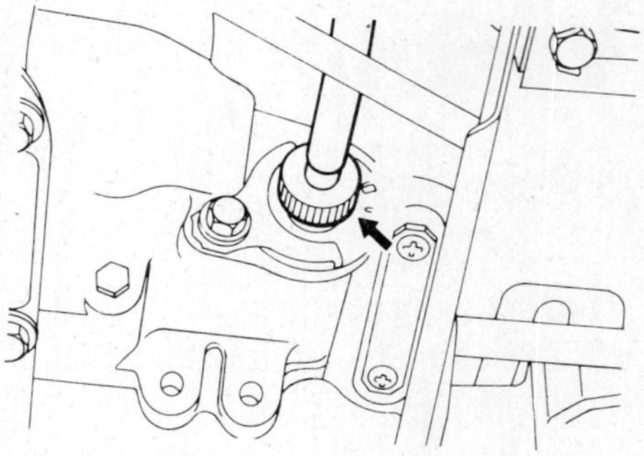

Question 7.
(Chrysler Corporation)

7. The unit shown in the picture is a
   A. transmission cooler line.
   B. speedometer cable.
   C. emergency-brake cable.
   D. pressure gauge hose.

8. Mechanic A says the transmission must be raised on the jack to take the load off the crossmember.
   Mechanic B says the transmission must be raised on the jack to lift the vehicle high enough for transmission removal.
   Who is correct?
   A. A only B. B only C. Both A and B D. Neither A nor B

9. Which of the following statements is correct?
   I. The torque converter must be removed together with the transmission.
   II. The torque converter must remain bolted to the flexplate during transmission removal.
   A. I only B. II only C. Both I and II D. Neither I nor II

10. What is the greatest danger during transmission removal?
    A. Cuts from sharp parts
    B. Burns from hot exhaust system parts
    C. Physical injury from a falling transmission
    D. Physical injury from a rotating driveline

15 Overhauling the Chrysler TorqueFlite A-727

YOUR LEARNING JOB

Transmission overhaul is the most complicated job in this manual. Step-by-step photographs are used to assist you in learning this procedure. Attention to detail is important throughout the overhaul procedure.

JOB COMPETENCY OBJECTIVES

This is a skill-building unit. When you have completed your assignments and exercises in this unit, you should be able to:

- ☐ Disassemble a TorqueFlite transmission.
- ☐ Inspect the transmission parts.
- ☐ Reassemble the TorqueFlite transmission with needed replacement parts.

15.1 TORQUEFLITE DESCRIPTION

Chrysler Corporation vehicles use automatic transmissions that are designated as TorqueFlites. TorqueFlite automatic transmissions differ from one model designation to another. The A-727 covered here is a heavier-duty transmission than the 904, 908, and 998 TorqueFlites that have been used over the years.

The A-727 TorqueFlite has been used in Chrysler Corporation cars and light trucks since 1962. It is also called the Torque-Command transmission by American Motors, which used the A-727 from 1972 to 1978. The A-727 also appears in International light and medium trucks since 1972 under the designation of T-407. The A-727 TorqueFlite consists of the following main parts:

- Torque converter
- Lockup torque converter (since 1978)

- Two multiple disc clutches
- One overrunning clutch
- Aluminum transmission case with integral bellhousing.
- Two servos and bands
- Compound planetary gearset
- Removable valve body

The TorqueFlite combines a torque converter with a fully automatic three-speed gear system. The bellhousing, or converter housing, and the transmission case are a single aluminum casting.

The torque converter has a front cover, impeller, turbine, stator, and stator overrunning clutch. The converter connects to the crankshaft through a flexplate. A lockup converter is used in Chrysler cars from 1978 and in American Motors cars from 1979.

The torque converter lockup has a lockup mechanism in the converter, a lockup module attached to the valve body, and a converter control valve in the valve body.

The compound planetary gear system has a common sun gear. The sun gear is connected to the front clutch by a driving shell. The driving shell is splined to the sun gear and to the front clutch retainer.

The hydraulic system has a single oil pump. The valve body contains all of the valves except the governor.

15.2 TRANSMISSION IDENTIFICATION

The TorqueFlite service identification number is located at the oil-pan flange on the right side of the transmission case. The service identification number has seven digits. A transmission identification number also appears on the right side at the bellhousing. Only the service identification number will give you information about the transmission. Use the proper service manual to decipher the service identification code.

15.3 BEFORE THE OVERHAUL

Be sure you are familiar with all of the assemblies that make up the automatic transmission. These assemblies are discussed in Parts II, III, and IV. This unit discusses the disassembly and reassembly of the TorqueFlite and Torque Command transmissions. Refer to the earlier units for teardown, inspection, and testing of the torque converter, planetary gearsets, apply devices, and the valve body.

Transmission specifications and parts are changed continually. For this reason, always refer to the proper service manual when performing any transmission overhaul.

Before proceeding with the overhaul, make sure that the transmission fluid has been drained. Plug all openings. Clean the transmission case, or housing, with a steam-cleaning machine. If a steam-cleaning machine is not available, use solvent, a wire brush, and shop towels. Do *not* wipe the inside of the transmission case with shop towels. Allow it to air dry. Any lint left inside the transmission case can cause problems in the transmission valves. Many times this lint can be blown out with dry compressed air.

After the transmission has been thoroughly cleaned, inspect the case visually for any signs of damage. A cracked or distorted transmission case can obviously cause transmission problems.

To better understand the parts and their locations in the transmission, refer to Figures 15-1 through 15-8.

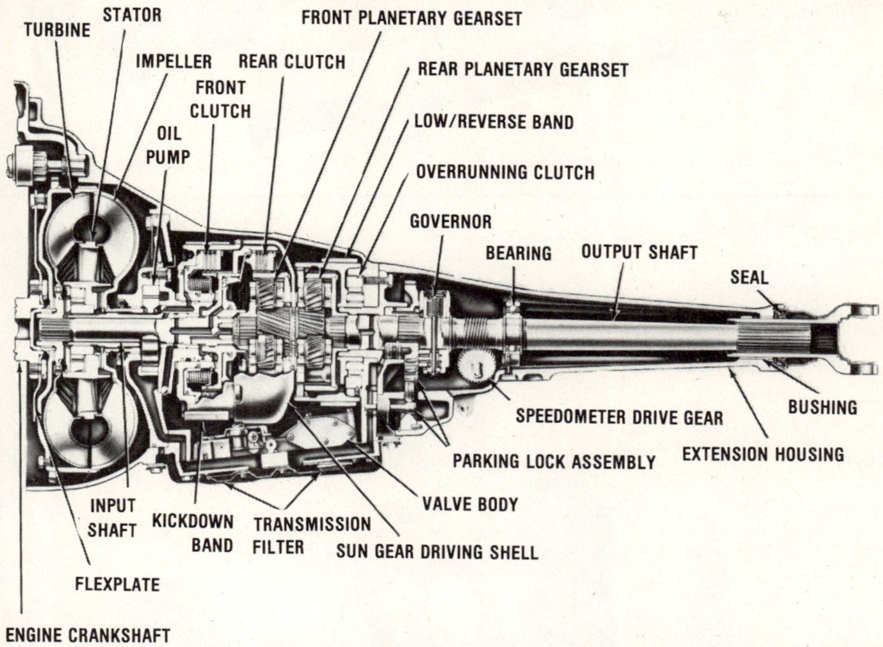

Figure 15-1. TorqueFlite transmission cutaway. (Chrysler Corporation)

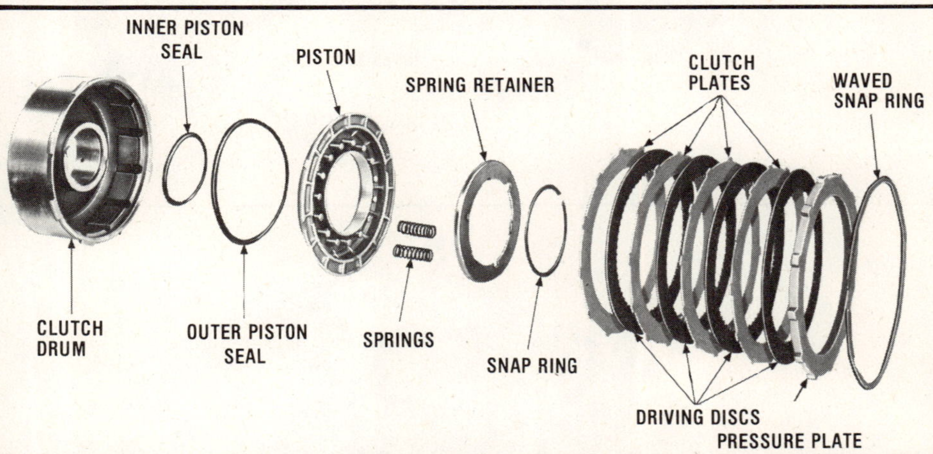

Figure 15-2. Front clutch disassembled. (Chrysler Corporation)

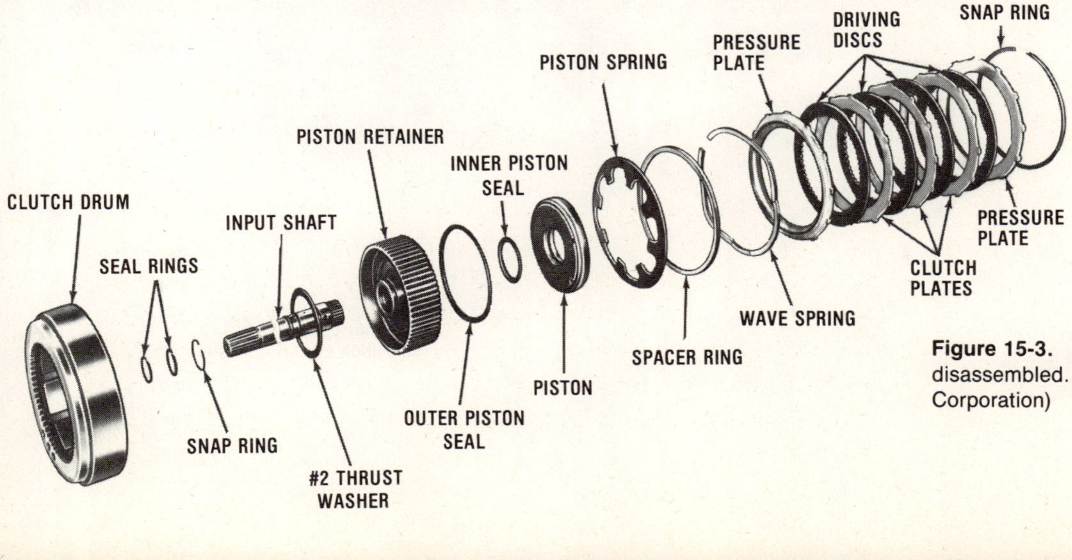

Figure 15-3. Rear clutch disassembled. (Chrysler Corporation)

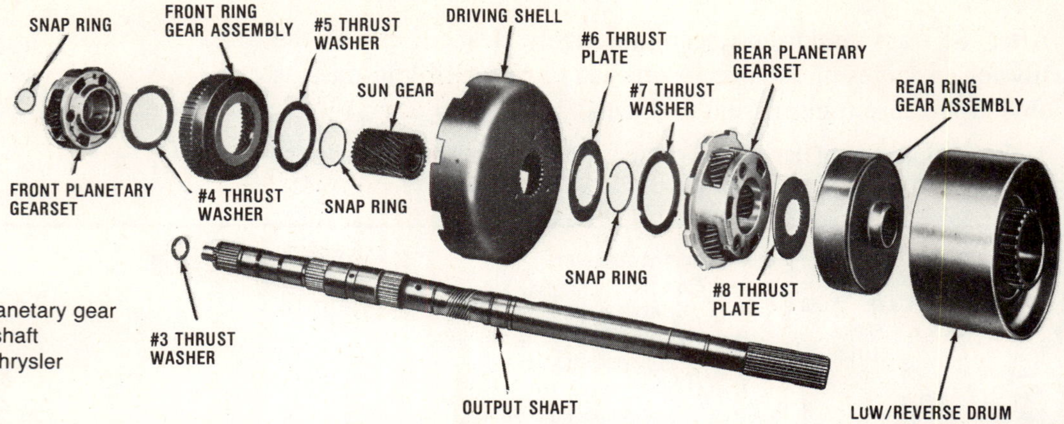

Figure 15-4. Planetary gear train and output shaft disassembled. (Chrysler Corporation)

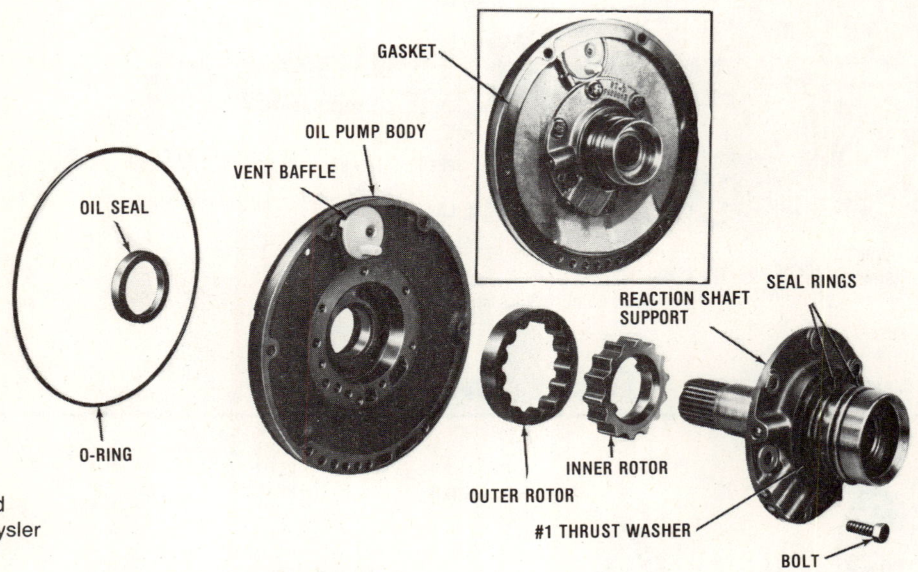

Figure 15-5. Oil pump and reaction shaft support. (Chrysler Corporation)

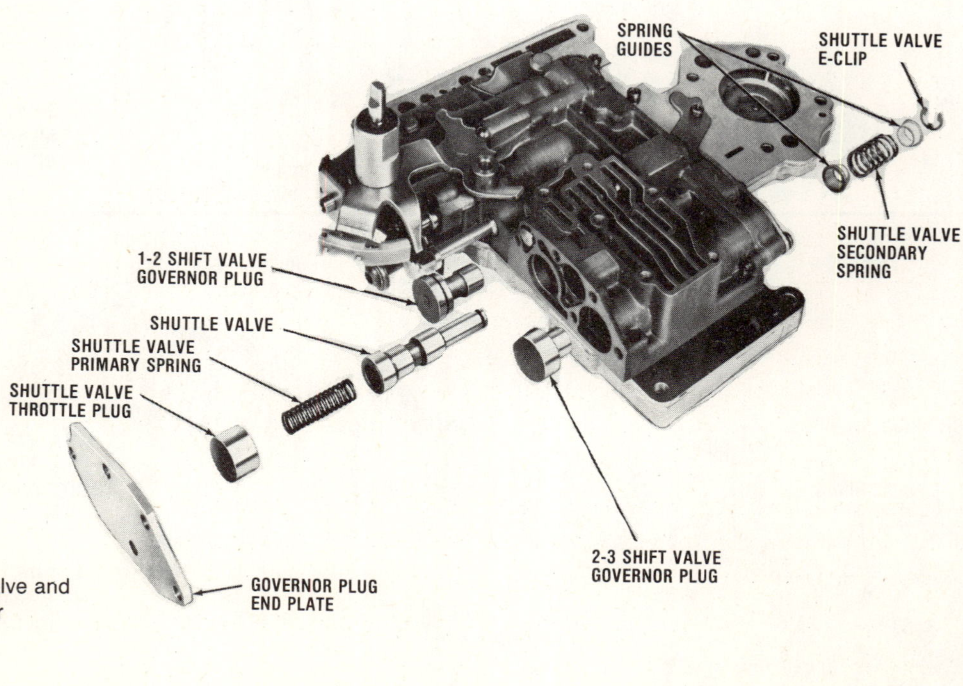

Figure 15-6. Shuttle valve and governor plugs. (Chrysler Corporation)

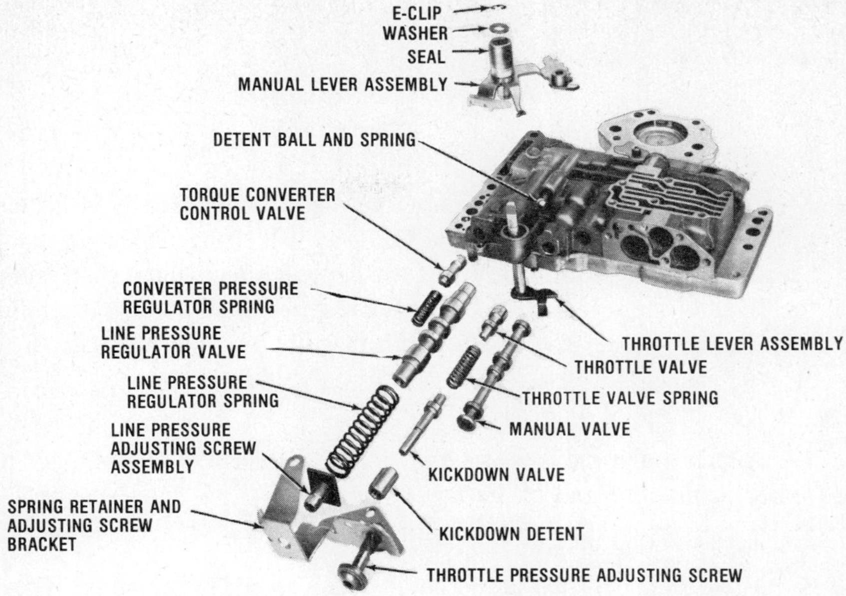

Figure 15-7. Pressure regulators and manual control. (Chrysler Corporation)

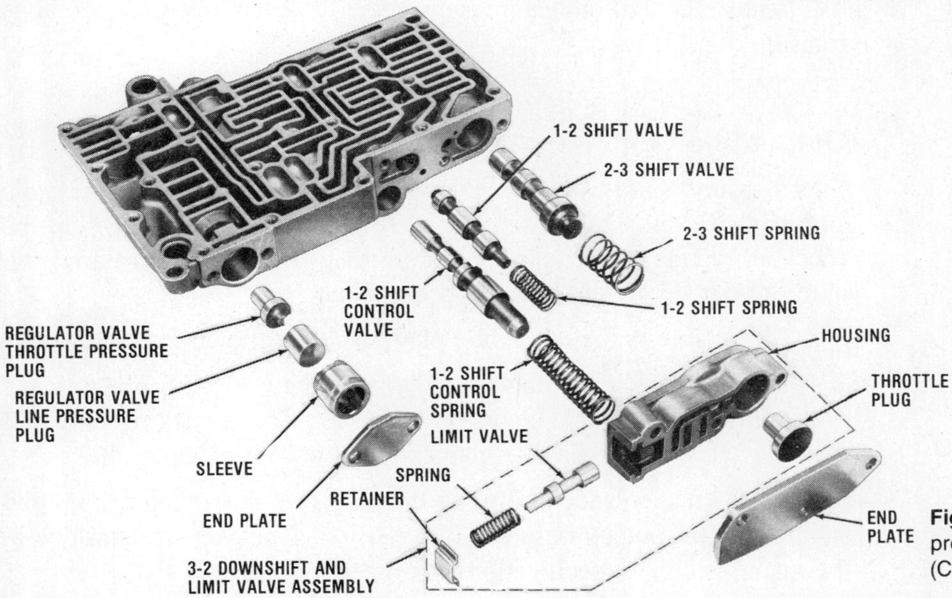

Figure 15-8. Shift valves and pressure regulator valve plugs. (Chrysler Corporation)

15.4 MEASURING END PLAY

End play is the front-to-rear movement along the centerline of a shaft. Each manufacturer has different input and output shaft end play specifications, so always refer to the proper service manual. Some manufacturers recommend simply checking and setting input shaft end play.

If there is not enough end play in the automatic transmission, the internal parts are too close together. This results in the transmission grabbing or locking up. If there is too much end play, the internal parts will rotate in a sloppy, or loose, manner. This causes hard shifts, noises, and excessive and uneven wear of the transmission parts.

End play measurements indicate whether thrust washers, bearings, or spacers need to be replaced with ones of a different size. Wear of these parts also changes end play.

Measurements for TorqueFlite end play are made at the input shaft and at the output shaft.

Figure 15-9. Measuring input end play with dial indicator. (Chrysler Corporation)

Input Shaft End Play Measurement

Measuring input shaft end play before disassembly usually indicates that a thrust washer change is necessary. End play will change when major transmission parts are replaced. The thrust washer that affects input shaft end play on the TorqueFlite is located between the reaction shaft and the front clutch retainer. The retainer shaft is the shaft that fits over the input shaft.

To measure input shaft end play, use the following procedure:

1. Attach a dial indicator as shown in Figure 15-9. The plunger must be seated against the end of the input shaft.

2. Zero the dial indicator.

3. Move the input shaft in and out to obtain an end play reading. The reading must be within the manufacturer's specifications.

4. Record the dial indicator reading for reference when reassembling the transmission.

Output Shaft End Play Measurement

This measurement is taken during disassembly of the planetaries, sun gear, and driving shell. Do *not* remove these parts from the output shaft before checking end play. Overhaul Steps 11 through 14 should not be disassembled until these end play measurements are taken.

To measure output shaft end play, use the following procedure:

1. Stand the assembly upright, as shown in Figure 15-10. The forward end of the output shaft should be on a wood block. This allows all parts to move forward against the selective snap ring at the front of the shaft.

2. Insert a feeler gauge between the rear ring-gear support hub and the shoulder on the output shaft. See Figure 15-10. The clearance must be within the manufacturer's specifications.

3. If the clearance is above the specifications, replace the thrust washers and/or any other necessary parts.

15.5 CLEANING AND INSPECTION

When disassembling and inspecting the parts of the TorqueFlite transmission, all parts should be cleaned in solvent as the disassembly procedure is performed. Wear safety glasses. Dry the parts with compressed air or allow them to air dry. *Never* use a shop towel to wipe off the parts. Lint can get into the transmission passages and cause valves to stick.

Wash all bearings in solvent and air dry. Rotate the bearings and listen carefully for noise and feel for tightness or roughness. Do *not* spin the dry bearings with compressed air. Lubricate the bearings with petroleum jelly or transmission fluid.

Check all needle bearings and needle-bearing bushings for signs of wear or roughness. If mating bearing or bushing parts need replacing, always replace both parts.

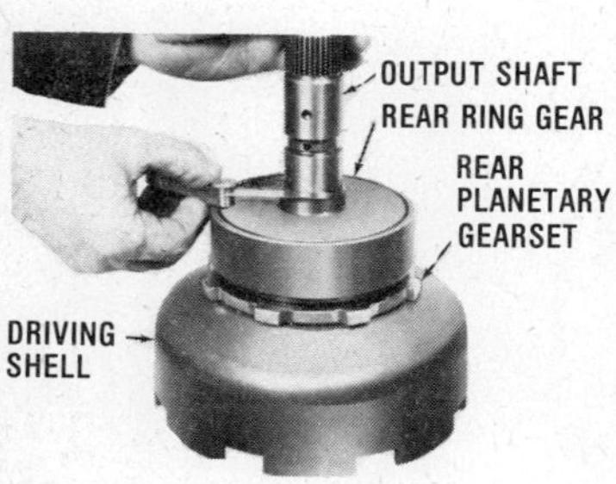

Figure 15-10. Measuring planetary gear train end play. (Chrysler Corporation)

Keep all clutch parts together with their respective clutch assemblies. Store front clutch parts separately from rear clutch parts. Do *not* mix them together.

Check the planetary gearsets carefully for signs of wear or chipped teeth. Revolve each planetary system to listen for noise or feel for roughness. Planetaries must be replaced as units.

Check all parts of the transmission for wear or damage. Replace parts when necessary.

Remove all traces of old gaskets. Do not damage sealing surfaces when scraping off old gaskets.

Clean all sealing surfaces with mineral spirits or carbon tetrachloride.

Lubricate each transmission part, after inspection, with petroleum jelly or transmission fluid.

Replace all oil seals and gaskets with new parts. Lubricate the lips of new seals with petroleum jelly.

Replace damaged snap rings and thrust washers with new parts.

15.6 TRANSMISSION OVERHAUL

1. Place the transmission on a transmission stand. Check and record the input shaft end play. Unbolt the oil pan. Remove the oil pan and gasket. Discard the gasket.

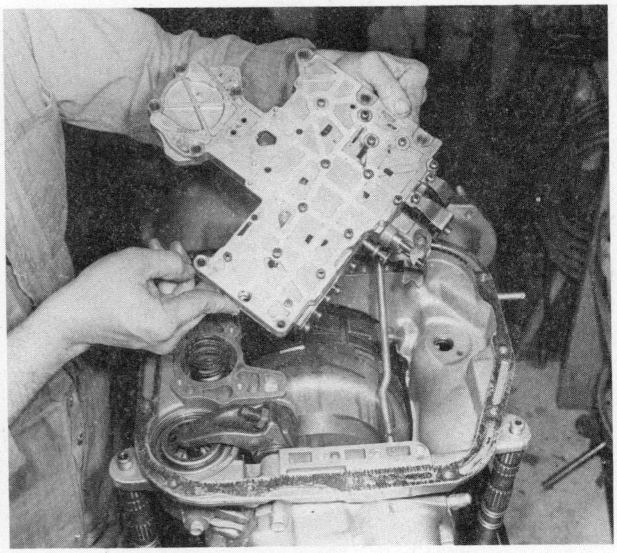

2. Remove and discard the fluid filter. Unbolt the valve body. Disconnect the parking lock rod from the manual lever. Remove the valve body from the transmission case.

3. Remove the spring from the accumulator piston. Remove the accumulator piston. On some versions, the spring will be under the piston.

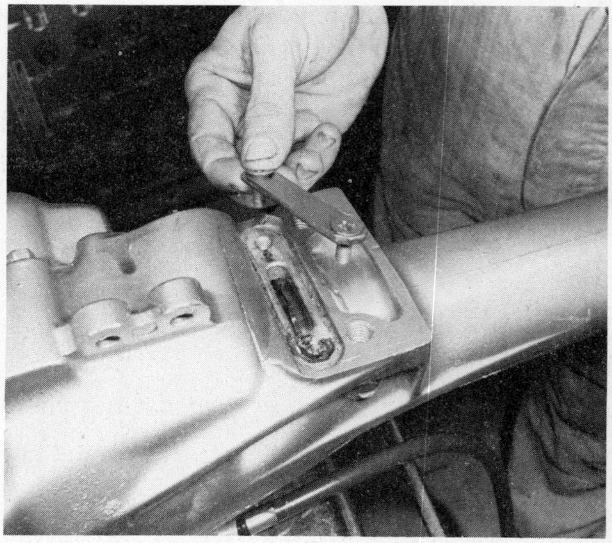

4. Remove the parking lock rod by pulling it forward and out of the transmission case. Unscrew and remove the plate (shown) from the bottom of the extension housing.

5. Remove the bolts that hold the extension housing to the transmission case. Use snap ring pliers to spread the snap ring holding the output shaft bearing. Hold the snap ring open and pull off the extension housing. It may be necessary to tap or wiggle the housing to pull the housing off the output shaft and bearing.

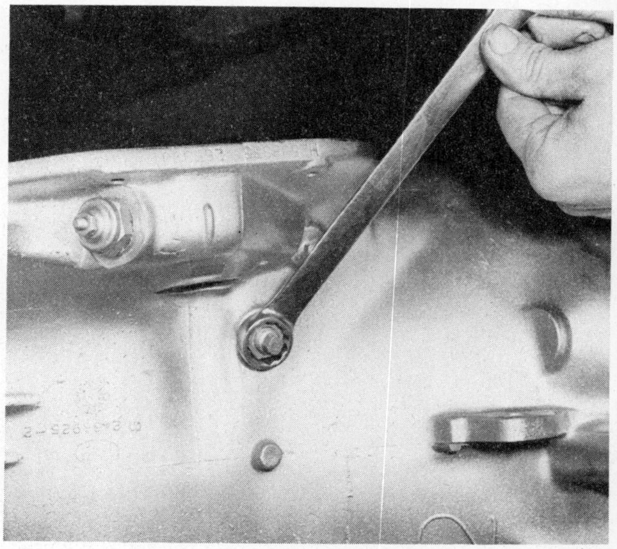

6. Tighten the adjusting screw on the front, or intermediate, band. When the band is tight, it prevents damage to the clutch discs when the oil pump is removed.

7. Remove the bolts that hold the oil pump housing in place. Use a slide hammer screwed into the threaded holes, as shown, to remove the pump housing.

8. When the pump housing is free, pull out on the housing to remove the pump and stator assembly.

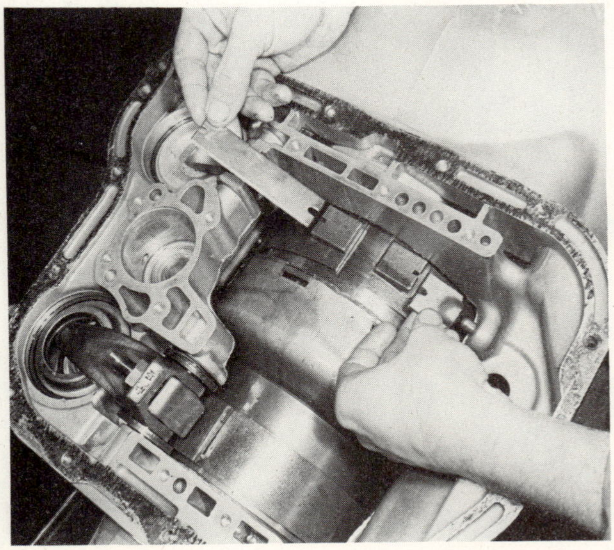

9. Loosen the adjusting screw on the front band. The adjusting screw is shown in step 6. Remove the band strut and anchor (shown).

10. Remove the front and rear clutch assemblies. Hold the clutch assemblies with one hand and pull straight out on the input shaft. The clutch assemblies are heavy, so use extreme care.

11. A thrust washer is located just behind the rear clutch assembly. It may come out when the clutches are removed. *Don't* lose it. Remove the snap ring on the output shaft.

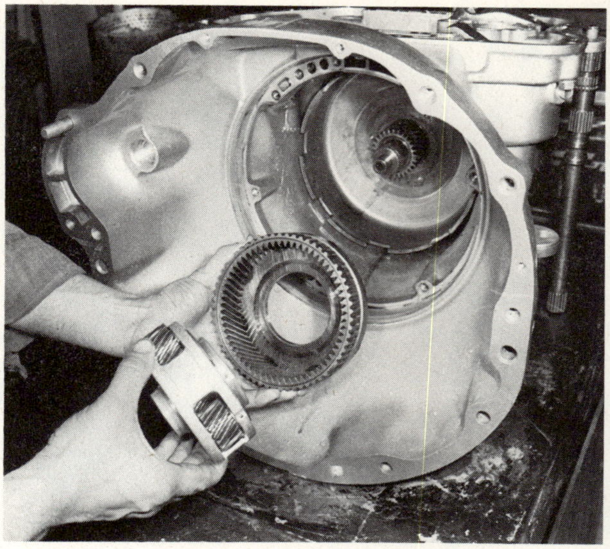

12. Remove the front planetary carrier and the ring gear. Chrysler calls the ring gear an annulus gear. Hold the driving shell while removing the planetary parts so that the driving shell remains in the transmission.

13. Remove the driving shell with the sun gear attached. Remove the thrust washer behind the driving shell.

14. Remove the snap ring that holds the rear planetary gearset in place. Remove the planetary carrier, ring gear, and low/reverse drum.

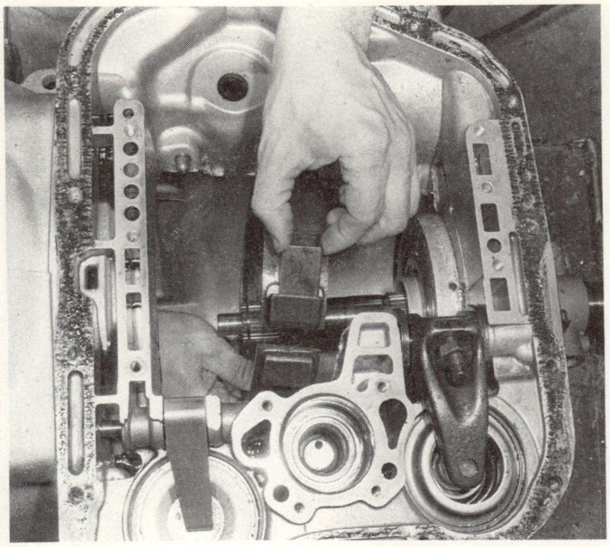

15. Loosen the rear, or low/reverse, band adjusting screw. Remove the band strut and anchor. Pull out the band from the front of the transmission case.

16. Remove the output shaft from the rear of the transmission case. The governor assembly will come out with the output shaft. Do *not* remove the output shaft support unless the clutch cam is to be replaced.

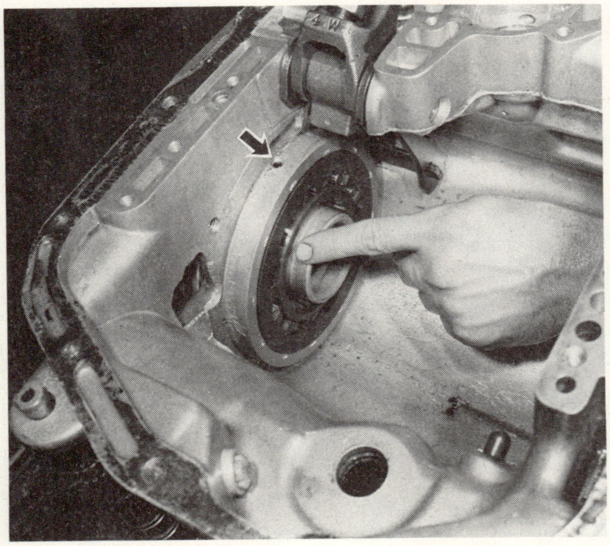

17. Check the overrunning clutch cam for damage. If it needs replacing, loosen the setscrew (arrow) and the output shaft support. Drive the cam out of the case.

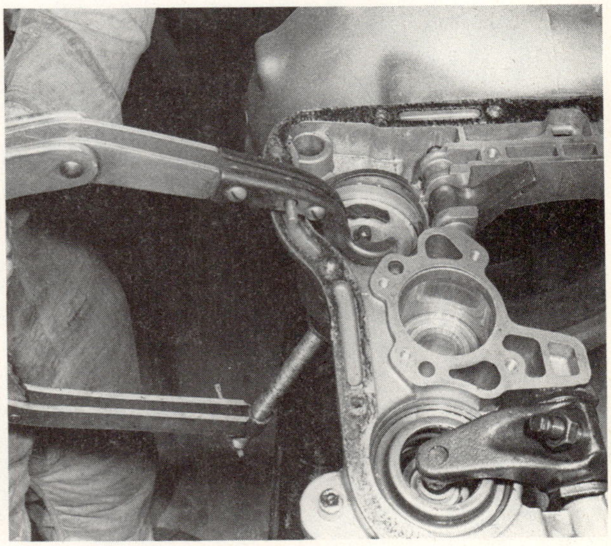

18. Hold down the front, or kickdown, servo with a valve spring compressor. Remove the snap ring from the bore. Loosen the compressor slowly and carefully. The spring pressure can cause the piston and spring to fly out and cause serious injury.

19. Remove the piston rod guide, springs, and piston rod. Be careful not to damage the rod or guide. Remove the piston. All seals must be replaced with new ones.

20. The rear, or low/reverse, servo can be removed without a special tool. Compress and remove the snap ring, spring retainer, servo piston, and plug assembly.

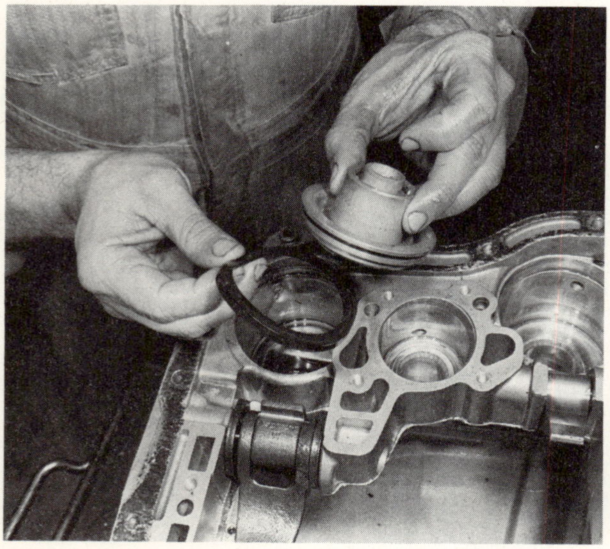

21. Remove and replace the rear servo piston seal ring with a new one. If the piston is to be disassembled, remove the snap ring at the bottom. This frees the plug from the piston body.

22. To reassemble the rear servo, position the piston and plug assembly in the bore. Place the piston, spring, and retainer in the bore. Depress the servo assembly while installing the snap ring.

23. The controlled load, or kickdown, front servo has a seal ring that attaches to the piston by the piston rod. Replace the three seals in the front servo with new ones. Assemble the front servo.

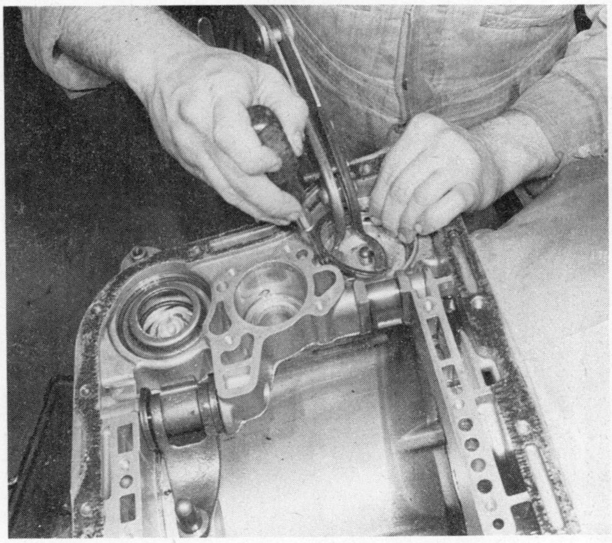

24. Place the front servo assembly into its bore. Hold the assembly in place with a valve spring compressor. Install the snap ring.

25. Inspect the rear ring gear, the #8 thrust washer, and the rear planetary gearset for wear and damage. The planetary gearset must be replaced as a unit.

26. Inspect the driving shell for wear or damage, especially the sun gear and bushings. Bushings can be pressed out and replaced. A bad sun gear requires that the entire planetary set be replaced. Check the drive shell, as shown, for signs of poor engagement.

27. Inspect both bands for signs of wear, cracks, burning, scoring, or other defects. However slight the imperfections, it is always best to replace the bands.

28. Inspect the thrust faces, as shown, of both planetary carriers for wear, cracks, scoring, or other damage. Inspect the planet gears for broken or worn teeth. It's also a good idea to clean out the valleys between the gear teeth. Contaminants will build up in these areas and create gear noise. If any damage is evident, replace the entire gearset.

29. Remove the oil seal rings on the governor housing and install new rings. The new rings can be broken or damaged easily, so use care when replacing them.

30. Pry out the C-clip from the end of the governor shaft. Remove the valve from the opposite side of the shaft.

31. After the governor valve has been removed, pressure is released on the snap ring that holds the governor weights in place. Remove the snap ring and governor weights.

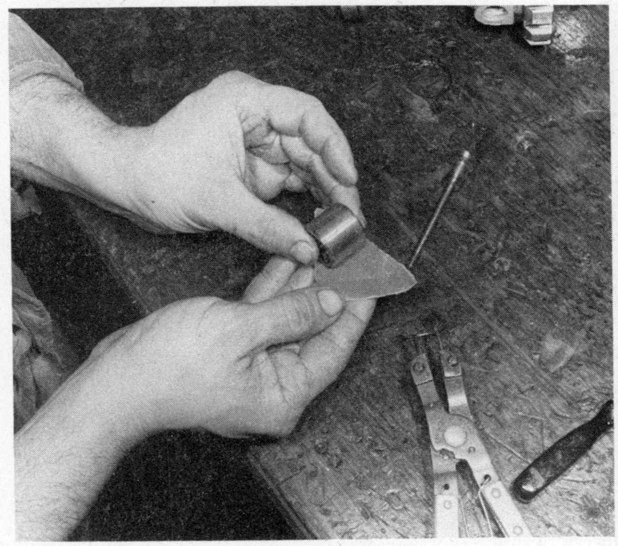

32. Inspect the governor weights for nicks, burrs, scoring, and other damage. Light scratches or small nicks and burrs can be removed by carefully sanding with crocus cloth. Replace any weights that are badly damaged.

33. Replace the weights and snap ring in the governor assembly. Turn the governor assembly over. Insert and seat the valve. Install the C-clip.

34. The rear bearing on the output shaft is held in place by snap rings on both sides of the bearing. Remove the rear snap ring only.

35. The front snap ring does not have to be removed. Pull the rear bearing off from the rear of the output shaft. Inspect the bearing for damage. Clean and lubricate the bearing or replace it with a new one if it is damaged.

36. Disassemble the front clutch pack by prying out the snap ring. Remove the pressure plate. Inspect the clutch plates. The clutch plate shown here has been scored and burned, and must be replaced.

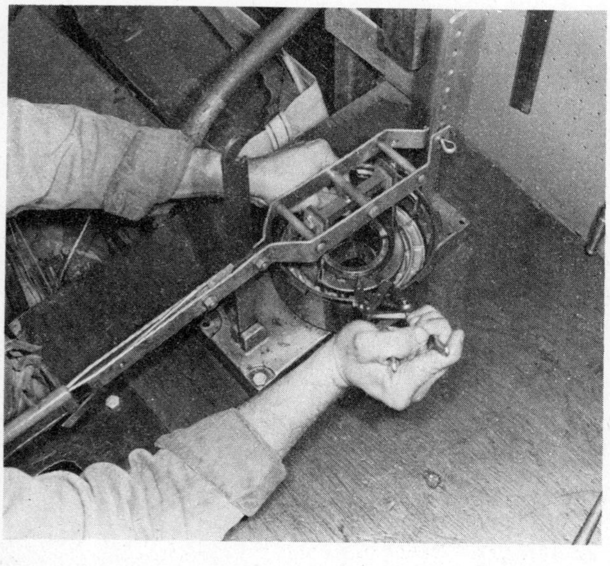

37. Use a special compressor tool to compress the front, or high/reverse, clutch springs. Remove the snap ring from the clutch hub. Release the compressor tool carefully, because the clutch pack is under high pressure. Clutch parts can fly loose and cause serious injury if not released properly.

38. Remove the spring retainer and inspect it for damage. The number of springs located under the spring retainer can vary from one vehicle to another. Replace any broken springs.

39. Remove the clutch piston. Remove the outer seal on the piston and the inner seal on the retainer. Replace both seals with new ones.

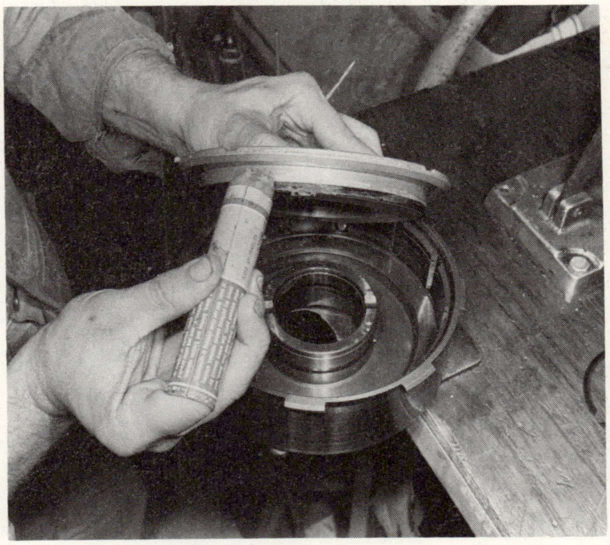

40. Lubricate the new seals with petroleum jelly, transmission fluid, or wax-type lubricant (shown). This allows easier piston replacement in the clutch drum without damage to the seals.

41. Install and seat the piston. Turn the piston to make sure it rotates freely. Place the springs on the piston shafts.

42. Using the special compressor tool, compress the spring retainer and springs. Install the snap ring. Install the clutch pack by starting with a friction plate, then a steel plate. Soak the friction and steel plates in transmission fluid for at least 15 minutes before installation. Install the pressure plate and snap ring. Check clearances against the manufacturer's specifications.

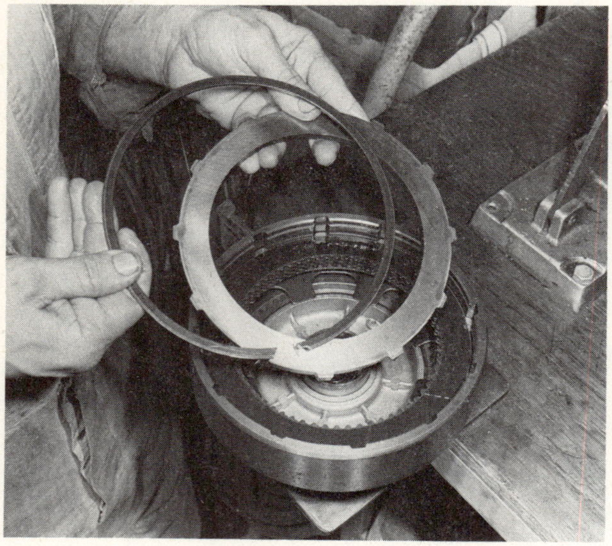

43. For the rear, or forward, clutch assembly, remove the large snap ring holding the pressure plate in the rear clutch retainer. Remove the friction and steel plates and the inner plate. Inspect the clutch assembly for damage or wear.

44. Pry out the wavy spring from the rear of the clutch drum. Remove the spacer ring and Belleville spring. Inspect these parts for wear and damage.

45. Push down on the piston retainer and remove the snap ring. Remove and replace the inner and outer piston seals with new ones. Lubricate the edges of the seals.

46. Use a feeler gauge or special seal tool, shown, to prevent damage to new seals. Press on the piston to seat it in the retainer.

47. Install the Belleville spring, spacer ring, and wavy spring. Install the clutch pack, beginning with a friction plate, then a steel plate. The inner lip of the pressure plate must face forward.

48. Turn the clutch drum over. Replace the two oil seal rings, shown, with new ones. Install the thrust washer. Set the assembly aside.

49. Remove the bolts that hold the stator support to the oil pump housing. Carefully lift the stator support off the oil pump assembly.

50. Apply a small amount of bluing on the inner and outer rotors of the oil pump. Scribe a mark across both rotors as a guide for correct tooth alignment when reassembling.

51. Inspect the inner and outer surfaces, as shown, for scoring or pitting. Check the rotor teeth for wear and damage. Slight nicks and burrs can be removed from machined surfaces with crocus cloth.

52. Use the scribed marks to install and align the teeth correctly. Lay a straightedge across the rotors and the pump. Place a feeler gauge under the straightedge. The clearance should be within the manufacturer's specifications.

53. Clearance must be measured between the rotor tip and the inner and outer teeth on the pump. Also check clearance between the rotor and the pump bore. If clearances are not within specifications, replace the oil pump assembly. Install new seals, shown, on the inner hub.

54. Place the stator support on the oil pump body. Install the attaching bolts finger-tight. Torque the bolts in crisscross sequence to specifications.

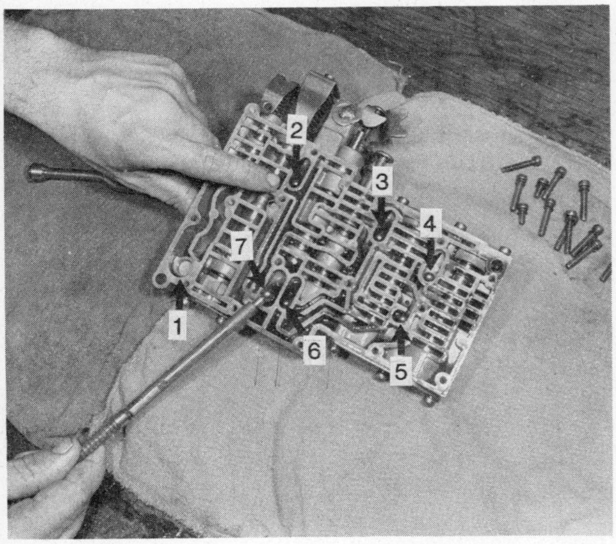

55. Remove the transfer plate and separator plate to disassemble the valve body. Remove the seven check balls and one spring from the valve body locations numbered in photo. Location 7 contains the largest check ball and the spring.

56. Disassemble and clean the valve body parts in clean solvent. Air dry on clean shop towels. Disassembly of valve body is discussed in Unit 12.

57. Lubricate the new oil rings on the output shaft support. Install the output shaft and governor assembly into the transmission case. Work carefully so that the oil seal rings are not damaged.

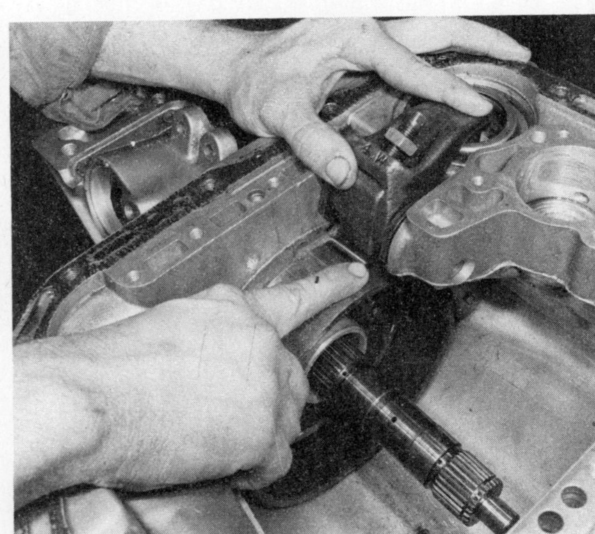

58. Install the rear, or low/reverse, band in the transmission case. Install the short strut, the link, and the anchor to the band. Tighten the adjusting screw to hold the assembly in place.

59. Place the rear planetary ring gear into the low/reverse drum. Install the assembly onto the output shaft. Install the #8 thrust washer (arrow). Install the rear planetary carrier.

60. Lubricate the #7 thrust washer with petroleum jelly to hold the washer in place. Install the washer.

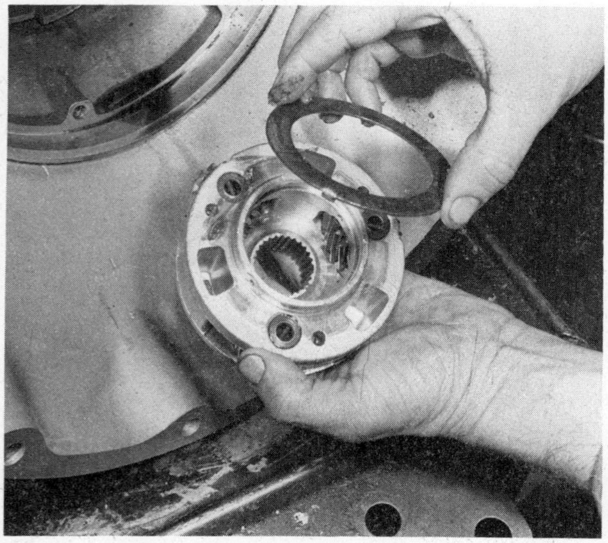

61. Fit the #4 thrust washer to the rear of the front planetary carrier. Place the carrier inside of the ring gear and over the output shaft. Install the snap ring and the #3 thrust washer.

62. Install the front, high/reverse, clutch assembly so that it engages with the rear clutch assembly. The rear clutch splines engage with the front ring gear splines. Check for proper installation by rotating the units, which should turn as one.

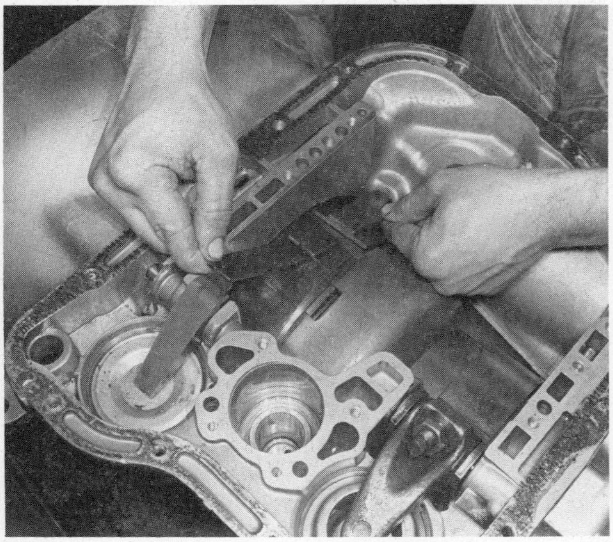

63. Make sure that the front clutch drive lugs engage the slots in the input shell. Install the front, or kickdown, band. Install the strut, link, and anchor. Tighten the adjusting screw to hold the assembly.

64. Install new O-rings on the accumulator piston. Place the piston and spring in the accumulator bore.

65. Adjust the low/reverse band by backing off the adjusting screw locknut five turns. Torque the screw to specifications. Back off on the screw according to specifications. Torque the locknut to specifications.

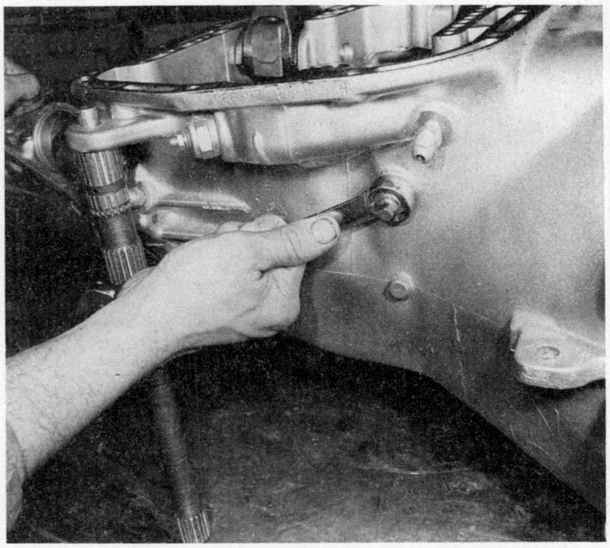

66. Adjust the kickdown band by backing off the adjusting screw locknut five turns. Torque the screw to specifications. Back off on the screw according to specifications. Torque the locknut to specifications.

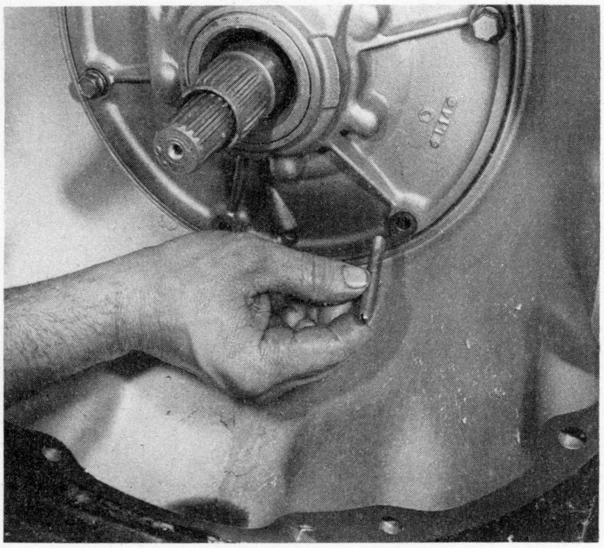

67. Place the oil pump assembly in position. Install two studs, as shown, to align the pump. Install the attaching bolts and torque to specifications.

68. If the oil pump seal requires replacement, use a seal puller to remove, as shown. Install a new seal with the lip facing inward.

69. Move the manual lever until the parking rod is at the rear of the transmission. Install the valve body while working the parking rod into the proper position. Torque the valve body attaching bolts to specifications from the center to the outside so the valve body won't warp. Reconnect the parking lock rod to the manual valve.

70. Install a new fluid filter. Install the oil pan and new gasket. Torque the pan bolts to specifications. The overhaul is completed.

NAME ______________________ SECTION ____________ DATE ____________

Job Sheet 15-1

Overhauling the Chrysler 727 TorqueFlite

Before you begin:

Read Topics 15.1 through 15.5.

Vehicle Make ____________________ Model ____________ Year ____________

Time Started ____________________ Time Finished ____________________

Job Completion Time ______________ Flat Rate Time ____________________

Tools Required

- Transmission stand
- Dial indicator and base
- Socket wrenches, extensions, handle
- Phillips screwdrivers
- Slotted screwdrivers
- Snap-ring pliers
- Safety glasses
- Engine valve-spring compressor
- Brass drift punch
- Soft-faced mallet
- Hydraulic press and support fixtures
- Crocus cloth
- Special tool to compress front clutch retainer
- Feeler gauges
- Torque wrench (lb.-inch) [Nm]
- Torque wrench (lb.-foot) [Nm]
- Magnet
- Seal puller

References

Manufacturer's service manual ______________________________

Repair manual ______________________________

Service bulletin ______________________________

Procedure

1. Attach a dial indicator to measure input shaft end play.
2. Zero the dial indicator.

In the spaces provided, write the manufacturer's specification and measured end play.

Manufacturer's Specified Input Shaft End Play .... ____________

Measured End Play .... ____________

Instructor Check: ____________

3. Unbolt the oil pan. Remove the oil pan and gasket. Discard the gasket.
4. Remove and discard the fluid filter. Unbolt the valve body. Disconnect the parking lock rod from the manual lever. Remove the valve body from the transmission case.
5. Remove the spring from the accumulator piston. Remove the accumulator piston.
6. Remove the parking lock rod. Unscrew and remove the plate from the bottom of the extension housing.
7. Remove the bolts that hold the extension housing to the transmission case. Open the snap ring holding the output shaft bearing and remove the extension housing.
8. Tighten the adjusting screw on the front, or intermediate, band.
9. Remove the bolts that hold the oil pump housing in place. Remove the pump housing.
10. Remove the pump and stator assembly.
11. Loosen the adjusting screw on the front band. Remove the band strut and anchor.
12. Remove the front and rear clutch assemblies.

SAFETY CAUTION: Use extreme care when removing the heavy clutch assemblies.

NOTE: A thrust washer is located just behind the rear clutch assembly. It may come out when the clutches are removed. Retain the thrust washer.

13. Stand the assembly upright. Place the forward end of the output shaft on a wood block.
14. Measure output-shaft end play between the rear ring-gear support hub and the shoulder on the output shaft.

In the spaces provided, write the manufacturer's specification and measured end play.

Manufacturer's Specified Output Shaft End Play ....________________

Measured End Play ....________________

Instructor Check: ________________

If output shaft end play is incorrect, replace thrust washers and other parts as necessary.

15. Remove the snap ring on the output shaft.
16. Remove the front planetary carrier and the ring gear.
17. Remove the driving shell with the sun gear attached. Remove the thrust washer behind the driving shell.
18. Remove the snap ring that holds the rear planetary gearset in place. Remove the planetary carrier, ring gear, and low/reverse drum.
19. Loosen the rear, or low/reverse, band adjusting screw. Remove the band strut and anchor. Pull out the band from the front of the transmission case.
20. Remove the output shaft and governor assembly from the rear of the transmission case.

NOTE: Do not remove the output shaft support unless the clutch cam is to be replaced.

21. If the overrunning clutch is to be replaced, loosen the setscrew and the output shaft support. Drive the assembly from the case.

22. Hold the front, or kickdown, servo down with a valve spring compressor. Remove the snap ring from the bore. Loosen the compressor slowly.

SAFETY CAUTION: Wear safety glasses. Loosen the compressor slowly and carefully. The spring pressure can cause the piston and spring to fly out and cause serious injury.

23. Remove the piston rod guide, springs, and piston rod. Remove the piston. Install new seals.
24. Remove the snap ring, spring retainer, servo piston, and plug assembly.
25. Replace the rear servo piston seal ring. If the piston is to be disassembled, remove the snap ring at the bottom.
26. Position the rear servo piston and plug assembly in the bore. Place the spring and retainer in the bore. Depress the servo assembly and install the snap ring.
27. Install three new seals in the front servo. Assemble the front servo.
28. Install the front servo assembly into its bore. Hold the assembly in place with a valve spring compressor. Install the snap ring.
29. Inspect the rear ring gear, the #8 thrust washer, and the rear planetary gearset. If wear or damage is seen on any part, replace the entire assembly.
30. Inspect the driving shell, sun gear, and bushings for wear or damage. If the sun gear is worn or damaged, replace the entire assembly. If a bushing is bad, press out the bushing and replace it.
31. Inspect both bands for signs of wear, cracks, burning, scoring, or other defects. Replace the bands if defects are found.
32. Inspect the thrust faces of both planetary carriers for wear, cracks, scoring, or other damage. Inspect the planet gears for broken or worn teeth. Clean the valleys between the gear teeth. If any damage is evident, replace the entire gearset.
33. Install new oil seal rings on the governor housing.

CAUTION: New rings can be broken or damaged easily, so use care when replacing them.

34. Remove the C-clip from the end of the governor shaft. Remove the valve from the opposite side of the shaft.
35. Remove the snap ring and governor weights.
36. Inspect the governor weights for nicks, burrs, scoring, and other damage. Use crocus cloth to carefully sand the weights. Remove light scratches, small nicks, or burrs. Replace any weights that are badly damaged.
37. Replace the governor weights and snap ring in the governor assembly. Turn the governor assembly over. Insert and seat the valve. Install the C-clip.
38. Remove the snap ring behind the rear output shaft bearing.
39. Pull the rear bearing off from the rear of the output shaft. Inspect the bearing. If damaged, replace it with a new bearing. If not damaged, clean and lubricate the old bearing.
40. Remove the snap ring and disassemble the front clutch pack. Remove the pressure plate. Inspect the clutch plates. Install new clutch plates if damage is seen.

41. Use the special tool to compress the front, or high/reverse, clutch springs. Remove the snap ring from the clutch hub. Carefully release the compressor tool.

SAFETY CAUTION: The clutch pack is under high pressure. Always wear safety glasses. Clutch parts can fly loose and cause serious injury if the tool is released suddenly.

42. Remove and inspect the clutch pack spring retainer. Replace it if damaged. Replace any broken springs.
43. Remove the front clutch piston. Remove the outer seal on the piston and the inner seal on the retainer. Install new seals.
44. Lubricate the new seals.
45. Install and seat the front clutch piston. Turn the piston to make sure it rotates freely. Place the springs on the piston shafts.
46. Compress the front clutch spring retainer and springs with the special tool. Install the snap ring. Soak the friction clutch discs in transmission fluid for at least 15 minutes before installation. Install the discs in this order: friction first, then steel, etc. Install the pressure plate and snap ring. Measure clearance.

In the spaces provided, write the manufacturer's specification and the measured clearance.

Manufacturer's Specified Clearance .... ____________________

Measured Clearance .... ____________________.

47. Remove the large snap ring (selective fit) holding the pressure plate in the rear clutch retainer. Remove the clutch discs and the inner plate. Inspect the clutch assembly. Replace any damaged friction or steel clutch plates.
48. Pry out the wavy spring from the rear clutch drum. Remove the spacer ring and Belleville spring. Inspect the parts. Replace any damaged parts.
49. Push down on the piston retainer and remove the snap ring. Install new inner and outer piston seals. Lubricate the edges of the seals with petroleum jelly.
50. Use a feeler gauge or special seal tool, shown, to prevent damage to new seals as the piston is installed. Press on the rear clutch piston to seat it in the retainer.
51. Install the Belleville spring, spacer ring, and wavy spring in the rear clutch. Soak the friction clutch discs in transmission fluid for at least 15 minutes. Install the clutch pack, beginning with a friction, then a steel. Install pressure plate so that the inner lip faces downward.
52. Turn the rear clutch drum over. Install two new oil seal rings. Install the thrust washer. Set the assembly aside.
53. Remove the bolts that hold the stator support to the oil pump housing. Carefully lift the stator support off the oil pump assembly.
54. Apply Prussian blue and gently mark both rotors of the oil pump.
55. Inspect the inner and outer surfaces of the oil pump rotors for scoring or pitting. Check the rotor teeth for wear and damage. Sand with crocus cloth to remove slight nicks and burrs. Replace both rotors if defects are found.
56. Align the rotor teeth correctly. Measure the clearance between the rotors and the pump.

In the spaces provided, write the clearances.

Manufacturer's Specified Clearance .... ____________________

NAME ____________________ SECTION ____________ DATE ____________

Measured Clearance ....____________

57. Measure the clearance between the rotor tip and the inner and outer teeth on the pump. In the spaces provided, write the clearances.

Measured Clearance ....____________

Manufacturer's Specified Clearance ....____________

Measure the clearance between the rotor and the pump bore.

Measured Clearance ....____________

Manufacturer's Specificied Clearance ....____________

Instructor Check: ____________

If clearances are not within specifications, replace the oil pump assembly. Install new seals on the inner hub.

58. Place the stator support on the oil pump body. Install the attaching bolts finger-tight. Torque the bolts in crisscross sequence to the manufacturer's specifications.

Manufacturer's Specified Torque ....____________

59. Remove the transfer plate and separator plate to disassemble the valve body. Remove the seven check balls and one spring from the valve body.

60. Disassemble and clean the valve body parts in clean solvent. Allow to air dry.

61. Reinstall the valve body parts, separator plate, and transfer plate. Torque the valve body attaching screws to the manufacturer's specifications.

Manufacturer's Specified Torque ....____________

62. Lubricate the new oil rings on the output shaft support. Carefully install the output shaft and governor assembly into the transmission case.

CAUTION: Work carefully so that the oil seal rings are not damaged.

63. Install the rear, or low/reverse, band in the transmission case. Install the short strut, the link, and the anchor to the band. Tighten the adjusting screw to hold the assembly in place.

64. Place the rear planetary ring gear into the low/reverse drum. Install the assembly onto the output shaft. Install the #8 thrust washer (arrow) in step 66. Install the rear planetary carrier.

65. Lubricate the #7 thrust washer with petroleum jelly, to hold the washer in place. Install the washer.

66. Install the #6 thrust washer and snap ring on one end of the sun gear. Insert the sun gear through the driving shell and onto the output shaft. Install the #5 thrust washer, snap ring, and planetary ring gear in the driving shell.

67. Fit the #4 thrust washer to the rear of the front planetary carrier. Place the carrier inside of the ring gear and over the output shaft. Install the snap ring and the #3 thrust washer.

68. Install the front clutch assembly so that it engages with the rear clutch assembly.

Instructor Check: ____________

69. Engage the front clutch drive lugs to the slots in the input shell. Install the front, or kickdown, band. Install the strut, link, and anchor. Tighten the adjusting screw to hold the assembly.
70. Install new steel oil rings on the accumulator piston. Place the piston and spring in the accumulator bore.
71. Adjust the low/reverse band by backing off the adjusting screw locknut five turns. Torque the screw to the manufacturer's specifications. Back off on the screw according to specifications. Torque the locknut to the manufacturer's specifications.

Low/Reverse Band Adjusting Screw Torque Specification .. ________________

Turns Backed Off After Torquing .... ________________

Locknut Torque Specification .... ________________

72. Adjust the kickdown band by backing off the adjusting screw locknut five turns. Torque the screw to the manufacturer's specifications. Back off on the screw according to specifications. Torque the locknut to the manufacturer's specifications.

Kickdown Band Adjusting Screw Torque Specification ... ________________

Turns Backed Off After Torquing .... ________________

Locknut Torque Specification .... ________________

73. Place the oil pump assembly in position. Install two studs to align the pump. Install the attaching bolts and torque to the manufacturer's specification.

Attaching Bolt Torque Specification .... ________________

74. Remove the oil pump seal. Install a new seal with the lip facing inward.
75. Move the manual lever until the parking rod is at the rear of the transmission. Install the valve body while working the parking rod into the proper position. Be careful not to bind and bend the manual valve. Torque the valve body attaching bolts to the manufacturer's specifications from the center to the outside so the valve body won't warp. Reconnect the parking lock rod to the manual valve.

Valve Body Attaching Bolt Torque Specification .... ________________

Instructor Check: ________________

76. Install a new fluid filter. Install the oil pan and new gasket. Torque the pan bolts to the manufacturer's specifications.

Job Completion Report

1. Other than seals and gaskets, which parts needed replacement? ____________

____

____

2. What difficulties, if any, did you encounter during the overhaul? __________

____

____

INSTRUCTOR SIGNOFF ________________ **DATE** ____________

NAME ______________________ SECTION ______________ DATE ______________

Certification Practice

Test 15

DIRECTIONS:

The following questions are similar to those used on mechanic certification tests. Answer each question by circling the letter of the correct choice.

1. Mechanic A says the A-727 TorqueFlite transmission was used only in Chrysler Corporation vehicles.

   Mechanic B says the A-727 was used in American Motors and International vehicles.

   Who is correct?

   A. A only B. B only C. Both A and B D. Neither A nor B

2. The A-727 TorqueFlite contains all of the following EXCEPT
   A. a torque converter lockup (all years).
   B. two multiple-disc clutches.
   C. one overrunning clutch.
   D. two servos and bands.

3. Which of the following statments is correct?
   I. The transmission case should be steam cleaned and cleaned inside and out with clean shop towels before overhaul.
   II. The transmission case can be cleaned with solvent and a wire brush, and wiped inside and out with clean shop towels before overhaul.

   A. I only B. II only C. Both I and II D. Neither I nor II

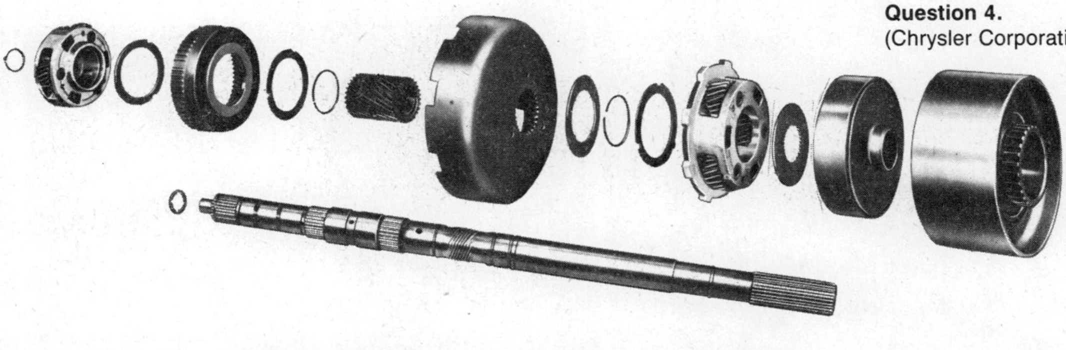

Question 4.
(Chrysler Corporation)

4. The A-727 TorqueFlite assembly shown in the picture is the
   A. front clutch.
   B. rear clutch.
   C. overrunning clutch.
   D. planetary gear train.

5. After washing the transmission bearings in solvent, all of the following should be done EXCEPT
   A. Allow the bearings to air dry.
   B. Spin the bearings with compressed air.
   C. Check the bearings by hand for tightness or roughness.
   D. Lubricate the bearings.

6. Which of the following should always be replaced during a transmission overhaul?
   A. Planetary gearsets
   B. Clutch friction and steel plates
   C. Gaskets and oil seals
   D. Thrust washers and snap rings

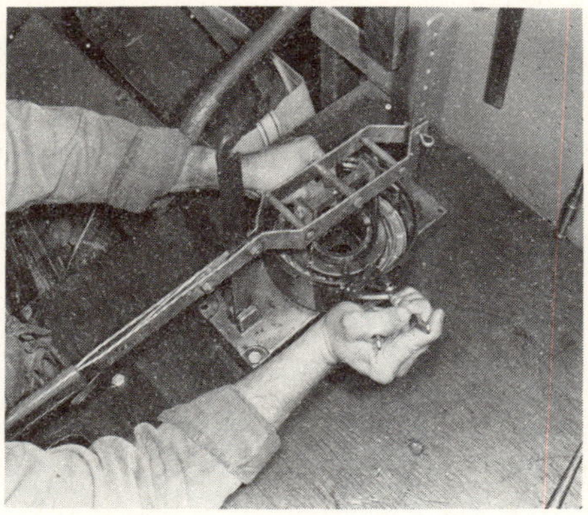

Question 7.

7. The tool shown in the picture is
   A. an engine valve-spring compressor.
   B. a special tool for removing the front clutch retainer.
   C. a hydraulic jack.
   D. a scissors-type transmission jack.

8. The A-727 TorqueFlite has how many check balls in the valve body under the transfer plate?
   A. 6
   B. 7
   C. 8
   D. 9

9. All of the following A-727 TorqueFlite oil pump clearances must be checked EXCEPT
   A. rotor faces to pump face.
   B. inner to outer rotor tip.
   C. outer rotor to pump bore.
   D. inner rotor to pump bore.

10. Mechanic A says that if mating parts in a transmission assembly are worn, the entire assembly should be replaced.
    Mechanic B says that only the worn part of a mating assembly should be replaced. Who is correct?
    A. A only B. B only C. Both A and B D. Neither A nor B

16 Overhauling the Ford AOD

YOUR LEARNING JOB

Overhauling a four-speed automatic transmission is more complicated than a three-speed overhaul. The main reason is because the four-speed automatic contains more parts. This unit covers the Ford AOD, a four-speed overdrive transmission that has parts similar to other four-speed automatics.

JOB COMPETENCY OBJECTIVES

This is a skill-building unit. When you have completed your assignments and exercises in this unit, you should be able to:

- ☐ Disassemble a Ford AOD transmission.
- ☐ Inspect the transmission parts.
- ☐ Reassemble the Ford AOD transmission with the needed replacement parts.

16.1 FORD AOD DESCRIPTION

This overdrive automatic transmission has been used on full-size Ford Motor Company automobiles since 1980. It has also been used on Ford light trucks since 1981. The Ford AOD consists of the following main parts:

- Torque converter
- Direct drive clutch
- Four multiple-disc clutches
- Two overrunning clutches
- Two servos and bands
- Compound planetary gearset
- Removable valve body
- Aluminum transmission case with integral bellhousing and removable extension housing.

The Ford AOD combines a torque converter with a fully automatic four-speed gear system. The bellhousing and the transmission case are a single removable aluminum casting. The extension housing is removable.

The torque converter is a three-element unit. The direct drive (third gear) clutch used with the direct drive input shaft locks up third gear. Fourth gear locks up by way of the overdrive band.

The compound planetary gear system uses a front and a rear sun gear.

The hydraulic system has a single oil pump. The valve body is removable.

This transmission differs from other automatics because it uses mechanical rather than hydraulic methods to drive third and fourth gears. The direct drive input shaft couples the engine to the direct clutch drive. This bypasses the torque converter. The input shaft is then driven by the torque converter cover through a damper assembly.

The Ford AOD is an all-metric transmission.

16.2 BEFORE THE OVERHAUL

Be sure you are familiar with all of the assemblies that make up the automatic transmission. These assemblies are discussed in Parts II, III, and IV. This unit discusses the disassembly and reassembly of the Ford AOD transmission. Refer to earlier units for teardown, inspection, and testing of the torque converter, planetary gearsets, apply devices, and the valve body.

Transmission parts and specifications change often during their lifespan. For this reason, always refer to the proper service manual when performing any transmission overhaul.

Before proceeding with the overhaul, make sure that the transmission fluid has been drained. Plug all openings in the transmission. Clean the transmission case with a steam-cleaning machine. If a steam-cleaning machine is not available, use solvent, a wire brush, and shop towels. Do *not* wipe the inside of the transmission case with shop towels. Allow it to air dry. Any lint left inside of the transmission case can affect the operation of the automatic transmission.

After the transmission has been thoroughly cleaned, inspect the case for any signs of damage. A cracked or distorted transmission case can cause internal transmission problems.

To better understand the parts and their locations in the transmission, refer to Figures 16-1 through 16-3.

Clean and inspect the transmission parts during overhaul as discussed in 15.5. Be sure to check end play measurements. Refer to 15.4.

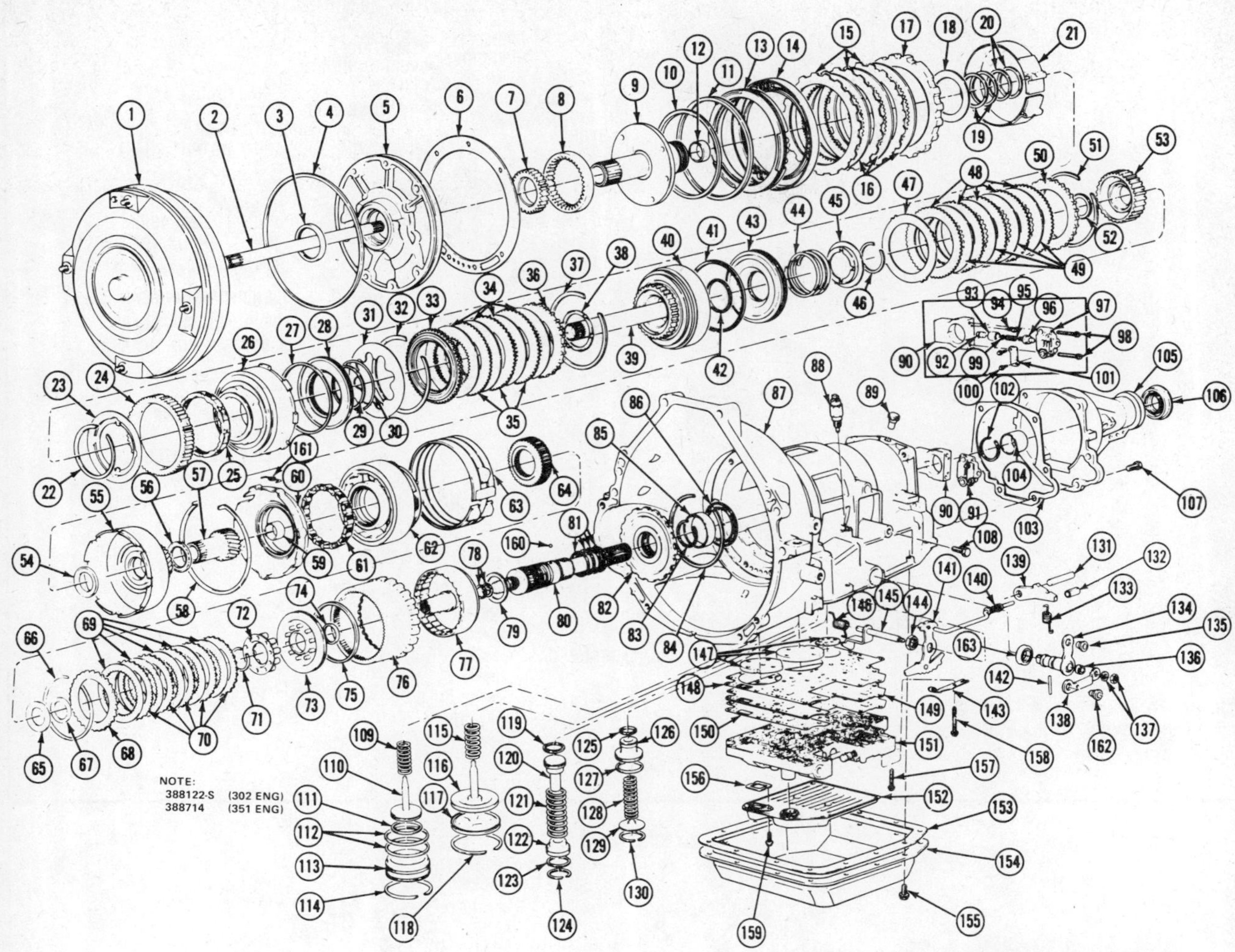

Figure 16-1. Ford AOD transmission, exploded view and parts identification chart. (Ford Motor Company)

1	Torque converter
2	Direct drive shaft
3	Front pump seal
4	Front pump "O" ring
5	Front pump body
6	Front pump gasket
7	Front pump drive gear
8	Front pump driven gear
9	Stator support — front pump
10	Interm. clutch piston inner lip seal
11	Interm. clutch piston outer lip seal
12	Front pump bushing
13	Interm. clutch piston
14	Interm. clutch piston return springs & retainer
15	Interm. clutch external spline steel plates (sel.)
16	Interm. clutch internal spline friction plates
17	Interm. clutch pressure plate
18	#1 thrust washer (front pump) selective
19	Stator support seal rings (rev. clutch)
20	Stator support seal rings (fwd. clutch)
21	Overdrive band
22	Interm. OWC retaining snap
23	Interm. OWC retaining plate
24	Interm. OWC outer race
25	Interm. one-way clutch assy.
26	Reverse clutch drum
27	Reverse clutch piston seal (outer)
28	Reverse clutch piston
29	Reverse clutch piston seal (inner)
30	Thrust ring
31	Reverse clutch piston return spring
32	Retaining snap ring
33	Reverse clutch front pressure plate
34	Reverse clutch internal spline friction plate
35	Reverse clutch external spline steel plate
36	Reverse clutch rear pressure plate
37	Reverse clutch retaining ring (sel.)
38	#2 thrust washer (rev. clutch)
39	Turbine shaft
40	Forward clutch cylinder & turbine shaft
41	Forward clutch piston seal (outer)
42	Forward clutch piston seal (inner)
43	Forward clutch piston
44	Forward clutch piston return spring
45	Return spring retainer
46	Retaining snap ring
47	Waved Spring
48	Forward clutch external spline steel plate
49	Forward clutch internal spline friction plate
50	Forward clutch pressure plate
51	Retaining snap ring (selective)
52	#3 needle bearing (fwd. clutch)
53	Forward clutch hub
54	#4 needle bearing
55	Reverse sun gear & drive shell assy.
56	#5 needle bearing
57	Forward sun gear
58	Center support retaining ring
59	Forward sun gear precision bushing
60	Center support planetary
61	Planetary OWC cage spring & roller assy.
62	Planetary assy.
63	Reverse band
64	Direct clutch hub
65	#7 needle bearing (direct clutch inner)
66	Retaining snap ring (selective)
67	Thrust spacer
68	Direct clutch pressure plate
69	Direct clutch internal spline plates
70	Direct clutch external spline plates
71	Retaining snap ring
72	Return spring & retainer
73	Direct clutch piston
74	Direct clutch piston seal (inner)
75	Direct clutch piston seal (outer)
76	Ring gear & park gear
77	Direct cylinder
78	Output shaft small (2) steel seal rings (direct clutch)
79	#8 needle bearing (direct clutch outer)
80	Output shaft
81	Output shaft large (4) steel seal rings
82	Output shaft hub
83	Retaining snap ring (O.P.S. hub to O.P.S.)
84	Retaining snap ring (O.P.S. hub to ring gear)
85	Rear case bushing
86	#9 needle bearing (rear case)
87	Case assy.
88	Neutral start switch
89	Vent cap
90	Governor counterweight
91	Body assy. — governor
92	Plug governor
93	Sleeve governor
94	Screen assy. — gov. oil
95	Spring gov. valve
96	Valve governor
97	Body governor
98	Bolt (governor body to counterweight)
99	Clip — governor spring retainer
100	Bolt (governor cover to governor body)
101	Cover — governor valve body
102	Retaining snap ring (governor assy. to O.P.S.)
103	Extension housing gasket
104	Extension housing bushing
105	Extension housing
106	Extension housing seal
107	Bolt (ext. hsg. to case) M6-12.5 x 30 (6 req'd)
108	Pipe plug — 1 8-27 dry seal
109	Overdrive servo piston return spring
110	Overdrive servo piston
111	Overdrive servo piston seal
112	Overdrive servo cover seal rings
113	Overdrive servo cover
114	Retaining snap ring (O.D servo to case)
115	Reverse servo piston return spring
116	Reverse servo piston (selective)
117	Reverse servo cover
118	Retaining snap ring (rev. servo to case)
119	3-4 accumulator valve seal
120	3-4 accumulator valve
121	3-4 accumulator valve return spring
122	3-4 accumulator cover
123	3-4 accumulator cover seal
124	Retaining snap ring (3-4 accum. to case)
125	2-3 accumulator valve seal (small)
126	2-3 accumulator valve
127	2-3 accumulator valve seal (large)
128	2-3 accumulator valve return spring
129	2-3 accumulator cover
130	Retaining snap ring (2-3 accum. to case)
131	Park pawl shaft
132	Guide cup
133	Park pawl return spring
134	Manual lever
135	Grommet
136	Throttle lever oil seal
137	Attaching nut & lock washer — M8 x 1.25
138	Throttle lever (outer)
139	Park pawl
140	Park pawl actuating rod
141	Manual lever (inner)
142	Roll pin — 1.8 x 0.95 grooved
143	Detent spring
144	Attaching nut (manual LVR) — M14 x 1.5 HEX
145	Throttle lever (inner)
146	Throttle torsion spring
147	Valve body reinforcement plate
148	Separator plate gasket (upper)
149	Separator plate
150	Separator plate gasket (lower)
151	Valve body (main control)
152	Filter & Grommet assy. — oil pan
153	Oil pan gasket
154	Oil pan
155	Bolt (oil pan to case) — M8-1.25 x 15 (17 req'd)
156	Oil filter gasket
157	Bolt (valve body to case) M6-1.0 x 30 (8 req'd)
158	Bolt (valve body to case) M6-1.0 x 40 (17 req'd)
159	Bolt (screen to valve body) M6-1.0 x 16 (3 req'd)
160	Ball (governor drive)
161	Spring (anti clunk)
162	Grommet
163	Oil seal assy.

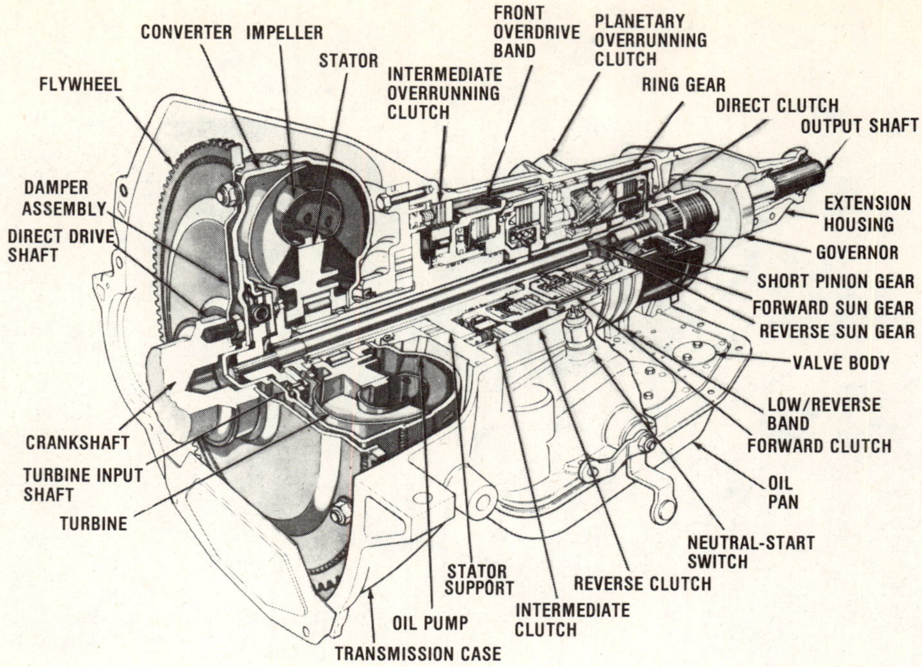

Figure 16-2. Ford AOD transmission cutaway. (Ford Motor Company)

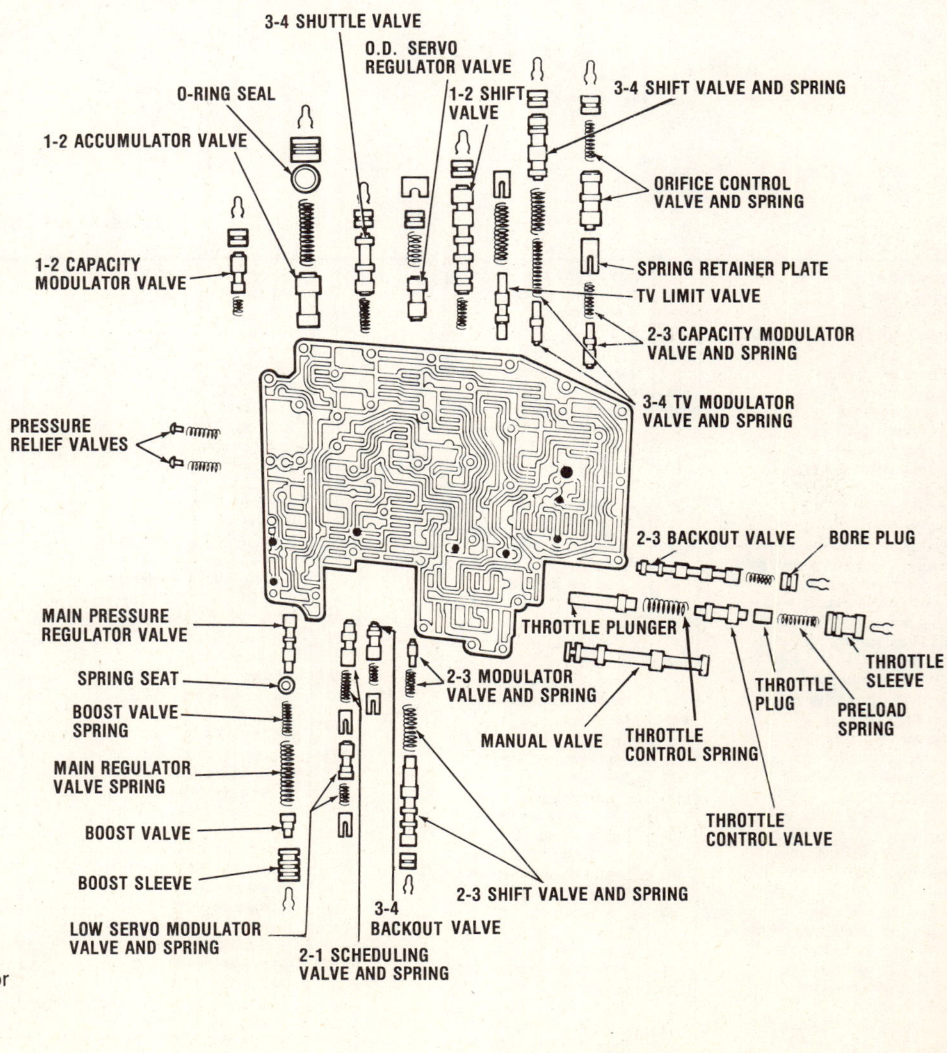

Figure 16-3. Ford AOD valve body disassembled. (Ford Motor Company)

1. Place the transmission in a transmission stand. Remove the torque converter. Refer to 14.2. Remove pan bolts. Lift off the oil pan and gasket. Discard the gasket and use a new one for reassembly.

2. Remove the bolts holding the transmission filter in place. Remove the filter, grommet, and gasket from the valve body. Discard the filter and gasket. Install a new filter and gasket during reassembly.

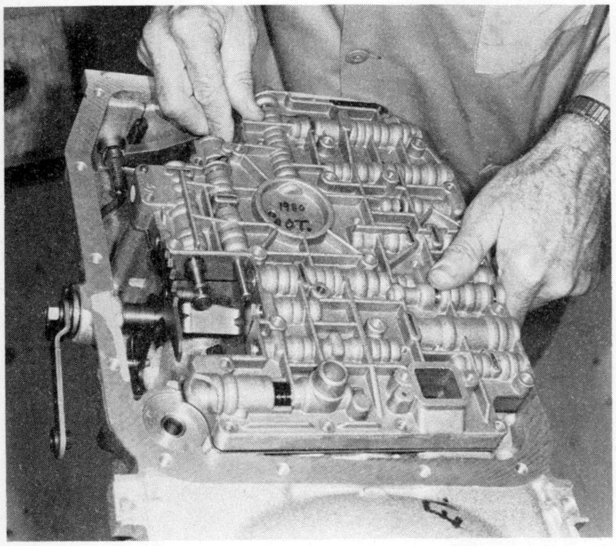

3. Remove the detent spring/bolt and valve body bolts. Lift off the valve body. Remove the separator plate gasket from the transmission case.

4. Use a wooden handle or dowel to push down on the overdrive servo cover. Pry out the snap ring. Remove the cover, piston, and spring. If the cover sticks, apply air pressure to the overdrive servo release passage in the transmission case. CAUTION: Hold the servo cover securely with a rag and a block of wood when applying air pressure. The cover can fly loose and cause serious physical injury.

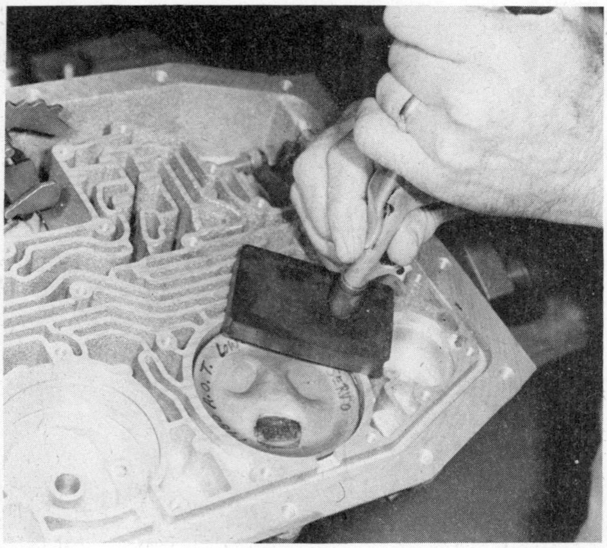

5. Push down on the low/reverse servo cover. Pry out the snap ring. If the cover does not pop up, tap lightly or apply air pressure. CAUTION: Hold the servo cover securely with a rag and a block of wood when applying air pressure. The cover can fly loose and cause serious physical injury.

6. Remove the low/reverse cover, piston, and spring. Inspect both servo assemblies and the servo bores. Replace the seal rings as necessary.

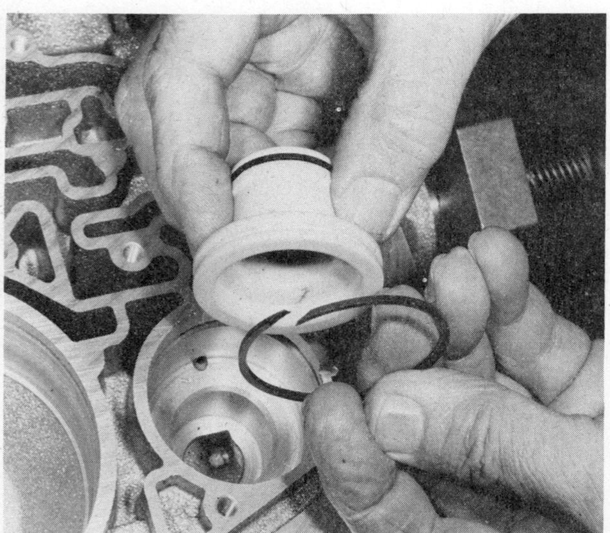

7. Push down on the 2-3 accumulator cover. Remove the snap ring. The cover and piston can be removed with little resistance. Remove the 3-4 accumulator in the same way. The 3-4 accumulator may have a spring, so use caution. The cover and piston may fly out. Discard and replace the seals.

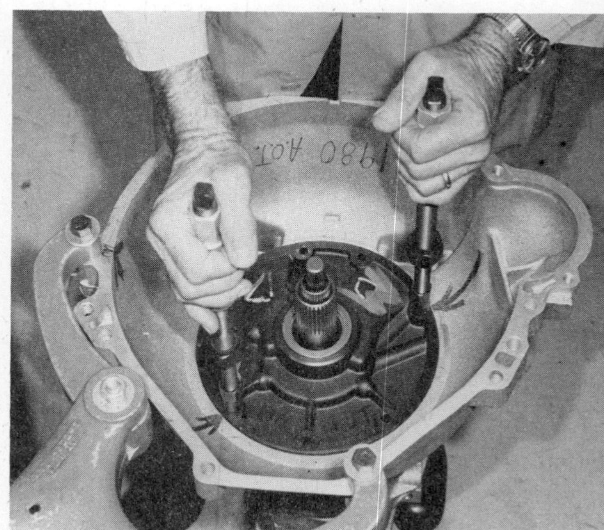

8. Use a seal remover to pull out the oil pump seal. Discard the seal and replace it with a new one. Remove the pump bolts. Thread a pair of metric slide hammers into the pump holes at opposite ends, as shown. Pull out the oil pump with the slide hammers. Remove and discard the pump gasket, and install a new gasket during reassembly.

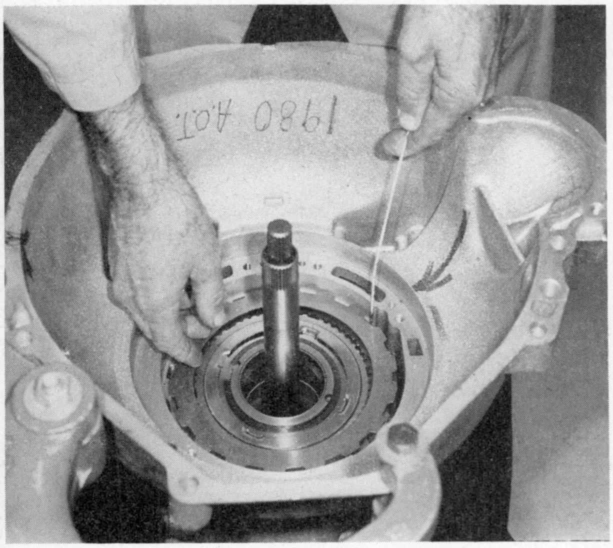

9. Use a piece of coat hanger, or stiff wire, to lift the intermediate clutch plates from the case.

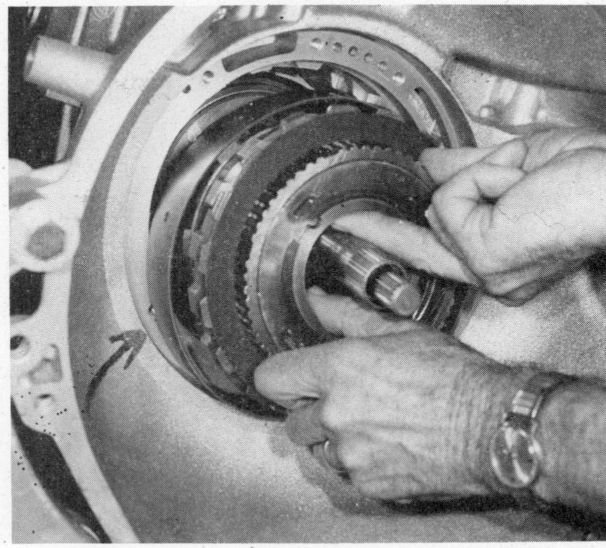

10. Rotate the transmission case so that it is in a horizontal position. Grasp the intermediate clutch assembly, as shown. Pull forward to free the clutch assembly. Remove the overdrive band.

11. Remove the intermediate clutch assembly. Be careful—the clutch assembly is heavy. Watch closely so that the #9 thrust washer and the needle bearings from the geartrain are not lost.

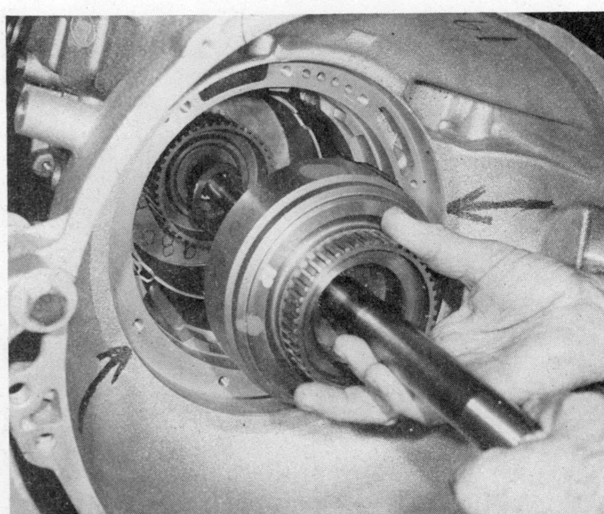

12. Remove the forward and reverse clutch assemblies. Needle bearings and a thrust bearing are located behind these assemblies. Inspect each part of the assemblies as they are removed.

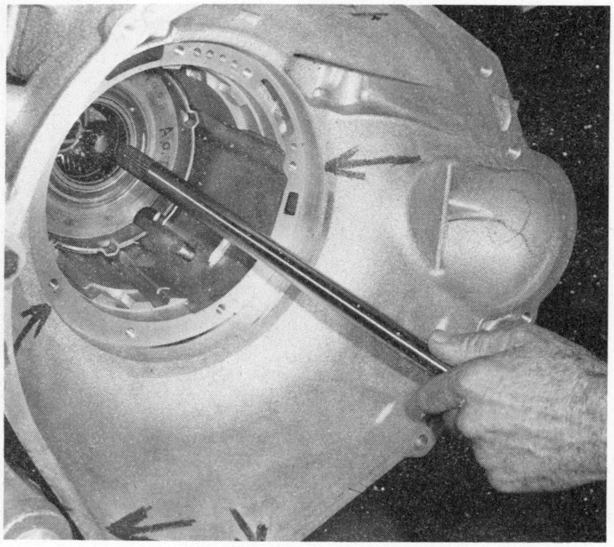

13. Pull out the direct drive shaft. NOTE: Because both ends of the shaft are identical, it is possible to replace the shaft facing the wrong way. Mark the shaft for proper reassembly direction.

14. Two springs are now removed. Locate and remove the center support retainer and note the position of the tabs for reassembly. Also, locate the anti-clunk spring (shown) and note the location for reassembly. Remove the springs with a piece of coat hanger, stiff wire, or a small screwdriver.

15. Remove the center support and low/reverse planetary gearset. Separate the direct clutch hub at the rear from the low/reverse planetary gearset.

16. Remove the low/reverse band. Reach inside and remove the needle bearing and spacer from the direct clutch drum. Inspect all needle bearings as they are removed.

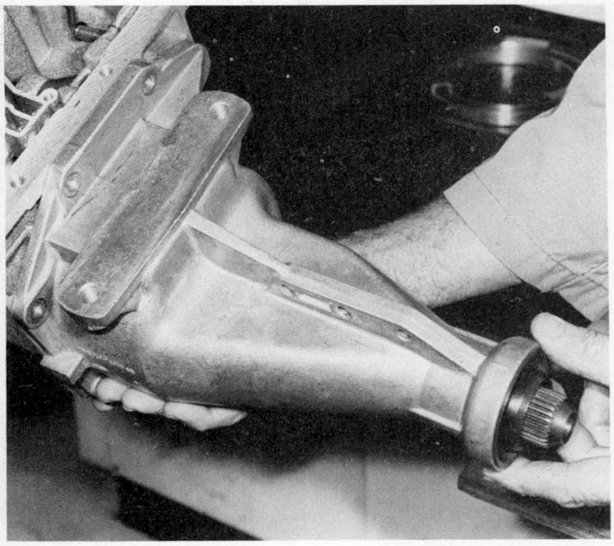

17. Unbolt and remove the extension housing. If necessary, tap the housing to break the seal for removal. Discard the housing gasket and replace it with a new one. Clean the mating flanges.

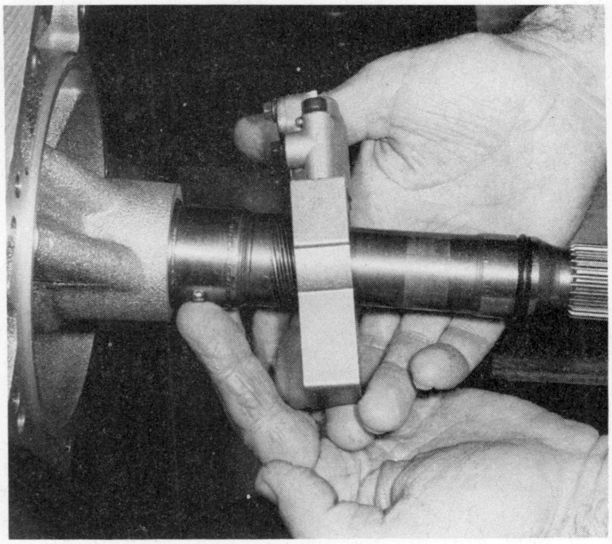

18. Remove the retaining snap ring and the governor assembly from the output shaft. Remove the governor drive ball from depression in the shaft. A magnet may be necessary to remove the ball.

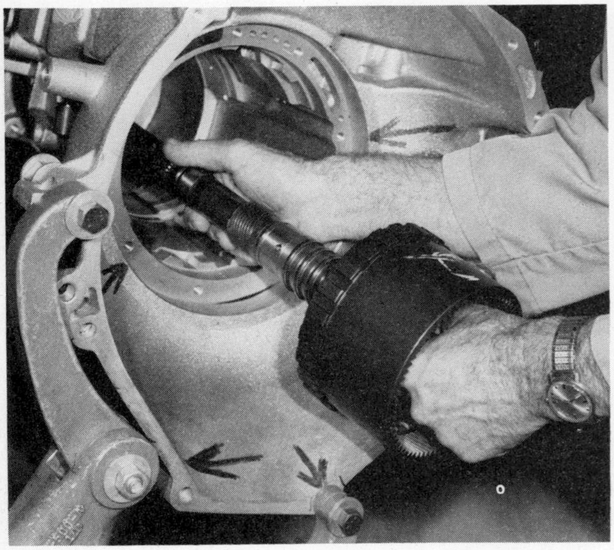

19. Remove the output shaft, ring gear, and direct clutch assembly from the front of case. Work carefully to avoid damaging the seal ring grooves and the case mating surface.

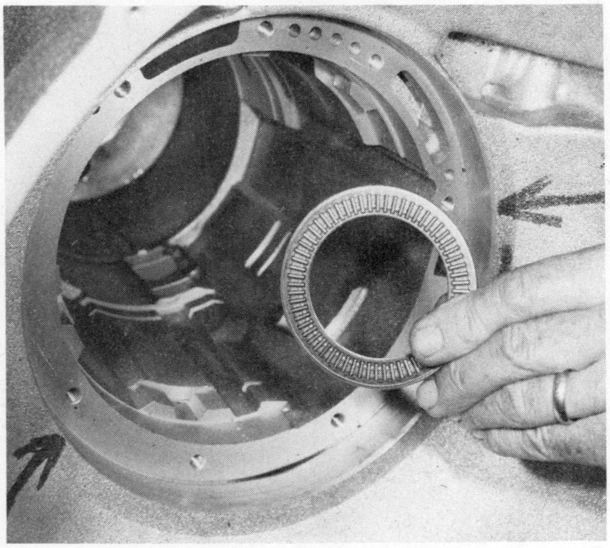

20. Remove the output-shaft needle bearings. Unless the rear case bushing must be removed because of damage, this completes transmission case disassembly. The rear case bushing is removed with a slide hammer.

21. To disassemble the oil pump, carefully pry open the tabs holding the spring retainer assembly. Remove the spring retainer assembly. Note the position of the bleed hole.

22. The spring retainer has springs that are permanently attached. If the springs are damaged, the entire spring retainer assembly must be replaced. The retainer snaps back for reassembly.

23. Rotate, lift, and remove the intermediate clutch piston from the oil pump assembly. Replace the inner and outer lip seals on the piston during reassembly.

24. Adjust transmission end play to specifications with the nylon stator-support thrust washer. The thrust washer comes in five color-coded thicknesses. The tang must fit into the locator slot.

25. To install the clutch piston, coat the seals and pump sealing area with petroleum jelly. Use a lip seal protector tool to prevent damage to the new seals.

26. The center support and low/reverse planetary assemblies contain an overrunning roller clutch. Rotate the center support counterclockwise to remove the overrunning clutch.

27. Inspect the overrunning clutch assembly. If a roller is lost or damaged, the entire clutch assembly must be replaced. To replace the overrunning clutch support, rotate counterclockwise.

28. To disassemble the output shaft and direct clutch assembly, pry out the waved snap ring. Remove the shaft from the ring gear. Install new sealing rings on the output shaft.

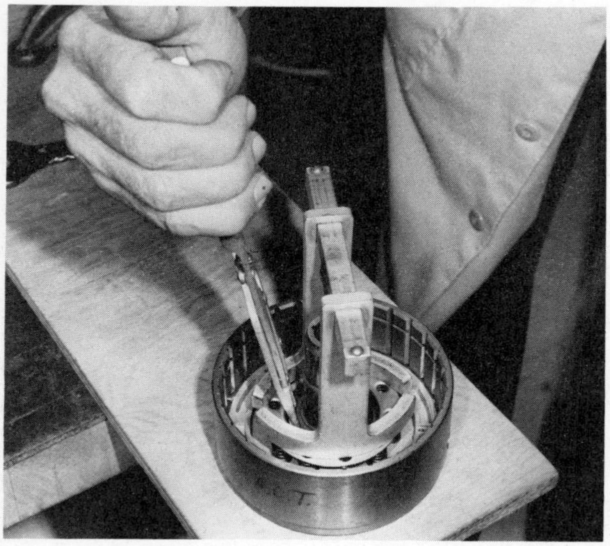

29. Remove the direct clutch assembly from the ring gear. Pry out the snap ring. Remove the pressure plate and clutch pack. Install a piston compressor, shown, and remove the snap ring. Component assembly sequences may differ, so note correct sequence during disassembly.

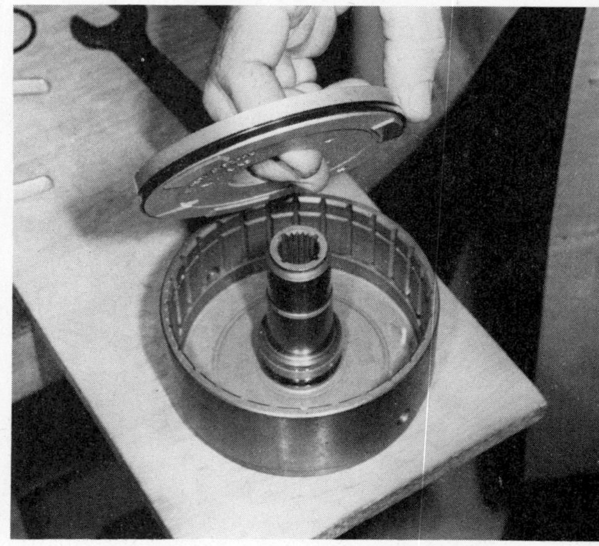

30. Remove the spring retainer assembly and piston from the drum. Lubricate and install new piston seals. Be sure the check ball in the piston is free.

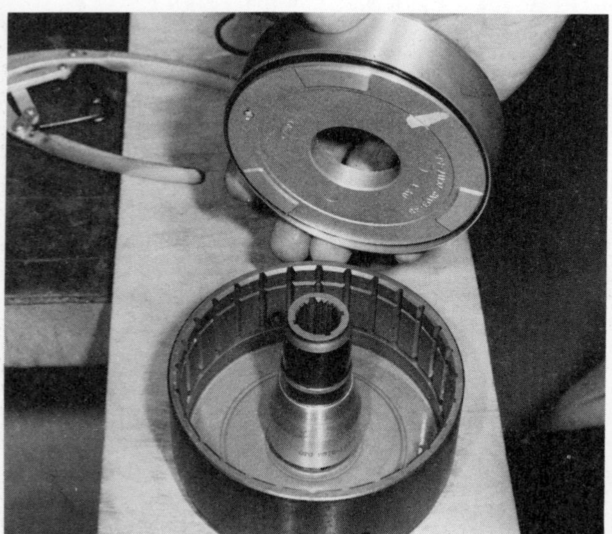

31. Lubricate the outer piston seal with petroleum jelly. Fit a lip seal protector over the piston. Install the piston into the clutch housing and remove the seal protector.

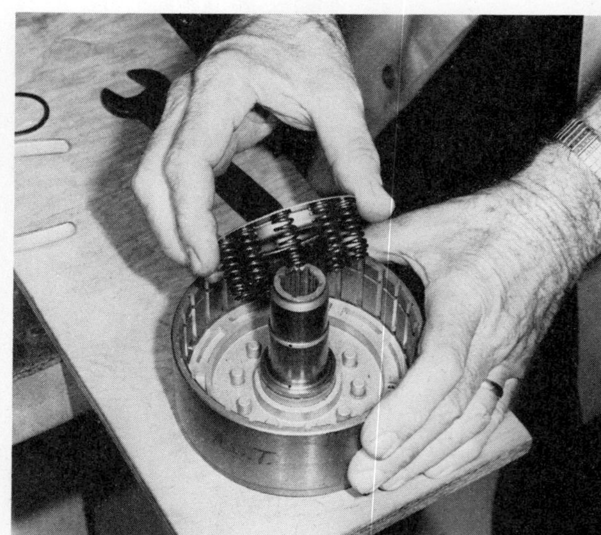

32. Place the piston spring and retainer assembly into the drum.

33. Install the compresser tool. Compress the spring retainer and install the snap ring. Make sure the snap ring seats completely in its groove.

34. Install the clutch pack, pressure plate, and snap ring. Place the needle bearings and clutch housing inside the ring gear. Replace the thrust washer and spacer, as shown.

35. Make sure the thrust washer tabs face away from the clutch drum. Install the direct clutch hub. Rotate the clutch hub to seat it properly.

36. Disassemble each remaining clutch drum in the same way. Pry the snap rings loose from the clutch drum grooves, and remove the pressure plates and clutch packs.

37. Inspect each clutch pack for burned or scored plates. Install new ones where needed. Compress the retainers and remove the pistons. Replace all seals. Reassemble.

38. Replace the pressure plates. Install the retaining snap rings. Component assembly sequences may differ, so refer to sequence noted in step 29.

39. Snap rings are of a selective fit. This means that different size snap rings can be used to gain proper clearances, or tolerances. To measure tolerance, insert a feeler gauge between the snap ring and the pressure plate. Compare this reading to specifications. Replace the snap ring with one of a different size, if needed.

40. The reverse clutch drum has an overruning clutch attached. To disassemble, remove the snap ring and the retaining plate.

41. Rotate the outer race counterclockwise and remove it from the overrunning clutch assembly. Replace the entire overrunning clutch assembly if the rollers are lost or damaged.

42. Place the outer race over the overrunning clutch, with the chamfered side facing the drum. Install the retaining plate and snap ring over the overrunning clutch assembly.

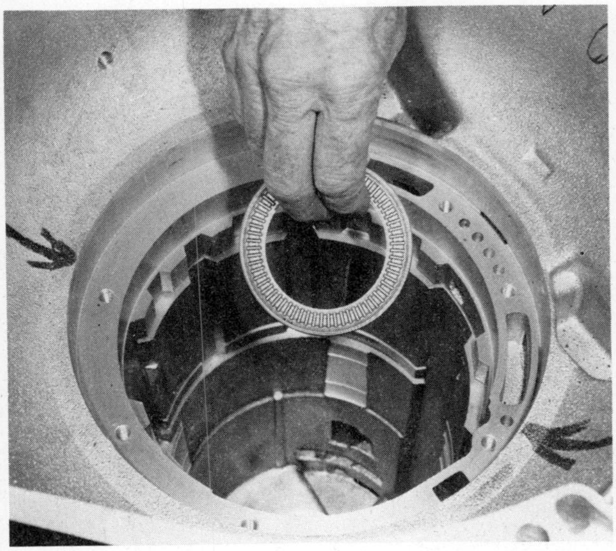

43. To reassemble, lubricate the race side of #9 needle bearing with petroleum jelly. Install the bearings in the transmission case. It helps to rotate the case to a vertical position.

44. Install the output shaft, ring gear, and direct clutch assembly into the transmission case. Use extreme caution not to damage the bushing or seal rings during installation.

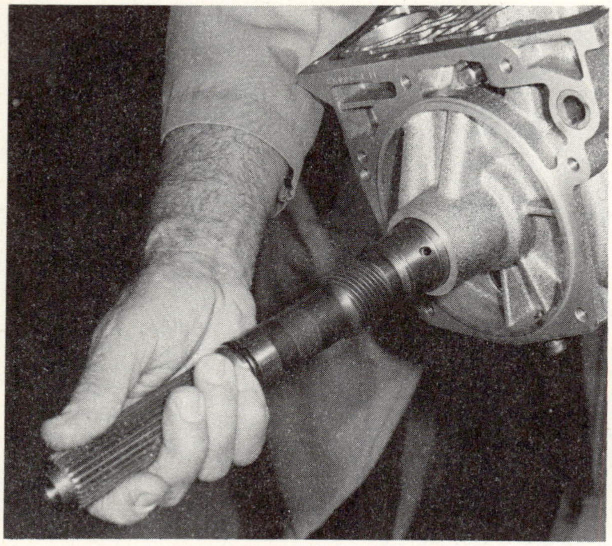

45. Support the output shaft as it is inserted through the case. This prevents damage to the bushing, to the seal ring grooves, and to the case mating surfaces.

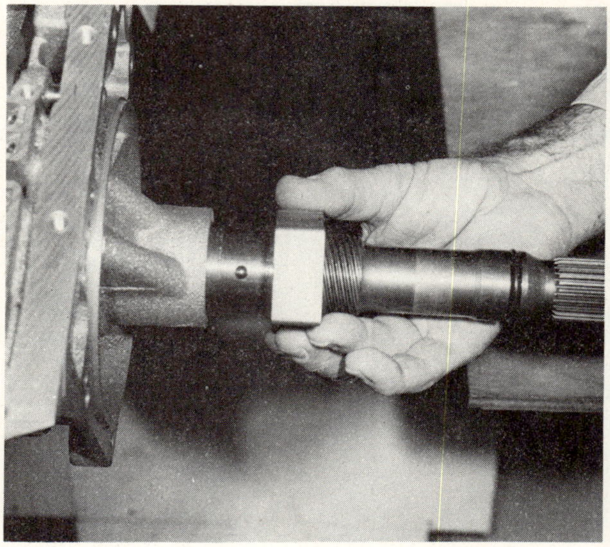

46. Install the governor drive ball into the output shaft depression. Slide the governor assembly onto the shaft and over the drive ball until the governor seats against the case casting.

47. Replace the retaining snap ring. Make sure the snap ring seats correctly in the shaft groove.

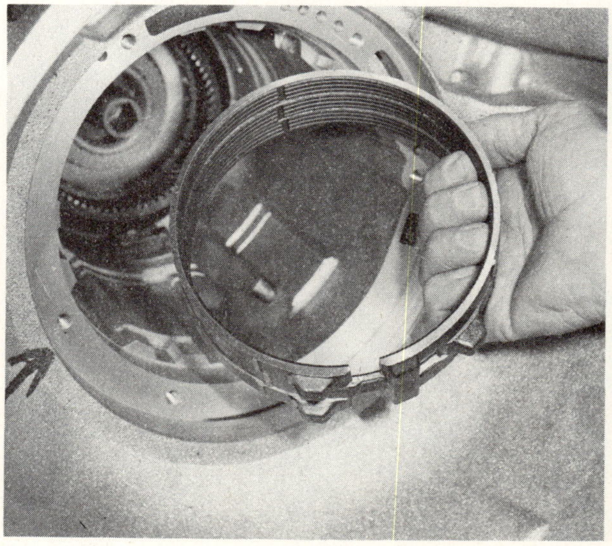

48. Band surfaces should be carefully inspected before installing them in the case. If any signs of wear are evident, replace the band without question.

49. Install the low/reverse band in the case. Locate the band positioning ears (arrows) against the transmission case anchor pins. Make sure the band is fully seated.

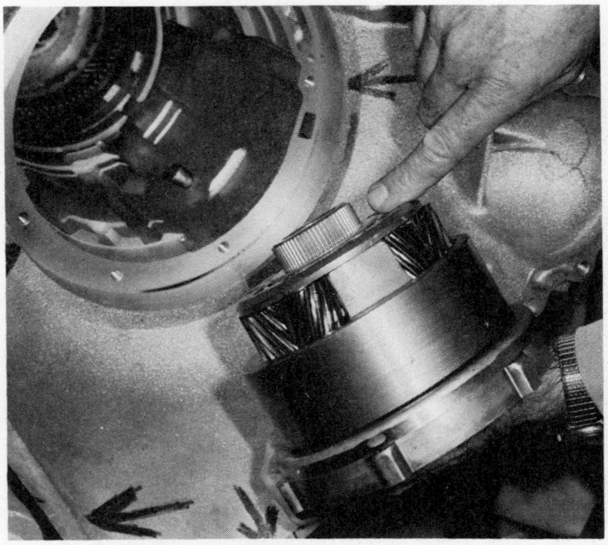

50. Before installing the center support and the low/reverse planetary gearset, make sure the needle bushing is positioned as shown.

51. Install the center support and the low/reverse planetary gearset. Apply constant forward pressure to mesh it with ring gear and direct clutch hub.

52. Installing the center support anti-clunk spring is a critical step. Apply thumb pressure clockwise to the center support to verify spring tension.

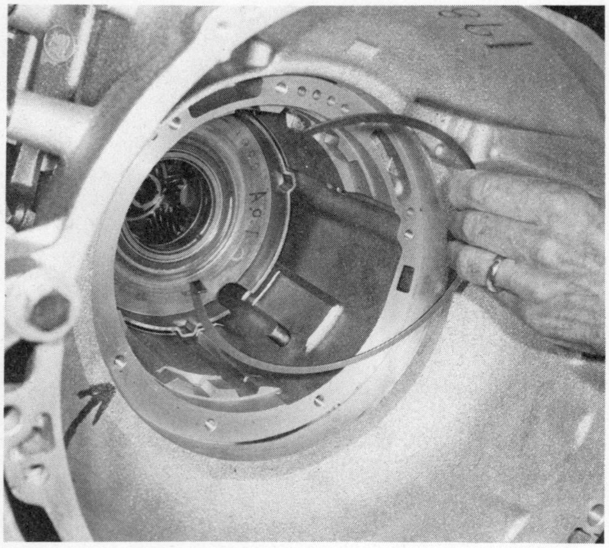

53. Install the center-support snap ring in the case groove next to the low/reverse planetary gearset. Make sure the snap ring seats properly in its groove.

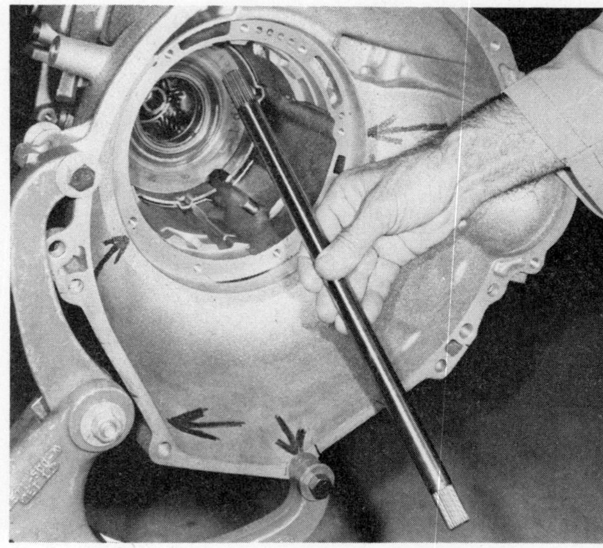

54. Install the direct drive shaft. Though the shaft is equally splined at each end, it must be replaced in the same direction it occupied before removal.

55. Lubricate the inner race of the #5 needle bearing with petroleum jelly. Slide the bearing onto the sun gear shaft, with the tabs facing away from the gear. Insert the sun gear shaft into the drive shell.

56. Assemble the forward and reverse clutch assemblies. Install the needle bearing and forward clutch hub.

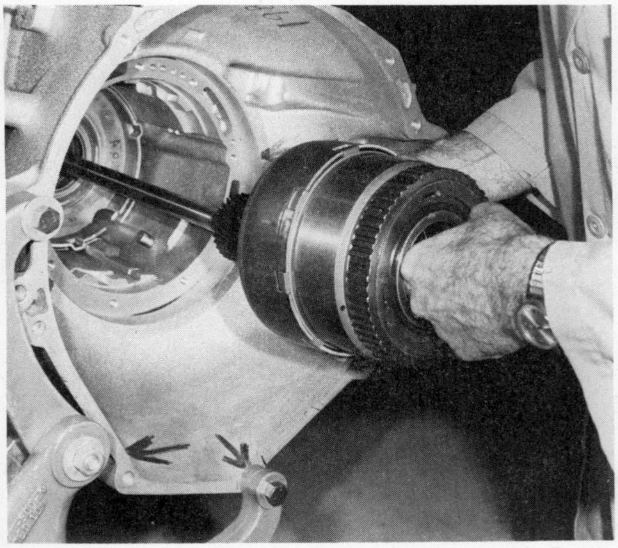

57. Assemble the forward and reverse sun-gear shell assemblies, and the forward and reverse clutch packs, into a single unit. Install these assemblies onto the direct drive shaft.

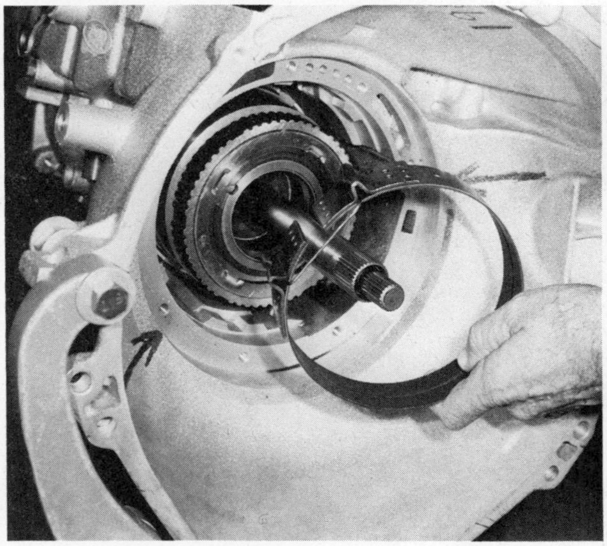

58. The overdrive band surface is delicate. Cut a piece of welding rod to size and bend it to form legs. Insert the rod into the band ears and install the band. Remove the welding rod from the band ears.

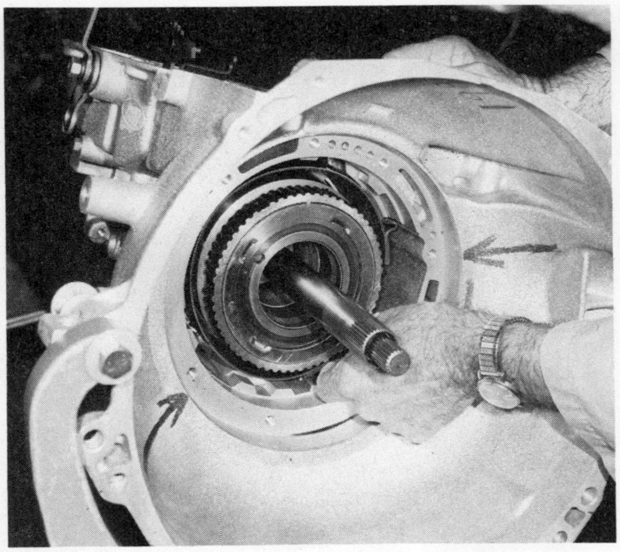

59. Position the band against the anchor pin in the case. Make sure the band is properly seated before continuing with the reassembly.

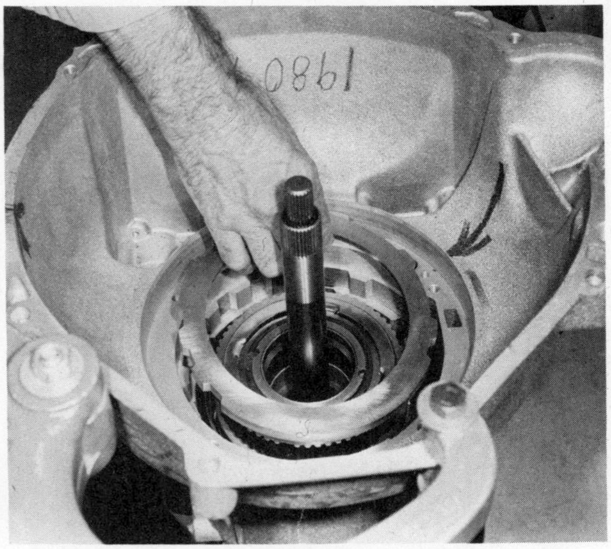

60. Be sure to install the pressure plate before reassembling the intermediate clutch pack. Alternately install three friction and three steel clutch plates.

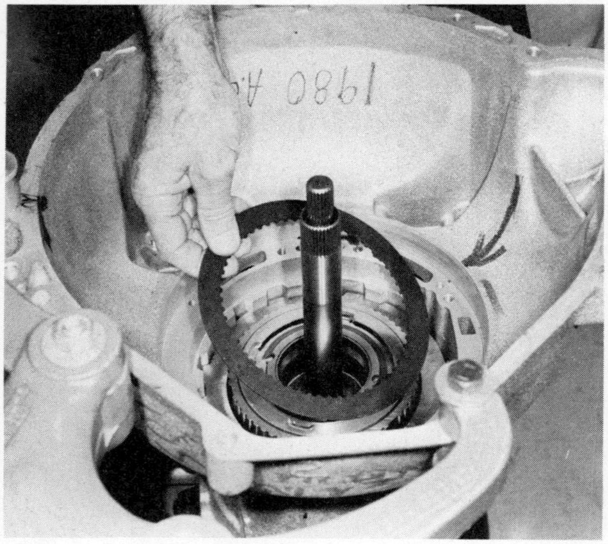

61. The outer steel separator plate on the intermediate clutch pack is selective. Intermediate clutch clearance should be checked at this point, using a special Ford tool.

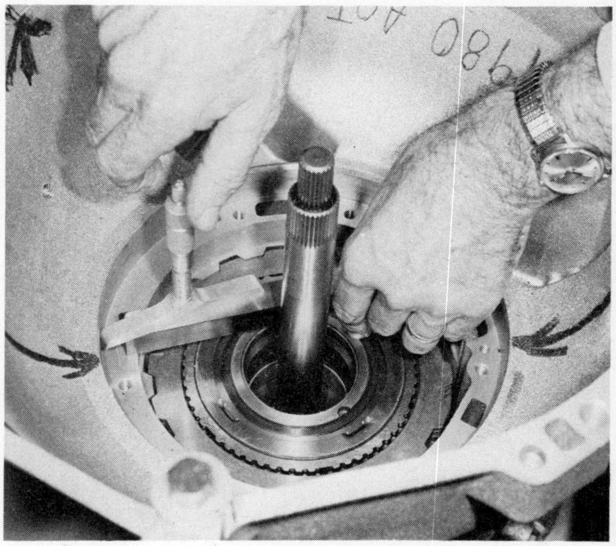

62. Place the special end-play checking tool across the oil pump case mounting. Use a micrometer to measure the depth. Repeat the measurement 180 degrees opposite the first reading. Record the average reading.

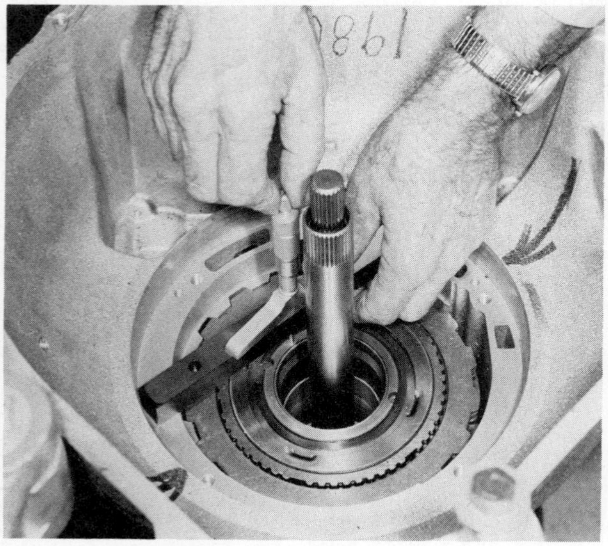

63. Also check the depth at the thrust face of the reverse clutch drum. Repeat the measurement 180 degrees opposite the first reading. Record the average reading. Compare the averages to specifications to see if adjustment is necessary.

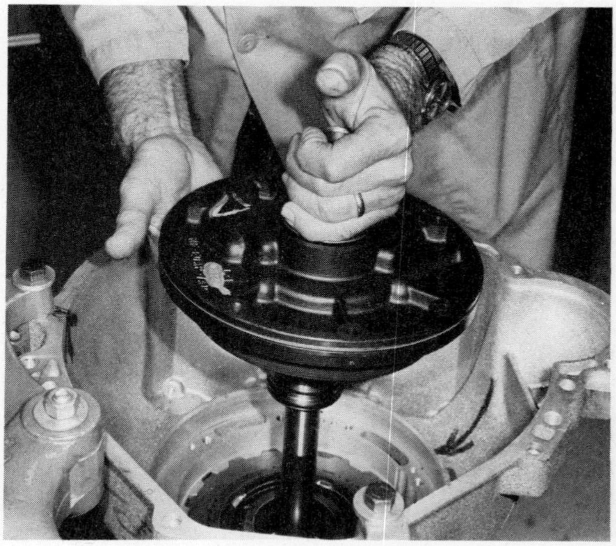

64. If necessary, change the separator plate or thrust washer on the oil pump. Install a new pump gasket. Lower the pump into the case. Torque the bolts to specifications.

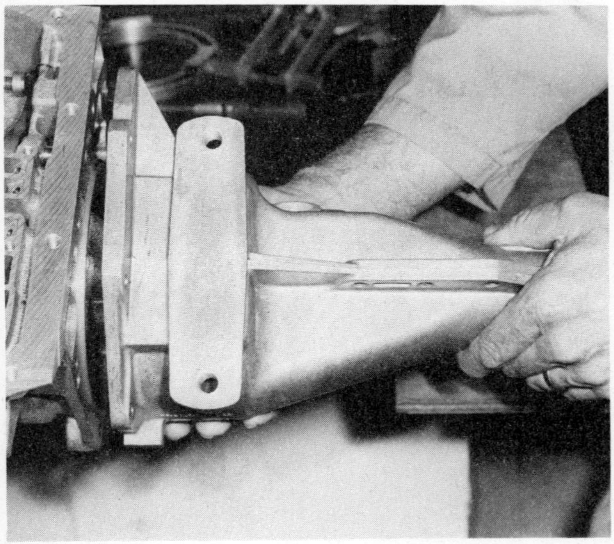

65. Install the new pump seal with a seal driver and a hammer. Replace the extension housing on the output shaft with a new seal and gasket. Torque the bolts to specifications.

66. With new seals on the 2-3 accumulator piston, install the piston and spring in the case. Replace the piston cover and snap ring. Repeat the procedure with the 3-4 accumulator.

67. Check the piston-rod bore to make sure that the low/reverse band is properly aligned. Then install the low/reverse band spring, piston, and cover.

68. Center the overdrive band under the piston hole with a screwdriver. Install the piston, spring, and cover. Tap cover in place and install snap ring.

69. Mate the manual valve with the inner manual-shaft shift linkage. Mate the TV plunger with the inner TV lever. Place the valve body in the transmission case.

70. Reconnect the inner linkage spring to the separator plate notch. Torque the valve body bolts to specifications. The four front, one center, and three rear bolts are short.

71. Replace the detent spring and bolt. Torque to specifications. Install the filter seal and gasket. Replace the filter on the valve body. Torque the bolts to specifications.

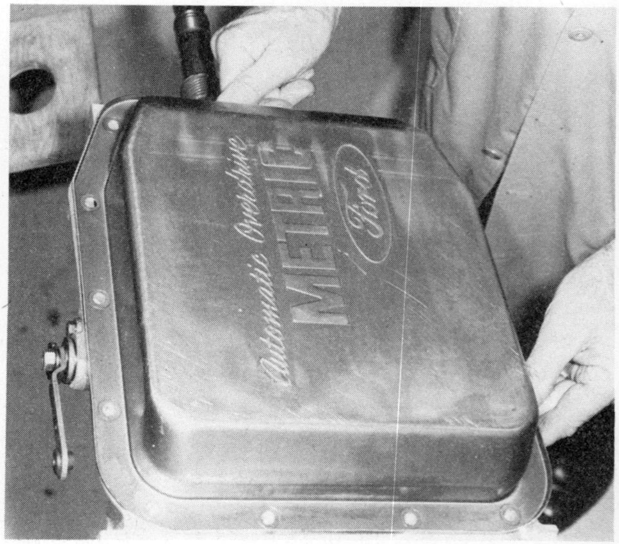

72. Install the pan with a new gasket. Tighten the attaching bolts to specifications. Replace the torque converter. The AOD overhaul is complete.

NAME ______________________ SECTION ____________ DATE ____________

Job Sheet 16-1

Overhauling the Ford AOD Transmission

Before you begin:

Read Topics 16.1 through 16.3.

Vehicle Make ____________________ Model ____________ Year ____________

Time Started ____________________ Time Finished ____________________

Job Completion Time ____________________ Flat Rate Time ____________________

Tools Required

- Metric sockets
- Socket extensions, handle
- Metric open-end or combination wrenches
- Screwdrivers
- Safety glasses
- Air blow gun
- Seal puller
- Slide hammers with metric bolt ends
- Stiff wire or small screwdriver
- Pencil magnet
- Ball-peen hammer
- Automatic transmission clutch piston compressor
- Snap-ring pliers
- Feeler gauges
- Longnose pliers
- Welding rod
- End-play straightedge tool
- Depth micrometer
- Seal driver
- Soft-faced mallet
- Torque wrench

References

Manufacturer's service manual ________________________________

Repair manual ________________________________

Service bulletin ________________________________

Procedure

1. Attach a dial indicator to measure input shaft end play. Seat the plunger against the end of the input shaft.
2. Zero the dial indicator.
3. Move the input shaft in and out to obtain an end play reading.

In the spaces provided, write the manufacturer's specification and measured end play.

Manufacturer's Specified Input Shaft End Play .... ____________________

Measured Input Shaft End Play .... ____________________

Instructor Check: ____________________

4. Place the transmission in a transmission stand. Remove the torque converter. Remove the pan bolts. Lift off the oil pan and gasket. Remove and discard the gasket.
5. Remove the bolts holding the transmission filter in place. Remove the filter, grommet, and gasket from the valve body. Discard the filter and gasket.
6. Remove the detent spring and bolt and the valve body bolts. Note the position of the 8 short bolts. Lift off the valve body. Remove the separator plate from the transmission case.
7. Push down carefully on the overdrive servo cover and pry out the snap ring. Remove the cover, piston, and spring. Apply air pressure to the servo release passage in the case.

SAFETY CAUTION: Use safety glasses. Hold the servo cover securely with a rag and a block of wood when applying air pressure. The cover can fly loose and cause serious physical injury.

8. Push down on the low/reverse servo cover. Pry out the snap ring. If the cover does not pop up, tap lightly or apply air pressure.
9. Remove the low/reverse cover, piston, and spring. Inspect both servo assemblies and the servo bores. Replace the seal rings as necessary.
10. Push down on the 2–3 accumulator cover. Remove the snap ring. Remove the cover and piston. Remove the 3–4 accumulator cover and piston in the same way. Discard the old seals.

SAFETY CAUTION: The 3–4 accumulator may have a spring. The cover and piston may fly out.

11. Remove the oil pump seal. Discard the old seal. Remove the pump bolts. Pull out the oil pump with slide hammers. Remove and discard the old pump gasket.
12. Remove the intermediate clutch plates from the case.
13. Rotate the transmission case so that it is in a horizontal position. Remove the clutch assembly. Remove the overdrive band.
14. Remove the intermediate clutch assembly. Retain the #9 thrust washer and the needle bearings from the geartrain.
15. Remove the forward and reverse clutch assemblies. Retain the needle bearings and thrust bearing. Inspect each part as it is removed.
16. Pull out the direct drive shaft.

NOTE: Because both ends of the shaft are identical, it is possible to replace the shaft facing the wrong way. Mark the shaft for proper reassembly direction.

17. Locate and remove the center support retainer and note the position of the tabs for reassembly. Remove the anti-clunk spring and note the location for reassembly.
18. Measure output shaft end play.

In the spaces provided, write the manufacturer's specification and measured output shaft end play.

Manufacturer's Specified Output Shaft End Play .... ____________________

NAME ______________________ SECTION ____________ DATE ____________

Measured Output Shaft End Play .... ________________

Instructor Check: ________________

If end play is not to specifications, replace selective fit thrust washers and/or other parts as required.

19. Remove the center support and low/reverse planetary gearset. Separate the direct clutch hub at the rear from the low/reverse planetary gearset.
20. Remove the low/reverse band. Remove the needle bearing and spacer from the direct clutch drum. Inspect all needle bearings as they are removed.
21. Unbolt and remove the extension housing. Discard the old housing gasket. Clean the mating flanges.
22. Remove the retaining snap ring and the governor assembly from the output shaft. Remove the governor drive ball from the depression in the shaft.
23. Remove the output shaft, ring gear, and direct clutch assembly from the front of case.

CAUTION: Work carefully to avoid damaging the seal ring grooves and the case mating surface.

24. Remove the output-shaft needle bearings. If damaged, remove the rear case bushing.
25. Disassemble the oil pump. Remove the spring retainer assembly. Note the position of the bleed hole.
26. If the springs on the retainer are damaged, replace the entire spring retainer assembly.
27. Rotate, lift, and remove the intermediate clutch piston from the oil pump assembly. Remove and discard the inner and outer lip seals on the piston.
28. Adjust transmission end play to specifications with the selective fit nylon stator-support thrust washer. Fit the tang into the locator slot.
29. Coat the clutch piston seals and pump sealing area with petroleum jelly. Install new seals.
30. Rotate the low/reverse planetary assembly center support counterclockwise to remove the overrunning clutch.
31. Inspect the overrunning clutch assembly. Replace the entire clutch assembly if defects are found.
32. Remove the waved snap ring from the output shaft and direct clutch assembly. Remove the shaft from the ring gear. Install new sealing rings on the output shaft.
33. Remove the direct clutch assembly from the ring gear. Pry out the snap ring. Remove the pressure plate and clutch pack. Install a piston compressor and remove the snap ring.
34. Remove the spring retainer assembly and piston from the drum. Lubricate and install new piston seals. Be sure the check ball in the piston is free.
35. Lubricate and install the outer piston seal.
36. Place the piston spring and retainer assembly into the drum.
37. Install the compressor tool. Compress the spring retainer and install the snap ring.
38. Install the clutch pack, pressure plate, and snap ring. Place the needle bearings and clutch housing inside of the ring gear. Replace the thrust washer and spacer.

39. Make sure the thrust washer tabs face away from the clutch drum. Install and seat the direct clutch hub.
40. Disassemble each remaining clutch drum in the same way. Pry the snap rings loose from the clutch drum grooves, and remove the pressure plates and clutch packs.
41. Inspect each clutch pack for burned or scored plates. Install new friction and steel plates where needed. Compress the retainers and remove the pistons. Install all new seals. Reassemble the clutch packs.
42. Replace the pressure plates. Install the retaining snap rings.
43. Check the multiple-disc clutch clearances. Replace the selective-fit snap ring to increase or decrease clearance to specifications.

Instructor Check: ___________________

44. Remove the snap ring and the retaining plate to disassemble the overrunning clutch from the reverse clutch drum.
45. Remove the outer race from the overrunning clutch assembly. Replace the entire overrunning clutch assembly if defects are found.
46. Place the outer race over the overrunning clutch, with chamfered side facing the drum. Install the retaining plate and snap ring over the overrunning clutch assembly.
47. Lubricate the race side of #9 needle bearing with petroleum jelly. Install the bearings in the transmission case.
48. Install the output shaft, ring gear, and direct clutch assembly into the transmission case.

CAUTION: Do not damage the bushing or seal rings during installation.

49. Support and insert the output shaft through the case.
50. Install the governor drive ball into the output shaft depression. Seat the governor assembly against the case casting.
51. Replace the retaining snap ring.
52. Replace defective bands.
53. Install the low/reverse band in the case. Locate the band positioning ears against the transmission case anchor pins.
54. Position the needle bushing on the center support and the low/reverse planetary gearset.
55. Install the center support and mesh the low/reverse planetary gearset.
56. Install the center support anti-clunk spring. Check that it applies spring tension.
57. Install and seat the center-support snap ring in the case groove next to the low/reverse planetary gearset.
58. Install the direct drive shaft facing in the correct direction.
59. Lubricate the inner race of the #5 needle bearing with petroleum jelly. Slide the bearing onto the sun gear shaft, with the tabs facing away from the gear. Insert the sun gear shaft into the drive shell.
60. Assemble the forward and reverse clutch assemblies. Install the needle bearing and forward clutch hub.
61. Assemble the forward and reverse sun-gear shell assemblies, and the forward and reverse clutch packs into a single unit. Install these assemblies onto the direct drive shaft.

62. Bend and insert a welding rod into the band ears and install the band. Remove the welding rod from the band ears.

63. Position the band against the anchor pin in the case. Seat the band in the case.

64. Install the pressure plate. Install the friction and steel plates.

65. Check intermediate clutch pack clearance. Install a selective-fit outer steel separator plate to increase or decrease clearance to specifications.

Instructor Check: ____________________

66. Place the end-play checking tool across the oil pump case mounting. Use a depth micrometer to measure end play. Repeat the measurement 180 degrees opposite the first reading. Record the average reading.

Manufacturer's Specified End Play ....____________________

Measured End Play 1.____________________ 2.____________________

Average End Play ....____________________

Instructor Check: ____________________

67. Check the depth at the thrust face of the reverse clutch drum. Repeat the measurement 180 degrees opposite the first reading. Record the average reading.

Manufacturer's Specified End Play ....____________________

Measured End Play 1.____________________ 2.____________________

Average End Play ....____________________

Instructor Check: ____________________

68. If necessary, install a new separator plate or thrust washer on the oil pump. Install a new pump gasket. Lower the pump into the case. Torque the bolts to specifications.

Manufacturer's Specified Torque ....____________________

69. Install the new pump seal. Install a new seal and gasket on output shaft extension housing. Install the housing and torque the bolts to specifications.

Manufacturer's Specified Torque ....____________________

70. Install new seals on the 2–3 accumulator piston. Install the 2–3 accumulator piston and spring in the case. Replace the piston cover and snap ring. Repeat the procedure with the 3–4 accumulator.

71. Check the piston-rod bore to make sure that the low/reverse band is properly aligned. Install the low/reverse band spring, piston, and cover.

72. Center the overdrive band under the piston hole with a screwdriver. Install the piston, spring, and cover. Tap the cover in place, and install the snap ring.

Instructor Check: ____________________

73. Connect the manual valve to the inner manual-shaft shift linkage. Connect the TV plunger to the inner TV lever. Place the valve body in the transmission case. Install the retaining bolts finger tight.

74. Reconnect the inner linkage spring to the separator plate notch. Torque the valve body bolts to specifications.

Manufacturer's Specified Torque .... ____________

75. Replace the detent spring and bolt. Torque to specifications.

Manufacturer's Specified Torque .... ____________

Install a new filter seal and gasket. Install a new filter on the valve body. Torque the bolts to specifications.

Manufacturer's Specified Torque .... ____________

Instructor Check: ____________

76. Place a new gasket on the pan. Install the pan. Tighten the pan bolts to specifications.

Manufacturer's Specified Torque .... ____________

77. Replace the torque converter.

Job Completion Report

1. What difficulties, if any, did you encounter in the overhaul procedure?

____

____

____

____

____

____

INSTRUCTOR SIGNOFF ____________ **DATE** ____________

NAME ______________________ SECTION ____________ DATE ____________

Certification Practice

Test 16

DIRECTIONS:

The following questions are similar to those used on mechanic certification tests. Answer each question by circling the letter of the correct choice.

1. The Ford AOD has all of the following EXCEPT
   A. one direct drive clutch.
   B. two servos and bands.
   C. three overrunning clutches.
   D. four multi-disc clutch assemblies.

2. The Ford AOD drives fourth gear
   A. mechanically.
   B. hydraulically.
   C. electrically.
   D. hydropneumatically.

Question 3.

3. The part shown is the
   A. overdrive servo piston.
   B. low/reverse band piston.
   C. 2-3 accumulator piston.
   D. overdrive clutch.

4. Mechanic A says the Ford AOD has only metric fasteners.
   Mechanic B says the Ford AOD has only SAE fasteners.
   Who is correct?
   A. A only B. B only C. Both A and B D. Neither A nor B

5. A servo or piston can be removed from its bore by using
   A. heat.
   B. compressed air.
   C. brake fluid.
   D. Ford type CJ or DEXRON II transmission fluid.

6. Which of the following statements is correct?
   I. The Ford AOD oil pump is removed with slide hammers.
   II. The Ford AOD oil pump is removed with air pressure.
   A. I only B. II only C. Both I and II D. Neither I nor II

7. The setup shown is used for checking
   A. rotor wear.
   B. end play.
   C. pinion depth.
   D. accumulator bore dimension.

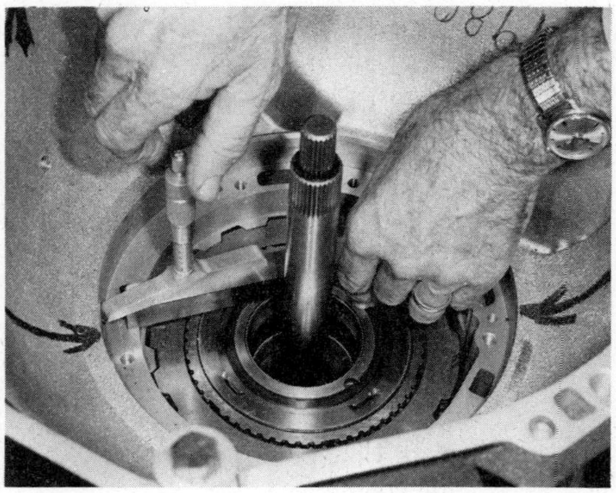

Question 7.

8. Mechanic A says seals and gaskets must always be replaced during transmission overhaul.
   Mechanic B says seals only need to be replaced if they are leaking.
   Who is correct?
   A. A only B. B only C. Both A and B D. Neither A nor B

9. End play on the Ford AOD is checked with
   A. feeler gauges.
   B. dial indicator.
   C. vernier calipers.
   D. depth micrometer.

10. Which part of the Ford AOD can be mistakenly installed facing the wrong way?
    A. Sun gear
    B. Valve Body
    C. Thrust bearing
    D. Direct-drive shaft

17 Installing the Transmission

YOUR LEARNING JOB

Placing the automatic transmission back in the vehicle is a heavy, difficult job. All parts must be aligned and connected properly. This unit deals with proper installation procedures.

JOB COMPETENCY OBJECTIVES

This is a skill-building unit. When you have completed your assignments and exercises in this unit, you should be able to:

☐ Install the torque converter.

☐ Install the automatic transmission.

☐ Replace the control linkages and systems.

17.1 TORQUE CONVERTER AND FLEXPLATE INSTALLATION

As with removing the transmission, installing the transmission has many variations. The installation procedure discussed here is for the Chrysler TorqueFlite.

First be sure that the converter drain plug is installed and torqued to specifications.

The transmission and the torque converter must be installed as an assembly. If not, the flexplate, pump bushing, and oil seal will be damaged.

Remember, the flexplate will *not* support a load. No transmission weight should be allowed to rest on the flexplate during installation.

To install the torque converter and flexplate, use the following procedure:

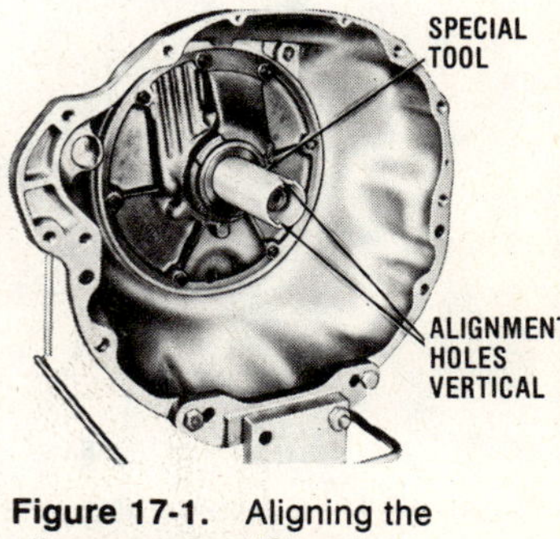

Figure 17-1. Aligning the oil pump rotor. (Chrysler Corporation)

1. Turn the pump rotors with a special tool until the two small holes in the handle are vertical. See Figure 17-1. Remove the tool.

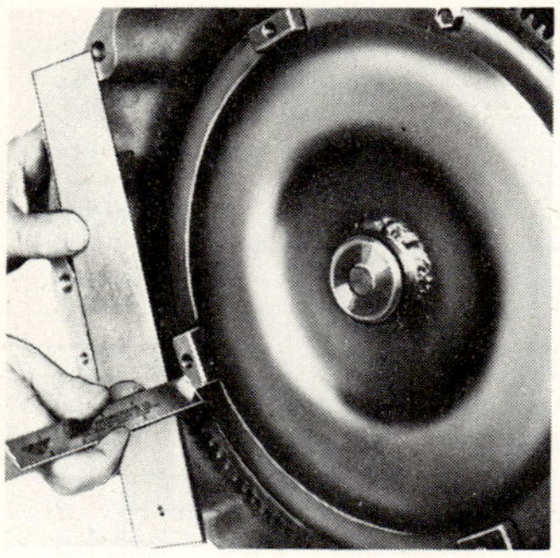

Figure 17-2. Measuring the torque converter for full engagement in transmission. (Chrysler Corporation)

2. Carefully slide the torque converter assembly over the input shaft and the reaction shaft. Make sure the converter hub slots are also vertical and fully engage the pump inner rotor lugs.

3. To test for full engagement, place a straightedge on the face of the transmission case. See Figure 17-2. The surface of the front cover lug of the converter should be at least 1/2 inch [12.7 mm] to the rear of the straightedge. The converter must be pushed all the way into the transmission when this reading is taken.

4. Attach a small C-clamp to the edge of the converter housing. This holds the converter in place during installation.

5. Inspect the flexplate for distortion or cracks. Replace if necessary.

6. Install the bolts that hold the flexplate to the crankshaft. Torque the bolts to specifications.

NOTE: If the flexplate was replaced, make sure both transmission dowel pins are in the engine block. The dowel pins must protrude far enough to hold the transmission in alignment.

7. Lubricate the converter hub bore in the crankshaft with multi-purpose grease.

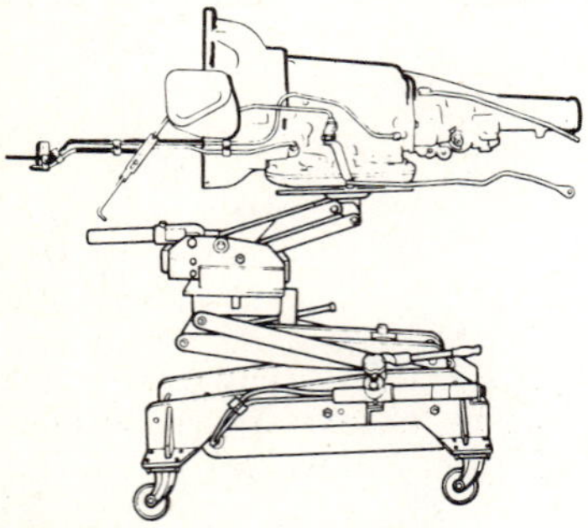

Figure 17-3. Use a suitable jackstand to raise the transmission back into position. (Chrysler Corporation)

17.2 INSTALLING THE TRANSMISSION

Remember, the automatic transmission is very heavy. Use every safety precaution, including safety glasses, to prevent serious personal injury. Always have a helper assist when installing the transmission.

Check that the engine safety support stand is solidly in position.

Be sure all linkages and wiring that are to be reconnected are in good condition. Replace any parts that do not look good. Make sure all linkages and other parts are held or wired out of the way for easier transmission installation.

1. Place the transmission on a suitable jack. See Figure 17-3. Use a safety strap or safety chain to hold the transmission on the jack.

2. Position the transmission assembly under the vehicle. Jack up the transmission, raising or tilting it until it is aligned with the engine.

3. Remove the small C-clamp holding the converter. Rotate the torque converter until the mark made during removal aligns with the mark on the flexplate. The offset holes are located in the inner circle of the flexplate. A stamped "v" identifies the offset hole in the converter front cover. Markings are shown in Figure 17-4.

Figure 17-4. Converter and flexplate markings. (Chrysler Corporation)

4. Carefully work the transmission assembly forward over the engine block dowels. Be sure the converter hub enters the crankshaft opening.

5. After the transmission is in position, install the lower bellhousing bolts.

6. Raise the transmission slightly, and install the upper bellhousing bolts.

7. Torque the bellhousing bolts to specifications. Tighten the bottom bolts first, then the top bolts.

NOTE: Reinstall the vibration dampener weight on the rear of the extension housing, if so equipped.

8. Install the crossmember to the frame.

9. Remove the engine support fixture.

10. Lower the transmission to install the extension housing mount to the crossmember. Torque bolts to specifications. See Figure 17-5.

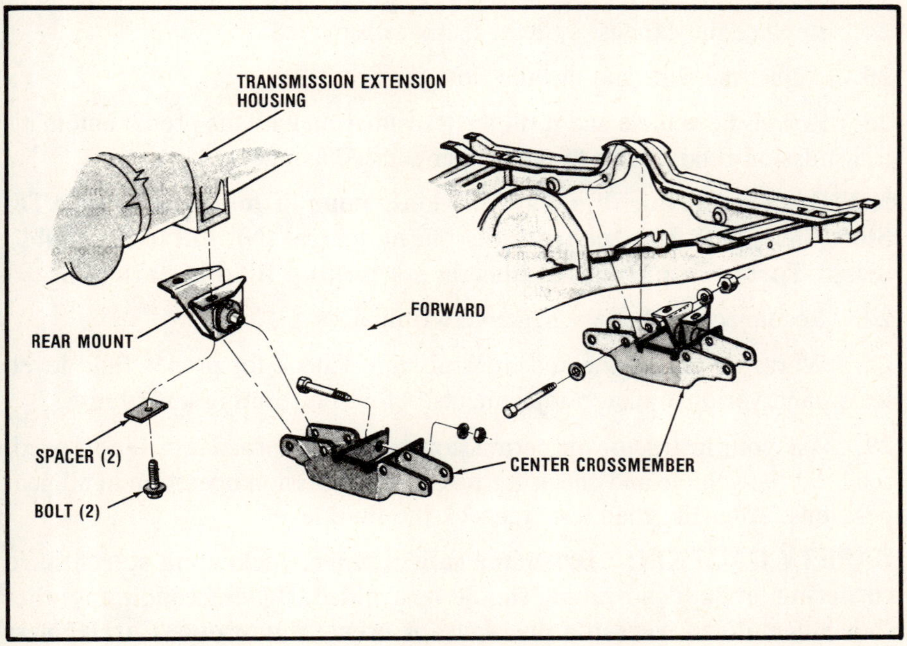

Figure 17-5. Center crossmember and rear engine mount. (Chrysler Corporation)

NOTE: Crossmember connections vary according to the vehicle being repaired. Check the proper service manual.

11. Remove the safety strap or safety chain from the transmission jack. Remove the transmission jack.

12. Install the oil filler tube, taking care not to dislodge dirt or other foreign material in the opening.

13. Install the speedometer sleeve and cable. The sleeve and pinion should have stayed with the transmission.

14. Connect the throttle rod to the transmission lever.

15. Connect the gearshift rod and torque shaft assembly to the transmission lever and frame.

16. Reconnect the combination backup light and neutral starting switch.

17. Lubricate the slip joint inside and outside. Carefully slide the slip joint end of the driveshaft into the extension housing and onto the output shaft splines.

18. Align the marks at the rear universal joint and the companion flange.

19. Rotating the engine clockwise with a socket wrench on the vibration damper, align the matchmarks made during removal. Install the bolts through the flexplate and into the torque converter and torque to specifications.

20. Install the torque converter access cover.

21. Install the starter motor. Torque the bolts to specifications. Install the starter motor wiring.

22. Install the cooler line bracket, attach the cooler lines, and torque to specifications.

23. If necessary, install the struts from the engine to the transmission.

NOTE: Tighten the bolts holding the strut to the transmission before tightening the strut to the engine.

24. Replace the exhaust system, if it was removed.

25. Adjust the shift and throttle linkage. See 3.9.

26. Lower the vehicle and refill the transmission with the proper automatic transmission fluid. Refer to the proper service manual.

NOTE: On TorqueFlite transmissions, pour in four quarts of ATF. Start the engine with the transmission in neutral (N). Fill to the "add" mark. TorqueFlite transmissions do *not* pick up fluid in park (P).

27. Reconnect the battery negative terminal.

28. Warm up the engine and transmission. Check for proper fluid level, leaks, and various linkage adjustments before performing a road test.

29. Ask your instructor for permission to conduct a road test. If approved, road test the vehicle and check for proper transmission operation in all gear positions. After the road test, recheck the fluid level.

SAFETY CAUTION: Drive in a safe manner. Jackrabbit starts, hard cornering, or sudden braking should be avoided. Driving conditions other than normal can worsen a problem, or create a new one. Carelessness during the test drive can result in an accident. Remember, you are responsible for someone else's vehicle. Drive only in a manner that is required to check or diagnose proper transmission operation.

NAME ______________________ SECTION ____________ DATE ____________

Job Sheet 17-1

Installing the Transmission

Before you begin:

Read Topics 17.1 through 17.2.

Vehicle Make ____________________ Model ____________ Year ____________

Time Started ____________________ Time Finished ____________________

Job Completion Time ______________ Flat Rate Time ____________________

Tools Required

- Safety glasses
- Special oil-pump rotor alignment tool
- Small C-clamp
- Sockets, extensions, handle
- Torque wrench
- Transmission jack and safety straps or chains
- Open-end or combination wrenches
- Pliers
- Screwdrivers
- Transmission funnel

References

Manufacturer's service manual ______________________________

Repair manual ______________________________

Service bulletin ______________________________

Procedure: Install Torque Converter and Flexplate

1. If recommended, align the pump rotors or gears with the necessary tool.

Instructor Check: ____________________

Remove the tool.

2. Replace the torque converter assembly on the transmission. Engage the drive lugs to the pump.

Instructor Check: ____________________

3. Attach a small C-clamp to the edge of the converter housing to hold the converter to the bellhousing during installation.
4. Inspect the flexplate for distortion or cracks. Replace the flexplate if defects are found. Torque the bolts that hold the flexplate to the crankshaft to the specified torque.

Manufacturer's Specified Torque .... ____________________

NOTE: If the flexplate is replaced, make sure both transmission dowel pins are still in the engine block. The dowel pins must protrude far enough to hold the transmission in alignment during installation.

5. Lubricate the converter hub bore in the crankshaft with multi-purpose grease.

Instructor Check: ____________________

Procedure: Install Transmission

1. Place the transmission on a suitable jack. Secure the transmission to the jack with safety straps or chains.

SAFETY CAUTION: The transmission is extremely heavy. If it falls, it will damage the transmission and/or cause serious physical injury. Make sure the transmission is securely mounted before moving the jack. Have a helper assist in holding and positioning the transmission during installation.

2. Position the transmission assembly under the vehicle. Jack up the transmission, raising or tilting it until it is aligned with the engine.
3. Rotate the torque converter to align the matchmarks made on the converter and on the flexplate before removal.

Instructor Check ____________________

4. Remove the C-clamp holding the converter to the bellhousing. Carefully move the transmission assembly forward over the engine block dowels. Be sure the converter hub enters the crankshaft opening.
5. Push the transmission until the bellhousing is against the engine mating surface. Install the lower bellhousing bolts finger-tight.
6. Raise the transmission slightly, and install the upper bellhousing bolts.
7. Torque the bottom bolts to specifications. Torque the top bolts to specifications.

In the space provided, write the correct torque specification.

Manufacturer's Specified Torque .... ____________________

NOTE: Reinstall the vibration damper weight on the rear of the extension housing, if so equipped.

8. Install the crossmember to the frame.
9. Lower the transmission to install the extension housing mount to the crossmember. Torque bolts to specifications.

Manufacturer's Specified Torque .... ____________________

NOTE: Crossmembers connections vary according to the vehicle being repaired. Check the proper service manual.

10. Remove the safety strap or safety chain from the transmission jack. Remove the transmission jack.
11. Remove the engine support fixture.

Instructor Check: ____________________

NAME ____________________ SECTION ______________ DATE ______________

12. Install the oil filler tube. Be careful not to dislodge any dirt into the filler tube hole.
13. Install the speedometer. The sleeve and pinion should have stayed with the transmissions. If not, install the sleeve and pinion. Install the cable.
14. Connect and adjust all throttle, kickdown, and gearshift linkage.
15. Reconnect all electrical wires to the proper connections.
16. Lubricate the slip joint on the inside and the outside. Carefully slide the slip joint end of the driveshaft into the extension housing and onto the output shaft splines.
17. Rotate the driveline to align the matchmarks made on the rear universal joint and the companion flange before disassembly. Reconnect the driveline and torque the bolts or nuts to specifications.

Manufacturer's Specified Torque .... ______________

18. Rotate the engine clockwise to align the matchmarks made on the flexplate and torque converter before removal. Install the bolts and torque to specifications.

Manufacturer's Specified Torque .... ______________

19. Install the torque converter access cover.
20. Install the starter motor. Torque the bolts to specifications.

Manufacturer's Specified Torque .... ______________

Install the starter motor electrical connections.

Instructor Check: ______________

21. Install the cooler line bracket, attach the cooler lines, and torque to specifications.

Manufacturer's Specified Torque .... ______________

22. Install struts from the engine to the transmission, if so equipped.

NOTE: Tighten the bolts holding the strut to the transmission before tightening the strut to the engine.

23. Replace the exhaust system, if it was removed.

Instructor Check: ______________

24. Lower the vehicle. Refill the transmission with the proper automatic transmission fluid.

NOTE: On TorqueFlite transmissions, add four quarts of ATF. Start the engine with the transmission in neutral (N). Fill the transmission to the "add" mark. TorqueFlite transmissions will *not* pick up fluid in park (P).

25. Reconnect the grounded battery cable to the battery.
26. Warm up the engine and transmission. Check the fluid level and check for leaks and for linkage adjustments before the road test.

Instructor Check: ______________

27. Ask your instructor's permission to conduct a road test. If approved, road test the vehicle and check for proper transmission operation in all gear positions.

SAFETY CAUTION: Drive in a safe manner. Jackrabbit starts, hard cornering, or sudden braking should be avoided. Driving conditions other than normal can worsen a problem, or create a new one. Carelessness during the test drive can result in an accident. Remember, you are responsible for someone else's vehicle. Drive only in a manner that is required to check or diagnose proper transmission operation.

Instructor Check: ____________________

Job Completion Report

1. Was it necessary to replace the flexplate? ______________________________
2. What difficulties, if any, were encountered during transmission installation?

____

____

____

INSTRUCTOR SIGNOFF ______________________ **DATE** ____________

NAME ______________________ SECTION ____________ DATE ____________

Certification Practice

Test 17

DIRECTIONS:

The following questions are similar to those used on mechanic certification tests. Answer each question by circling the letter of the correct choice.

1. Mechanic A says that the torque converter should be bolted to the flexplate before the transmission is installed.

   Mechanic B says that the torque converter and transmission should be installed as a unit.

   Who is correct?

   A. A only B. B only C. Both A and B D. Neither A nor B

2. During installation, a transmission cannot be pushed far enough forward to touch the engine mating surface. The most likely cause is

   A. incorrectly installed transmission internal parts.
   B. a misaligned flexplate.
   C. an improperly installed torque converter.
   D. movement of the engine crankshaft.

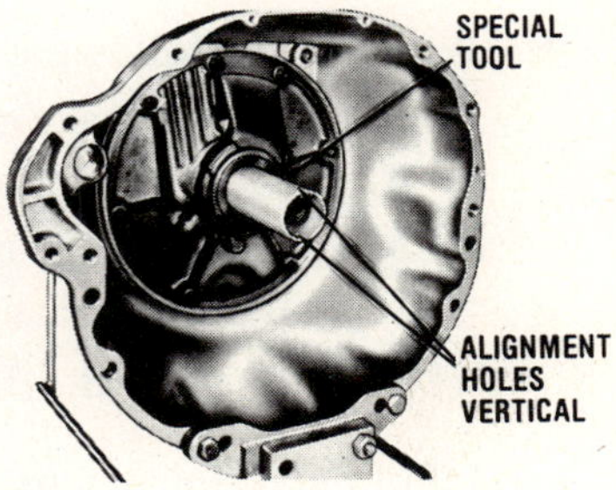

Question 3.
(Chrysler Corporation)

3. The tool shown is used to

   A. align TorqueFlite oil pump rotors before installation.
   B. remove oil pumps.
   C. align the transmission to the engine during transmission installation.
   D. install the front seal.

4. Which of the following statements is correct?

   I. A small C-clamp is used to hold the torque converter to the transmission during installation.
   II. The small C-clamp must be removed before the transmission is pushed all the way against the engine.

   A. I only B. II only C. Both I and II D. Neither I nor II

5. All of the following should be done before attempting to install the transmission EXCEPT

   A. Make sure the engine is properly supported.
   B. Make any necessary linkage or wiring repairs.
   C. Wire or hold all linkage and parts out of the way.
   D. Reconnect the battery.

6. Mechanic A says that the transmission jack can be removed after the crossmember is securely bolted to the vehicle chassis.

   Mechanic B says that the transmission jack can be removed after the transmission is securely bolted to the crossmember.

   Who is correct?

   A. A only B. B only C. Both A and B D. Neither A nor B

7. What additional equipment should be used with the tool shown?
   A. Lever handle
   B. Hydraulic line
   C. Air hose
   D. Safety straps or chains

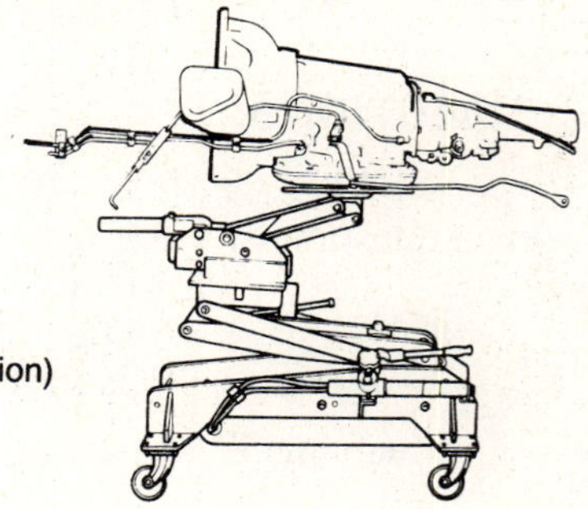

Question 7.
(Chrysler Corporation)

8. When should the battery ground cable be reconnected to the battery?
   A. Before the starter is installed
   B. Before the cooler lines are installed
   C. After all other mechanical work is completed
   D. After the road test

9. Which of the following statements is correct?
   I. Matchmarks made on the torque converter and flexplate must be aligned during transmission installation.
   II. Matchmarks made on the driveline and differential companion flange must be aligned during transmission installation.

   A. I only B. II only C. Both I and II D. Neither I nor II

10. All of the following are part of transmission installation procedures EXCEPT
    A. checking engine problems.
    B. refilling the transmission with the proper fluid.
    C. reconnecting the driveline.
    D. road testing the vehicle.

18

Turbo Hydra-Matic 200, 200C, 200-4R, 325, 325-4L

YOUR LEARNING JOB

This series of General Motors transmissions are similar. The biggest differences are in the 200-4R and in the 325. This unit describes the differences and illustrates the various assemblies found in this General Motors family of transmissions.

JOB COMPETENCY OBJECTIVES

This is a knowledge-building unit. When you have completed your assignments and exercises in this unit, you should be able to:

- ☐ Explain the major assemblies shared by each transmission.
- ☐ Describe the differences in the overdrive transmissions.
- ☐ Locate the transmission identification numbers.

18.1 TURBO HYDRA-MATIC 200/200C DESCRIPTION

All General Motors automobiles and the GMC Caballero and GMC Sprint have used the 200 and 200C transmission. Beginning in 1976, these transmissions have been used with four-cylinder, V-6, V-8, and diesel engines. They appear in subcompacts as well as full-sized General Motors cars. The 200 and 200C transmissions consist of the following main parts:

- Torque converter
- Lockup torque converter (200C)
- Three multiple-disc clutches
- One overrunning clutch
- Intermediate band and servo
- One-piece aluminum transmission case with integral bellhousing and extension housing.
- Compound planetary gearset
- Integral valve body

The 200/200C are all-metric transmissions. Both transmissions combine a torque converter with a fully automatic three-speed gear system. The bellhousing, transmission case, and extension housing are a single aluminum casting.

The torque converter is a three-element unit that connects to the crankshaft through a flexplate or flywheel. A lockup converter is used on the 200C, which is the biggest difference between the two transmissions.

The lockup torque converter has a lockup mechanism in the converter, a converter apply valve in the oil pump, and a solenoid. Numerous other switches and valves also control the operation of the lockup converter.

The compound planetary gear system has two sun gears. The front sun gear meshes with the rear sun gear. The front sun gear is also meshed to the input shell.

The hydraulic system has one oil pump. The bottom half of the valve body is removable. The top half of the valve body is cast into the transmission case. Because the entire valve body is not removable, this type of valve body assembly is referred to as an integral valve body. Integral valve bodies are common in General Motors vehicles.

Numerous changes have been made to the 200/200C over the years. Always refer to the proper service manual when servicing one of these transmissions.

Figures 18-1 through 18-5 show the major parts of the 200/200C transmissions.

18.2 TURBO HYDRA-MATIC 200-4R DESCRIPTION

The 200-4R is an overdrive transmission that has been used in full-size General Motors cars since 1982. It is used with V-6 and V-8 gasoline and V-8 diesel engines. The 200-4R transmission consists of the following main parts:

- Torque converter
- Lockup torque converter
- Five multiple-disc clutches
- Two overrunning clutches
- One servo and band
- Compound planetary gearset
- Overdrive planetary gearset
- Integral valve body
- Aluminum transmission case with integral bellhousing and extension housing.

The working parts of the 200-4R are similar to those used in the 200C. The basic difference is that the 200-4R has two extra multiple-disc clutches, an extra overrunning clutch, and an additional overdrive planetary gearset. The overdrive parts are located at the forward end of the transmission case. Figure 18-6 shows where the added parts are located in the 200-4R.

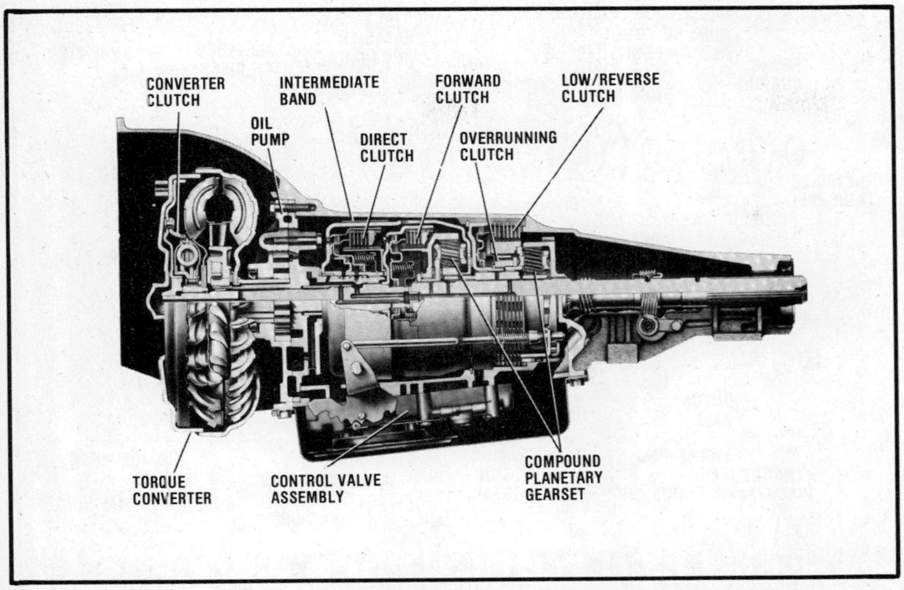

Figure 18-1. Turbo Hydra-Matic 200C automatic transmission cutaway. (Oldsmobile Division-GMC)

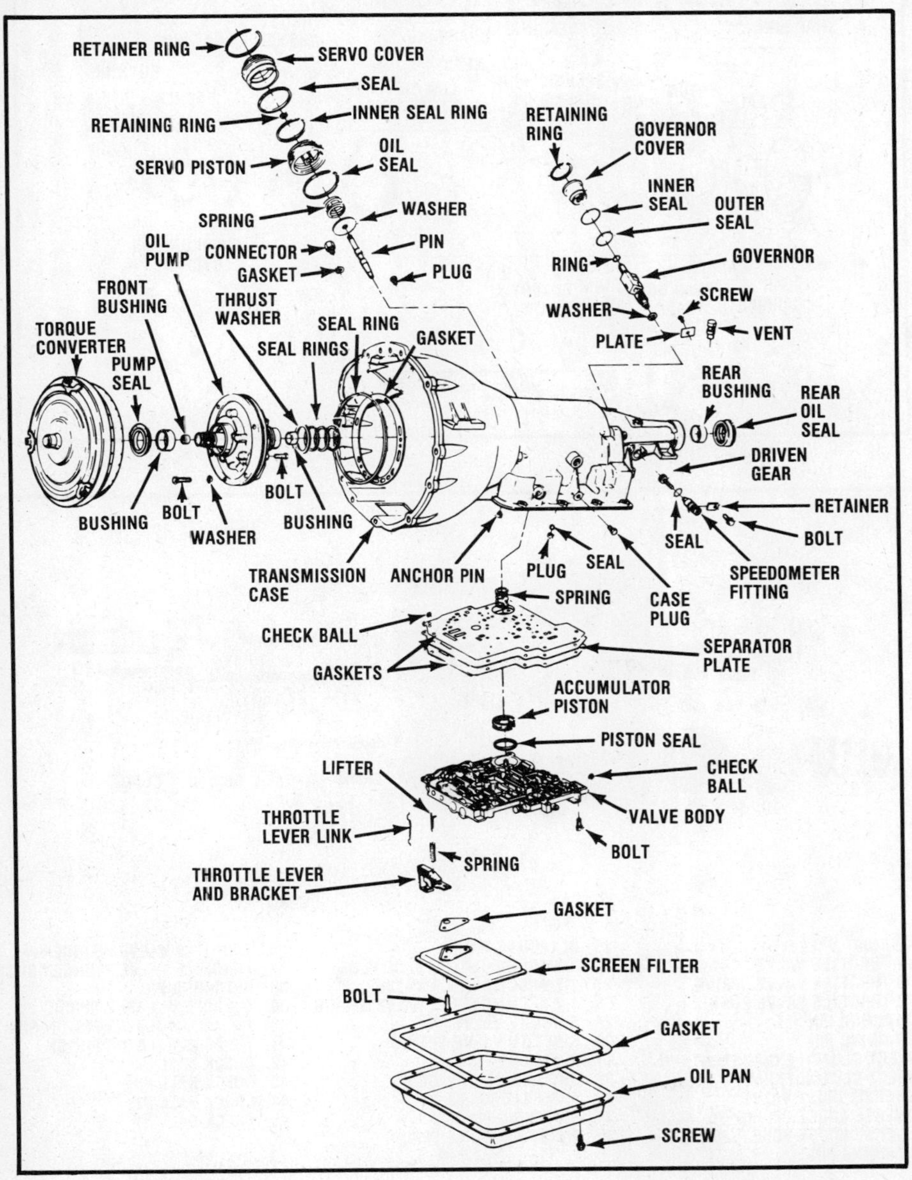

Figure 18-2. Turbo Hydra-Matic 200 transmission case, governor, servo, torque converter, oil pump, and valve body. (Pontiac Motor Division-GMC)

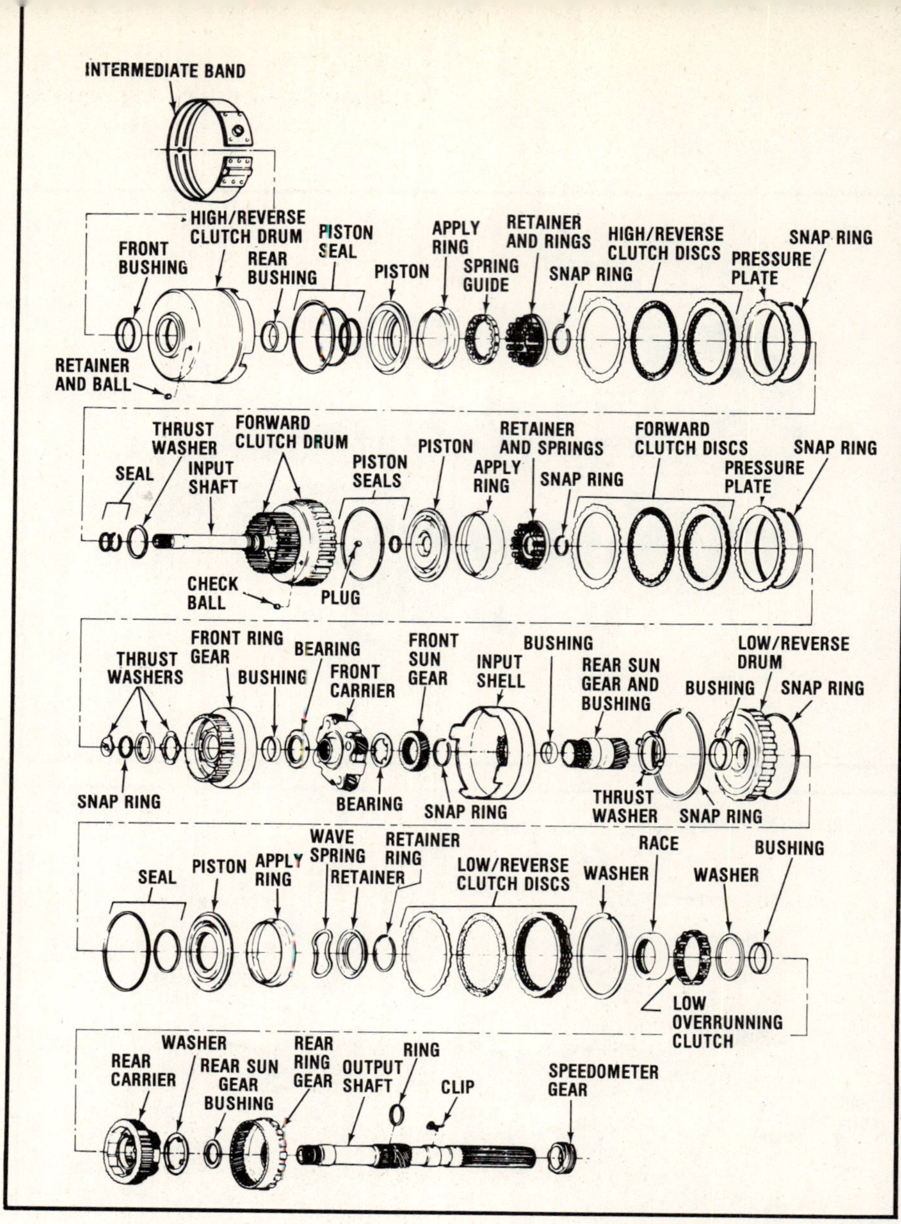

Figure 18-3. Turbo Hydra-Matic 200 clutches and planetary gearsets. (Pontiac Motor Division-GMC)

Figure 18-4. Turbo Hydra-Matic 200C control valve assembly. (Oldsmobile Division-GMC)

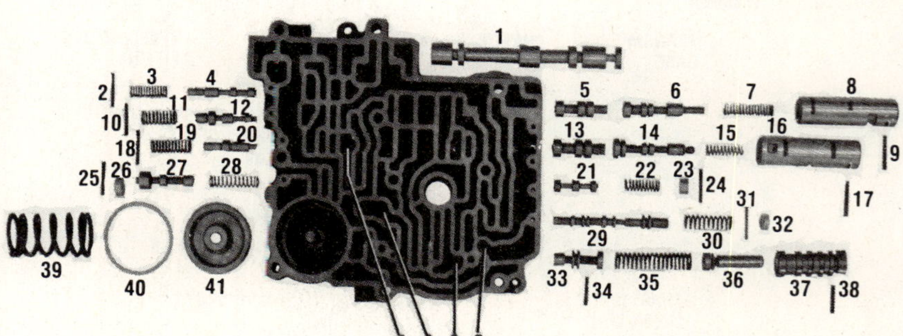

1. MANUAL VALVE
2. RETAINING PIN
3. INTERMEDIATE BOOST SPRING
4. INTERMEDIATE BOOST VALVE
5. 2-3 SHIFT VALVE
6. 2-3 THROTTLE VALVE
7. 2-3 THROTTLE VALVE SPRING
8. 2-3 THROTTLE VALVE BUSHING
9. RETAINING PIN
10. RETAINING PIN
11. LOW OVERRUNNING CLUTCH SPRING
12. LOW OVERRUNNING CLUTCH VALVE
13. 1-2 SHIFT VALVE
14. 1-2 THROTTLE VALVE
15. 1-2 THROTTLE VALVE SPRING
16. 1-2 THROTTLE VALVE BUSHING
17. RETAINING PIN
18. RETAINING PIN
19. DIRECT CLUTCH EXHAUST SPRING
20. DIRECT CLUTCH EXHAUST VALVE
21. REVERSE BOOST VALVE
22. REVERSE BOOST SPRING
23. REVERSE BOOST BORE PLUG
24. RETAINING PIN
25. RETAINING PIN
26. 1-2 ACCUMULATOR BORE PLUG
27. 1-2 ACCUMULATOR VALVE
28. 1-2 ACCUMULATOR VALVE SPRING
29. SHIFT TV VALVE
30. SHIFT TV VALVE SPRING
31. RETAINING PIN
32. SHIFT TV BORE PLUG
33. THROTTLE VALVE
34. RETAINING PIN
35. THROTTLE VALVE SPRING
36. THROTTLE VALVE PLUNGER
37. THROTTLE VALVE PLUNGER BUSHIN
38. RETAINING PIN
39. 1-2 ACCUMULATOR SPRING
40. 1-2 ACCUMULATOR PISTON SEAL
41. 1-2 ACCUMULATOR PISTON
42. CHECK BALL #4
43. CHECK BALL #3
44. CHECK BALL #2
45. CHECK BALL #1

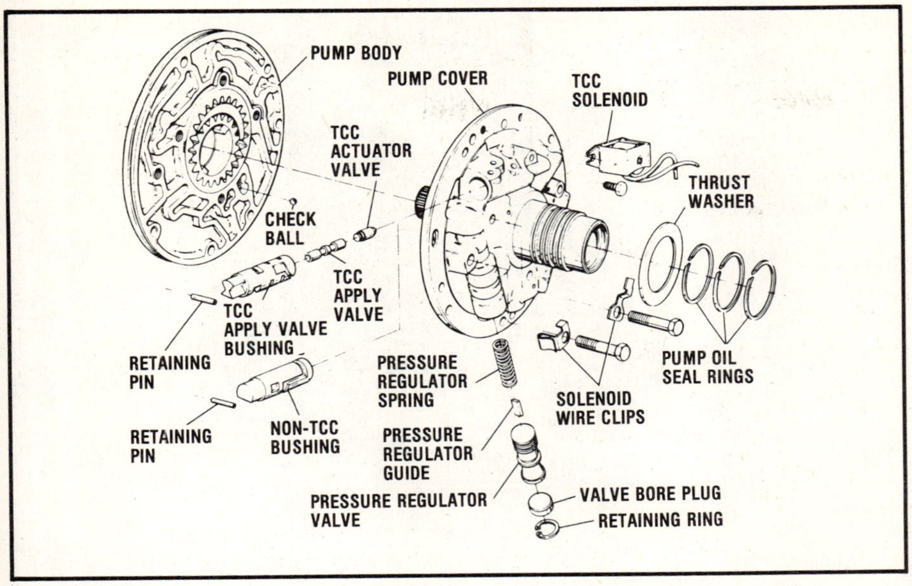

Figure 18-5. Turbo Hydra-Matic 200C oil pump assembly. (Oldsmobile Division-GMC)

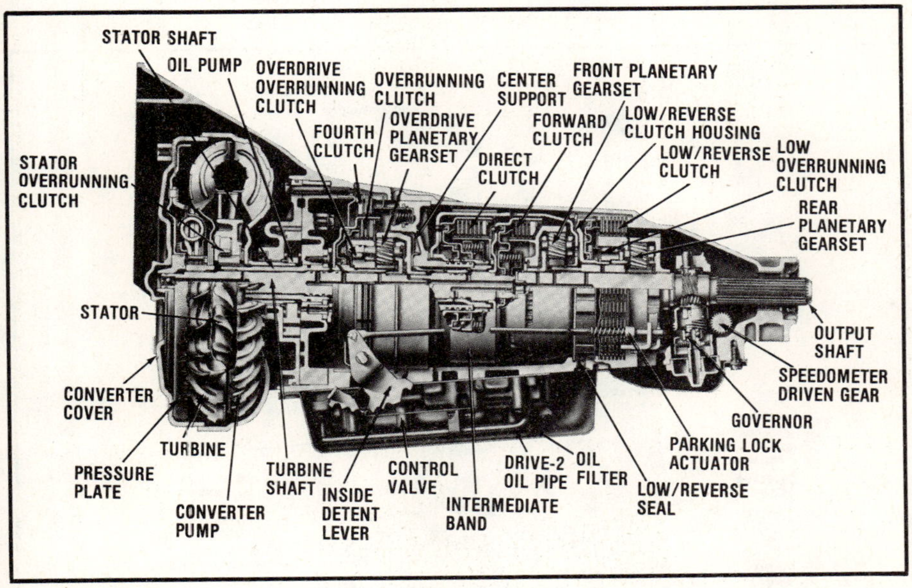

Figure 18-6. Turbo Hydra-Matic 200-4R automatic transmission cutaway. (Oldsmobile Division-GMC)

18.3 TURBO HYDRA-MATIC 325/325-4L DESCRIPTION

The 325 is a front-drive transmission used in Cadillac Eldorados, Buick Rivieras, and Oldsmobile Toronados. It is *not* a transaxle. Transaxles are discussed in Unit 25. The internal working parts are similar to those found in the 200C. The biggest differences are in the transmission case and the connection between the oil pump and the input shaft. See Figure 18-7.

The oil pump and input shaft use sprockets to drive a chain. This allows the transmission to redirect power to the front driving wheels.

The 325-4L is similar to the 325. However, like the 200-4R, it contains extra parts to accommodate an overdrive planetary gearset. See Figure 18-8.

With the exception of the chain and sprocket system, other overhaul procedures are similar to those for the 200C and 200-4R.

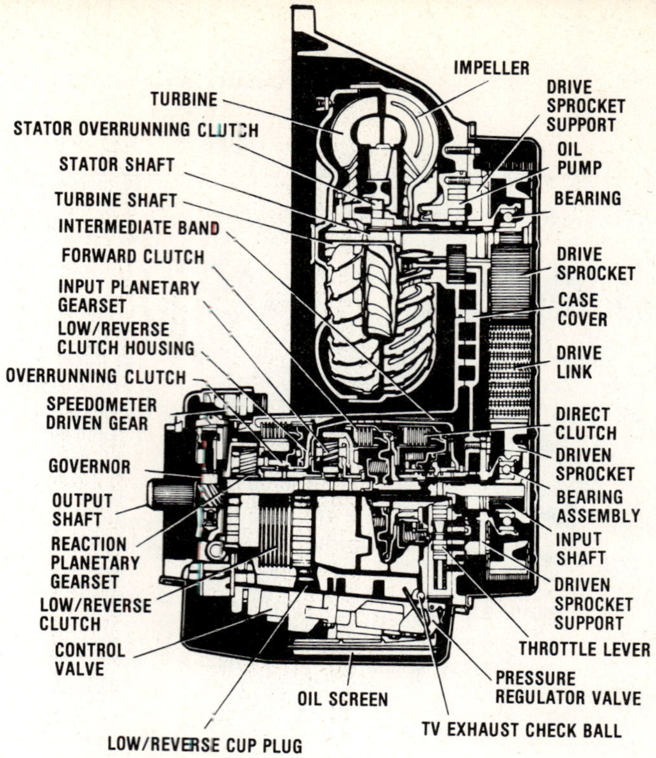

Figure 18-7. Turbo Hydra-Matic 325 automatic transmission cutaway. (Oldsmobile Division-GMC)

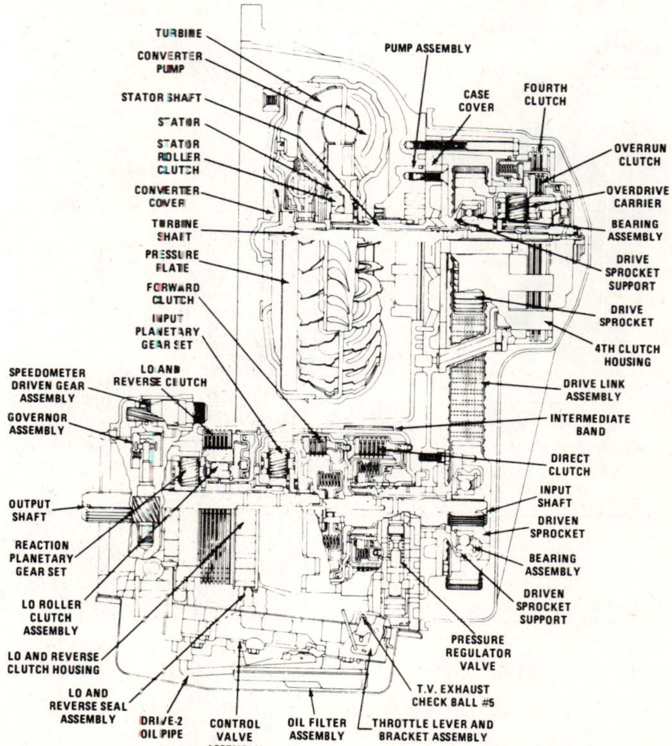

Figure 18-8. Turbo Hydra-Matic 325-4L automatic transmission cutaway. (Oldsmobile Division-GMC)

18.4 TRANSMISSION IDENTIFICATION

The word "metric" is stamped into the bottom of the 200/200C/200-4R oil pan. The transmission codes are stamped on a plate at the right side of the transmission case. On the 325 and 325-4L, the identification plate is located on the left side of the converter housing. Check the proper service manual to decipher the code.

NAME ____________ SECTION ____________ DATE ____________

Certification Practice

Test 18

DIRECTIONS:

The following questions are similar to those used on mechanic certification tests. Answer each question by circling the letter of the correct choice.

1. The GM Turbo Hydra-Matic 200/200C transmission has been used since 1976 with all of the following engines EXCEPT
   A. inline four-cylinder.
   B. inline six-cylinder.
   C. V-6.
   D. V-8.

2. Which of the following statements is correct?
   I. The GM 200 transmission has a torque converter lockup mechanism.
   II. The GM 200C transmission has a torque converter lockup mechanism.
   A. I only B. II only C. Both I and II D. Neither I nor II

3. The part indicated in the picture is a
   A. converter clutch.
   B. oil pump.
   C. forward clutch.
   D. direct clutch.

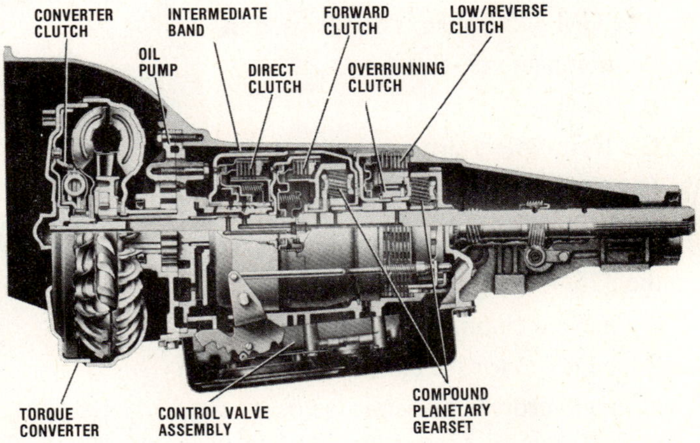

Question 3.
(Oldsmobile Division-GMC)

4. Mechanic A says the GM 200 transmission has some SAE fasteners.
   Mechanic B says the GM 200C transmission has some SAE fasteners.
   Who is correct?
   A. A only B. B only C. Both A and B D. Neither A nor B

5. The GM 200-4R transmission has all of the following EXCEPT
   A. two extra multiple-disc clutches.
   B. one extra overrunning clutch.
   C. one additional overdrive planetary gearset.
   D. two sprockets and a drive chain.

6. The GM 200-4R transmission is used with all of the following EXCEPT
   A. V-6 engines.
   B. V-8 engines.
   C. diesel engines.
   D. compact and intermediate-size vehicles.

7. Which transmission is shown in the picture?
   A. GM 200
   B. GM 200C
   C. GM 200-4R
   D. GM 325

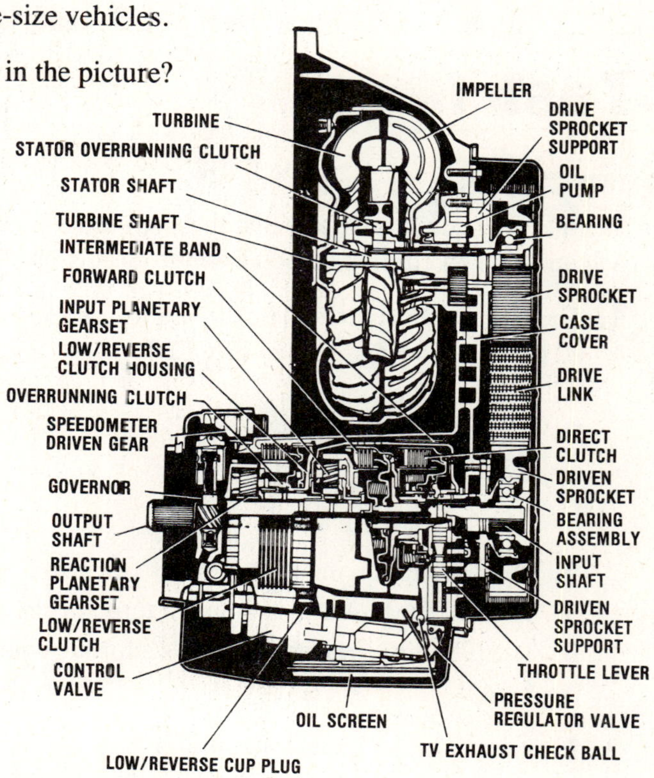

Question 7.
(Oldsmobile Division-GMC)

8. The GM 325 transmission is used in
   A. compact vehicles.
   B. trucks.
   C. front-wheel-drive vehicles.
   D. rear-wheel-drive vehicles.

9. The main difference between the GM 325-4L and the GM 325 transmissions is that the 325-4L has
   A. a drive chain.
   B. a lockup torque converter.
   C. an overdrive planetary gearset.
   D. a differential.

10. All of the following statements about identification are correct EXCEPT
    A. The word "metric" is stamped into the bottom of the oil pan on the GM 200/200C/200-4R.
    B. The GM 200/200C/200-4R have an identification number stamped on a plate at the right side of the transmission case.
    C. The GM 325/325-4L have an identification plate on the left side of the converter housing.
    D. The word "metric" is stamped into the bottom of the oil pan on the GM 325/325-4L.

19 Turbo Hydra-Matic 250, 250C, 350, 350C, 375B

YOUR LEARNING JOB

These General Motors transmissions are considered to be medium-duty. As with the transmissions discussed in Unit 18, these five transmissions are very similar. This unit covers the basic operation of this family of transmissions and illustrates the differences.

JOB COMPETENCY OBJECTIVES

This is a knowledge-building unit. When you have completed your assignments and exercises in this unit, you should be able to:

- ☐ Describe the main parts that are similar in each transmission.
- ☐ Explain the differences between each transmission.
- ☐ Identify each transmission.

19.1 TURBO HYDRA-MATIC 250/250C/350/350C/375B DESCRIPTION

This family of transmissions has been used in most General Motors cars of all sizes and light trucks since 1969. These transmissions consist of the following main parts:

- Torque converter
- Lockup torque converter (since 1980 on 250C and 350C)
- Three multiple-disc clutches (four on the 350, 350C, and 375B)
- One overrunning clutch (two on the 350, 350C, and 375B)
- Servo and intermediate band
- Compound planetary gearset
- Integral valve body
- Aluminum bellhousing and transmission case with removable extension housing.

These transmissions combine a torque converter with a fully automatic three-speed gear system. The bellhousing and the transmission case are a single aluminum casting.

The torque converter is a three-element unit that connects to the crankshaft through a flexplate or flywheel. The only difference between the 250/350 and the 250C/350C transmissions is the use of a lockup torque converter.

The lockup torque converter is controlled by a converter clutch apply valve located in the auxiliary valve body. The apply valve is controlled by a solenoid. Various vacuum and electrical controls also affect the operation of the lockup system.

The compound planetary gear system shares a common sun gear. The sun gear is connected to the front clutch by a driving shell.

The hydraulic system has a single oil pump and an integral valve body assembly.

The main difference between the 350 and 375B is that the 375B has a longer output shaft. Also, because the 375B is a heavier-duty transmission, it has five direct clutch plates instead of four.

Figures 19-1 through 19-6 illustrate the similarities and differences between each of these transmissions.

19.2 TRANSMISSION IDENTIFICATION

The identification number for the 250 transmission is stamped into the governor cover on the extension housing. On 350s prior to 1973, the identification number is stamped on the intermediate accumulator cover. Later 350s and 375Bs have the identification number stamped on the governor cover on the extension housing. Refer to the proper service manual to decipher these codes.

On some 250 and 350 transmissions, governor and accumulator covers may have been changed so that no number is evident. Also, identification plates on 350 and 375B can be missing.

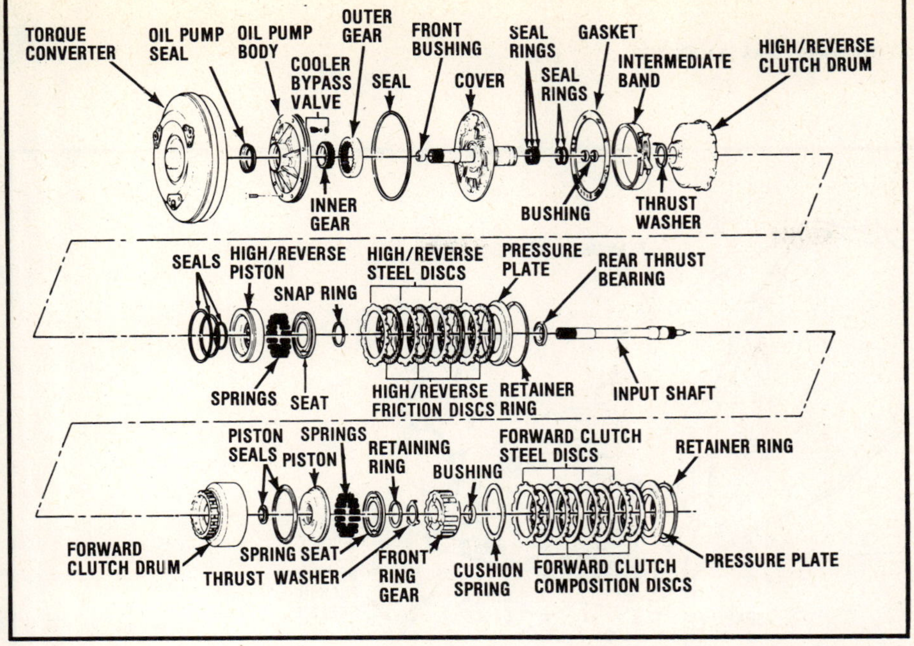

Figure 19-1. Turbo Hydra-Matic 250 oil pump and clutches. (Chevrolet Motor Division-GMC)

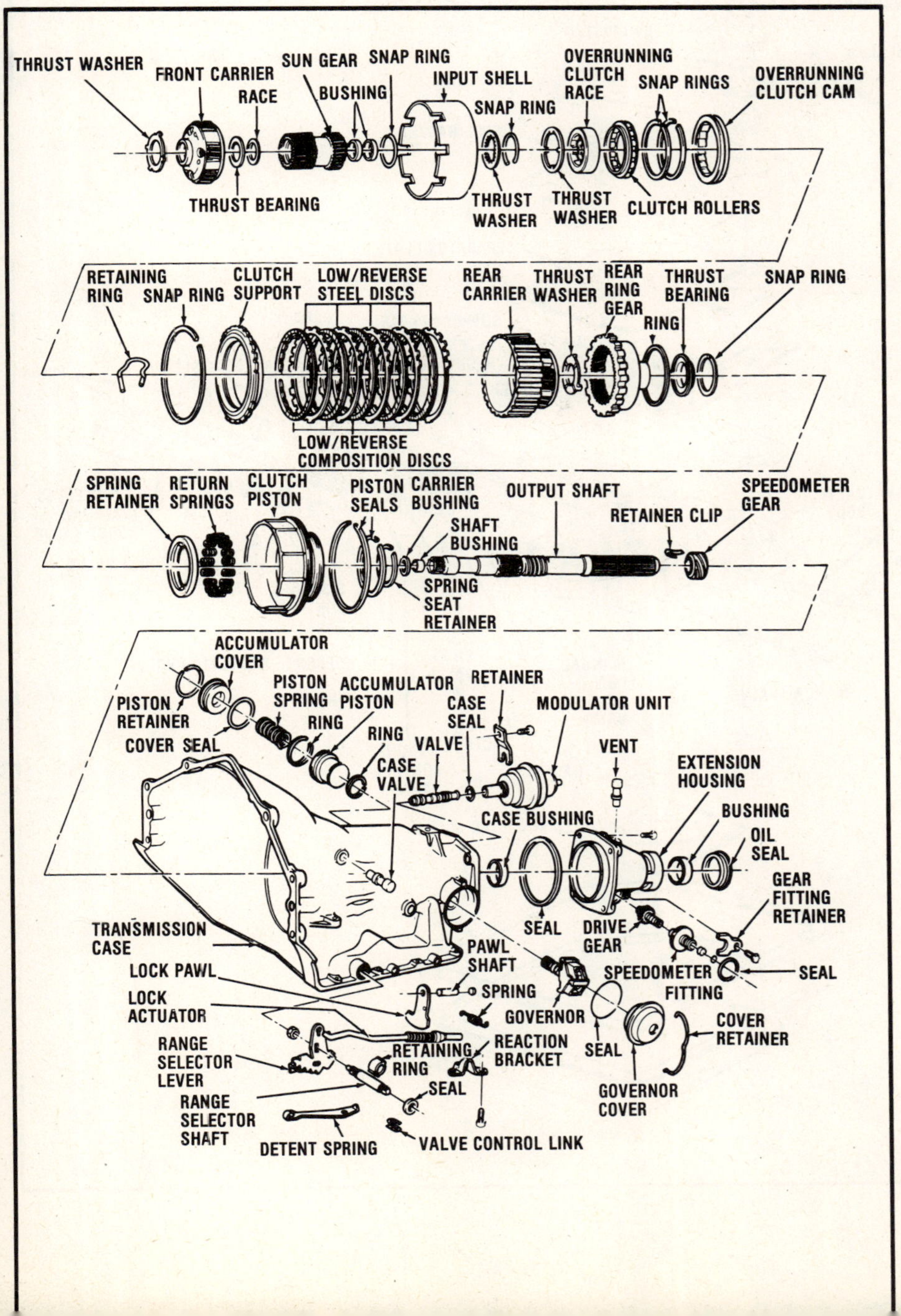

Figure 19-2. Turbo Hydra-Matic 250 transmission case and planetary gearsets. (Chevrolet Motor Division-GMC)

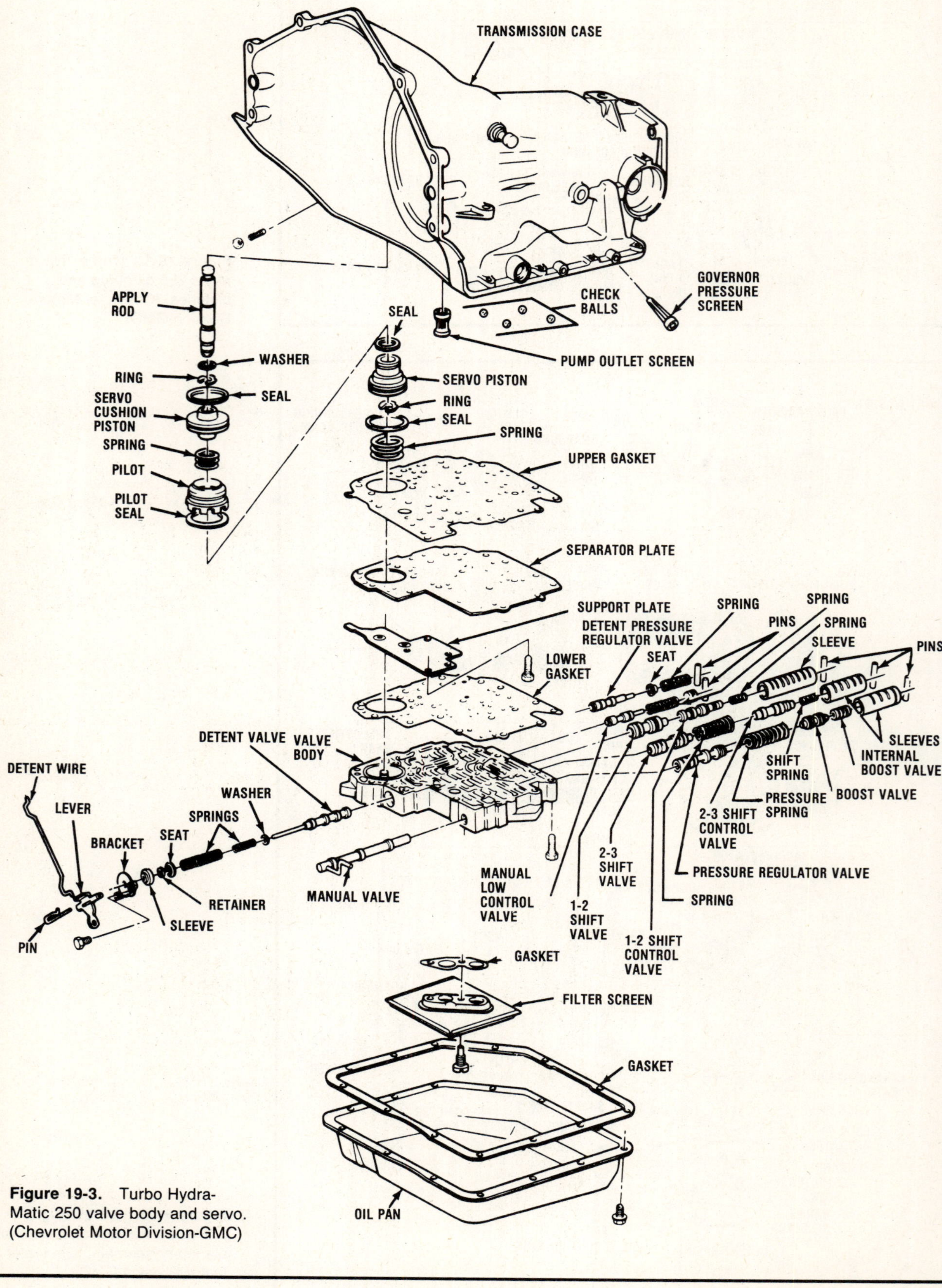

Figure 19-3. Turbo Hydra-Matic 250 valve body and servo. (Chevrolet Motor Division-GMC)

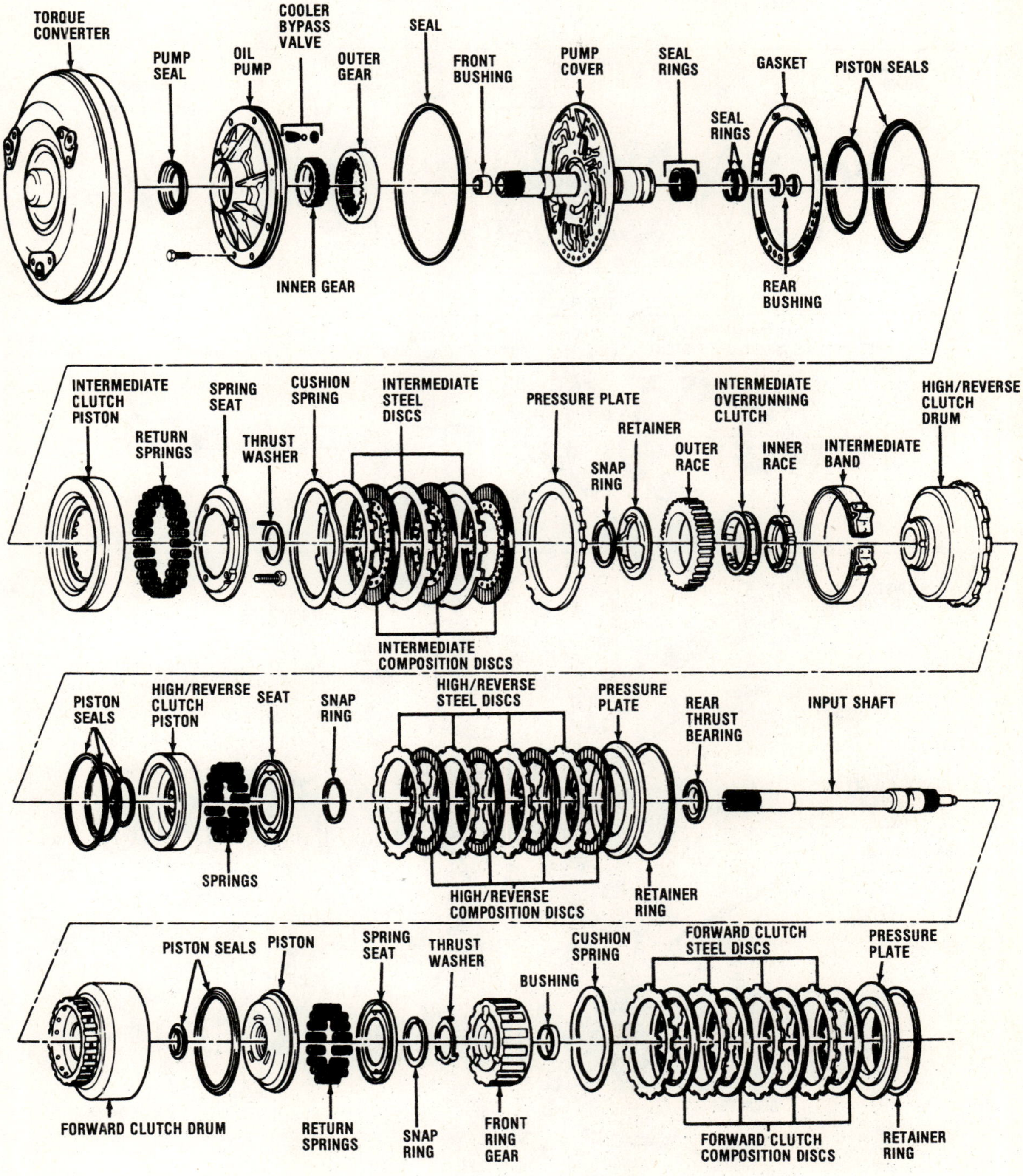

Figure 19-4. Turbo Hydra-Matic 350 oil pump and clutches. (Chevrolet Motor Division-GMC)

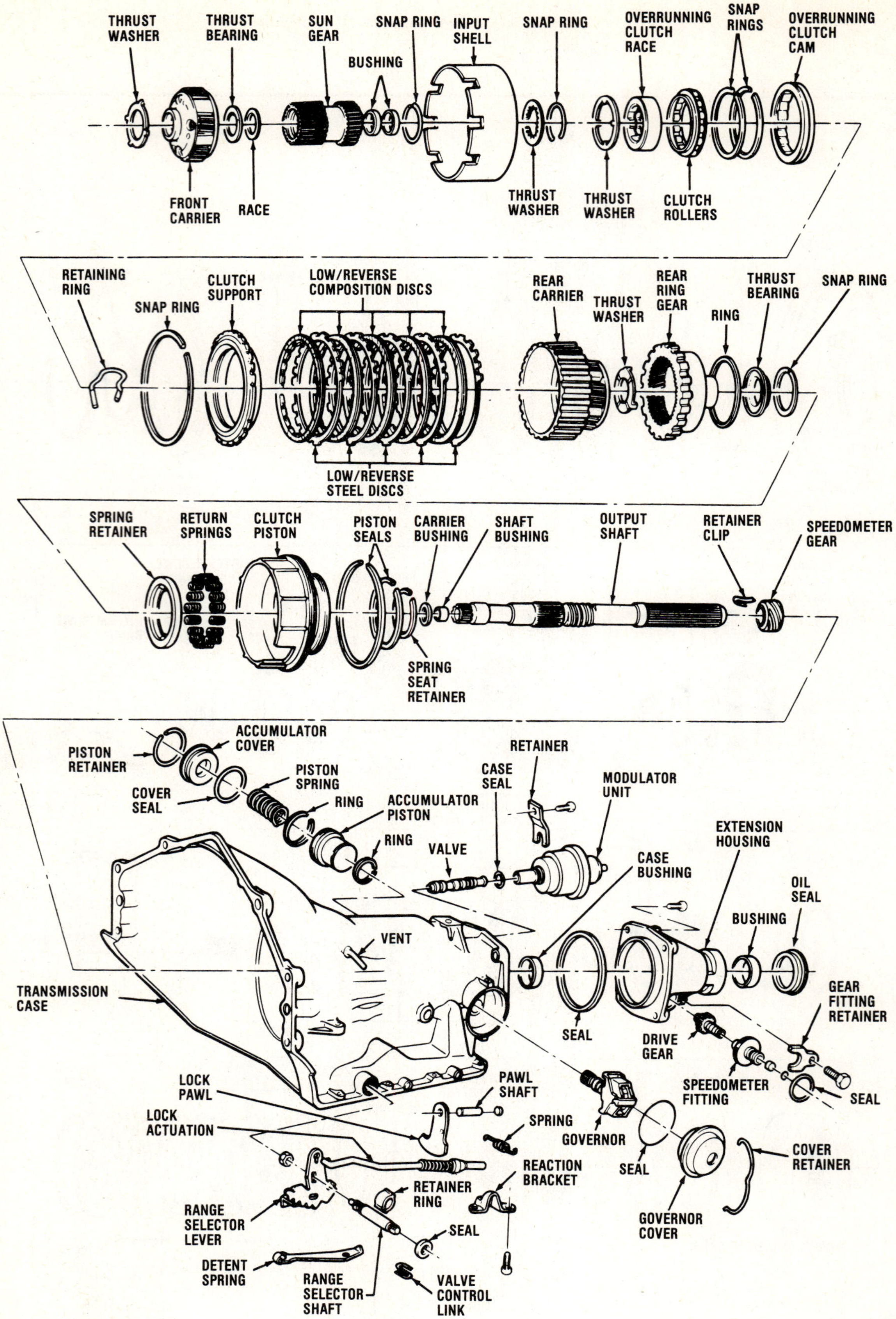

Figure 19-5. Turbo Hydra-Matic 350 transmission case and planetary gearset. (Chevrolet Motor Division-GMC)

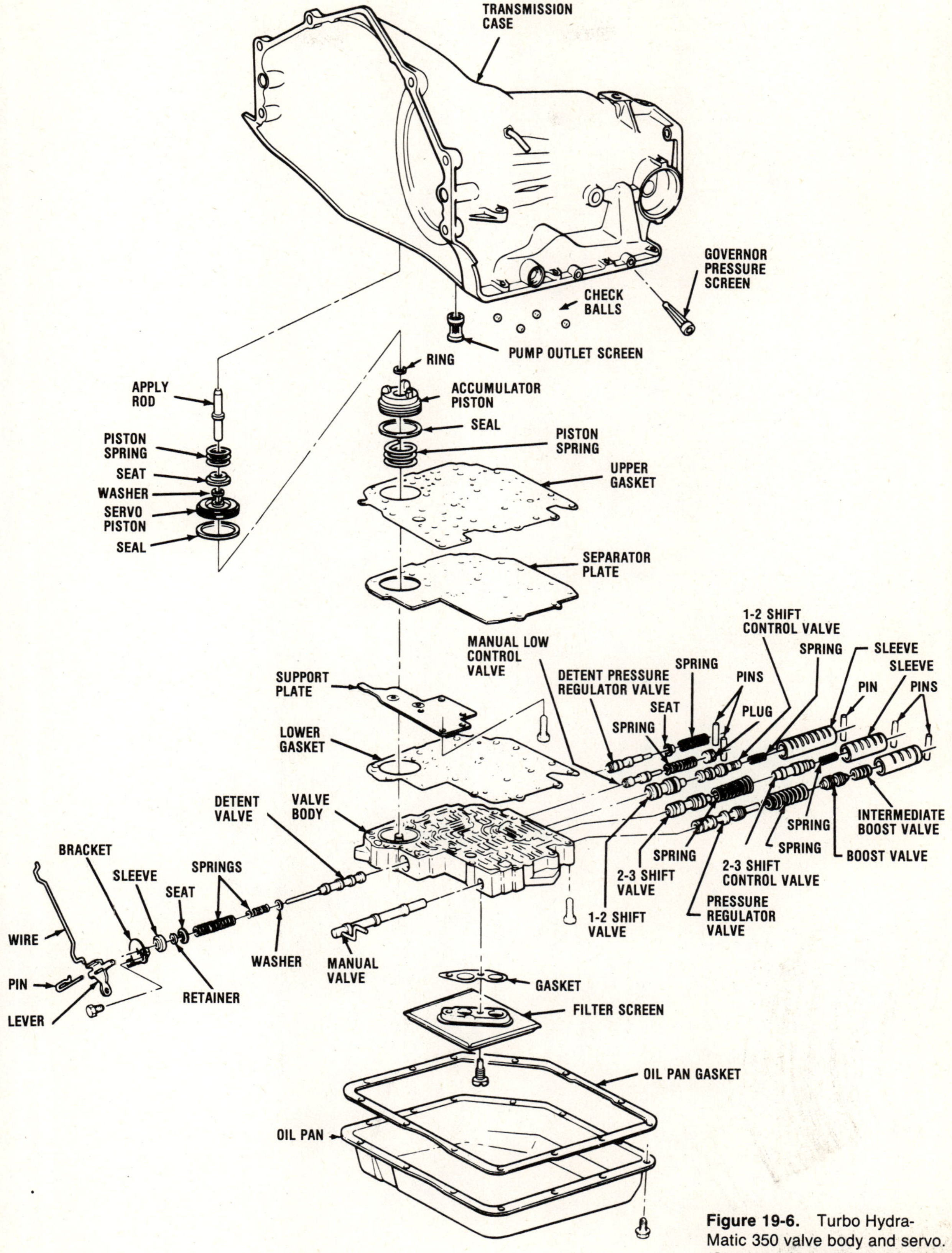

Figure 19-6. Turbo Hydra-Matic 350 valve body and servo. (Chevrolet Motor Division-GMC)

NAME ______________________ SECTION ____________ DATE ____________

Certification Practice

Test 19

DIRECTIONS:

The following questions are similar to those used on mechanic certification tests. Answer each question by circling the letter of the correct choice.

1. Which of these transmissions has a lockup torque converter mechanism?
   A. GM 250/350
   B. GM 250C/350C
   C. GM 375B
   D. All of the above

2. The lockup torque converter clutch apply valve is controlled by a
   A. solenoid.
   B. regulating valve.
   C. stator.
   D. change in fluid pressure.

3. Which of the following statements is correct?
   I. The GM 250/250C transmissions have three multiple-disc clutches.
   II. The GM 350/350C transmissions have four multiple-disc clutches.
   A. I only B. II only C. Both I and II D. Neither I nor II

4. Mechanic A says that the GM 350/350C/375B transmissions all have one overrunning clutch.
   Mechanic B says that the GM 250/250C transmissions both have two overrunning clutches.
   Who is correct?
   A. A only B. B only C. Both A and B D. Neither A nor B

5. Compared to a GM 350 transmission, a GM 375B transmission
   A. has a longer output shaft.
   B. is a heavier-duty transmission.
   C. has five direct clutch plates instead of four.
   D. has a lockup torque converter.

6. The most common place to locate the identification number for the GM 250/250C/350/350C/375B transmission family is on the
   A. governor cover on the extension housing.
   B. intermediate accumulator cover.
   C. transmission case.
   D. transmission pan.

7. The drawing shows the pump and clutch assemblies for

A. the GM 250 transmission.

B. the GM 350C transmission.

C. the GM 375B transmission.

D. none of the above.

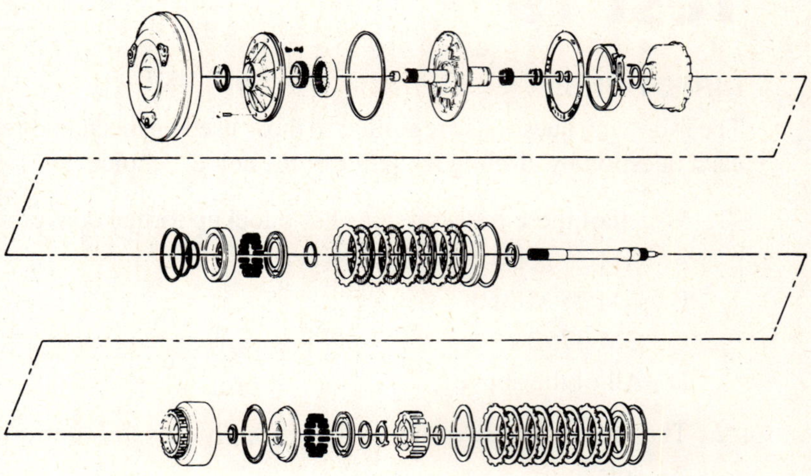

Question 7.
(Chevrolet Motor Division-GMC)

20 Turbo Hydra-Matic 375, 400, 425

YOUR LEARNING JOB

This General Motors transmission family shares most of the same parts. Overhaul work on each of these transmissions is almost identical. This unit discusses the similarities and differences between these three transmissions.

JOB COMPETENCY OBJECTIVES

This is a knowledge-building unit. When you have completed your assignments and exercises in this unit, you should be able to:

- ☐ Describe the main parts that are common to each transmission.
- ☐ Explain differences in the Turbo Hydra-Matic 425.
- ☐ Locate the transmission identification code on these transmissions.

20.1 TURBO HYDRA-MATIC 375/400/425 DESCRIPTION

This family of transmissions was introduced in 1964 with the Turbo Hydra-Matic 400. These transmissions have been used with all General Motors cars with big engines and in light trucks. These transmissions were last used in GM automobiles in 1982 but are still found in light trucks. The 425 is a front-drive transmission, *not* a transaxle. It was last used in 1978 Cadillac Eldorados, Buick Rivieras, and Oldsmobile Toronados. The 400 transmission was also used in Jeeps, Jaguars, and Rolls-Royces. The 375/400/425 transmissions consist of the following main parts:

- Torque converter
- Three multiple-disc clutches
- Two overrunning clutches
- Two servos and bands
- Compound planetary gearset
- Integral valve body
- Aluminum bellhousing and transmission case with removable extension housing.

The Turbo Hydra-Matic 375/400/425 transmissions combine a torque converter with a fully automatic three-speed gear system. The bellhousing and the transmission case are a single aluminum casting.

The torque converter is a three-element unit that connects to the crankshaft through a flexplate.

The compound planetary gear system shares a common sun gear. The planetaries are operated by four different shafts: input shaft, mainshaft, sun gear shaft, and output shaft.

The hydraulic system has a single oil pump. The bottom half of the valve body is removable. The top half of the valve body is cast into the transmission case.

The main difference in the 375 is that it has a longer output shaft than the 400. Otherwise, the internal parts are the same.

The 425 is a heavy-duty front-drive transmission. It uses a similar sprocket and chain system as the 325, which is discussed in 18.3. The other internal parts of the 425 are the same as those found in the 375/400.

Figures 20-1 through 20-6 illustrate the differences between the three transmissions.

Numerous changes have been made in this family of transmissions over the years. Be sure to check the proper service manual before making any repairs or an overhaul.

20.2 TURBO HYDRA-MATIC 375/400/425 IDENTIFICATION

The 375 and 400 have the transmission identification plate mounted on the right side of the case, just above the valve body. The 425 identification plate is located on the left side of the bellhousing. To decipher the code on the identification plate, refer to the proper service manual.

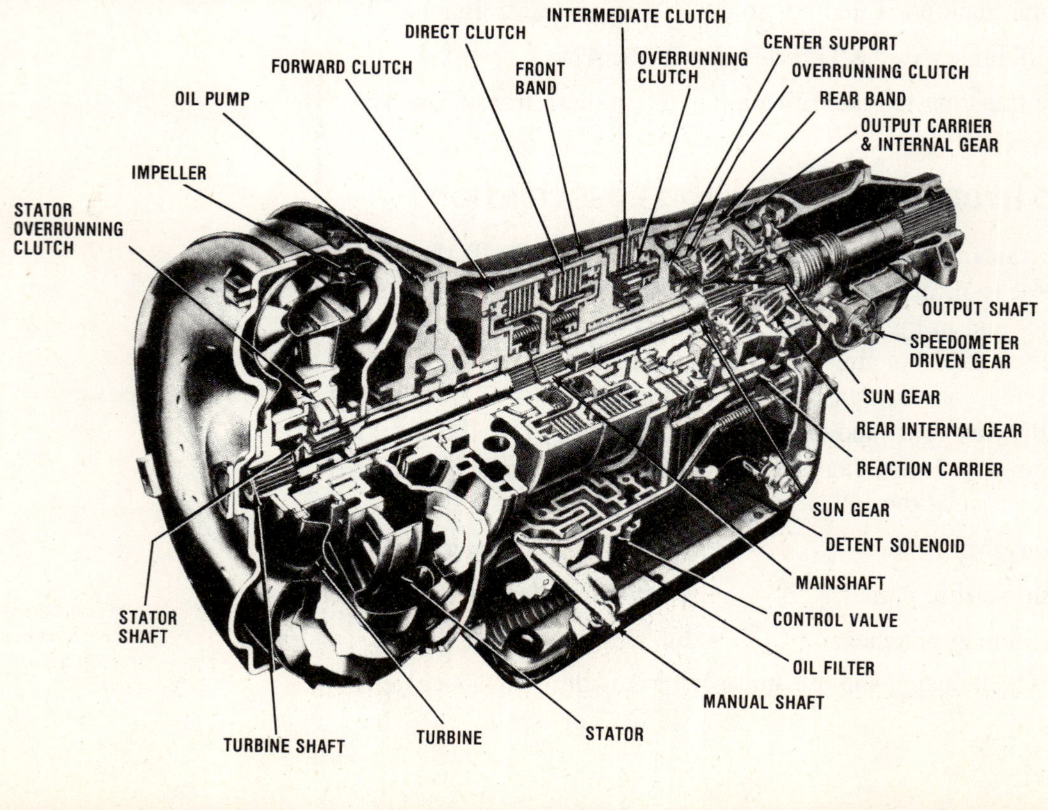

Figure 20-1. Turbo Hydra-Matic 400 automatic transmission cutaway. (Chevrolet Motor Division-GMC)

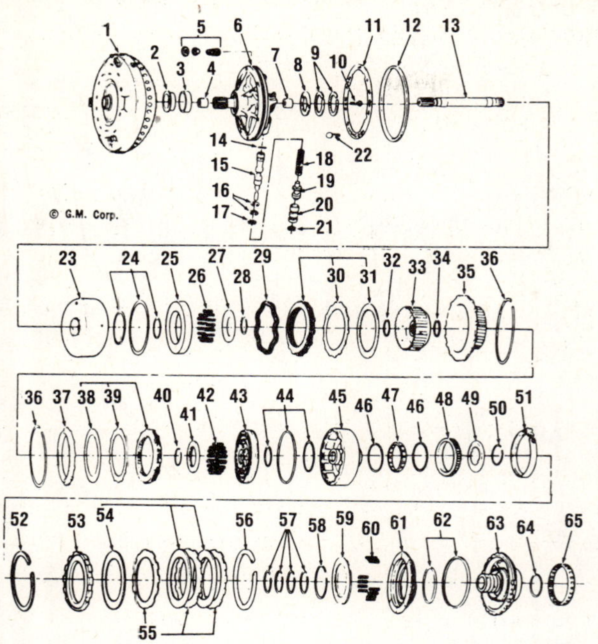

1. TORQUE CONVERTER
2. OIL SEAL
3. BUSHING
4. BUSHING
5.&6. OIL PUMP
7. BUSHING
8. THRUST WASHER
9. SEAL RING
10. BOLT
11. GASKET
12. SEAL
13. TURBINE SHAFT
14. PIN
15. PRESSURE REGULATOR VALVE
16. SPACER
17. WASHER
18. REGULATOR SPRING
19. VALVE AND BUSHING
20. VALVE AND BUSHING
21. RETAINING RING
22. BOLT
23. FORWARD CLUTCH HOUSING
24. SEALS
25. FORWARD CLUTCH PISTON
26. RELEASE SPRING
27. SPRING RETAINER
28. SNAP RING
29. WAVE PLATE
30. DRIVEN CLUTCH PLATE
31. DRIVE CLUTCH PLATE
32. THRUST WASHER
33. FORWARD CLUTCH HUB
34. THRUST WASHER
35. DIRECT CLUTCH HUB
36. SNAP RING
37. BACKING PLATE
38. DRIVING PLATE
39. DRIVEN PLATE
40. SNAP RING
41. SPRING RETAINER
42. RELEASE SPRING
43. DIRECT CLUTCH PISTON
44. SEALS
45. DIRECT CLUTCH HOUSING
46. INTERMEDIATE SPRAG
47. INTERMEDIATE SPRAG
48. OUTER RACE
49. CLUTCH RETAINER
50. SNAP RING
51. FRONT BAND
52. SNAP RING
53. BACKING PLATE
54. DRIVING CLUTCH PLATE
55. REACTION PLATE
56. SNAP RING
57. SNAP RING
58. SNAP RING
59. SPRING RETAINER
60. RETURN SPRING
61. INTERMEDIATE PISTON
62. SEALS
63. CENTER SUPPORT
64. THRUST WASHER
65. LOW OVERRUNNING CLUTCH

Figure 20-2. Turbo Hydra-Matic 400 oil pump and clutches. (Chevrolet Motor Division-GMC)

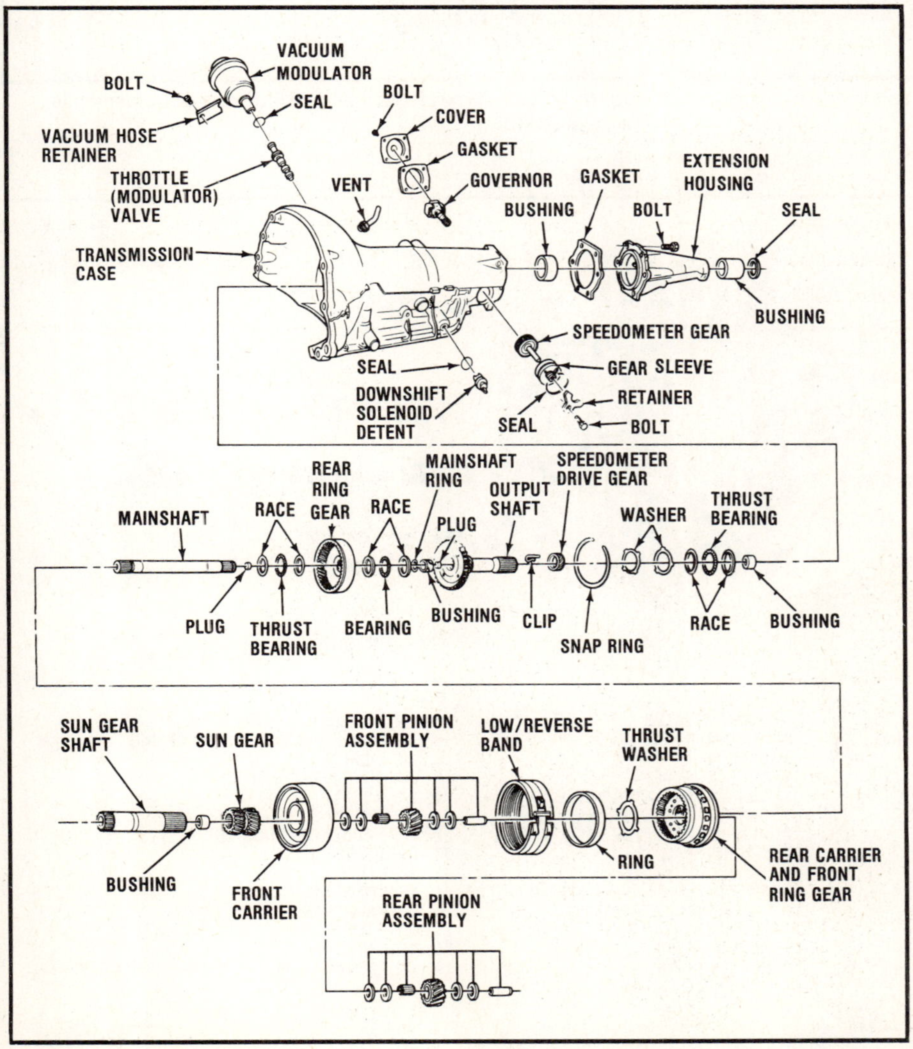

Figure 20-3. Turbo Hydra-Matic 400 transmission case and planetary gearset. (Chevrolet Motor Division-GMC)

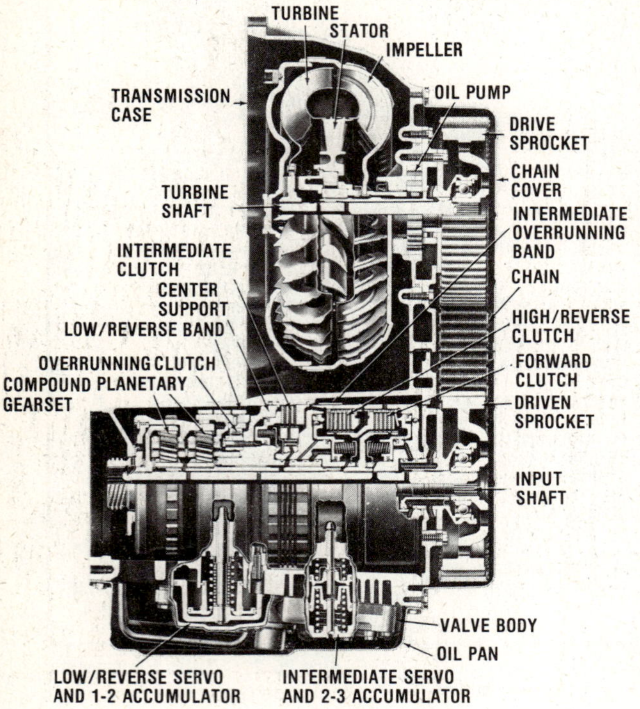

Figure 20-4. Turbo Hydra-Matic 425 automatic transmission cutaway. (Hydra-Matic)

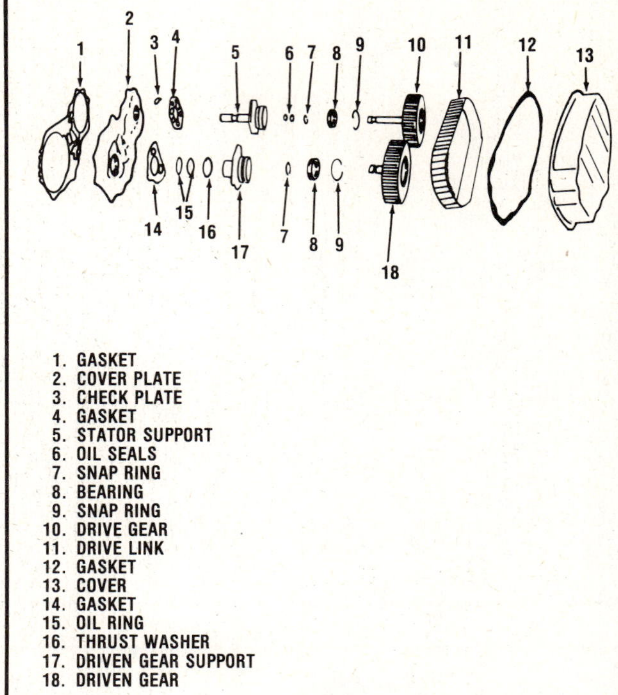

Figure 20-5. Turbo Hydra-Matic 425 drive link. (Oldsmobile Division-GMC)

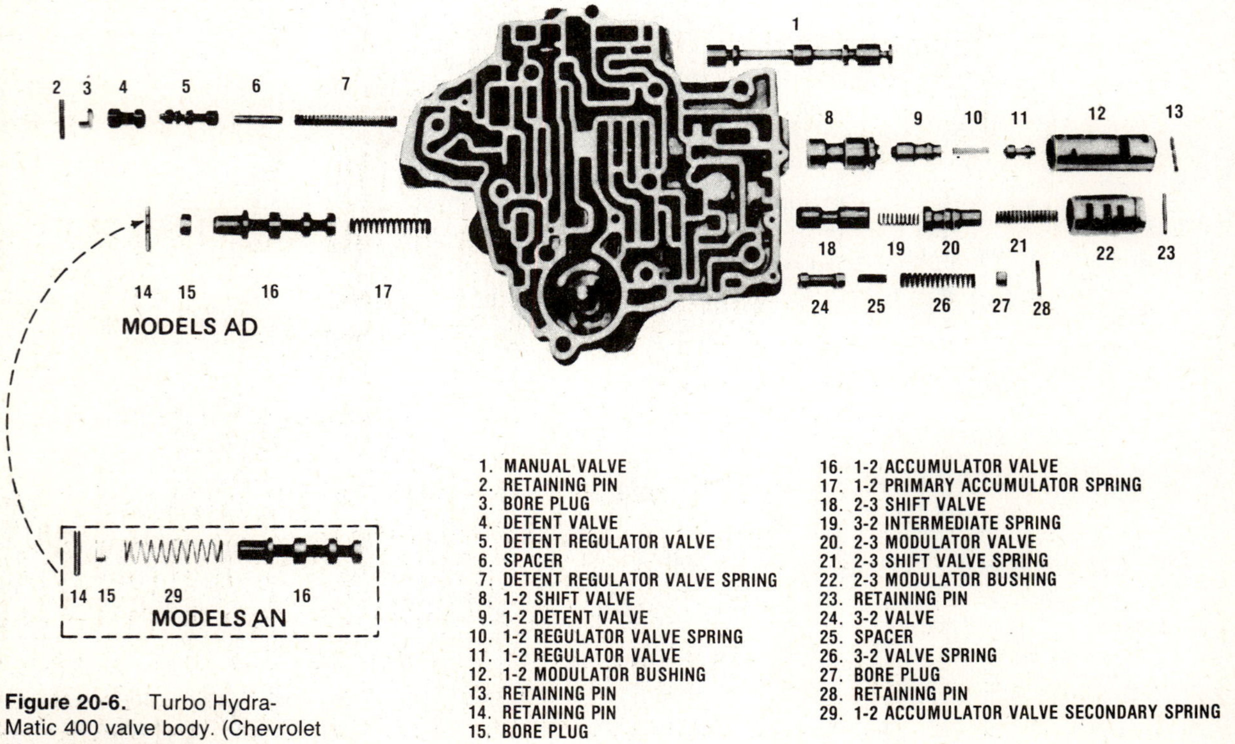

Figure 20-6. Turbo Hydra-Matic 400 valve body. (Chevrolet Motor Division-GMC)

NAME ______________ SECTION ______________ DATE ______________

Certification Practice

Test 20

DIRECTIONS:

The following questions are similar to those used on mechanic certification tests. Answer each question by circling the letter of the correct choice.

1. The GM 425 transmission is used in
   A. vans.
   B. light trucks.
   C. foreign vehicles.
   D. front-wheel-drive vehicles.

2. The GM 400 transmission was used in
   A. BMW.
   B. Peugeot.
   C. Mercedes-Benz.
   D. Rolls-Royce.

3. The GM 375/400/425 transmission family has all of the following EXCEPT
   A. one simple planetary gearset.
   B. two overruning clutches.
   C. two servos and bands.
   D. three multiple-disc clutches.

4. Compared to the GM 375 transmission, the GM 400 has
   A. a shorter output shaft.
   B. a longer output shaft.
   C. two sprockets and a chain.
   D. two chains and a sprocket.

5. The transmission shown is
   A. a GM 375.
   B. a GM 400.
   C. a GM 425.
   D. none of the above.

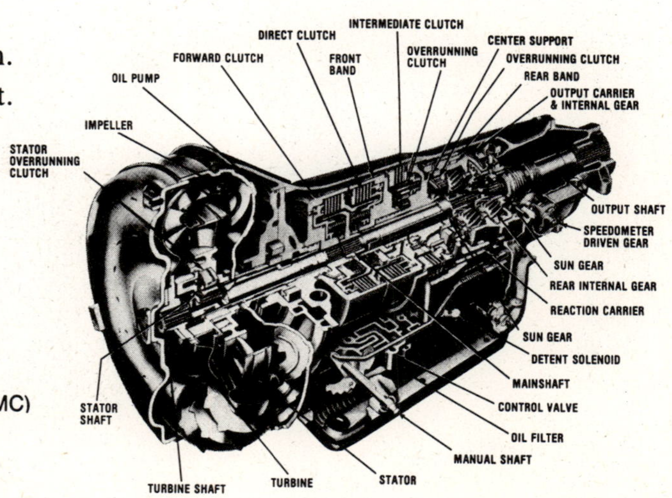

Question 4.
(Chevrolet Motor Division-GMC)

6. The transmission shown is
   A. a GM 375.
   B. a GM 400.
   C. a GM 425.
   D. none of the above.

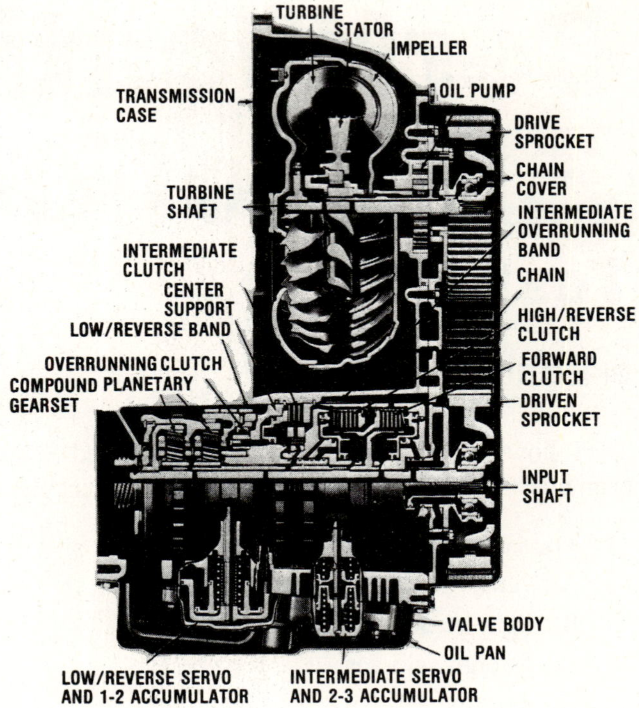

Question 6.
(Hydra-Matic)

7. Which of the following statements is correct?
   I. The GM 375 and 400 identification plate is on the right side of the case.
   II. The GM 425 identification plate is on the left side of the bellhousing.
   A. I only B. II only C. Both I and II D. Neither I nor II

21

Ford C-3, C-4, C-5

YOUR LEARNING JOB

The C-3, C-4, and C-5 transmissions are very popular transmissions in Ford automobiles. It's important to know the difference between these models. One is for light-duty purposes and the other two are for medium-duty purposes. This unit discusses the differences and similarities of these transmissions.

JOB COMPETENCY OBJECTIVES

This is a knowledge-building unit. When you have completed your assignments and exercises in this unit, you should be able to:

- ☐ Describe the main parts of the three transmissions.
- ☐ Explain the differences between the C-3/C-4 and C-5 transmissions.
- ☐ Locate the transmission identification numbers.

21.1 FORD C-3/C-4/C-5 DESCRIPTION

The C-4, which has been used since 1964, was discontinued after 1982. It has been used on compacts, full-size cars, and light trucks. The C-4 was used with four-cylinder, six-cylinder, and eight-cylinder engines.

The C-5 was introduced in 1982 as a replacement for the C-4. It is used primarily with six-cylinder engines. The C-5 is found in Ford compact, intermediate, and full-size automobiles, as well as in light trucks.

The C-3 is a lighter-duty version of the C-4. The C-3 first appeared in 1974 and is used in subcompacts, compacts, and a few full-size automobiles. It is used with four-cylinder and six-cylinder engines. The C-3, C-4, and C-5 transmissions consist of the following main parts:

- Torque converter
- Torque converter lockup (C-5 only)
- Two multiple-disc clutches
- Aluminum transmission case with removable bellhousing and extension housing.
- One overrunning clutch
- Two servos and bands
- Compound planetary gearset
- Removable valve body

The C-3/C-4 design combines a torque converter with a fully automatic three-speed gear system. The bellhousing, transmission case, and extension housing are separate aluminum castings.

The torque converter is a three-element unit that connects to the crankshaft through the flexplate.

The compound planetary gear system shares a common sun gear. The sun gear is connected to the driving shell and the output shaft through the rear planetary gearset.

The hydraulic system has a single oil pump. The valve body is removable.

Some 1980 and later C-4 transmissions use an altitude-compensating vacuum diaphragm (modulator). An altitude-compensating vacuum diaphragm (modulator) compensates for, or takes care of shifting problems caused by, low atmospheric pressures at high elevations.

The C-3 has provision for only one band adjustment. The C-4 and C-5, however, have two adjustable bands.

Figures 21-1 through 21-3 illustrate the various parts of the C-3/C-4 transmissions.

Numerous changes have been made to the C-3/C-4 family of transmissions over the years. Always refer to the proper service manual.

The C-5 transmission has internal parts similar to those of the C-3/C-4. The biggest differences are a modified valve body for smoother shifts and a lockup clutch in the converter. The lockup clutch is a centrifugal type. Weights and friction shoes are used to activate it. The C-5 torque converter is locked up in *all* forward gears instead of just high. Figures 21-4 and 21-5 show the main parts of the C-5.

21.2 C-3/C-4/C-5 IDENTIFICATION

Two identification numbers are used on Ford vehicles. On the doorpost on the driver's side, a plate gives the transmission model code. The transmissions also have metal tags that further identify the transmission. The tag on the C-3 is located on the lower left extension-housing attaching bolt. The tag on the C-4 and C-5 transmissions is located under the attaching bolt at the intermediate servo cover. Refer to the service manual to decipher the codes.

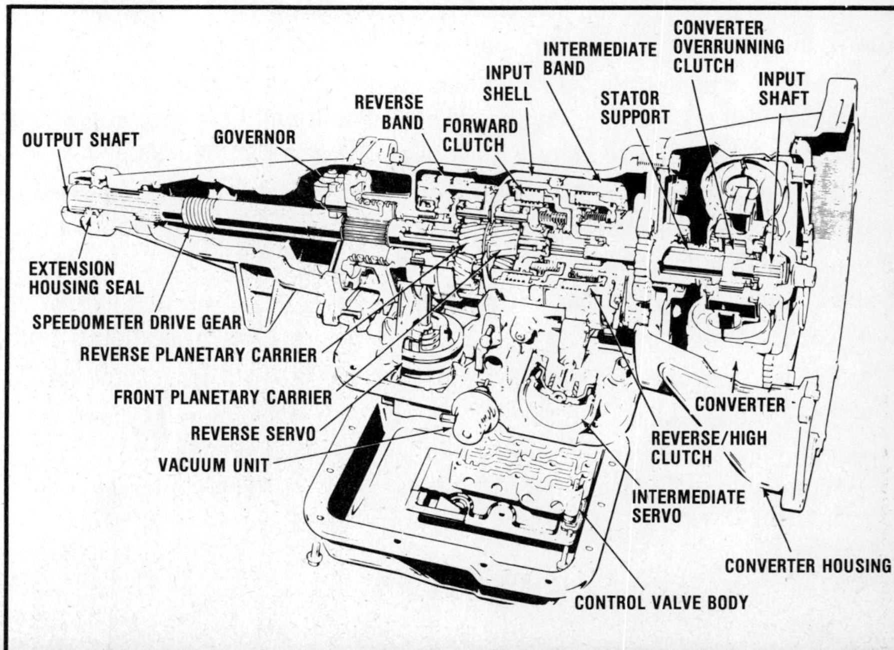

Figure 21-1. Ford C-3 automatic transmission cutaway. (Ford Motor Company)

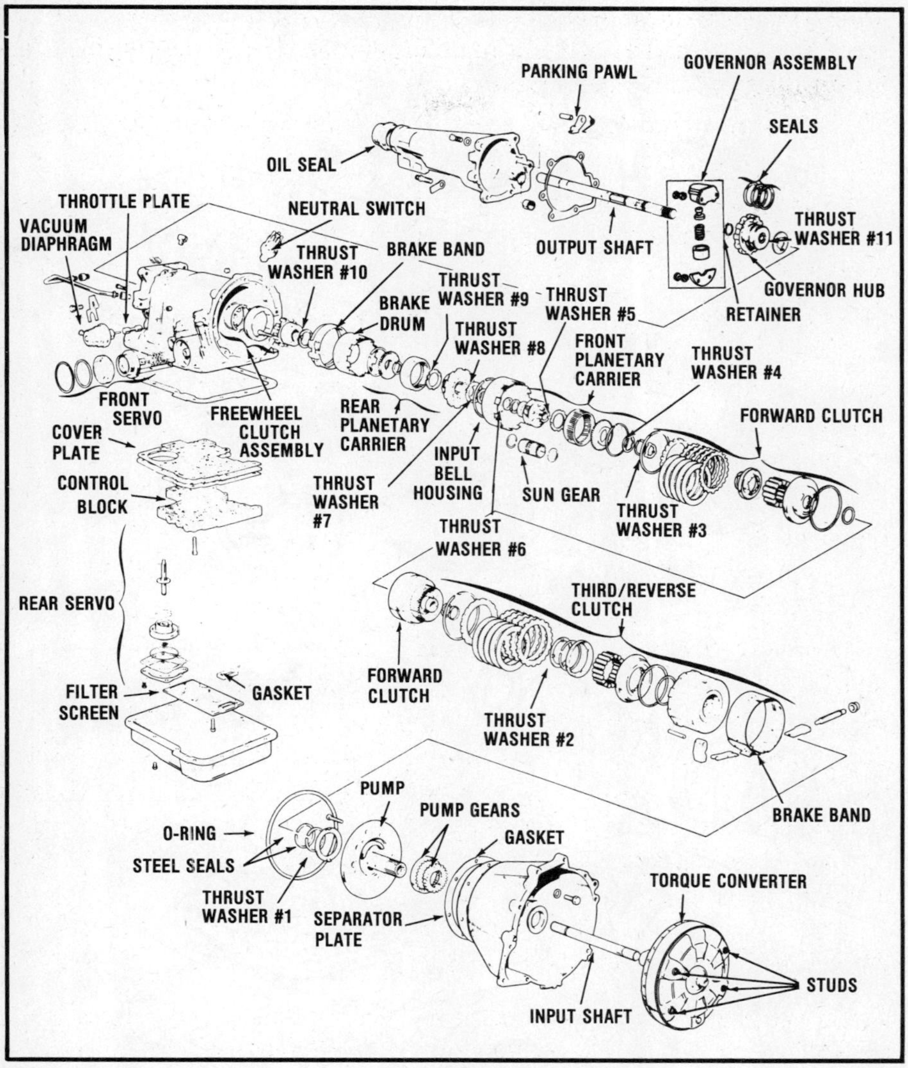

Figure 21-2. Ford C-3 automatic transmission, exploded view. (Ford Motor Company)

Figure 21-3. Ford C-3 control valve body. (Ford Motor Company)

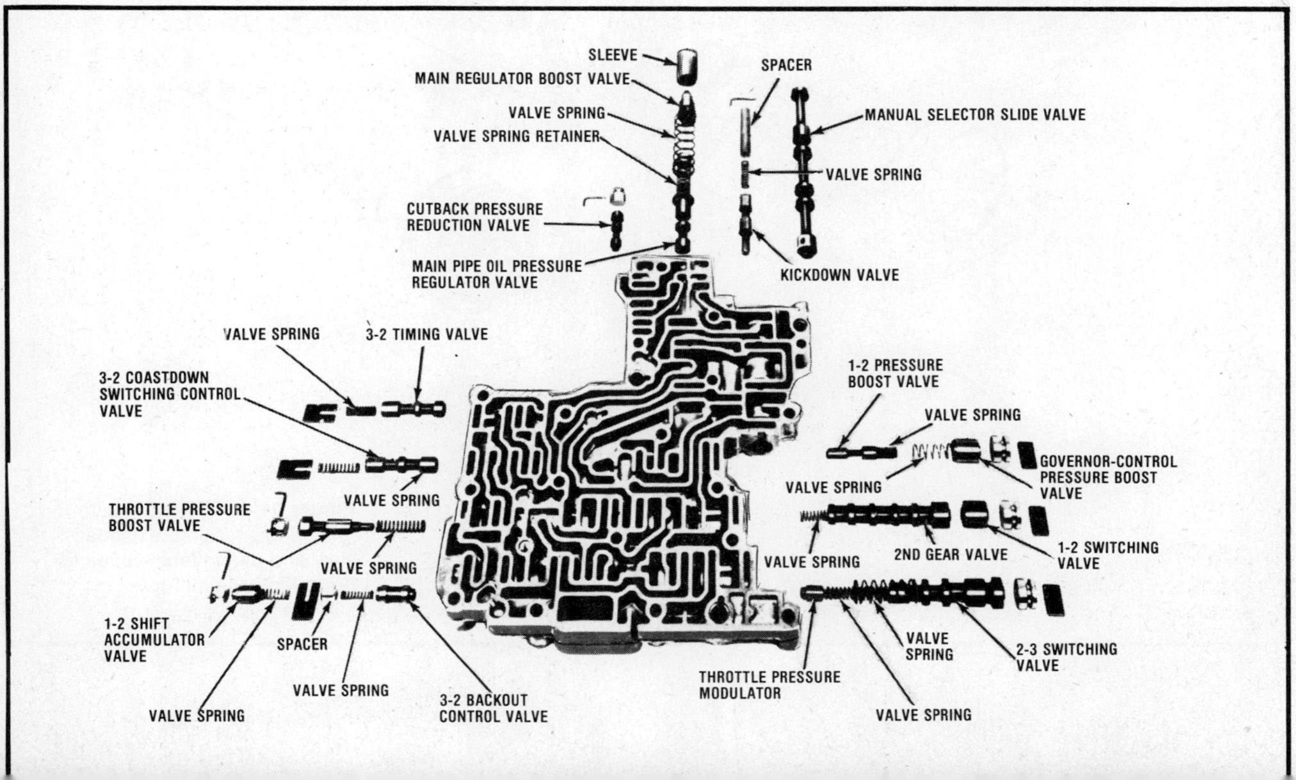

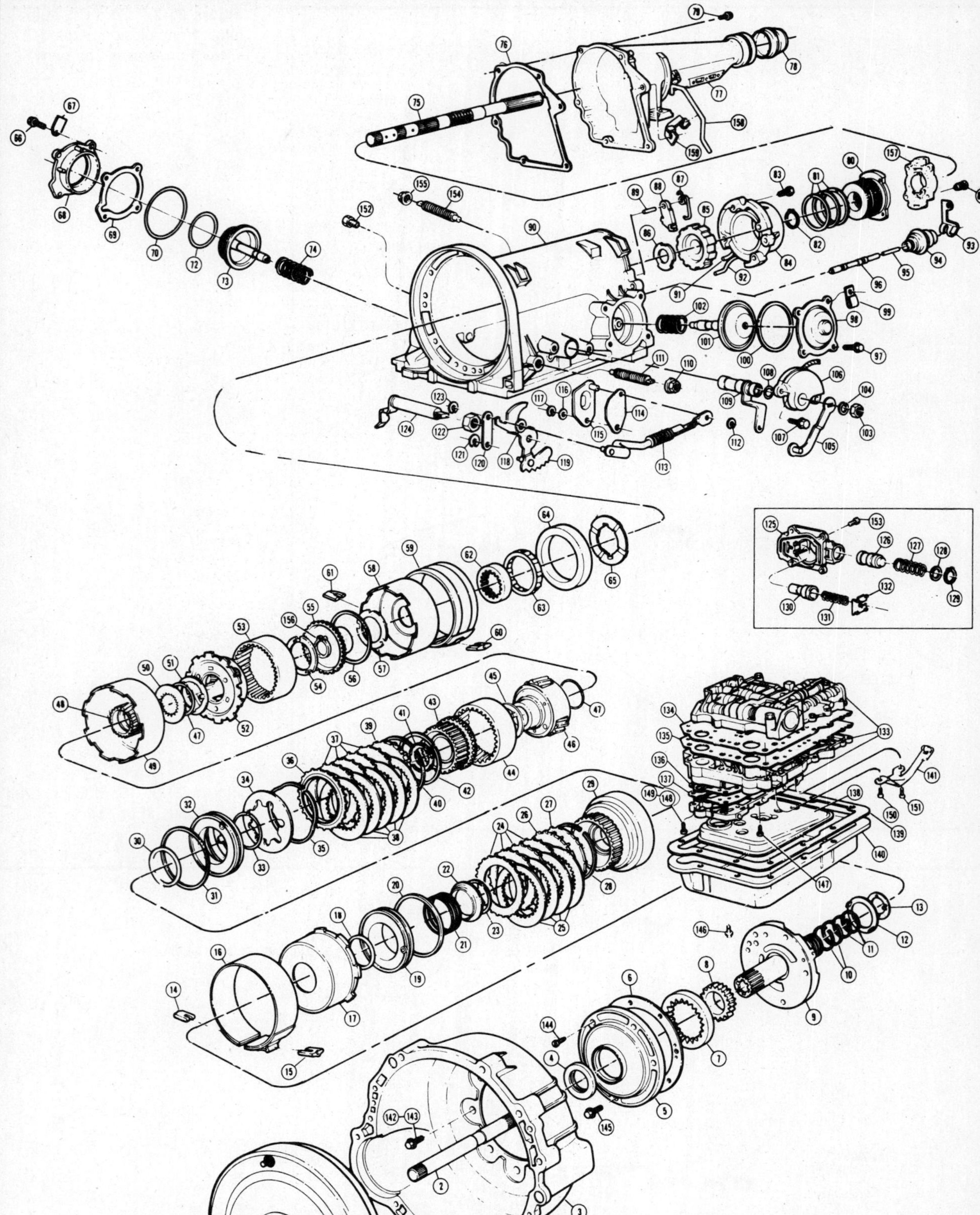

Figure 21-4. Ford C-5 automatic transmission, exploded view and parts identification list. (Ford Motor Company)

1. CONVERTER ASSEMBLY
2. SHAFT – INPUT
3. HOUSING CONVERTER
4. SEAL ASSY. – FRONT OIL PUMP
5. BODY – FRONT PUMP
6. GASKET – FRONT OIL PUMP
7. GEAR – FRT. OIL PUMP DRIVEN
8. GEAR – FRT. OIL PUMP DRIVE
9. STATOR SUPPORT – FRT. OIL PUMP
10. SEAL – FWD. CL. CYL.
11. SEAL – REV. CL. CYL.
12. WASHER – FRT. PUMP SUPT. THRUST (SEL.) – NO. 1
13. WASHER – REV. CL. THRUST – NO. 2
14. STRUT – INTERMEDIATE BRAKE
15. STRUT – INTERMEDIATE BRAKE (ANCHOR INT. BAND)
16. BAND ASSY. – INTERMEDIATE
17. DRUM ASSY. – INTERMEDIATE BRAKE
18. SEAL – REV. CL. INNER
19. PISTON ASSY. – HIGH CLUTCH
20. SEAL – HIGH CL. OUTER
21. SPRING – HIGH CL. PISTON
22. RETAINER – CL. PISTON SPRING
23. RING – 2¼ RETAINING
24. PLATE – CL. EXT. SPLINE – (STEEL)
25. PLATE ASSY. – CL. INT. SPLINE (FRICTION)
26. PLATE – CLUTCH PRESSURE
27. SPRING – REV. CL. PRESS. PLATE DISC
28. SNAP RING (SELECTIVE)
29. CYLINDER – FWD. CLUTCH
30. O-RING – CL. PISTON OIL
31. SEAL – FWD. CL. OUTER
32. PISTON ASSY. – FWD. CLUTCH
33. RING – FWD. CL. PST. SPRING PRESS
34. SPRING – FWD. CL. PISTON DISC
35. RING – RET. WAVE INT.
36. PLATE – FWD. CL. PRESS.
37. PLATE – CL. EXT. SPLINE (STEEL)
38. PLATE ASSY. CL. INT. SPLINE (FRICTION)
39. PLATE – CLUTCH PRESS.
40. RING – RETAINING EXT. (SEL.)
41. WASHER – FWD. CL. HUB – THRUST – NO. 3
42. RETAINER – 4¼ SNAP
43. HUB & BSHG. ASSY. – FWD. CL.
44. GEAR – OUTPUT SHAFT RING
45. WASHER – REV. PLANET PINION – THRUST – NO. 4
46. PLANET ASSY. – FWD.
47. RETAINER – 1-13/16 SNAP (2-REQ'D.)
48. GEAR – SUN
49. SHELL – INPUT
50. WASHER – INPUT SHELL THRUST – NO. 5
51. WASHER – REV. PLT. CARRIER – THRUST (FRT.) – NO. 6
52. PLANET ASSY. – REVERSE
53. GEAR – OUTPUT SHAFT RING
54. WASHER – REV. DRUM THRUST – NO. 7
55. HUB – OUTPUT SHAFT
56. RING – 4-21/32 RETAINING
57. WASHER – REV. DRUM THRUST – NO. 8
58. DRUM ASSY. – REV. BRAKE
59. BAND ASSY. – REVERSE
60. STRUT – REV. BRAKE (ACTUATES REV. BND.)
61. STRUT – REV. BAND (ANCHOR)
62. RACE – OVERRUN CLUTCH – INNER
63. SPRING & ROLLER ASSY. – OWC
64. RACE – OVERRUN CLUTCH – OUTER
65. WASHER – REV. DRUM TO CASE – THRUST – NO. 9
66. BOLT – 5/16-18 x 3/4 HEX. WSHR. HEAD (7D027 TO 7005)
67. TAG – SERVICE IDENTIFICATION
68. COVER – INTERM. BAND SERVO
69. GASKET – INTERM. BAND SERVO
70. SEAL – INTERM. BAND SERVO COVER – LARGE
71. DELETE
72. SEAL – INTERM. BAND SERVO PISTON – SMALL
73. PISTON ASSY. INTERM. BAND SERVO
74. SPRING – INTERM. BAND SERVO PISTON
75. SHAFT ASSY. – OUTPUT
76. GASKET – EXTENSION
77. EXTENSION ASSY.
78. SEAL ASSY. – EXT. OIL
79. BOLT – 3/8-16 x 1-3/8 HEX. WSHR. HEAD (6-REQ'D.)
80. BODY – GOV. OIL COLLECTOR
80A. SCREEN ASSY. – GOV. OIL
81. RING – GOVERNOR SEAL
82. RING – 1¼ RETAINING (7C053 TO 7060)
83. BOLT – 5/16-18 HEX. HD. (7C232 TO 7005) (4-REQ'D.)
84. SLEEVE – OIL DISTRIBUTOR
85. GEAR – OUTPUT SHAFT PARKING
86. WASHER – OUTPUT SHAFT THRUST – RR – NO. 10
87. SPRING – P.P. RETURN
88. PAWL ASSY. – PARKING
89. PIN – 1/2 x 1.57 STRAIGHT
90. CASE ASSEMBLY
91. TUBE – OIL DISTRIBUTOR
92. TUBE – OIL DISTRIBUTOR
93. CLIP – THROT. CNTL. VALVE DIAPHRAGM
94. DIAPHRAGM ASSY. – T.V. CONTROL
95. ROD – T.V. CONTROL
96. VALVE – THROTTLE CONTROL
97. BOLT – 5/16-18 x 3/4 HEX. HD. (4-REQ'D.)
98. COVER – REV. BAND SERVO
99. CLIP – ELECTRICAL WIRING HARNESS
100. SEAL – REV. BND. SERVO PISTON COVER
101. PISTON ASSY. – REV. BAND SERVO
102. SPRING – REV. SERVO PISTON
103. NUT – 5/16-24 x 4 HEX (7D261 TO 7A394)
104. WASHER – 5/16 (7D261 TO 7A394)
105. LEVER ASSY. – DWNSHIFT CONTROL – OUTER
106. SWITCH ASSY. – NEUTRAL START
107. SCREW & WSHR. ASSY. (7A247 TO 7005) (2 REQ'D.)
108. SEAL – 3/8 O-RING
109. LEVER ASSY. – MANUAL CONTROL
110. NUT – 1/2-28 HEX. SEALER
111. SCREW – 1/2-28 x 2.88 ADJUSTING
112. RING – 1/4 RET. TYPE RB EXT. (7A117 TO 7A232)
113. ROD ASSY. – PARK LEVER ACTUATING
114. SPACER – PARKING LEVER
115. LEVER ASSY. – PARKING ACTUATING
116. WASHER – .380-.385 x .62 O.D. (7A117 & 7F286 TO 7005)
117. RING RETAINING (7A117 TO 7005)
118. ROLLER – PARK LEVER ACTUATING ROD
119. LEVER ASSY. – MAN. VLV. DETENT – INNER
120. LINK – P.P. TOGGLE OPER. LEVER
121. RING – 1/4 RET. TYPE RB EXT. (7C495 TO 7005)
122. NUT – 9/16-18 HEX. (7A115 TO 7A256)
123. RING – 1/4 RET. TYPE RB EXT. (7A232 TO 7C495)
124. LEVER ASSY. – DWNSHFT. DETENT – INNER
125. BODY – GOVERNOR VALVE
126. VALVE – GOVERNOR PRIMARY
127. SPRING – GOV. PRIMARY VALVE
128. WASHER – 1/2 FLAT
129. RETAINING RING – INTERNAL
130. VALVE – GOV. SECONDARY
131. SPRING – GOV. SECONDARY VALVE
132. RETAINER – GOV. SEC. VALVE SPRING
133. CONTROL ASSY. – MAIN
134. PLATE – CONTROL VLV. BODY SEP.
135. GAKSET – CONTROL VLV. BODY SEP.
136. PLATE – 3-2 TIMING BODY SEP.
137. GASKET – 3-2 TIMING BODY SEP.
138. SCREEN & GROMMET ASSY. – OIL PAN
139. GASKET – OIL PAN
140. PAN ASSY. – OIL PAN
141. SPRING ASSY. – MAN. VLV. DETENT
142. BOLT – 3/8-16 x 1.75 HEX. WSHR. HD. (7-REQ'D.)
143. BOLT – 7/16-14 x 1.12 HEX. WSHR. HD. (5-REQ'D.)
144. SCREW – 10-24 x .68 CROSS RECESSED HD. (2-REQ'D.)
145. BOLT – 3/8-16 x 1.38 HEX. WSHR. HD (7-REQ'D.)
146. BOLT – 5/16-18 x .54 HEX. WSHR. HD. (11-REQ'D.)
147. SCREW – 1/4-20 x 1 HEX WSHR. HD. (7-REQ'D.)
148. SCREW – 1/4-20 x 1½ HEX. WSHR. HD. (1-REQ'D.)
149. SCREW – 1/4-20 x 2.08 HEX. WSHR. HD. (2-REQ'D.)
150. SCREW – 1/4-20 x 1½HEX. WSHR. HD. (1-REQ'D.)
151. SCREW – 10-24 x .88 HEX. WSHR. HD. (1-REQ'D.)
152. PLUG – 1/8-27 DRY SEAL TAPPED THREAD
153. BOLT 1/4-20 x .68 HEX. WSHR. HD. (4-REQ'D.)
154. SCREW 1 2-16 x 2.88 ADJUSTING
155. NUT 1 2-16 HEX SEALER
156. RING 1-3 16 RETAINER (7D163 TO 7060)
157. BODY ASSY. GOVERNOR
158. TUBE – VENT
159. CLAMP – VENT TUBE

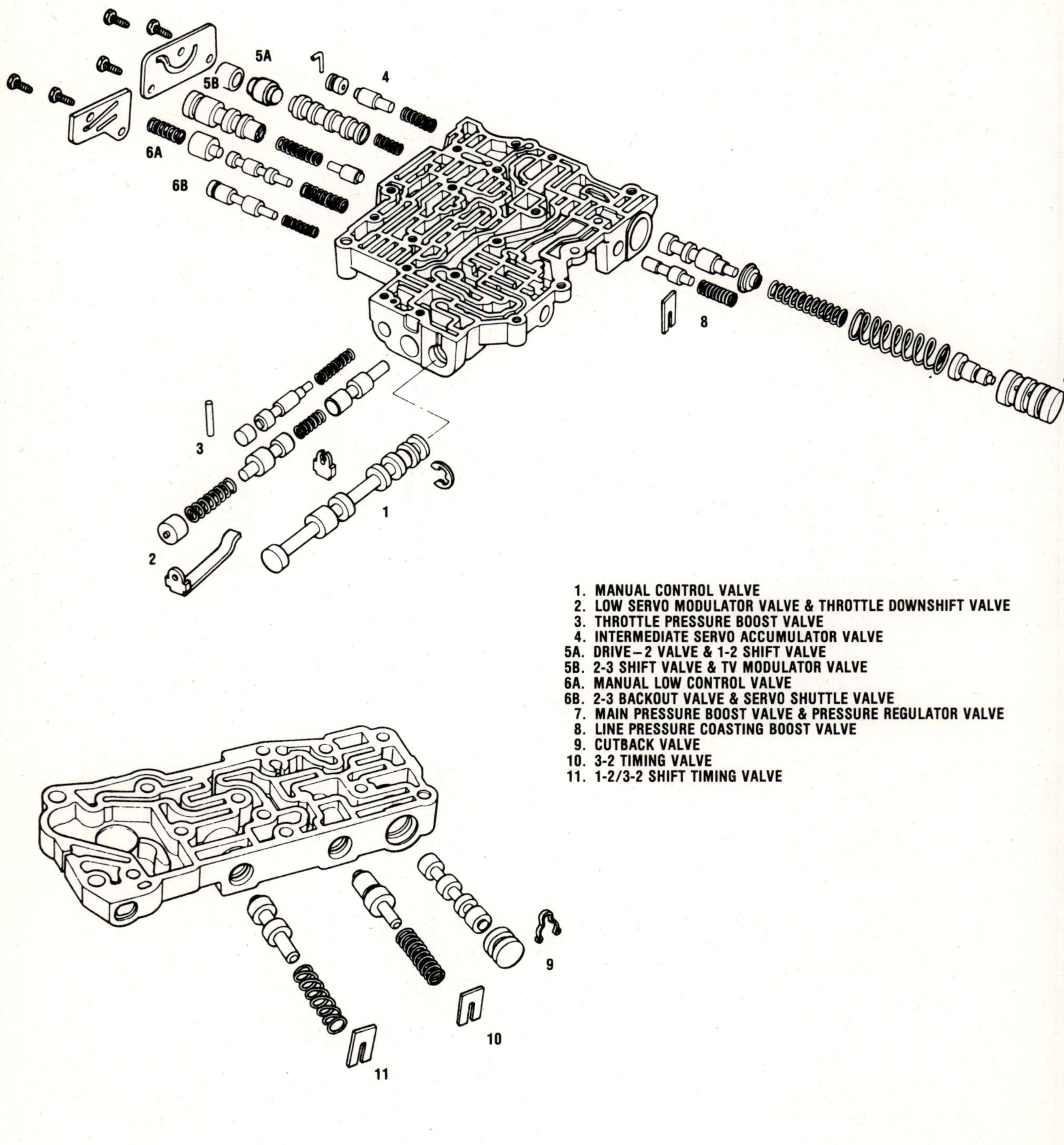

1. MANUAL CONTROL VALVE
2. LOW SERVO MODULATOR VALVE & THROTTLE DOWNSHIFT VALVE
3. THROTTLE PRESSURE BOOST VALVE
4. INTERMEDIATE SERVO ACCUMULATOR VALVE

5A. DRIVE–2 VALVE & 1-2 SHIFT VALVE
5B. 2-3 SHIFT VALVE & TV MODULATOR VALVE
6A. MANUAL LOW CONTROL VALVE
6B. 2-3 BACKOUT VALVE & SERVO SHUTTLE VALVE

7. MAIN PRESSURE BOOST VALVE & PRESSURE REGULATOR VALVE
8. LINE PRESSURE COASTING BOOST VALVE
9. CUTBACK VALVE
10. 3-2 TIMING VALVE
11. 1-2/3-2 SHIFT TIMING VALVE

Figure 21-5. Ford C-5 control valve body. (Ford Motor Company)

NAME ______________________ SECTION ____________ DATE ____________

Certification Practice

Test 21

DIRECTIONS:

The following questions are similar to those used on mechanic certification tests. Answer each question by circling the letter of the correct choice.

1. Unlike the GM automatic transmissions, the Ford C-3/C-4/C-5 transmissions have
   A. torque converter lockup mechanisms.
   B. compound planetary gearsets.
   C. overrunning clutches.
   D. removable bellhousings.

2. The C-5 is a replacement for
   A. the C-3 transmission.
   B. the C-4 transmission.
   C. the C-6 transmission.
   D. none of the above.

3. Mechanic A says the Ford C-3/C-4/C-5 transmissions have band adjustments.
   Mechanic B says the Ford C-3 has two band adjustments, while the C-4 and C-5 have only one.
   Who is correct?
   A. A only B. B only C. Both A and B D. Neither A nor B

4. The valve body shown is from which transmission?
   A. C-3
   B. C-4
   C. C-5
   D. C-6

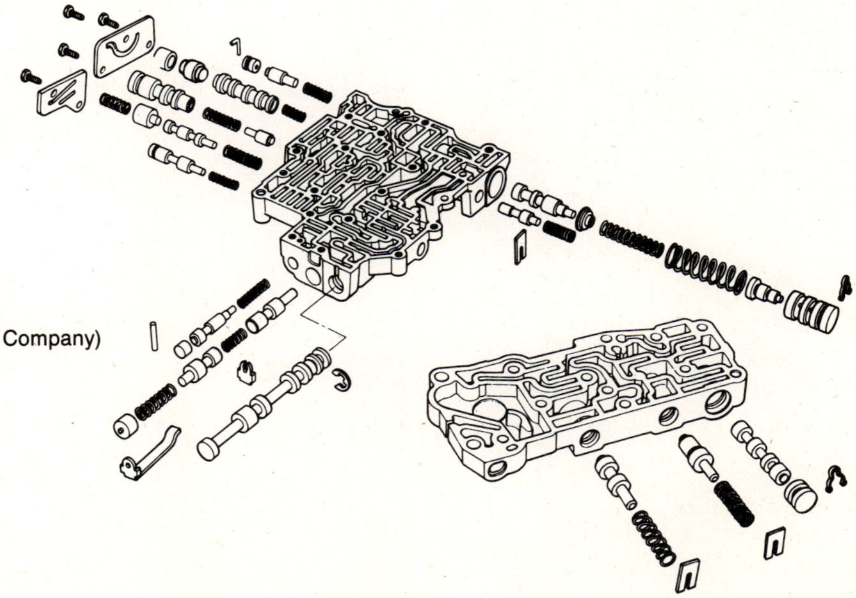

Question 4.
(Ford Motor Company)

5. The C-5 torque converter
   A. is a conventional type (non-locking).
   B. locks up in high gear at speeds above 35 mph.
   C. locks up in all forward gears.
   D. uses a piston and friction facings in the lockup mechanism.

6. Which of the following statements is correct?
   I. The Ford C-3/C-4/C-5 transmissions have metal identification tags attached to bolts on the transmission.
   II. Ford vehicles have a plate on the passenger's doorpost that lists a transmission code.

   A. I only   B. II only   C. Both I and II   D. Neither I nor II

7. All of the following statements about the Ford C-3/C-4/C-5 transmissions are correct EXCEPT
   A. The valve body is removable.
   B. Some of them have an altitude-compensating vacuum control diaphragm (modulator).
   C. The C-3 is a lighter-duty version of the C-4.
   D. The C-5 is used on four-cylinder vehicles.

22 Ford C-6

YOUR LEARNING JOB

The Ford C-6 automatic transmission has been one of the most popular transmissions used in almost two decades. It's one of the more basic automatic transmissions, and a good one on which to learn the job of overhauling. This unit discusses the C-6 and its operation, with illustrations of the major parts.

JOB COMPETENCY OBJECTIVES

This is a knowledge-building unit. When you have completed your assignments and exercises in this unit, you should be able to:

- ☐ Describe the major parts of the Ford C-6.
- ☐ Explain the planetary gear system.
- ☐ Locate the transmission identification tag.

22.1 FORD C-6 DESCRIPTION

This is a heavy-duty transmission that has been used successfully in Ford intermediate and full-size cars. It has also been used throughout the Ford light truck line. It is usually coupled to Ford's big V-8 engines, including the super V-8s of the late 1960s and early 1970s. Beginning in 1977, it was used with Ford's 300-cubic-inch 6-cylinder engine.

The C-6 is very similar in design to the smaller C-4. The biggest difference is that the C-6 uses a low/reverse clutch instead of a low/reverse band. Other than that, the hydraulic control system, clutch combinations, and geartrain are very similar. The C-6 transmission consists of the following main parts:

- Torque converter
- Three multiple-disc clutches
- One overrunning clutch

- One servo and band
- Compound planetary gearset
- Removable valve body
- Aluminum transmission case with integral bellhousing and removable extension housing.

The Ford C-6 combines a torque converter with a fully automatic three-speed gear system. The bellhousing and the transmission case are a single aluminum casting. The extension housing is removable.

The torque converter is a three-element unit that connects to the crankshaft through a flywheel.

The compound planetary gear system shares a common sun gear. The sun gear is connected to the input shell and to the output shaft through the rear planetary gearset.

The hydraulic system has a single oil pump. The valve body is removable.

Earlier C-6 transmissions had many differences from those manufactured today. Always refer to the proper service manual for the correct repair procedures.

Figures 22-1 through 22-4 illustrate the major parts of the C-6 transmission.

22.2 TRANSMISSION IDENTIFICATION

The C-6 requires locating two identification tags that will give the necessary information on the transmission. A plate on the driver's side doorpost gives a transmission identification. For further identification, a tag is located at the lower front bolt of the intermediate servo. To decipher the transmission codes, refer to the proper service manual.

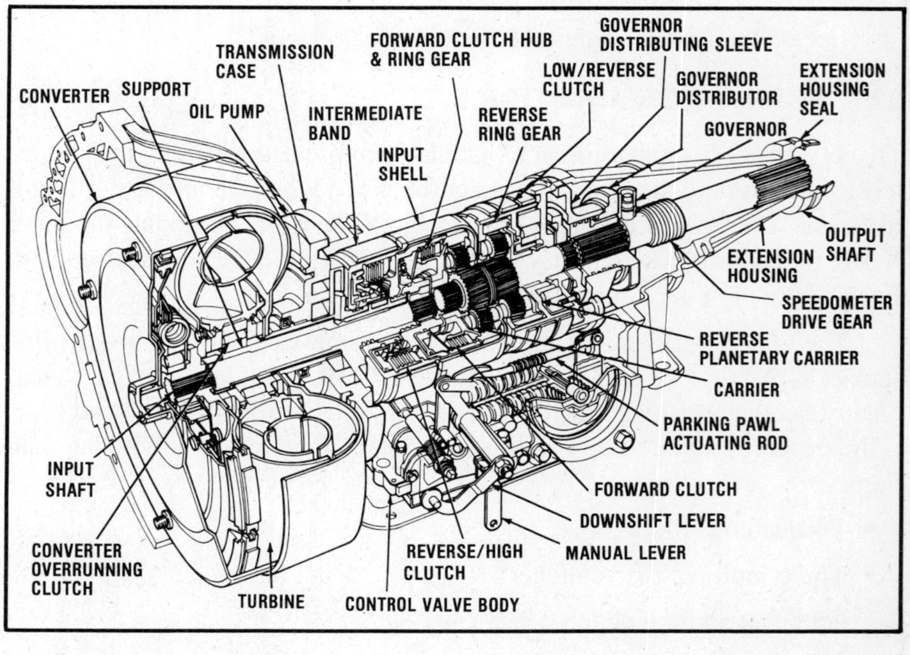

Figure 22-1. Ford C-6 automatic transmission, cutaway. (Ford Motor Company)

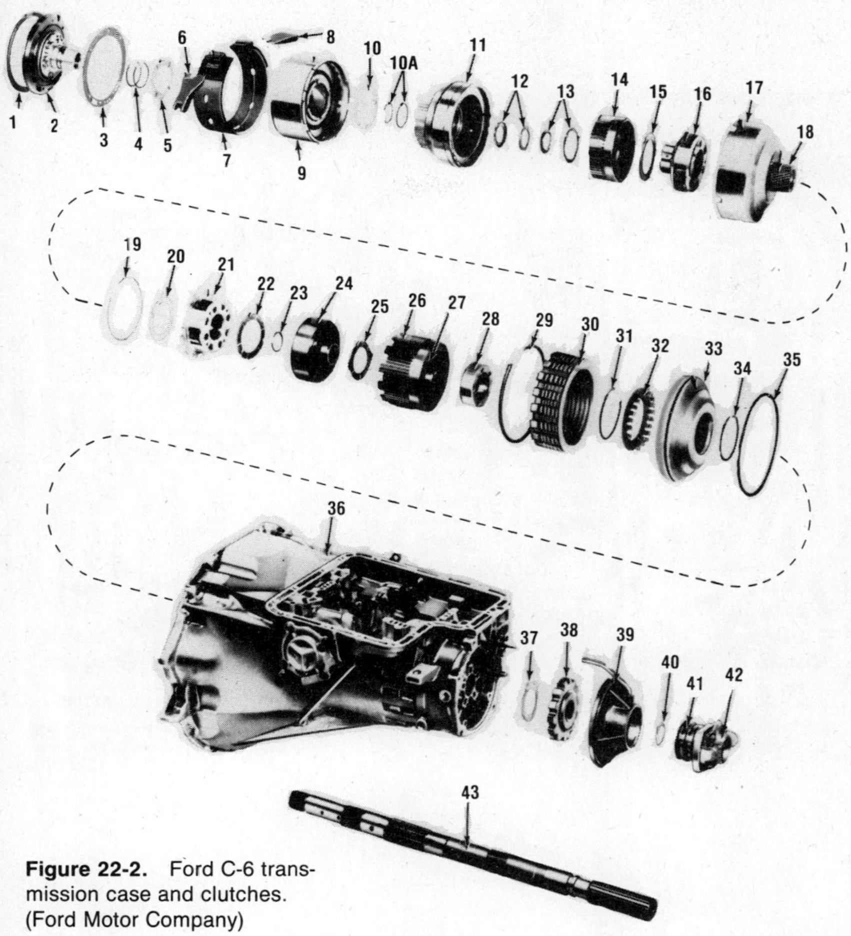

Figure 22-2. Ford C-6 transmission case and clutches. (Ford Motor Company)

1. OIL PUMP SEAL
2. OIL PUMP
3. GASKET
4. SEAL RINGS
5. #1 THRUST WASHER
6. INTERMEDIATE BAND STRUT
7. INTERMEDIATE BRAKE BAND
8. INTERMEDIATE BAND STRUT ANCHOR
9. REVERSE/HIGH CLUTCH
10. #2 THRUST WASHER
10A. FORWARD CLUTCH SEAL RING
11. FORWARD CLUTCH
12. #3 THRUST WASHER
13. #4 THRUST WASHER
14. FORWARD RING GEAR & HUB
15. #5 THRUST WASHER
16. FORWARD PLANETARY CARRIER
17. INPUT SHELL
18. #6 THRUST WASHER
19. SNAP RING
20. #7 THRUST WASHER
21. REVERSE PLANETARY CARRIER
22. #8 THRUST WASHER
23. REVERSE RING GEAR & HUB RETAINER
24. REVERSE RING GEAR & HUB
25. #9 THRUST WASHER
26. LOW/REVERSE CLUTCH HUB
27. OVERRUNNING CLUTCH
28. OVERRUNNING CLUTCH INNER RACE
29. SNAP RING
30. LOW/REVERSE CLUTCH
31. SNAP RING
32. LOW/REVERSE PISTON RETURN SPRING & RETAINER
33. LOW/REVERSE PISTON
34. INNER SEAL
35. OUTER SEAL
36. TRANSMISSION CASE
37. #10 THRUST WASHER
38. PARKING GEAR
39. GOVERNOR DISTRIBUTOR SLEEVE
40. SNAP RING
41. GOVERNOR COLLECTOR
42. GOVERNOR
43. OUTPUT SHAFT

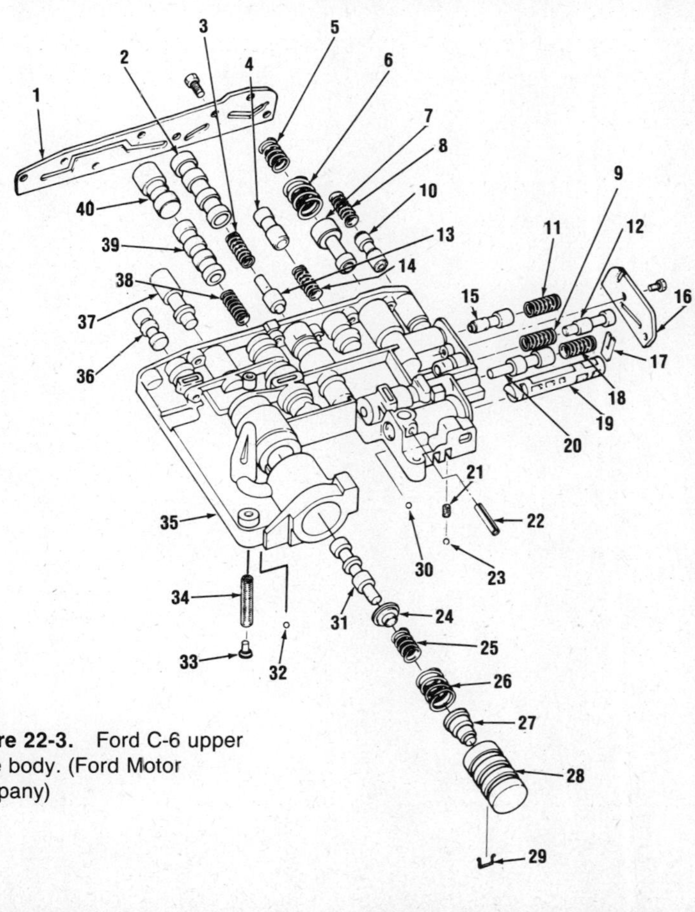

Figure 22-3. Ford C-6 upper valve body. (Ford Motor Company)

1. SHIFT VALVE PLATE
2. 2-3 SHIFT VALVE
3. THROTTLE MODULATING VALVE SPRING
4. 2-3 BACKOUT VALVE
5. SMALL INTERMEDIATE SERVO ACCUMULATOR SPRING
6. LARGE INTERMEDIATE SERVO ACCUMULATOR SPRING
7. INTERMEDIATE SERVO ACCUMULATOR VALVE
8. INTERMEDIATE SERVO MODULATOR VALVE SPRING
9. THROTTLE PRESSURE BOOST SPRING
10. INTERMEDIATE SERVO MODULATOR VALVE
11. 2-1 SCHEDULING VALVE SPRING
12. THROTTLE PRESSURE BOOST VALVE
13. THROTTLE MODULATOR VALVE
14. BACKOUT VALVE SPRING
15. 1-2 SCHEDULING VALVE
16. THROTTLE BOOST PLATE
17. RETAINER
18. DOWNSHIFT VALVE SPRING
19. MANUAL VALVE
20. THROTTLE DOWNSHIFT VALVE
21. THROTTLE PRESSURE RELIEF SPRING
22. MANUAL VALVE RETAINING PIN
23. THROTTLE PRESSURE RELIEF BALL
24. SPRING SEAT
25. INTERMEDIATE SPRING
26. PRESSURE REGULATOR SPRING
27. BOOSTER VALVE
28. BOOSTER VALVE SLEEVE
29. RETAINER
30. REVERSE CLUTCH CHECK BALL
31. MAIN REGULATOR VALVE
32. 2-3 SHIFT VALVE CHECK BALL
33. CONVERTER PRESSURE RELIEF VALVE
34. CONVERTER PRESSURE RELIEF SPRING
35. UPPER VALVE BODY
36. CUTBACK CONTROL VALVE
37. LINE PRESSURE COAST VALVE
38. 1-2 SHIFT VALVE SPRING
39. DRIVE-2 SHIFT VALVE
40. 1-2 SHIFT VALVE

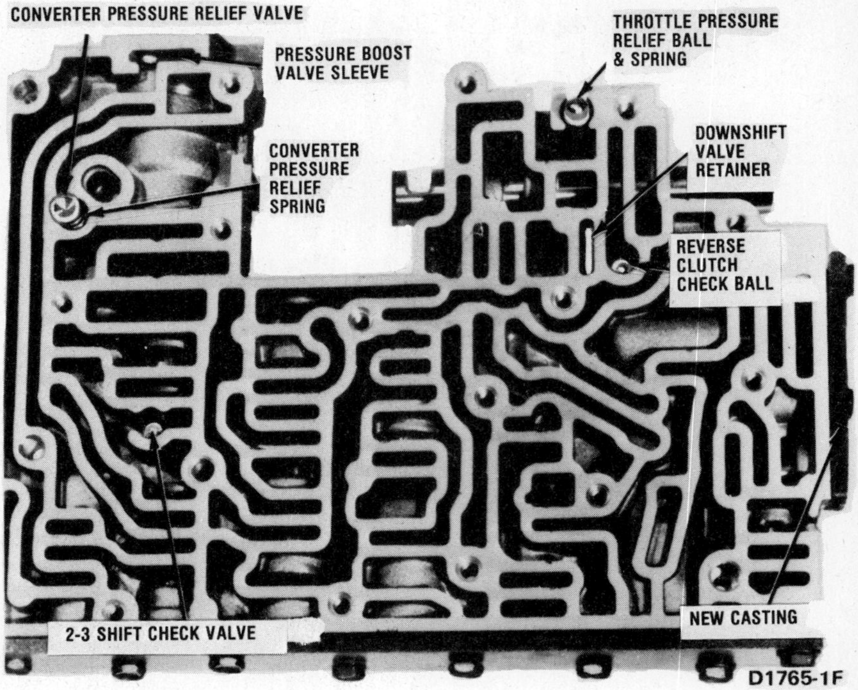

Figure 22-4. Ford C-6 converter pressure relief valve and check valve locations. (Ford Motor Company)

NAME ______________________ SECTION ____________ DATE ____________

Certification Practice

Test 22

DIRECTIONS:

The following questions are similar to those used on mechanic certification tests. Answer each question by circling the letter of the correct choice.

1. The Ford C-6 transmission is
   A. a light-duty transmission.
   B. a medium-duty transmission.
   C. a heavy-duty transmission.
   D. used only in light trucks.

2. Which of the following statements is correct?
   I. The Ford C-6 transmission has a low/reverse clutch.
   II. The Ford C-6 transmission has a low/reverse band.
   A. I only B. II only C. Both I and II D. Neither I nor II

3. The Ford C-6 transmission includes all of the following EXCEPT
   A. one overrunning clutch.
   B. one servo and band.
   C. three multiple-disc clutches.
   D. a lockup torque converter.

4. The sun gear is *not* connected to the
   A. input shell.
   B. front planetary gearset.
   C. output shell.
   D. rear planetary gearset.

5. The Ford C-6 transmission is used with
   A. four-cylinder engines.
   B. six-cylinder engines.
   C. V-8 engines.
   D. six-cylinder and V-8 engines.

6. Mechanic A says that the Ford C-6 transmission has a metal identification tag attached to the lower front servo cover bolt.
   Mechanic B says that Ford vehicles have a plate on the driver's doorpost that contains a transmission code.
   Who is correct?
   A. A only B. B only C. Both A and B D. Neither A nor B

7. The part of the C-6 transmission shown is
   A. the upper valve body.
   B. the lower valve body.
   C. the separator plate.
   D. none of the above.

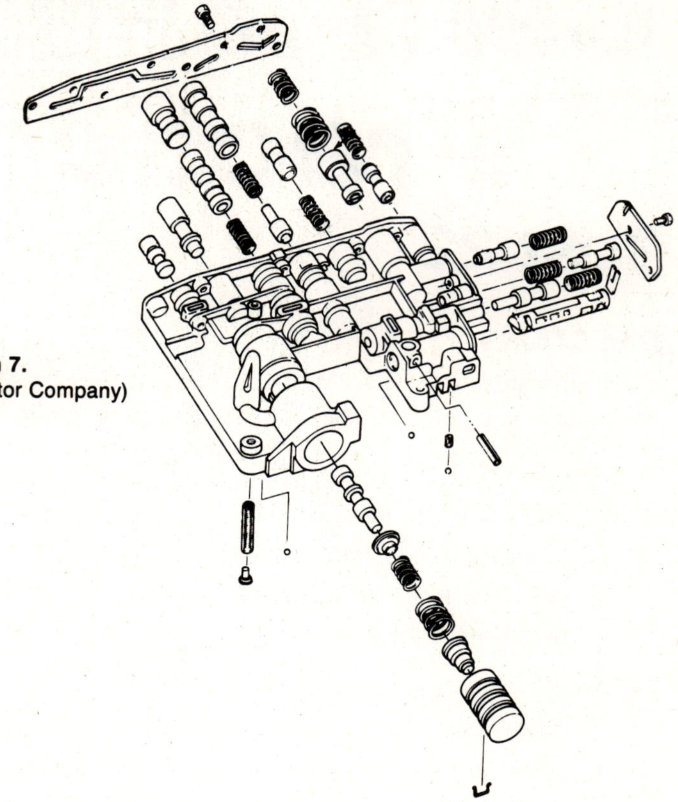

Question 7.
(Ford Motor Company)

23 Ford Jatco

YOUR LEARNING JOB

The Jatco is a truly international transmission used in Japanese and American cars and small trucks. The Jatco is similar to the Ford C-3/C-4 transmissions. This unit discusses the Jatco and illustrates the major parts of the transmission.

JOB COMPETENCY OBJECTIVES

This is a knowledge-building unit. When you have completed your assignments and exercises in this unit, you should be able to:

- ☐ Name vehicles that use the Jatco transmission.
- ☐ List the major parts of the Jatco transmission.
- ☐ Locate the identification plates.

23.1 JATCO DESCRIPTION

Commonly known as the Ford Jatco, this transmission was developed by Ford, Nissan, and Mazda. It has been used in Mazda-built cars and mini-trucks, including the Ford Courier, since 1972. Nissan also uses the Jatco in many of its Datsun automobiles. In 1977, Ford began using the transmission in Granada and Monarch automobiles with six-cylinder engines. The Jatco consists of the following main parts:

- Torque converter
- Two multiple-disc clutches
- One overrunning clutch
- One servo and band
- Compound planetary gearset
- Removable valve body
- Aluminum transmission case with removable bellhousing and extension housing.

The Jatco (Japan Automatic Transmission Company) transmission combines a torque converter with a fully automatic three-speed gear system. The bellhousing, transmission case, and extension housing are three separate aluminum castings.

The torque converter is a three-element unit that connects to the crankshaft through a flexplate or flywheel.

The compound planetary gear system shares a common sun gear. The sun gear is connected to the input drum and to the output shaft through the rear planetary gearset.

The hydraulic system has a single oil pump. The valve body is removable.

Figures 23-1 through 23-4 illustrate the major parts of the Jatco transmission.

23.2 JATCO TRANSMISSION IDENTIFICATION

The Jatco identification plate is attached to the left side of the transmission. It is located just above the manual control lever. When repairing or overhauling the Jatco, the identification tag must always be replaced. Otherwise, it is very difficult to identify the transmission.

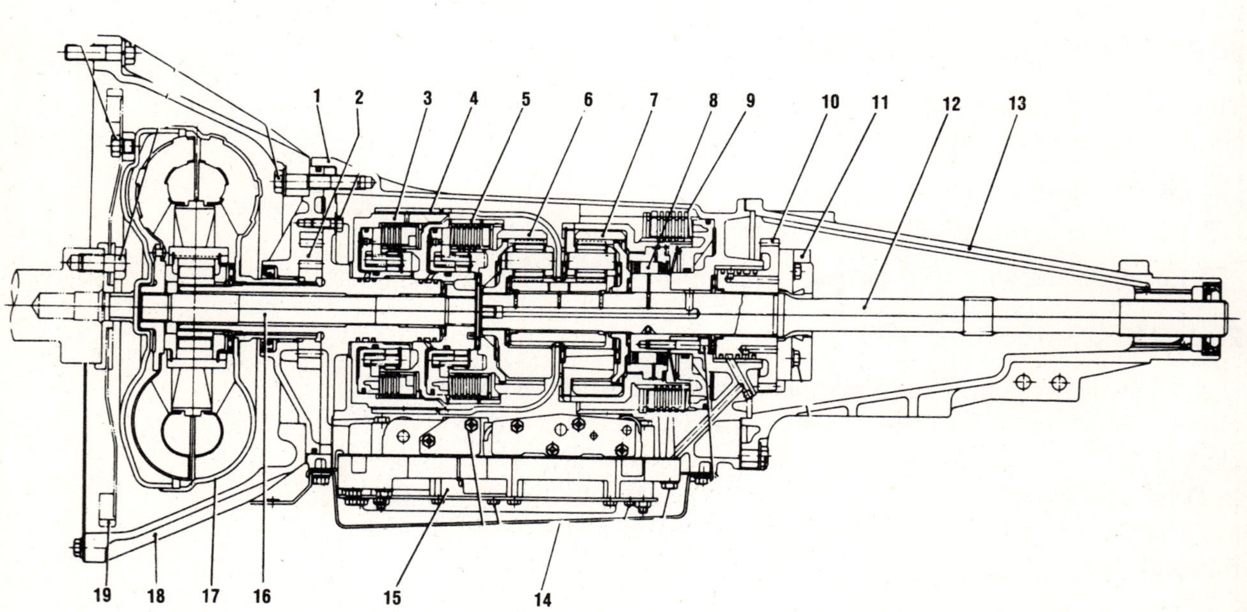

1. TRANSMISSION CASE
2. OIL PUMP
3. HIGH/REVERSE CLUTCH
4. BRAKE BAND
5. FORWARD CLUTCH
6. FRONT PLANETARY GEARSET
7. REAR PLANETARY GEARSET
8. OVERRUNNING CLUTCH
9. LOW/REVERSE BRAKE
10. OIL DISTRIBUTOR
11. GOVERNOR
12. OUTPUT SHAFT
13. EXTENSION HOUSING
14. OIL PAN
15. CONTROL VALVE BODY
16. INPUT SHAFT
17. TORQUE CONVERTER
18. CONVERTER HOUSING
19. FLEXPLATE

Figure 23-1. Jatco automatic transmission cutaway. (Nissan Motor Corporation)

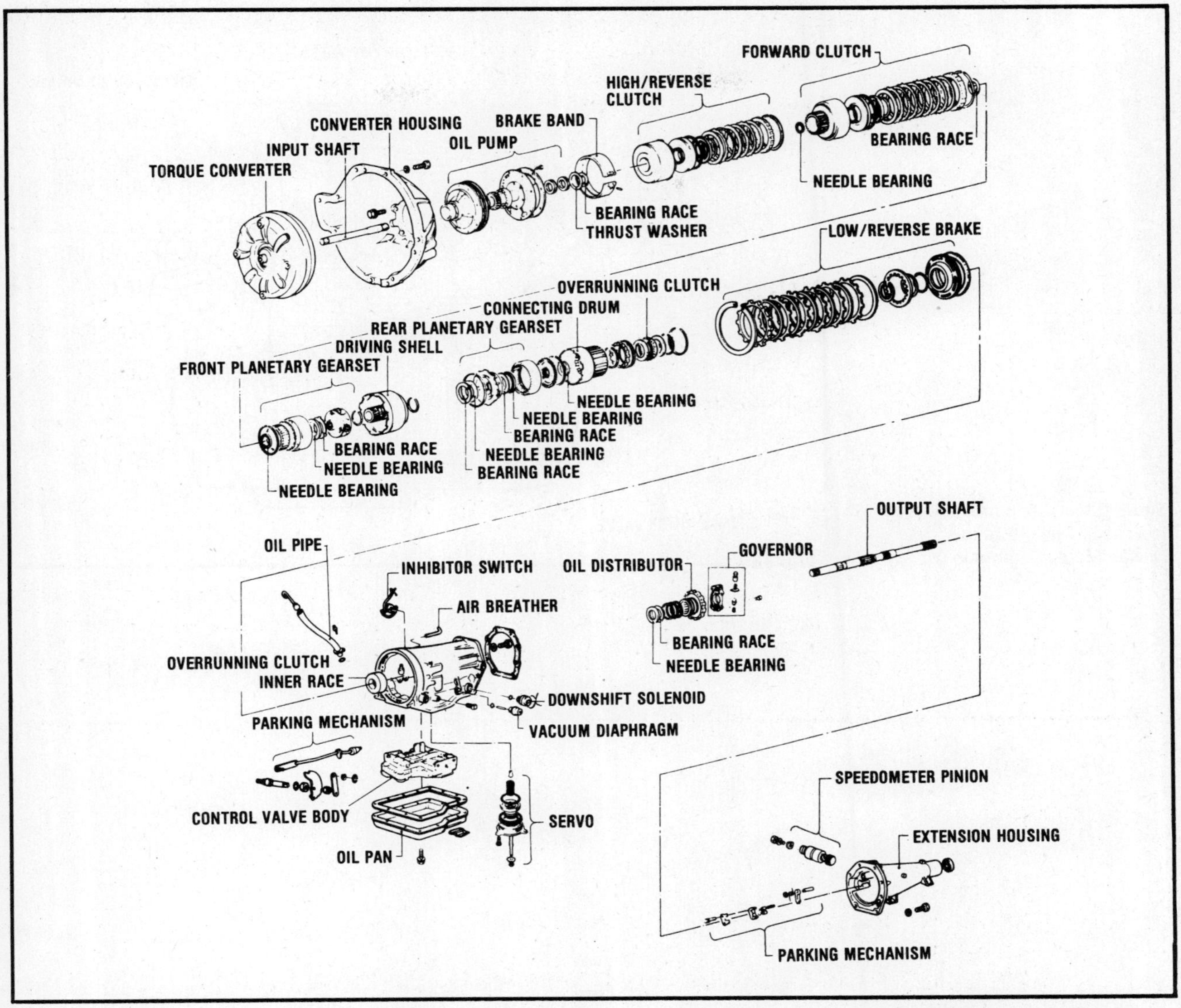

Figure 23-2. Jatco transmission case, oil pump, clutches, and planetary gearsets. (Nissan Motor Corporation)

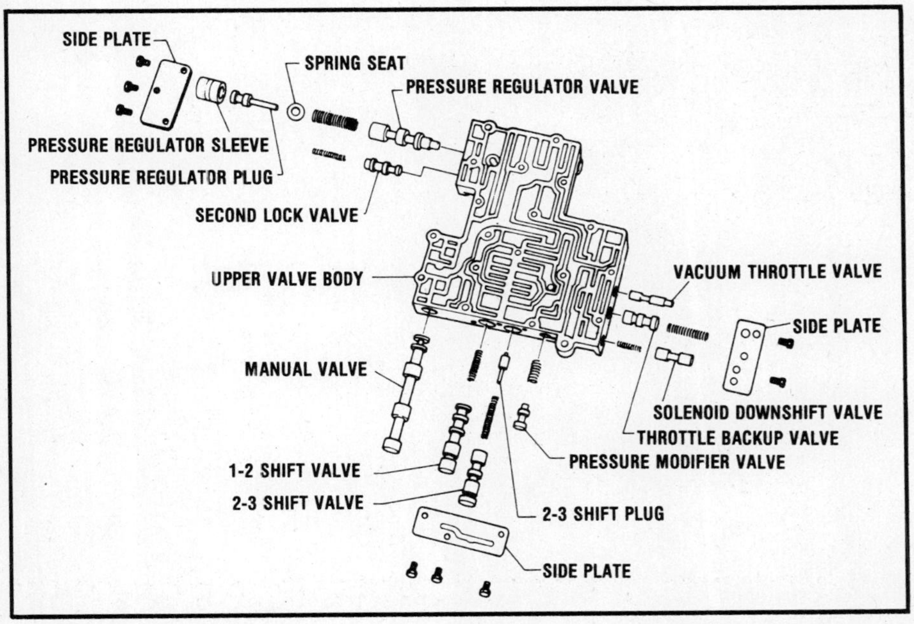

Figure 23-3. Jatco control valve body. (Nissan Motor Corporation)

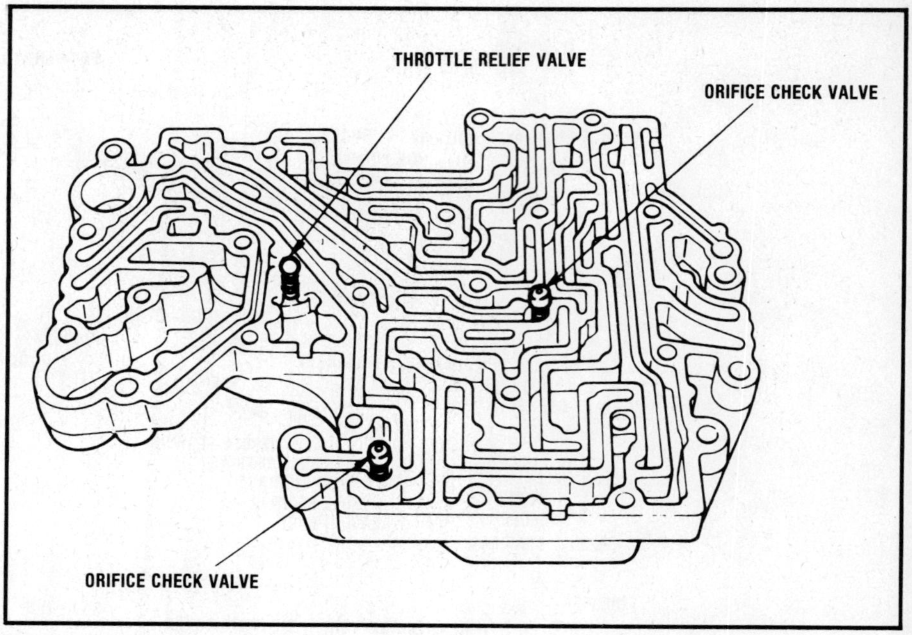

Figure 23-4. Jatco relief valve and check valve locations. (Nissan Motor Corporation)

NAME ______________________ SECTION ____________ DATE ____________

Certification Practice

Test 23

DIRECTIONS:

The following questions are similar to those used on mechanic certification tests. Answer each question by circling the letter of the correct choice.

1. Which of the following statements is correct?

   I. The Ford Jatco transmission is used in Mazda and Nissan vehicles.

   II. The Ford Jatco transmission is used in Ford vehicles.

   A. I only B. II only C. Both I and II D. Neither I nor II

2. The Ford Jatco transmission includes all of the following EXCEPT

   A. one overrunning clutch.
   B. two multiple-disc clutches.
   C. one servo and band.
   D. aluminum transmission case with integral bellhousing and valve body.

3. Mechanic A says the Ford Jatco transmission is a four-speed overdrive transmission. Mechanic B says the Ford Jatco transmission is a four-speed transmission with a lockup torque converter.

   Who is correct?

   A. A only B. B only C. Both A and B D. Neither A nor B

4. The Ford Jatco transmission sun gear is connected to

   A. the input drum and the rear planetary gearset.
   B. the front clutch by a driving shell.
   C. the driving shell and the output shaft.
   D. the input shell and the output shaft.

5. The valve body is

   A. integral.
   B. removable.
   C. jointed.
   D. sealed.

6. The Ford Jatco transmission identification plate is

   A. a removable metal tag on the lower front servo cover bolt.
   B. stamped into the right side of the bellhousing.
   C. stamped on the governor cover.
   D. on the left side of the transmission above the manual control lever.

7. The drawing shows the location of the
   A. shift valves.
   B. separator plate.
   C. check balls.
   D. pressure springs.

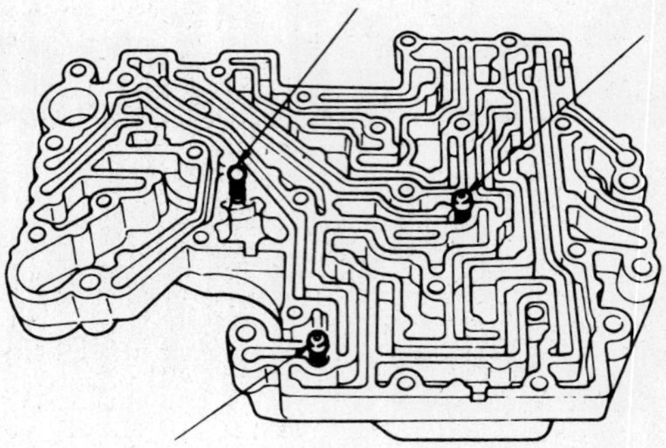

Question 7.
(Nissan Motor Corporation)

24 Borg-Warner A40

YOUR LEARNING JOB

Because it is one of the most common transmissions in Japanese cars, the A40 is important to the automatic transmission mechanic. This unit discusses the A40 and illustrates its major parts.

JOB COMPETENCY OBJECTIVES

This is a knowledge-building unit. When you have completed your assignments and exercises in this unit, you should be able to:

- ☐ Name vehicles that use the A40 transmission.
- ☐ Describe the major parts of the A40 transmission.
- ☐ Explain the differences between this transmission and others.

24.1 BORG-WARNER A40 DESCRIPTION

The A40 transmission was designed by Borg-Warner and is built by Aisin-Warner in Japan. The transmission has been used in Toyota Corollas, Coronas, Celicas, and pickup trucks. The transmission was used by Toyota from 1973 through 1981.

This transmission uses no bands, so no band adjustments are necessary. The Borg-Warner A40 consists of the following main parts:

- Torque converter
- Three multiple-disc clutches
- Two overruning clutches
- Compound planetary gearset
- Removable three-part valve body
- Aluminum transmission case with removable bellhousing and extension housing.

The Borg-Warner A40 combines a torque converter with a fully automatic three-speed gear system. The bellhousing, transmission case, and extension housing are separate aluminum castings.

The torque converter is a three-element unit that connects to the crankshaft through a flexplate.

The compound planetary gear system shares a common sun gear.

The hydraulic system has a single oil pump. The valve body is removable.

Figures 24-1 through 24-7 show main assemblies of the Borg-Warner A40.

1. THRUST BEARING RACE
2. THRUST BEARING
3. THRUST BEARING SEAT
4. SNAP RING
5. REAR CLUTCH FLANGE
6. CLUTCH DISC
7. CLUTCH PLATE
8. SNAP RING
9. CLUTCH-PISTON RETURN SPRING SEAT
10. COMPRESSION RING
11. O-RING
12. O-RING
13. REAR CLUTCH OUTER PISTON
14. O-RING
15. REAR CLUTCH INNER PISTON
16. REAR CLUTCH DRUM SUBASSEMBLY

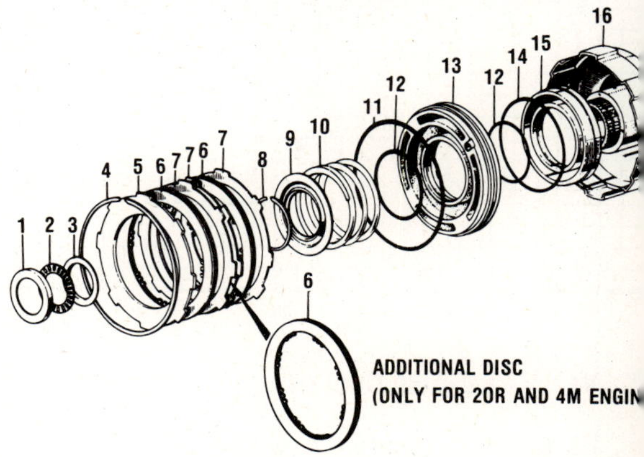

Figure 24-1. Borg-Warner A40 rear clutch. (Toyota Motor Sales U.S.A., Inc.)

1. INPUT SHAFT SUBASSEMBLY
2. FRONT CLUTCH PISTON
3. O-RING
4. O-RING
5. COMPRESSION SPRING
6. CLUTCH-PISTON RETURN SPRING SEAT
7. SNAP RING
8. THRUST BEARING RACE
9. THRUST BEARING
10. THRUST BEARING RACE
11. #1 CLUTCH PLATE
12. #1 CLUTCH DISC
13. FRONT CLUTCH HUB
14. SNAP RING
15. REAR CLUTCH HUB
16. SNAP RING

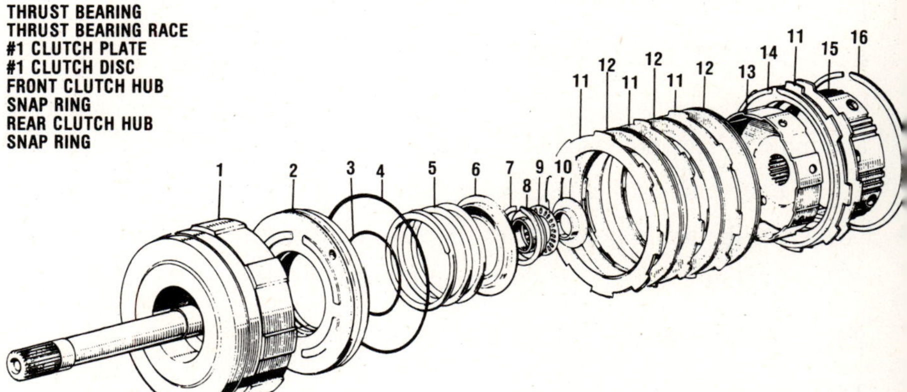

Figure 24-2. Borg-Warner A40 front clutch. (Toyota Motor Sales U.S.A., Inc.)

1. OIL SEAL RINGS
2. SNAP RING
3. SNAP RING
4. REAR CLUTCH FLANGE
5. #1 CLUTCH DISC
6. #3 CLUTCH PLATE
7. BRAKE RETURN SPRING SEAT
8. COMPRESSION SPRING
9. BRAKE PISTON
10. O-RING
11. O-RING
12. CENTER SUPPORT SUBASSEMBLY
13. SNAP RING
14. OVERRUNNING CLUTCH RETAINER
15. #2 BRAKE HUB
16. #1 OVERRUNNING CLUTCH
17. BUSHING
18. PLANETARY SUN GEAR
19. OIL SEAL RING

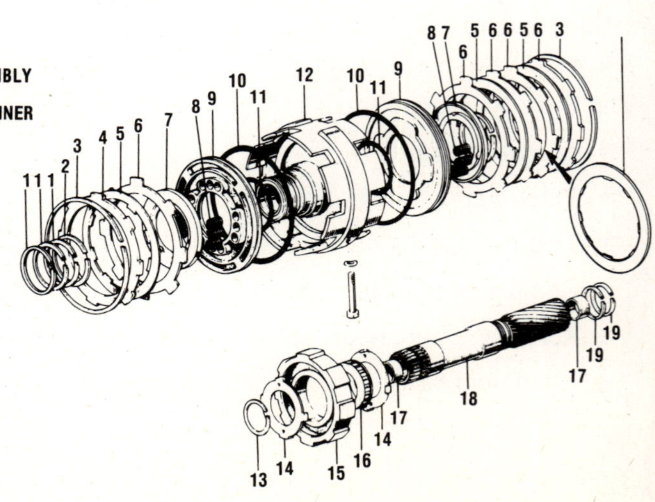

Figure 24-3. Borg-Warner A40 center support. (Toyota Motor Sales U.S.A., Inc.)

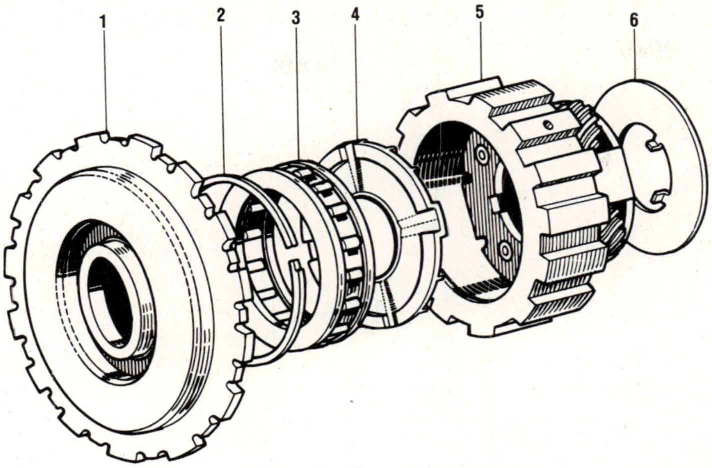

1. OVERRUNNING CLUTCH INNER RACE
2. SNAP RING
3. #2 OVERRUNNING CLUTCH
4. #2 THRUST WASHER
5. FRONT PLANETARY GEARSET
6. #1 THRUST WASHER

Figure 24-4. Borg-Warner A40 front planetary overrunning clutch. (Toyota Motor Sales U.S.A., Inc.)

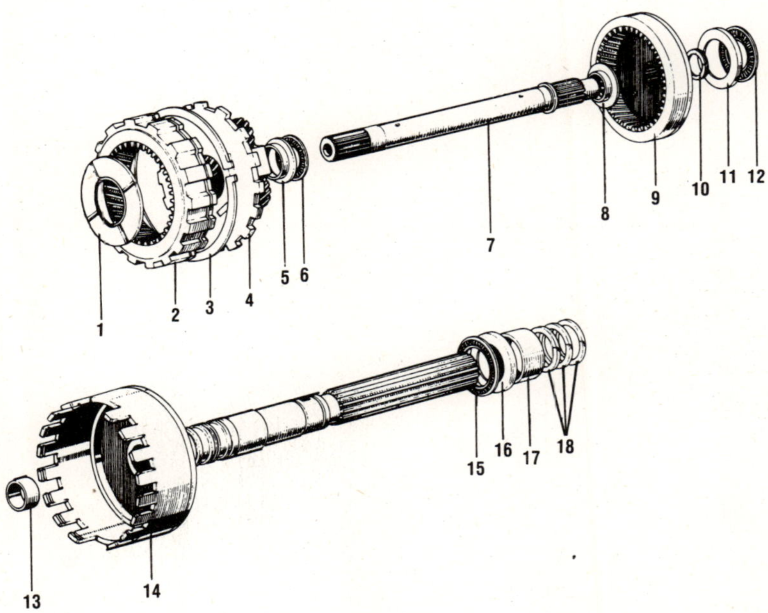

1. #2 THRUST WASHER
2. FRONT PLANETARY RING GEAR
3. SNAP RING
4. REAR PLANETARY GEARSET
5. THRUST BEARING RACE
6. THRUST BEARING
7. INTERMEDIATE SHAFT
8. THRUST BEARING RACE
9. REAR PLANETARY RING GEAR
10. SET RING
11. THRUST BEARING RACE
12. THRUST BEARING
13. BUSHING
14. OUTPUT SHAFT ASSEMBLY
15. THRUST BEARING
16. THRUST BEARING RACE
17. BUSHING
18. OIL SEAL RING

Figure 24-5. Borg-Warner A40 rear planetary gearset and output shaft. (Toyota Motor Sales U.S.A., Inc.)

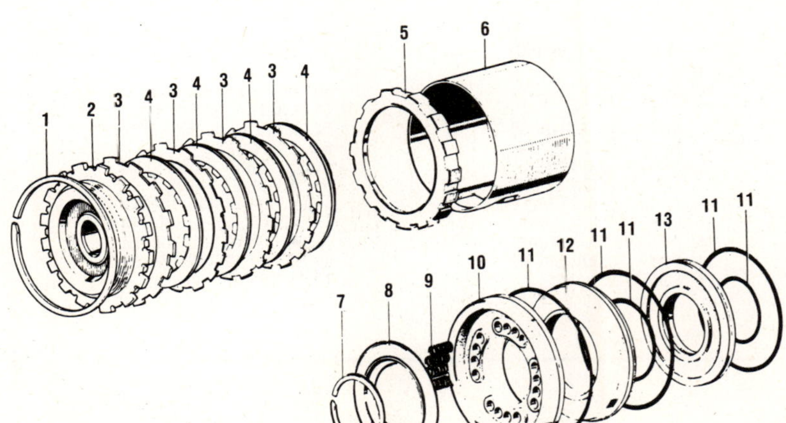

1. SNAP RING
2. BRAKE REACTION PLATE
3. #2 CLUTCH DISC
4. #2 CLUTCH PLATE
5. PRESSURE PLATE
6. BRAKE APPLY TUBE
7. SNAP RING
8. BRAKE RETURN SPRING SEAT
9. COMPRESSION SPRING
10. #3 BRAKE PISTON (OUTER)
11. O-RING
12. BRAKE REACTION SLEEVE
13. #3 BRAKE PISTON (INNER)

Figure 24-6. Borg-Warner A40 #3 brake assembly. (Toyota Motor Sales U.S.A., Inc.)

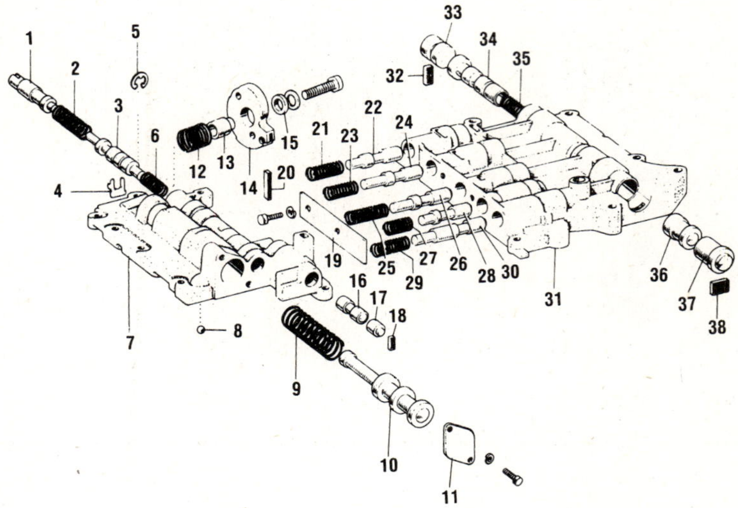

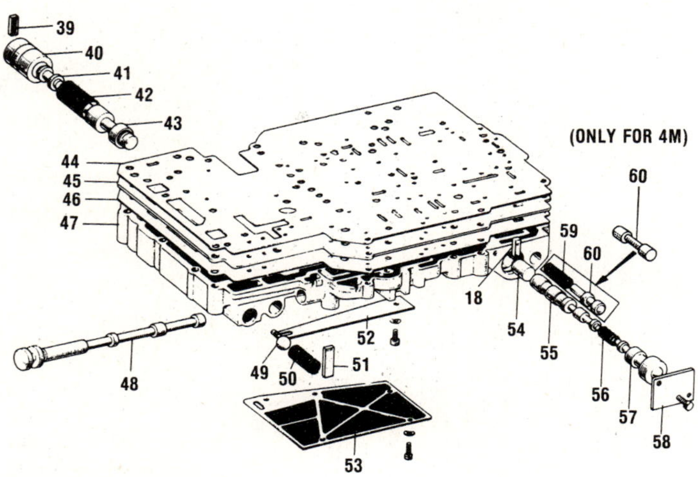

Figure 24-7. Borg-Warner A40 control valve body. (Toyota Motor Sales U.S.A., Inc.)

1. DOWNSHIFT PLUG
2. COMPRESSION SPRING
3. THROTTLE VALVE
4. KEEP PLATE
5. E-RING
6. COMPRESSION SPRING
7. FRONT UPPER VALVE BODY
8. CHECK BALL
9. COMPRESSION SPRING
10. SECONDARY REGULATOR VALVE
11. #3 VALVE BODY COVER
12. TORSION SPRING
13. THROTTLE VALVE CAM PIN
14. THROTTLE VALVE CAM
15. WASHER
16. CUTBACK VALVE
17. CUTBACK PLUG
18. SHIFT VALVE PLUG SEAT
19. #2 VALVE BODY COVER
20. DETENT SPRING SEAT
21. COMPRESSION SPRING
22. DETENT REGULATOR VALVE
23. COMPRESSION SPRING
24. INTERMEDIATE MODULATOR VALVE
25. COMPRESSION SPRING
26. SEQUENCE VALVE
27. COMPRESSION SPRING
28. GOVERNOR MODULATOR VALVE
29. COMPRESSION SPRING
30. LOW MODULATOR VALVE
31. REAR UPPER VALVE BODY
32. SHIFT VALVE PLUG SEAT
33. 2-3 SHIFT VALVE PLUG
34. 2-3 SHIFT VALVE
35. COMPRESSION SPRING
36. INTERMEDIATE SHIFT VALVE
37. INTERMEDIATE SHIFT VALVE PLUG
38. INTERMEDIATE SHIFT VALVE SEAT
39. PRESSURE REGULATOR VALVE SPRING SEAT
40. PRIMARY REGULATOR VALVE SPRING
41. PRIMARY REGULATOR VALVE PLUNGER
42. COMPRESSION SPRING
43. PRIMARY REGULATOR VALVE
44. #1 VALVE BODY GASKET
45. VALVE BODY PLATE
46. #2 VALVE BODY GASKET
47. LOWER VALVE BODY
48. MANUAL VALVE
49. CHECK BALL
50. COMPRESSION SPRING
51. CHECK VALVE SPRING SEAT
52. MANUAL DETENT SPRING
53. OIL FILTER
54. 1-2 SHIFT VALVE PLUG
55. 1-2 SHIFT VALVE
56. COMPRESSION SPRING
57. LOW COAST SHIFT VALVE
58. #1 VALVE BODY COVER
59. COMPRESSION SPRING
60. SEQUENCE VALVE

NAME ______________________ SECTION ____________ DATE ____________

Certification Practice

Test 24

DIRECTIONS:

The following questions are similar to those used on mechanic certification tests. Answer each question by circling the letter of the correct choice.

1. Which of the following statements is correct?
   I. The Borg-Warner A40 transmission is used in Toyota Corollas, Coronas, and Celicas.
   II. The Borg-Warner A40 transmission is used in pickup trucks.
   A. I only B. II only C. Both I and II D. Neither I nor II

2. The Borg-Warner A40 transmission includes all of the following EXCEPT
   A. one servo and band.
   B. two overrunning clutches.
   C. three multiple-disc clutches.
   D. removable valve body.

3. The assembly shown in the picture is
   A. a front disc clutch.
   B. a rear disc clutch.
   C. a center support.
   D. none of the above.

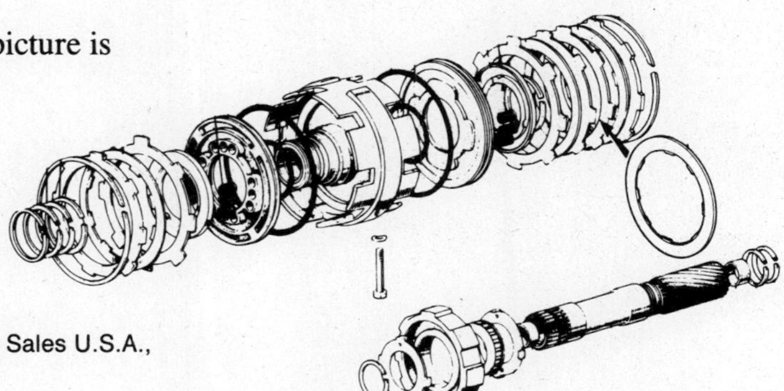

Question 3.
(Toyota Motor Sales U.S.A., Inc.)

4. The unit shown in the picture is
   A. a rear multiple-disc clutch.
   B. a front multiple-disc clutch.
   C. a center support assembly.
   D. an overrunning clutch.

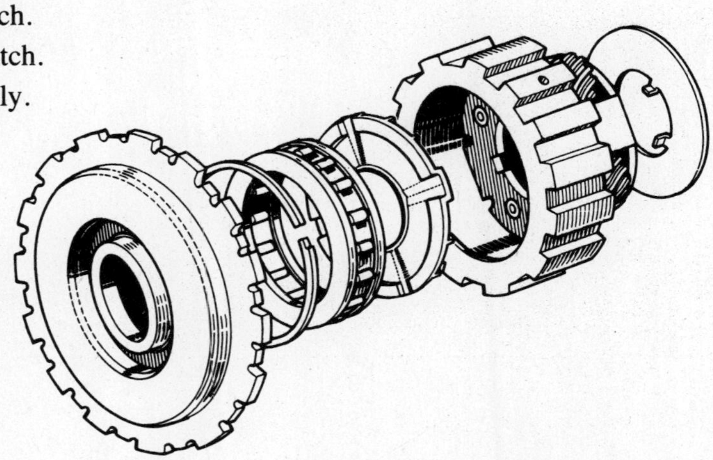

Question 4.
(Toyota Motor Sales U.S.A., Inc.)

5. The Borg-Warner A40 transmission is a
   A. 4-speed overdrive transmission.
   B. 4-speed transmission with lockup torque converter.
   C. 3-speed transmission.
   D. 3-speed transmission with lockup torque converter.
6. The hydraulic system has
   A. one oil pump.
   B. two oil pumps.
   C. a rear oil pump.
   D. no oil pump.
7. The Borg-Warner A40 transmission valve body has
   A. a one-piece casting.
   B. two parts.
   C. three parts.
   D. four parts.

25 Transaxle Differences

YOUR LEARNING JOB

Front-engine/front-wheel-drive has become an increasingly popular automotive design in recent years. Both domestic and foreign manufacturers have been turning to transaxles as a way to achieve major design and performance goals. Transaxle designs require some different techniques in transmission service. These differences are discussed in general terms in this unit.

JOB COMPETENCY OBJECTIVES

This is a knowledge-building unit. When you have completed your assignments and exercises in this unit, you should be able to:

- ☐ List reasons why transaxles are used on many modern cars.
- ☐ Describe how transaxles are different from transmissions.
- ☐ Describe how transaxles are similar to transmissions.
- ☐ Describe the general problems in working on transaxles.

25.1 TRANSAXLES

A transaxle is a combination of two drivetrain units: a transmission and a differential. The transmission and differential parts are combined in one unit, inside the same housing.

There are advantages to combining the units. The weight of the transaxle is less than the combined weight of a separate transmission and differential. The size of the transaxle is also smaller than the separate transmission and differential. In today's smaller cars, saving weight and space are important.

The engine and transaxle can be mounted together at the front or rear of the vehicle. See Figures 25-1 (A and B). If the engine and transaxle are located at the rear of the vehicle, the engine can be mounted behind the rear axle. See

Figure 25-1(B). The engine can also be mounted in front of the rear axle, as shown in Figure 25-1(C). In some vehicles, the engine is mounted at the front, and the transaxle is mounted separately, at the rear. See Figure 25-1(D).

The locations of the engine and transaxle, and the way the engine and transaxle are mounted—together or separately—affect the handling characteristics of a vehicle. One reason for a transaxle location is to put the weight of the powertrain directly over the driving wheels. This provides better traction.

Most cars with transaxles are front-wheel-drive (FWD) vehicles. In a FWD vehicle, the engine and drivetrain are combined at the front of the vehicle, and they drive the front wheels. In such a vehicle, the engine may be mounted transversely, or sideways. When the engine is mounted transversely, the transaxle is also mounted transversely. See Figure 25-1(A). On a few FWD vehicles, the engine is mounted front-to-back, and the transaxle is located to the rear and under the engine. See Figure 25-1(E).

Putting the engine and drivetrain in the front allows for more room inside the passenger compartment and in the trunk. There is no need for a large transmission hump or driveshaft tunnel. Many FWD cars do have smaller humps or tunnels, but their purpose is to stiffen the floorpan of the vehicle and provide a space for the exhaust system. Stiffening the floorpan gives the car body greater strength.

In most FWD vehicles, the powertrain is mounted transversely, or sideways. When the engine or drivetrain is mounted in a line from the front to the back of the car, it is called longitudinal, or front-to-back, mounting.

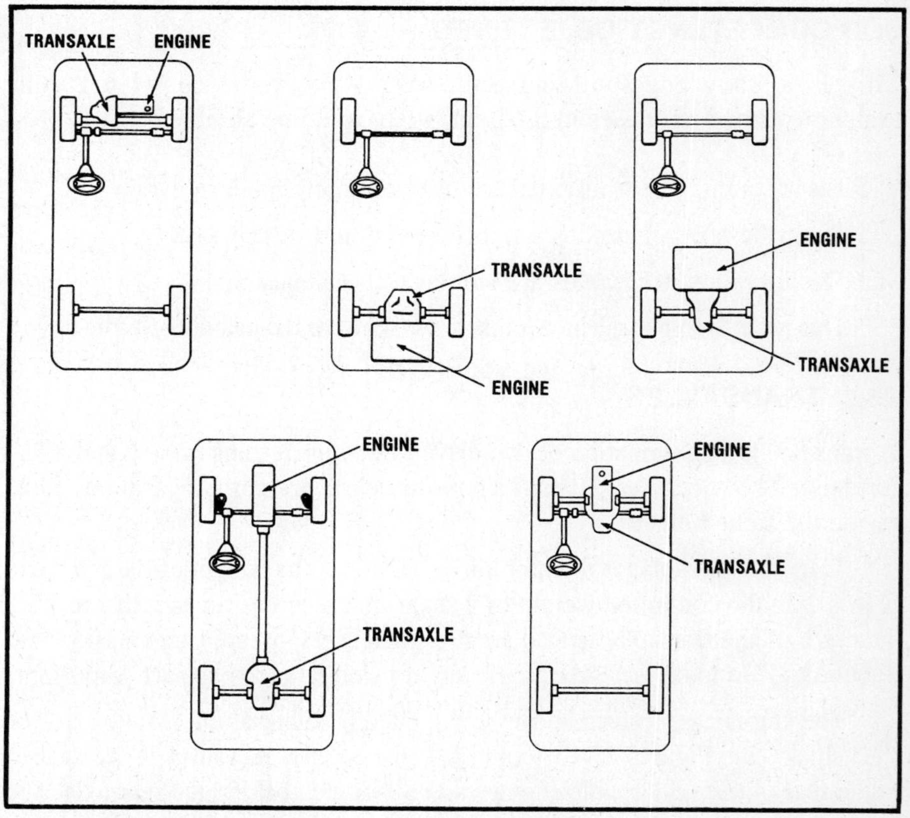

Figure 25-1. Engine and transaxle mountings.

Not all FWD vehicles use transaxles. Some use a transmission with a separate differential unit, such as the GM 325 and 325-4L transmissions used in the Cadillac Eldorado, Oldsmobile Toronado, and Buick Riviera automobiles.

Since most of the vehicles with transaxles are FWD with transversely mounted engines and drivetrains, the sections that follow will concentrate on that kind of vehicle.

Transaxle drivetrains are covered in *Manual Transmissions and Drivetrains,* another text/shop manual in the South-Western Auto Tech Series.

CHECKPOINT

1. **A transaxle is a combination of a ________________ and a ________________.**
2. **Combining the transmission and differential into one housing saves ______________ and ________________.**
3. **Transaxles may be located at the ________________ or ________________ of the vehicle.**
4. **Putting the weight of the powertrain directly over the ________________ ________________ results in better traction.**
5. **When an engine or drivetrain is mounted sideways, it is known as ________________ mounting.**

25.2 MOUNTING THE TRANSAXLE IN THE VEHICLE

In an FWD vehicle with a transversely mounted engine, the powertrain is mounted at the front of the car. The engine, transaxle, and driving axles are all in one relatively small space, so the area under the hood is crowded and often difficult to work in.

Also, when the engine is speeded up, it tends to twist slightly in the motor mounts. The engine will "torque-over," or twist, in the direction opposite to crankshaft rotation.

With a longitudinally mounted powertrain, torquing-over usually doesn't cause problems unless the engine or transmission mounts are loose or damaged. There is enough flex, or give, in the shift linkage to make up for slight twisting. In addition, the linkage tends to twist in the same direction as the powertrain.

In an FWD vehicle with a transversely mounted powertrain, torquing-over of the engine will pull on the shift linkage. This twisting could cause the transaxle to shift into another gear or damage the linkage. To prevent this, braces or struts are often connected between the body and the powertrain, parallel to the shift linkage. These braces or struts, shown in Figure 25-2, will sometimes be in the way when work is being done on a transaxle. Often they

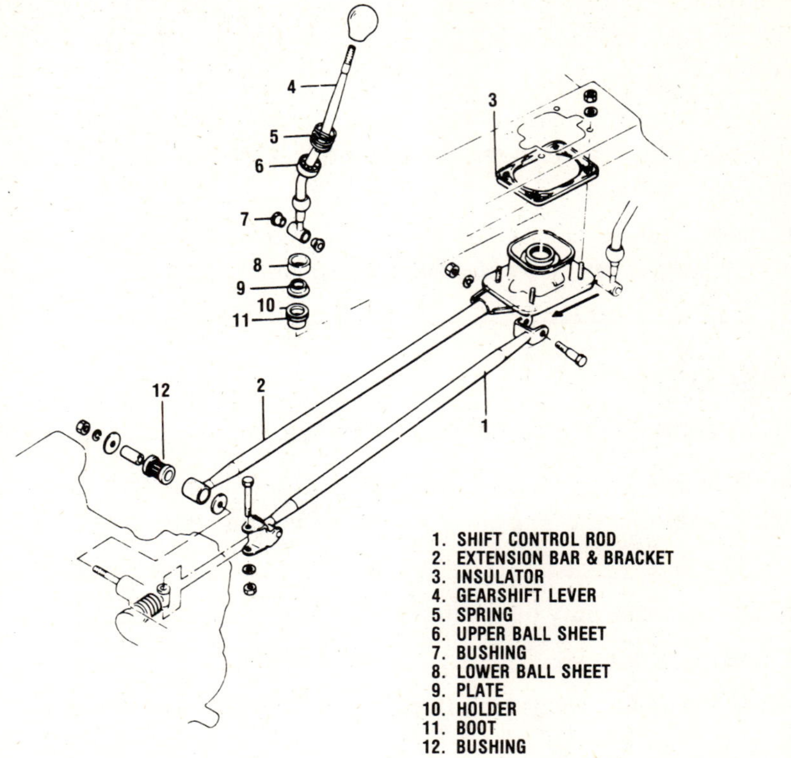

Figure 25-2. The upper rod is a brace used to prevent damage from engine torque. (Toyo Kogyo Co., Ltd.)

must be removed, and the engine supported, before trying to remove the transaxle. In some cases, it is easier to remove the entire engine/transaxle assembly to service the transaxle.

25.3 SHIFT LINKAGE

The shift linkage often consists of rods and levers (Figure 25-3) or a cable (Figure 25-4).

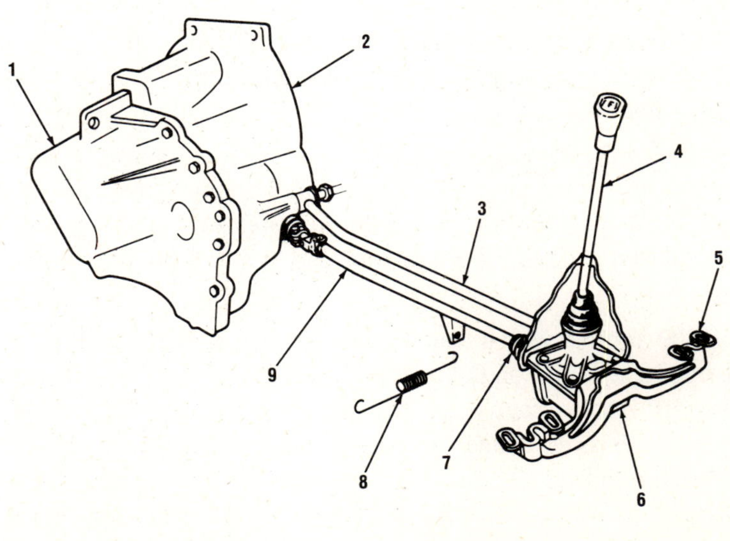

Figure 25-3. Rod-and-lever shift linkage control. (Ford Motor Company)

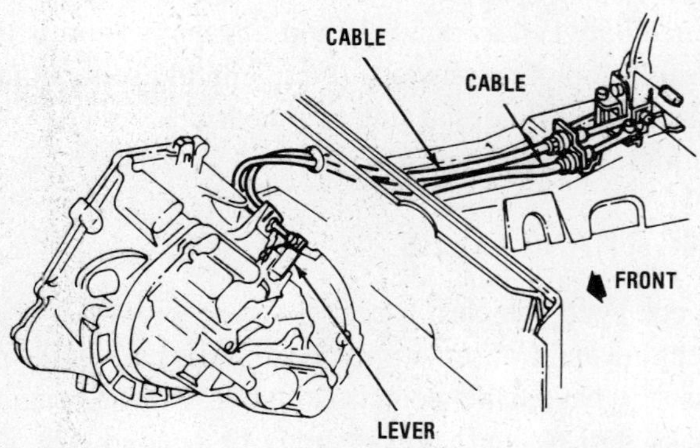

Figure 25-4. Cable shift linkage control. (Pontiac Motor Division-GMC)

The rods and levers are connected to the manual valve at the outside of the transmission. With a longitudinally mounted powertrain, the rods usually run parallel to, and next to, the transmission case. The linkage rods and levers transfer the motion of the shift lever to the manual valve arm. This linkage system is similar to that found in conventional rear-drive vehicles.

In a FWD vehicle with a transversely mounted powertrain, the transaxle is perpendicular, or at a 90-degree angle, to the front-to-back line of the car. The shift linkage must change the normal front-to-back motions of the shift pattern into side-to-side motions. The simplest solution is to use flexible cables to transfer the motions to the manual valve arm on the transaxle.

Rods and levers can also be used with transversely mounted powertrains to give a more normal ''feel'' to the motion of the gearshift lever. However, more complicated mechanical arrangements must be used.

CHECKPOINT

1. **The area under the hood is ________________ and often ________________ to work in on a FWD vehicle.**
2. **To prevent the engine from torquing over when it is accelerated, and possibly shifting the transaxle or damaging the shift linkage, ________________ or ________________ are used.**
3. **The shift linkage often consists of ________________ and ________________ or a ________________.**
4. **The shift linkage on a FWD vehicle with transversely mounted powertrain must change the front-to-back motion of the shift pattern into a ________________-to-________________ motion.**

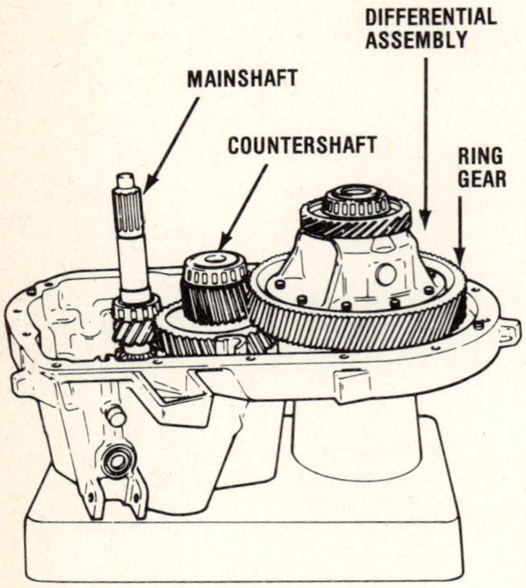

Figure 25-5. Differential assembly inside of a transaxle. Notice that the ring gear is a helical-cutgear, like other transmission gears. (Pontiac Motor Division-GMC)

25.4 INSIDE THE TRANSAXLE

The main difference between a transaxle and a transmission is that the differential and transmission of a transaxle are inside the same housing. Since there is no driveshaft, the differential ring gear is driven directly off the output shaft of the transmission. See Figure 25-5.

25.5 REMOVING TRANSAXLES

In some transaxle-equipped vehicles, especially those with rear- or mid-engine powertrains, the engine and transaxle must be removed as a unit. A transmission jack cannot usually handle the weight of the transaxle and engine together. Other types of jacks, hoists, and supports may be needed. On Volkswagen air-cooled models, it is necessary to remove the engine separately before the transaxle can be removed. An example of transaxle removal is discussed in Unit 26.

25.6 SERVICING AND REPAIRING TRANSAXLES

Transaxle disassembly is not simple. Once the transaxle is out of the vehicle, however, the servicing procedures are similar to those for conventional transmissions. The parts look the same, and many of the same checking and repair procedures are used.

Transaxle disassembly is not simple. However, in many cases it is similar to the disassembly of a conventional transmission. A transaxle overhaul is discussed in Unit 26.

CHECKPOINT

1. **The main difference between a transaxle and a transmission is that the ________________ and ________________ are inside one housing.**
2. **Since there is no driveshaft, the differential ring gear is driven directly off the ____________________________ ____________________ of the transmission.**
3. **The ________________ and ____________________ must be removed as a unit in some transaxle-equipped vehicles.**
4. **Once the transaxle is out of the vehicle, the ____________________ procedures are almost identical to those for conventional transmissions.**

NAME ______________________ SECTION ____________ DATE ____________

Certification Practice

Test 25

DIRECTIONS:

The following questions are similar to those used on mechanic certification tests. Answer each question by circling the letter of the correct choice.

1. Engines and transaxles are usually mounted in all of the following positions EXCEPT
   A. engine in front, transaxle in front.
   B. engine in rear, transaxle in front.
   C. engine in rear, in front of transaxle.
   D. engine in rear, behind transaxle.

2. Mechanic A says that putting the powertrain in front increases room inside the vehicle for passengers and luggage.
   Mechanic B says that putting the powertrain in front over the drive wheels increases traction on slippery roads.
   Who is correct?
   A. A only B. B only C. Both A and B D. Neither A nor B

3. The arrangement shown is a
   A. transverse FWD cable gearshift linkage.
   B. conventional rear-wheel drive cable gearshift linkage.
   C. transverse FWD rod-and-lever gearshift linkage.
   D. conventional rear-wheel drive rod-and-lever gearshift linkage.

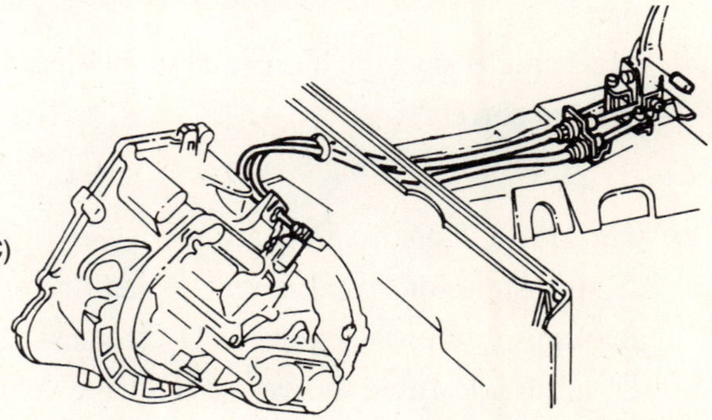

Question 3.
(Chevrolet Motor Division-GMC)

4. When an engine and/or transaxle is mounted at 90 degrees to the front-to-back centerline of the vehicle, it is called
   A. horizontal mounting.
   B. longitudinal mounting.
   C. transverse mounting.
   D. perpendicular mounting.

5. Which of the following statements is correct?
   I. A transaxle includes a transmission and a differential unit inside the same housing.
   II. Any transmission used in a FWD vehicle is a transaxle.
   A. I only B. II only C. Both I and II D. Neither I nor II

6. In a typical front-wheel-drive vehicle, the shift linkage must be protected from the
   A. power of the engine.
   B. effects of torque-over.
   C. effects of turning and steering.
   D. weight of the engine.

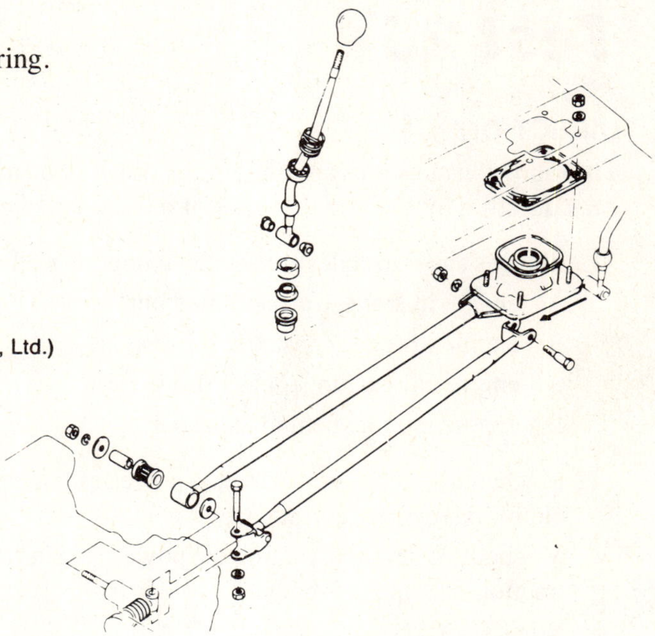

Question 7.
(Toyo Kogyo Co., Ltd.)

7. Which of the following statements about the picture is correct?
   I. These parts are used on FWD vehicles.
   II. These parts are used to protect a transaxle and linkage.
   A. I only B. II only C. Both I and II D. Neither I nor II

8. Mechanic A says the hump and tunnel in a FWD vehicle is to allow room for drivetrain parts.
   Mechanic B says the hump and tunnel in a FWD vehicle is to strengthen the body.
   Who is correct?
   A. A only B. B only C. Both A and B D. Neither A nor B

9. The GM 325 and 325-4L are
   A. transmissions used in rear-wheel-drive vehicles.
   B. transmissions used in FWD vehicles.
   C. transaxles used in rear-wheel drive vehicles.
   D. transaxles used in FWD vehicles.

10. When a powertrain is mounted front-to-back, it is called
   A. transverse mounting.
   B. longitudinal mounting.
   C. horizontal mounting.
   D. latitudinal mounting.

26 Overhauling the Turbo Hydra-Matic 125/125C

YOUR LEARNING JOB

The automatic transmission is only one part of the transaxle system. If the transmission needs repair, it's usually not necessary to disassemble the entire transaxle. However, transaxle removal, overhaul, and replacement is the most demanding job described in this manual. This unit discusses the overhaul procedure for the General Motors 125.

JOB COMPETENCY OBJECTIVES

This is a skill-building unit. When you have completed your assignments and exercises in this unit, you should be able to:

- ☐ Remove the transaxle.
- ☐ Remove and replace the necessary linkages and controls.
- ☐ Overhaul the GM 125 transaxle.
- ☐ Install the transaxle.

26.1 TURBO HYDRA-MATIC 125/125C DESCRIPTION

The Turbo Hydra-Matic 125 transaxle series has been used since the middle of 1979 on General Motors automobiles. It is found in almost all General Motors front-wheel-drive cars that use transaxles. The Turbo Hydra-Matic 125/125C series consists of the following main parts:

- Torque converter
- Lockup torque converter (125C)
- Three multiple-disc clutches
- One overrunning clutch
- One servo/band
- Compound planetary gearset
- Removable valve body
- Dual sprocket and drive link assembly
- One-piece aluminum case with removable case cover.

The Turbo Hydra-Matic 125/125C combines a torque converter with a fully automatic three-speed gear system and differential assembly. The transaxle case is a single aluminum casting. The transaxle case has a removable case cover, valve body cover, and oil pan.

The torque converter is a three-element unit. A lockup converter is used on the 125C, which is the major difference between the 125 and 125C.

The compound planetary gear system has two sun gears.

The hydraulic system has a single vane-type oil pump. The valve body is integrated with the removable case cover.

A dual sprocket and drive link assembly is the direct connection between the transmission and differential portions of the transaxle assembly.

26.2 TURBO HYDRA-MATIC 125/125C OPERATION

The torque converter on the 125 transaxle is on the side of the main transaxle body, instead of directly in front of it. A sprocket at the transmission input shaft and a sprocket at the differential output shaft are connected by a chain. See Figure 26-1.

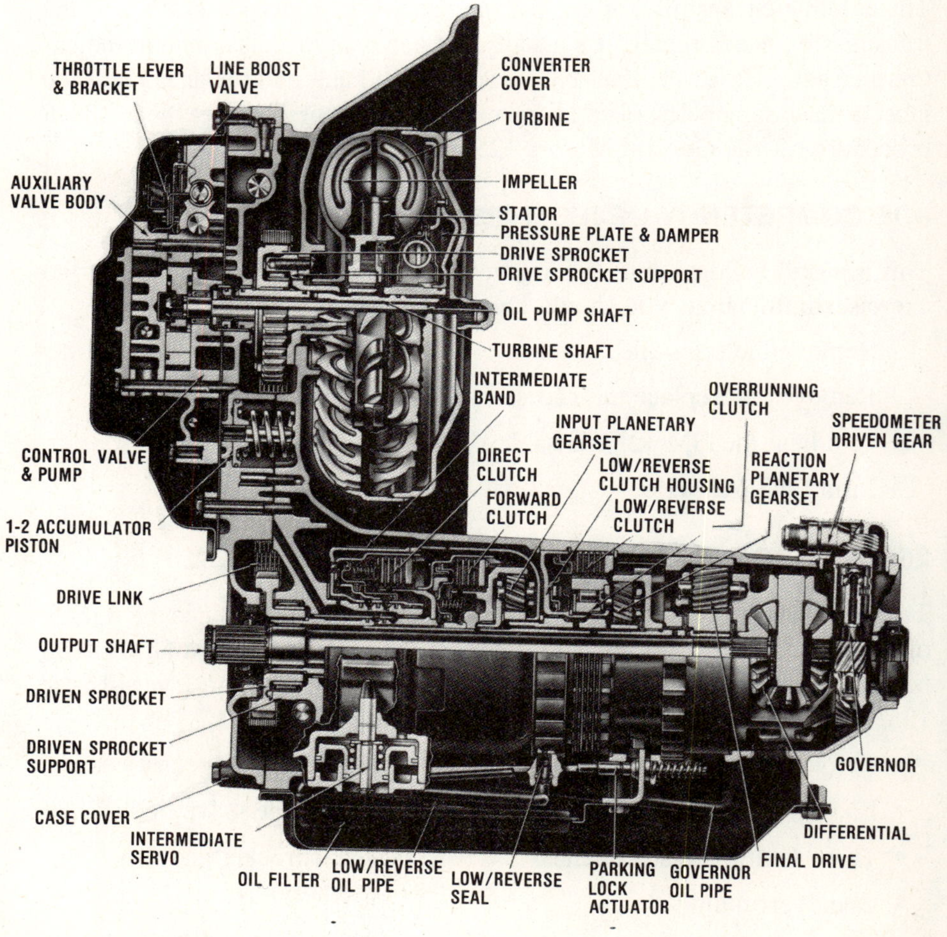

Figure 26-1. Turbo Hydra-Matic 125C automatic transmission cutaway. (Oldsmobile Division-GMC)

The reservoir in the oil pan is very shallow. For this reason, the valve body cover is used as a second reservoir to provide sufficient fluid for proper operation. A thermostatic valve allows fluid to flow between the two reservoirs, depending on transaxle temperature.

Engine torque is routed through two 90-degree turns to reach the transaxle—once at the output of the turbine shaft and again at the transmission input shaft. The differential operates in the same way as a conventional rear-drive differential. Differentials are covered in *Manual Transmissions and Drivetrains*, another text/shop manual in this Auto Tech Series.

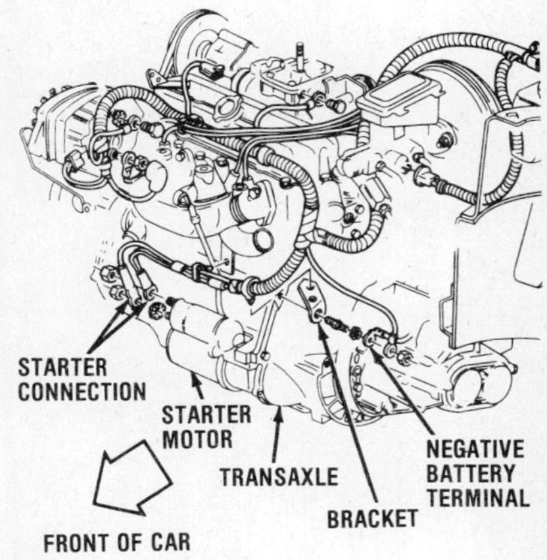

Figure 26-2. Disconnecting starter wiring. (Chevrolet Motor Division-GMC)

26.3 TURBO HYDRA-MATIC 125/125C IDENTIFICATION

Two identification numbers are found on the 125/125C transaxles. One is the vehicle identification code. The other, more important identification number, is the transaxle model code. The transaxle model code is located at the top of the removable case cover. Use the proper service manual to decipher the transaxle model code. Many changes occur in the life of a transaxle, so the transaxle model code number is important in obtaining the proper parts.

26.4 TRANSAXLE REMOVAL

Removing the transaxle is a complicated procedure that requires considerable time. It involves removing braces and supporting the engine.

SAFETY CAUTION: ***Never* attempt to remove the transaxle unless the vehicle is properly supported. Special hoists and transaxle jacks must be used. Trying to remove the transaxle while using jackstands is dangerous. Wear safety glasses when removing the transaxle. Serious damage and personal injury can result when using other than the recommended equipment.**

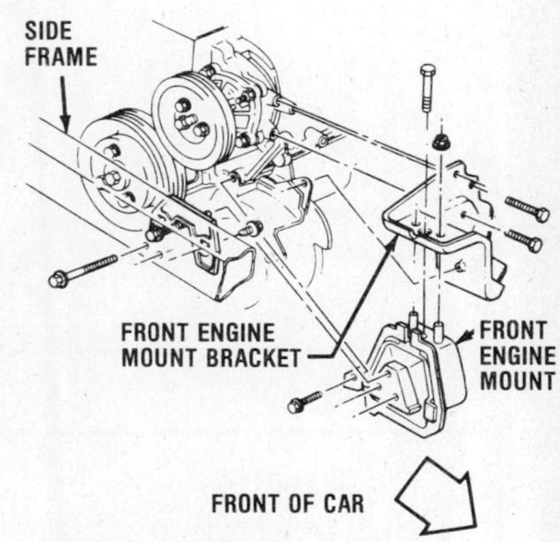

Figure 26-3. Securing the engine mount. (Chevrolet Motor Division-GMC)

The following transaxle removal procedure is for the Chevrolet Cavalier. Other 125/125C transaxle procedures are similar:

1. Disconnect the negative battery cable from the transaxle. See Figure 26-2.

2. Insert a 1/4 x 2-inch-long [6.35 x 50.8 mm] bolt at the front, right-hand engine mount. See Figure 26-3. This prevents the engine mount from becoming misaligned during transaxle removal.

3. Remove the carburetor air cleaner.

4. Disconnect the TV cable at the carburetor.

5. Remove the bolt holding the TV cable to the transaxle. Pull up on the TV cable cover at the transaxle until the cable can be seen. Disconnect the TV cable from the transaxle rod. See Figure 26-4.

6. Remove the bolt holding the engine wiring harness to the top of the transaxle.

7. Remove the hose from the air management valve. This allows the engine wiring harness to be pulled up and out of the way.

8. Install the special engine support fixture shown in Figure 26-5. Raise the engine just enough to take pressure off the engine mounts.

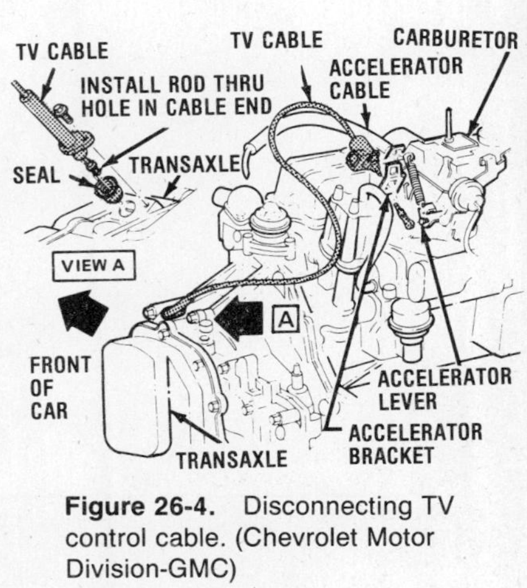

Figure 26-4. Disconnecting TV control cable. (Chevrolet Motor Division-GMC)

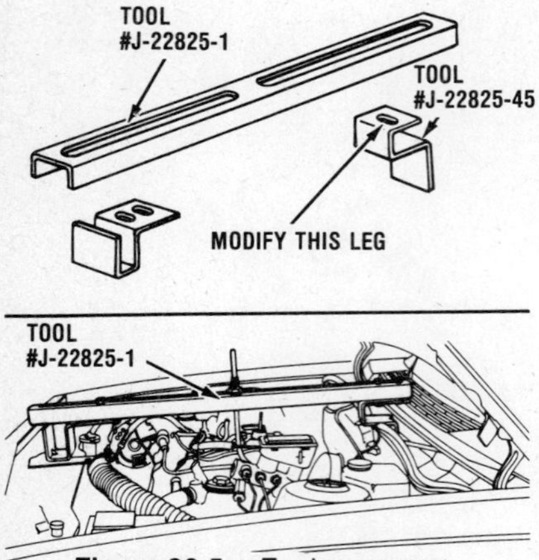

Figure 26-5. Engine support fixture. (Chevrolet Motor Division-GMC)

SAFETY CAUTION: The engine support fixture *must* be located in the center of the cowl. The fasteners must be torqued to specifications before supporting the engine. The fixture is *not* intended to support the entire weight of the engine and transaxle. Serious personal injury can result when the support fixture is improperly used.

Place a transaxle jack under the transaxle assembly to support it.

9. Remove the top transaxle mount and bracket assembly, shown in Figure 26-6. The engine may have to be raised slightly to remove the mounting assembly.

10. Disconnect the shift control linkage. See Figure 26-7.

11. Remove the top bolts that hold the transaxle to the engine. Loosen—do *not* remove—the bolt near the starter that holds the transaxle to the engine. Figure 26-8.

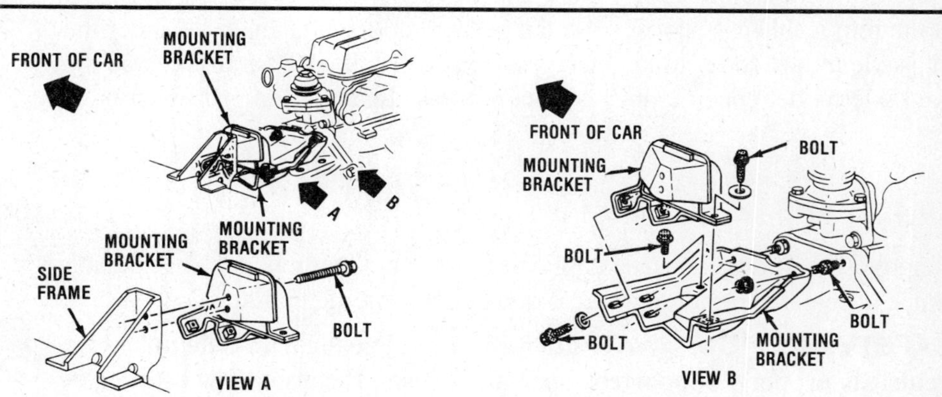

Figure 26-6. Transaxle mounting brackets. (Chevrolet Motor Division-GMC)

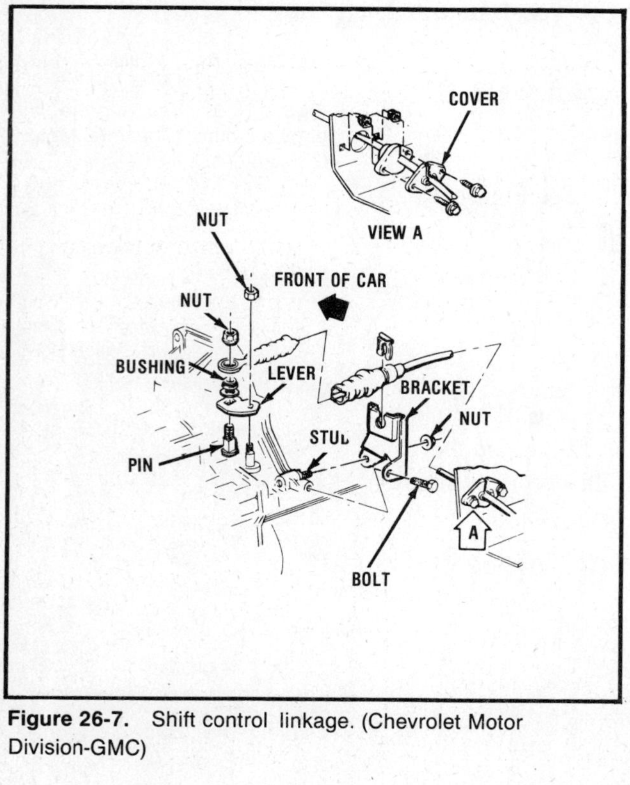

Figure 26-7. Shift control linkage. (Chevrolet Motor Division-GMC)

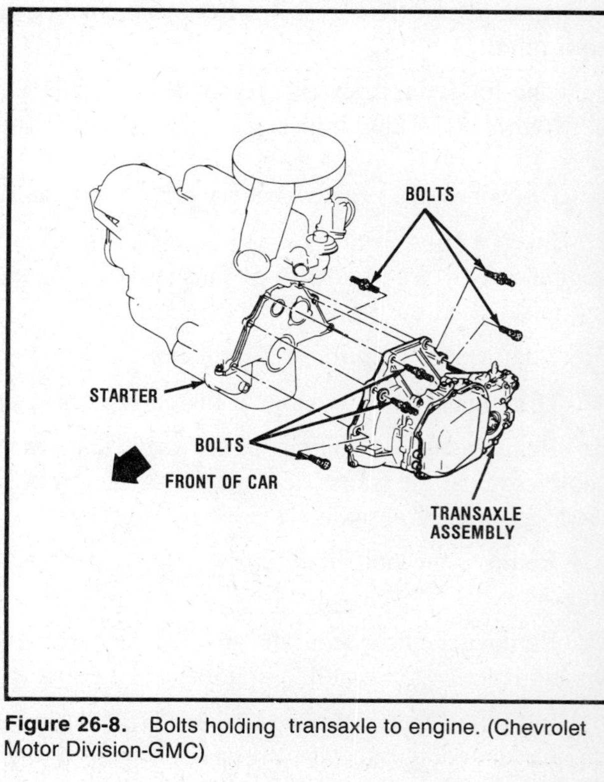

Figure 26-8. Bolts holding transaxle to engine. (Chevrolet Motor Division-GMC)

12. Unlock the steering column.

13. Raise the vehicle on a hoist.

14. Remove both front wheel and tire assemblies.

15. Remove the cotter pin and loosen the castellated ball-joint nut. Separate the ball joint from the control arm. See Figure 26-9.

16. Repeat step 15 at the opposite front wheel.

17. Remove the bolt holding the stabilizer bar to the lower left-hand control arm, shown in Figure 26-10.

18. Remove the bolts holding the front left-hand suspension support assembly. See Figure 26-11.

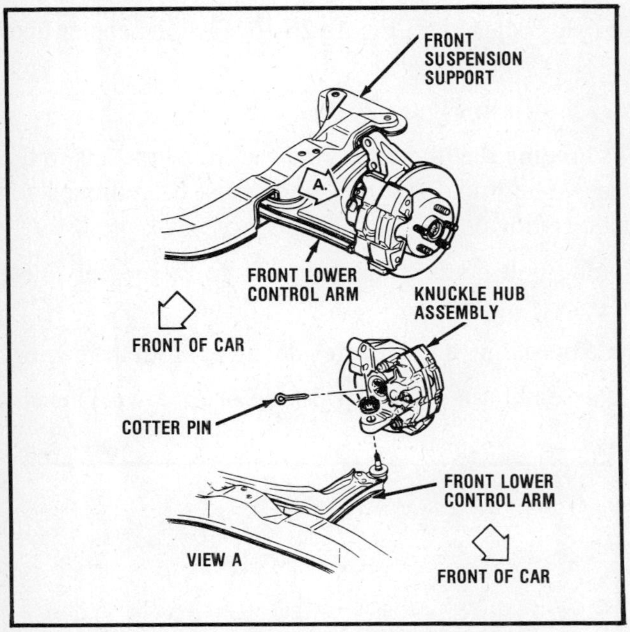

Figure 26-9. Separating ball joint from control arm. (Chevrolet Motor Division-GMC)

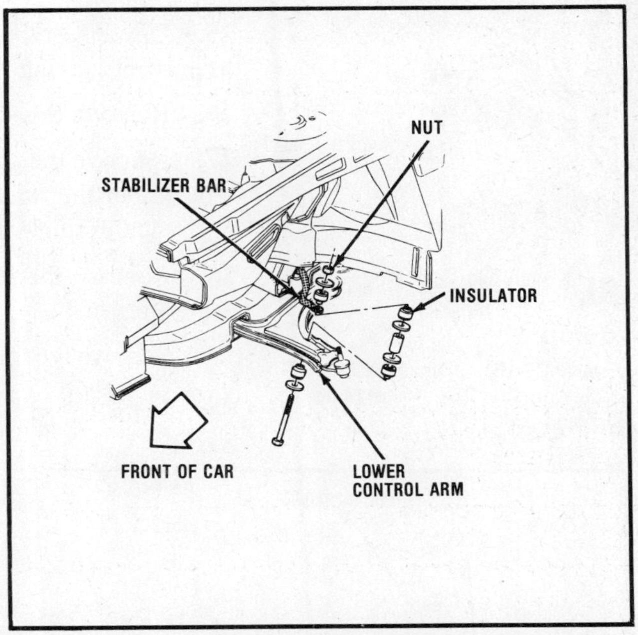

Figure 26-10. Stabilizer bar. (Chevrolet Motor Division-GMC)

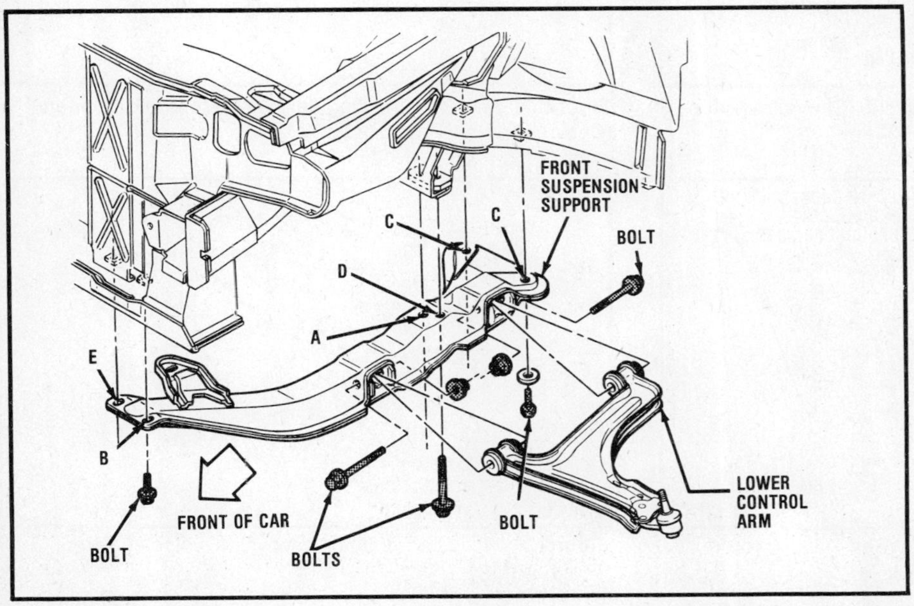

FRONT SUSPENSION SUPPORT ATTACHING BOLT/SCREW SEQUENCE:

1. INSTALL CENTER SCREW LOOSELY INTO HOLE [A].
2. INSTALL TIE BAR SCREW LOOSELY INTO THE SMALL SLOTTED OUTBOARD HOLE.
3. INSTALL AND TORQUE TO SPECIFICATIONS BOTH BOLTS IN THE REAR HOLES.
4. INSTALL AND TORQUE TO SPECIFICATIONS SECOND CENTER HOLE [D] BOLT/SCREW.
5. TORQUE CENTER HOLE [A] BOLT TO SPECIFICATIONS.
6. INSTALL OTHER TIE BAR BOLT (FRONT HOLE [E]) AND TORQUE TO SPECIFICATIONS.
7. TORQUE SECOND BOLT/SCREW IN FRONT HOLE [B].

Figure 26-11. Front suspension support. (Chevrolet Motor Division-GMC)

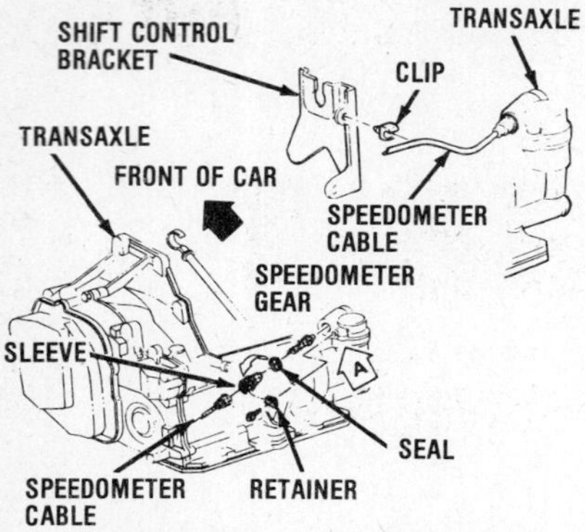

Figure 26-12. Disconnecting the speedometer cable. (Chevrolet Motor Division-GMC)

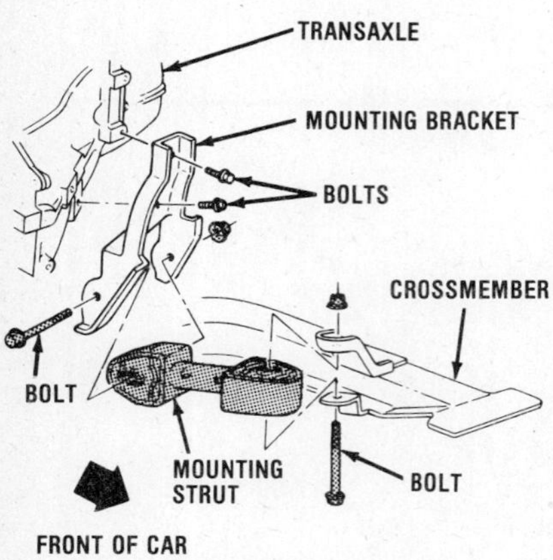

Figure 26-13. Transaxle mounting strut. (Chevrolet Motor Division-GMC)

19. Use a special axle shaft removal tool and a slide hammer. Position the axle shaft removal tool behind the axle shaft cones. Pull the cones away from the transaxle. Remove the axle shafts. Plug the bores to reduce fluid leakage.

20. Remove the nut holding the control cable bracket to the transaxle. Remove the stud (Figure 26-7) holding the engine to the transaxle.

21. Disconnect the speedometer cable shown in Figure 26-12.

22. Disconnect the transaxle strut at the transaxle. See Figure 26-13.

23. Remove the screws (Figure 26-14) holding the torque converter shield. Remove the shield.

24. Remove the bolts that hold the torque converter to the flexplate. Figure 26-15.

25. Disconnect the transaxle cooler lines, Figure 26-16. Plug the cooler lines to prevent leaking.

26. Remove the starter, shown in Figure 26-17.

27. Remove the screws holding the brake line and fuel line brackets to the left side of the underbody. The brake and fuel lines must be removed for clearance during transaxle removal.

28. Remove the remaining bolt holding the transaxle to the engine. Figure 26-8.

29. Remove the transaxle by sliding it to the left side, away from the engine.

30. Place the transaxle assembly in a support device or on a workbench.

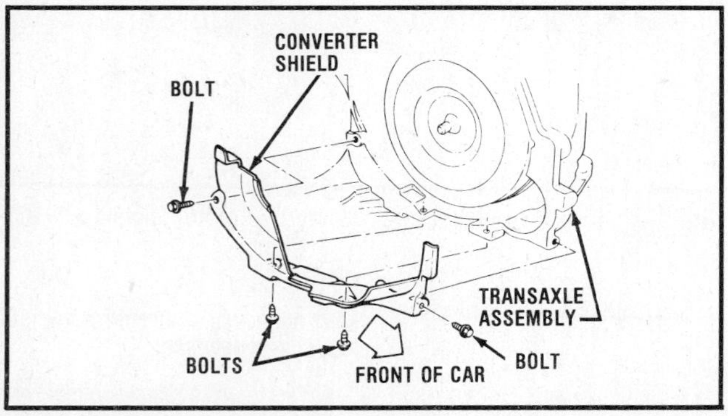

Figure 26-14. Transaxle converter shield. (Chevrolet Motor Division-GMC)

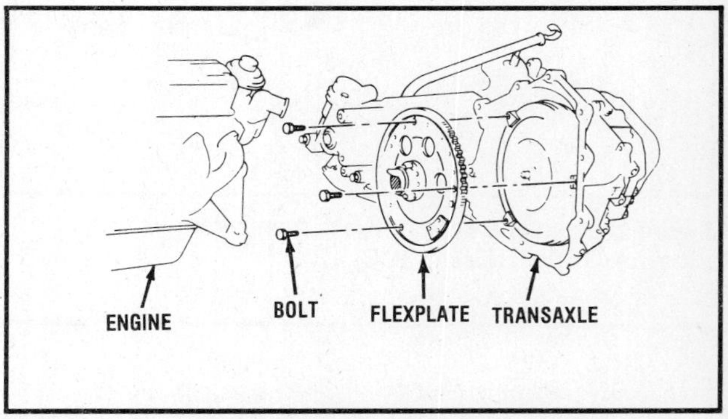

Figure 26-15. Bolts holding flexplate to torque converter. (Chevrolet Motor Division-GMC)

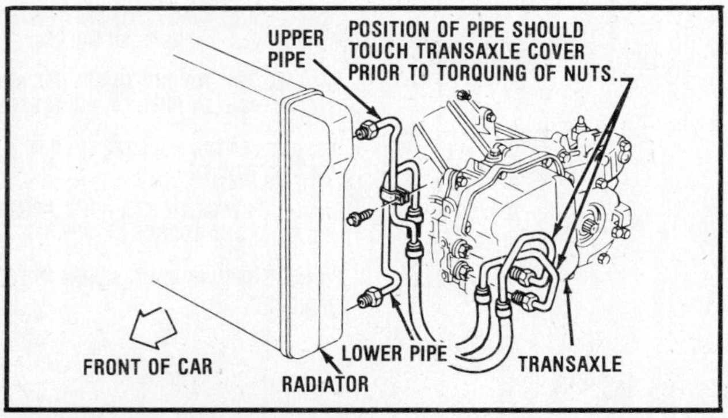

Figure 26-16. Transaxle cooler lines. (Chevrolet Motor Division-GMC)

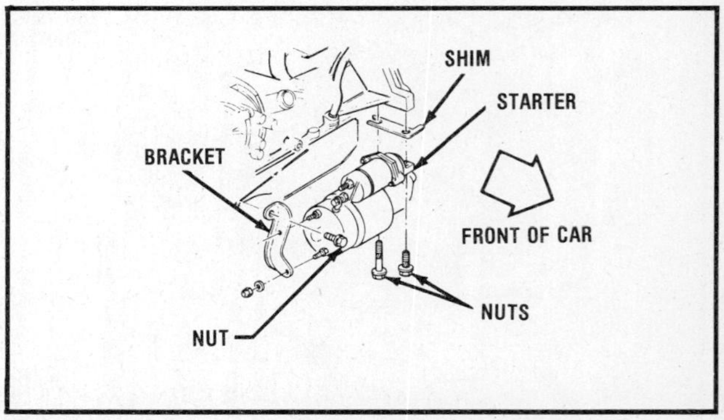

Figure 26-17. Starter motor removal. (Chevrolet Motor Division-GMC)

NAME ____________________ SECTION ____________ DATE ____________

Job Sheet 26-1

Removing the GM 125 Transaxle

Before you begin:

Read Topics 26.1 through 26.4.

Vehicle Make ____________________ Model ____________ Year ____________

Time Started ____________________ Time Finished ____________________

Job Completion Time ____________________ Flat Rate Time ____________________

Tools Required

- Hoist
- Safety glasses
- Special transaxle jack
- Metric sockets
- Socket extensions, universal joints, handle
- Metric open-end or combination wrenches
- Pliers
- Screwdrivers
- Special engine support fixture
- Torque wrench
- Ball joint separator tool
- Axle shaft removal tool
- Slide hammer

References

Manufacturer's service manual ____________________

Repair manual ____________________

Service bulletin ____________________

Procedure

SAFETY CAUTION: ***Never*** **attempt to remove the transaxle unless the vehicle is properly supported. Special hoists and transaxle jacks must be used. Trying to remove the transaxle while using only jackstands is dangerous. Serious damage and personal injury can result when using other than the recommended equipment.**

1. Disconnect the negative battery cable from the transaxle.
2. Insert a 1/4 x 2-inch-long [6.35 x 50.8 mm] bolt at the front, right-hand engine mount.
3. Remove the carburetor air cleaner.
4. Disconnect the TV cable at the carburetor.
5. Remove the bolt holding the TV cable to the transaxle. Pull up on the TV cable cover at the transaxle until the cable can be seen. Disconnect the TV cable from the transaxle rod.
6. Remove the bolt holding the engine wiring harness to the top of the transaxle.

7. Remove the hose from the air management valve. This allows the engine wiring harness to be pulled up and out of the way.
8. Install the manufacturer's recommended engine support fixture. Torque the bolts to the manufacturer's specifications. In the space provided, write the correct torque.

Manufacturer's Specified Torque .... ________________

Raise the engine just enough to take pressure off the engine mounts.

SAFETY CAUTION: The special engine support fixture *must* be located in the center of the cowl. The fasteners must be torqued to specifications before supporting the engine. The fixture is *not* intended to suport the entire weight of the engine and transaxle. Serious personal injury can result when the support fixture is improperly used.

Instructor Check: ________________

Place a transaxle jack under the transaxle assembly to support it.

9. Remove the top transaxle mount and bracket assembly. Raise the engine slightly and remove the mounting assembly.
10. Disconnect the shift control linkage.
11. Remove the upper bolts that hold the transaxle to the engine. Loosen—do *not* remove—the bolt near the starter that holds the transaxle to the engine.
12. Unlock the steering column.
13. Raise the vehicle on a hoist.
14. Remove both front wheel and tire assemblies.
15. Remove the cotter pin and loosen the castellated ball-joint nut at the left front wheel. Separate the ball joint from the control arm.
16. Repeat step 15 at the right front wheel.
17. Remove the bolt holding the stabilizer bar to the lower left-hand control arm.
18. Remove the bolts holding the front left-hand suspension support assembly.
19. Use a special axle shaft removal tool and a slide hammer. Position the axle shaft removal tool behind the axle shaft cones. Pull the cones away from the transaxle. Remove the axle shafts. Plug the transaxle bores to reduce fluid leakage.
20. Remove the nut holding the control cable bracket to the transaxle. Remove the stud holding the engine to the transaxle.
21. Disconnect the speedometer cable.
22. Disconnect the transaxle strut at the transaxle.
23. Remove the screws holding the torque converter shield. Remove the shield.
24. Remove the bolts that hold the torque converter to the flexplate.
25. Disconnect and plug the transaxle cooler lines.
26. Remove the starter.
27. Remove the screws holding the brake line and fuel line brackets to the left side of the underbody. Remove the brake and fuel lines. Plug or cap the connecting lines.

Instructor Check: ________________

NAME ______________________ SECTION ____________ DATE ____________

28. Remove the remaining bolt holding the transaxle to the engine.
29. Remove the transaxle by sliding it to the left side, away from the engine.
30. Place the transaxle assembly in a support device or on a workbench.

Job Completion Report

1. What difficulties, if any, were encountered during the removal procedure?

____

____

____

____

____

____

INSTRUCTOR SIGNOFF ______________________ DATE ____________

26.5 BEFORE THE OVERHAUL

Be sure you are familiar with all of the assemblies that make up the automatic transmission. These assemblies are discussed in Parts II, III, and IV. This unit discusses the disassembly and reassembly of the General Motors 125 transaxle. Refer to earlier units for teardown, inspection, and testing of the torque converter, planetary gearsets, apply devices, and the valve body.

Only the overhaul of the transmission part of the 125 transaxle is discussed in this unit.

Transaxle specifications are changed often during their lifespans. For this reason, always refer to the proper service manual when performing any transaxle overhaul.

Before proceeding with the overhaul, make sure that the transaxle fluid has been drained. Plug all openings in the transaxle case. Clean the case with a steam-cleaning machine. If a steam-cleaning machine is not available, use solvent, a wire brush, and shop towels. Do *not* wipe the inside of the transaxle case with shop towels. Allow it to air dry. Any lint left inside of the case can affect the operation of the automatic transmission.

After the transaxle case has been thoroughly cleaned, inspect the case for any signs of damage. A cracked or distorted case can cause internal problems.

For cleaning and inspection procedures, refer to 15.5.

Be sure to check for end play. Refer to 15.4.

26.6 TURBO HYDRA-MATIC 125 OVERHAUL

To better understand the parts and their locations in the 125 transaxle, refer to Figures 26-18 and 26-19. The following overhaul procedure was done on a four-cylinder Pontiac Phoenix. There may be differences from other 125 overhauls, so be sure to refer to the proper service manual.

NOTE: Reassembly notes are given in the disassembly procedure. Reassembly is essentially the reverse of disassembly. Refer to the manufacturer's service manual for specific instructions.

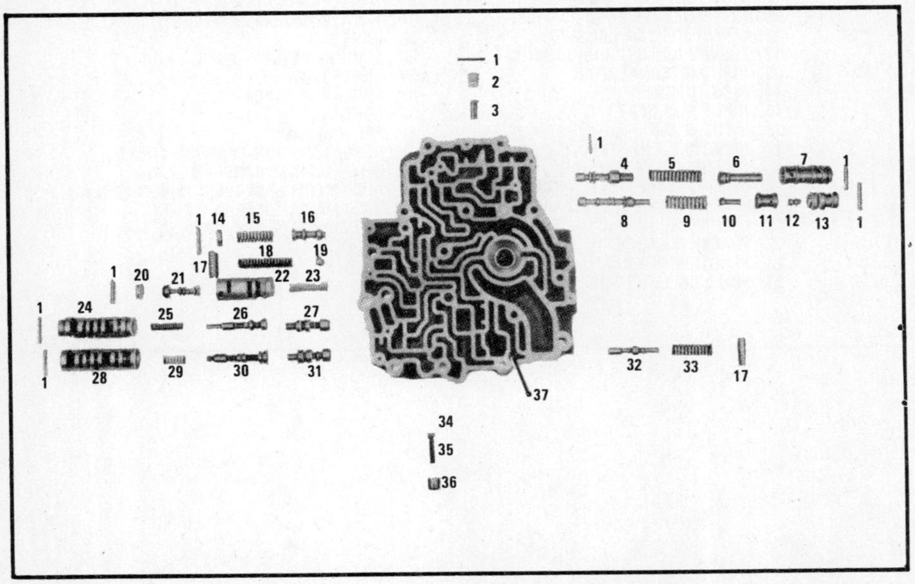

1. PIN
2. LINE BOOST VALVE PLUG
3. LINE BOOST VALVE
4. THROTTLE VALVE
5. THROTTLE VALVE SPRING
6. THROTTLE VALVE PLUNGER
7. TV PLUNGER BUSHING
8. PRESSURE REGULATOR VALVE
9. PRESSURE REGULATOR SPRING
10. REVERSE BOOST VALVE
11. REVERSE BOOST VALVE BUSHING
12. TV BOOST VALVE
13. TV BOOST VALVE BUSHING
14. VALVE BORE PLUG
15. SHIFT TV SPRING
16. SHIFT TV VALVE
17. SPRING RETAINING SLEEVE
18. PRESSURE RELIEF SPRING
19. PRESSURE RELIEF BALL
20. VALVE BORE PLUG
21. 1-2 ACCUMULATOR VALVE
22. 1-2 ACCUMULATOR BUSHING
23. 1-2 ACCUMULATOR SPRING
24. 2-3 THROTTLE VALVE BUSHING
25. 2-3 THROTTLE VALVE SPRING
26. 2-3 THROTTLE VALVE
27. 2-3 SHIFT VALVE
28. 1-2 THROTTLE VALVE BUSHING
29. 1-2 THROTTLE VALVE SPRING
30. 1-2 THROTTLE VALVE
31. 1-2 SHIFT VALVE
32. 3-2 CONTROL VALVE
33. 3-2 VALVE SPRING
34. LOW BLOW-OFF BALL
35. LOW BLOW-OFF SPRING & PLUG
36. #1 CHECK BALL

Figure 26-18. Turbo Hydra-Matic 125C control valve body. (Oldsmobile Division-GMC)

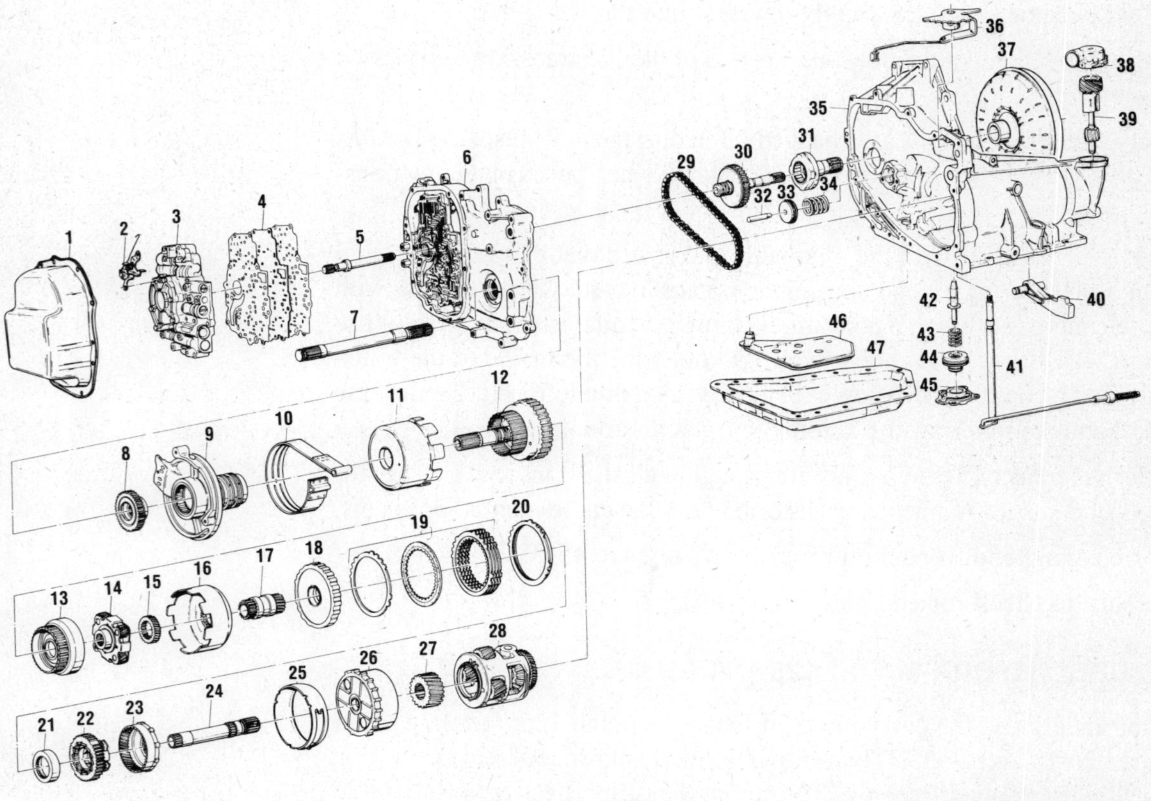

1. VALVE BODY COVER
2. THROTTLE LEVER BRACKET & LINK
3. CONTROL VALVE & OIL GASKET
4. SPACER PLATE & GASKETS
5. OIL PUMP SHAFT
6. CASE COVER
7. OUTPUT SHAFT
8. DRIVEN SPROCKET
9. DRIVEN SPROCKET SUPPORT
10. INTERMEDIATE BAND
11. DIRECT CLUTCH HOUSING
12. FORWARD CLUTCH HOUSING
13. INPUT INTERNAL GEAR
14. INPUT CARRIER
15. INPUT SUN GEAR
16. INPUT DRUM
17. REACTION SUN GEAR
18. LOW/REVERSE HOUSING
19. LOW/REVERSE CLUTCH PLATES
20. LOW/REVERSE CLUTCH BACKING PLATE
21. LOW OVERRUNNING CLUTCH RACE
22. REACTION CARRIER
23. REACTION INTERNAL GEAR
24. FINAL DRIVE SUN-GEAR SHAFT
25. FINAL DRIVE INTERNAL-GEAR SPACER
26. FINAL DRIVE INTERNAL GEAR
27. FINAL DRIVE SUN GEAR
28. DIFFERENTIAL CARRIER
29. DRIVE LINK
30. DRIVE SPROCKET & TURBINE SHAFT
31. DRIVE SPROCKET SUPPORT
32. 1-2 ACCUMULATOR PIN
33. 1-2 ACCUMULATOR PISTON
34. 1-2 ACCUMULATOR SPRING
35. CASE
36. MANUAL DETENT LEVER & ROD
37. CONVERTER
38. GOVERNOR COVER
39. GOVERNOR
40. PARKING PAWL
41. ACTUATOR ROD & MANUAL SHAFT
42. INTERMEDIATE BAND APPLY PIN
43. INTERMEDIATE SERVO CUSHION SPRING
44. INTERMEDIATE SERVO PISTON
45. INTERMEDIATE SERVO COVER
46. OIL FILTER
47. OIL PAN

Figure 26-19. Turbo Hydra-Matic 125 automatic transmission major parts. (Hydra-Matic)

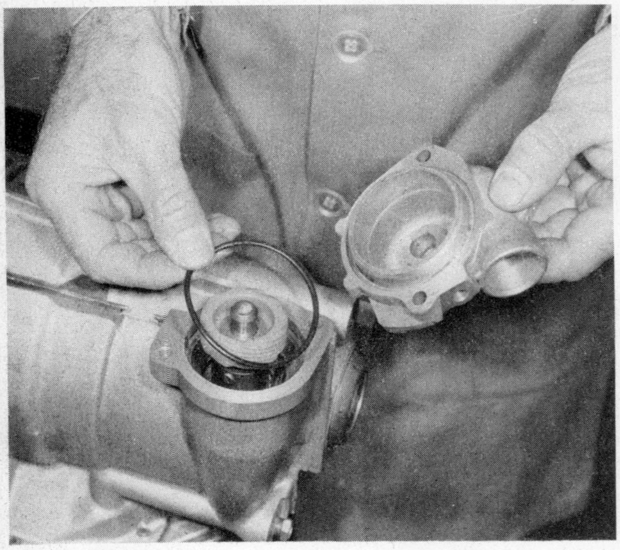

1. Place the transaxle in a holding fixture. Remove the torque converter. Unbolt the governor cover and discard the O-ring seal.

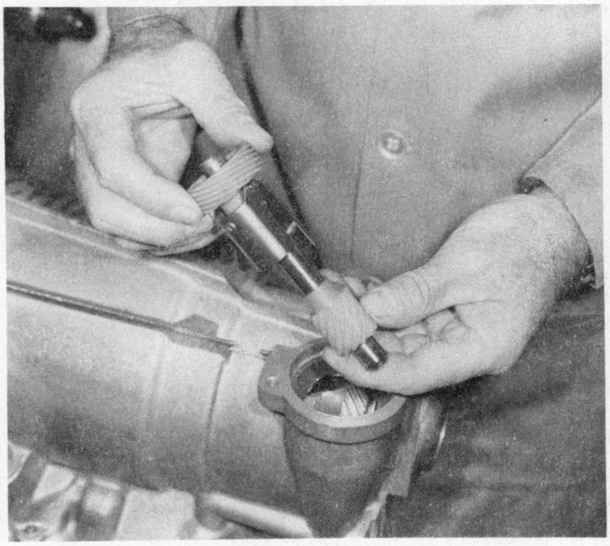

2. Remove the thrust washer between the governor cover and the speedometer drive gear. Lift the governor and drive gear assembly from transaxle case.

3. Rotate the transaxle in the holding fixture to position the oil pan so that it faces upward. Unbolt and remove the oil pan. Discard the old oil pan gasket.

4. Swivel the oil strainer to unhook it from the dipstick stop. Pull up on the opposite end of the oil strainer to free it from the transaxle case. Do *not* reuse the strainer.

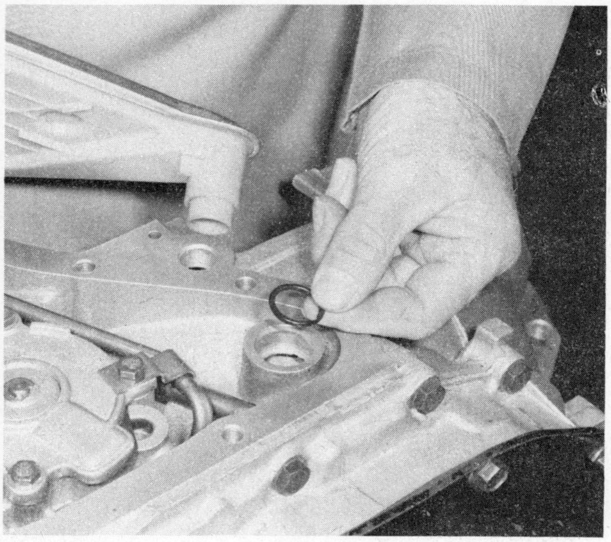

5. The plastic and metal oil strainer assembly uses an O-ring seal on the nipple. If the seal does not come off with the strainer, remove it from the bore. Discard the seal.

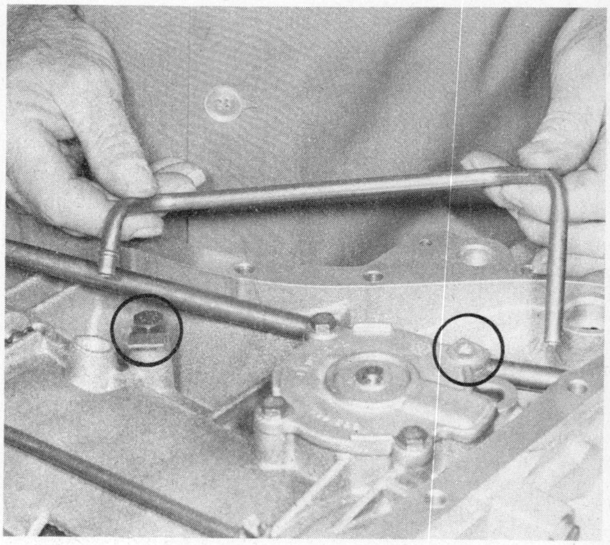

6. Unbolt the retaining brackets (circled) and remove the low/reverse oil pipe by pulling it from the case. Remove the O-ring seal and backup ring from the bore (arrow).

7. Unbolt the intermediate servo cover. The retaining bracket is held on by one cover bolt. Remove the servo cover from the case. Peel the old gasket from the servo bore flange and discard it.

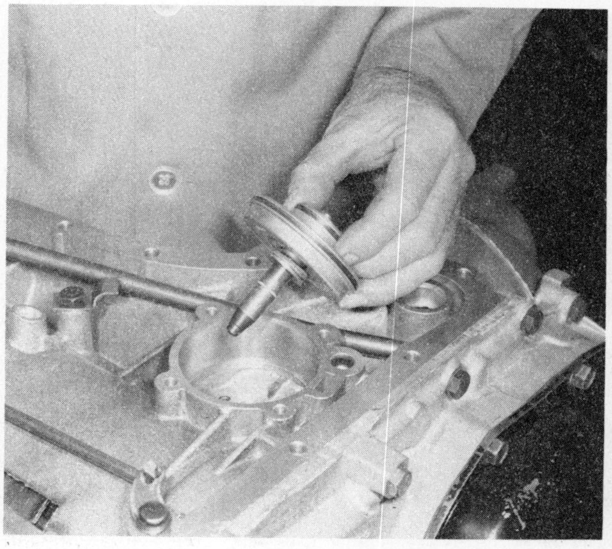

8. Remove the intermediate servo assembly from the case. Replace the oil seal rings with a new set. Check the band apply pin with a special tool recommended by GM.

9. Remove the third accumulator check valve. Early production models like this one use a cup plug to hold the valve in place. A screw extractor is required for removal.

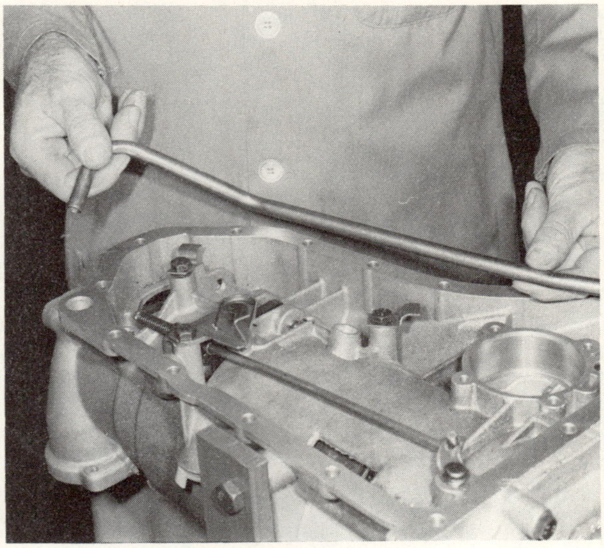

10. Remove the oil pipe by pulling the curved end out of the case. Rotate the pipe as the straight end is pulled from the case. Do *not* bend the pipe.

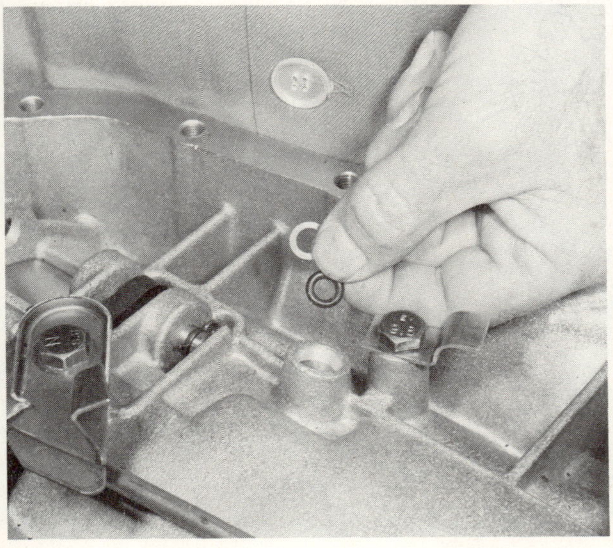

11. The low/reverse seal assembly is under the cup plug assembly. Use a modified screw extractor to remove the seal assembly from the case.

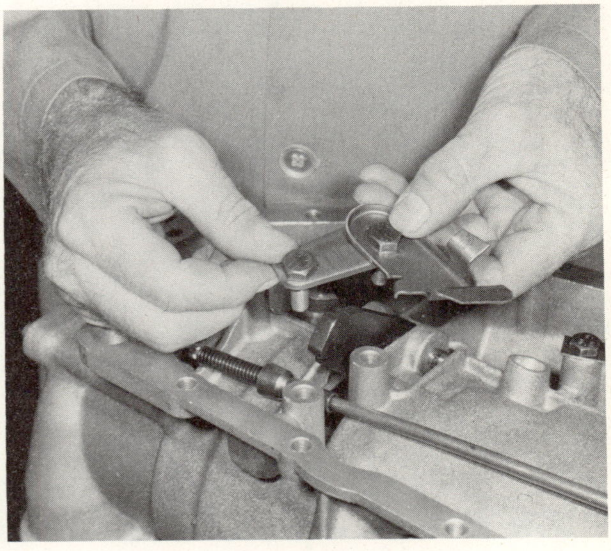

12. Unbolt and remove the dipstick stop and parking lock bracket.

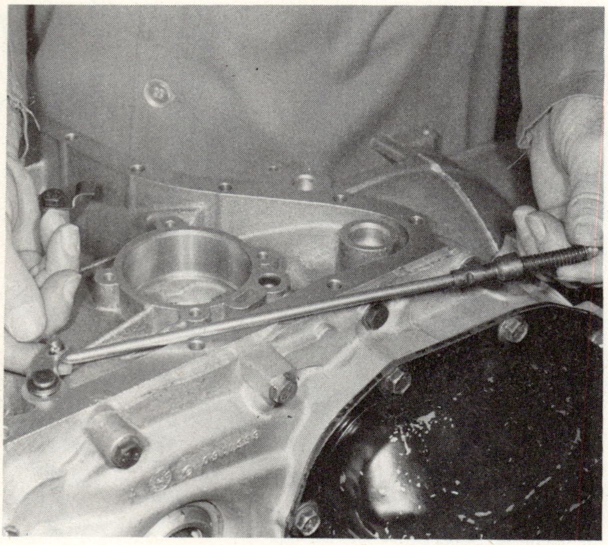

13. Pull the actuator rod end from the case. Swivel the actuator rod, as shown, to align the tanged end with the slot in the attaching arm. Depress the rod end and remove it from the attaching arm and case.

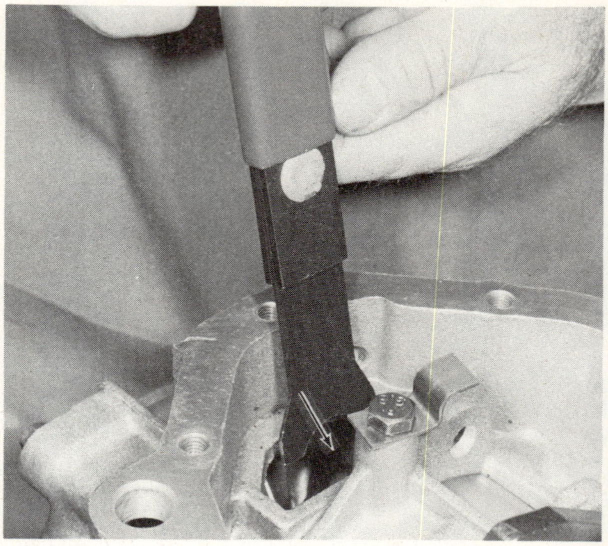

14. Rotate the final drive assembly until the C-ring can be seen in the access slot (arrow). Align the C-ring so that its open side faces the window. This special tool frees the ring from the output shaft.

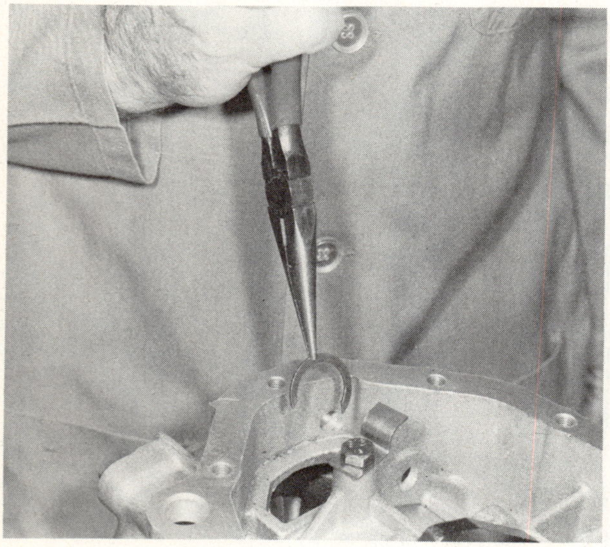

15. Turn the output shaft approximately 180 degrees. Reach into the access window with needlenose pliers and grasp the C-ring. Pull the C-ring off the shaft. Discard the C-ring and install a new C-ring during reassembly.

16. Once the C-ring is removed, the output shaft can be withdrawn from the case without difficulty.

17. Examine the output shaft for burrs, scoring, and other damage. Note that the ends are different. When the shaft is reinstalled, the short, splined end goes in first.

18. Rotate the transaxle so that the case cover is facing upward. Remove the valve body cover and gasket. Discard the gasket. Clean the cover in fresh solvent.

19. Two screws hold the throttle lever and bracket assembly to the control valve body. Remove the lever and bracket assembly with the TV cable link (arrow) attached.

20. Except for the bolt shown, the pump cover bolts also hold the valve body to the case. Remove all valve body bolts, but leave the pump cover bolt (shown) untouched.

21. Lift the valve body straight up and off the transaxle case. Place the valve body (machined side up) on a workbench. Note the difference in bolt lengths as you remove them.

22. To check the pump gears, remove the remaining cover bolt. Lift off the cover. The pump is of the vane type.

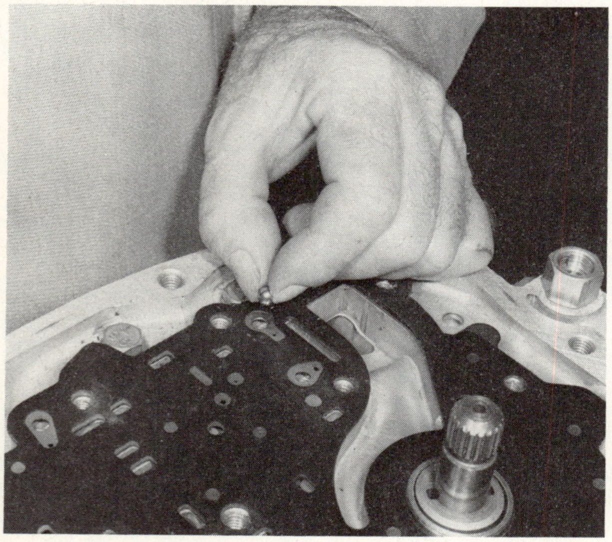

23. Remove the #1 check ball on the spacer plate. The spacer plate has gaskets on both sides. Keep the #1 check ball separate from others in the body.

24. Lift the spacer plate and gasket assembly from the transaxle case. Discard the old gaskets. Clean the spacer plate in fresh solvent before reassembly.

25. Remove the oil pump shaft from the transaxle case. Inspect the shaft closely.

26. The transaxle case contains five check balls located in the circled areas. Note the locations of each ball. Remove the balls. Do *not* mix these check balls with the #1 check ball.

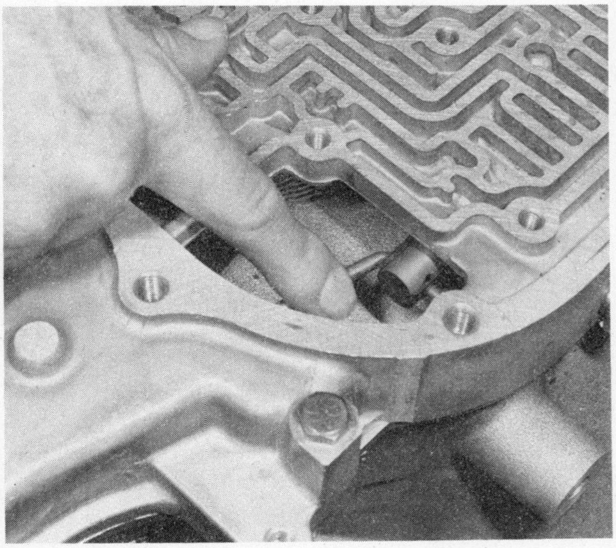

27. Disconnect the manual valve rod from the manual valve by depressing the rod. The spring clip on the rod will pop off the valve, separating the two parts.

28. Remove the two case cover screws and washers from the converter housing side. Then remove the remaining case cover screws. Dip the screws in ATF before reinstalling them to prevent galling the aluminum threads.

29. Lift the case cover from the transaxle case. Invert and place the cover on a workbench. The 1-2 accumulator will face up. The 1-2 accumulator pin may fall out. Remove and discard the gasket.

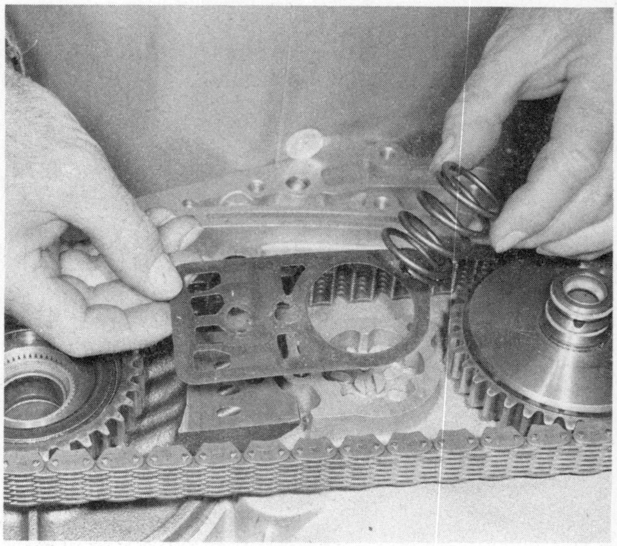

30. Remove the 1-2 accumulator spring and the center cover gasket. Discard the gasket. Inspect the oil seal rings on the drive sprocket shaft.

31. The drive sprocket thrust washer may remain attached to the case cover. If the washer is not on the drive sprocket shaft, check the inside of the cover for the nylon washer shown here.

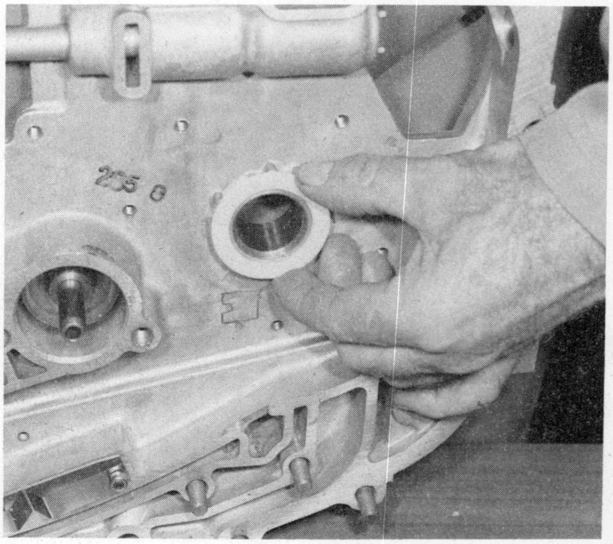

32. When reassembling the case cover to the case, attach the drive sprocket thrust washer to the cover, as shown, for easier installation. The washer presses into place.

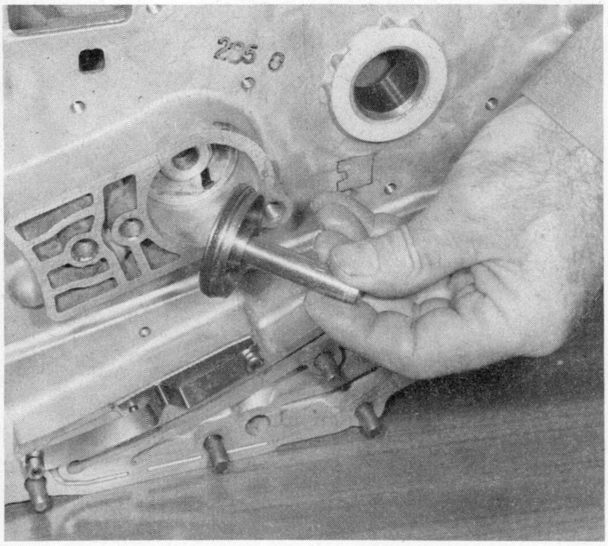

33. The 1-2 accumulator pin should fall out of the case cover when the cover is removed. If not, replace the oil seal rings. Then reinstall the pin in the cover bore.

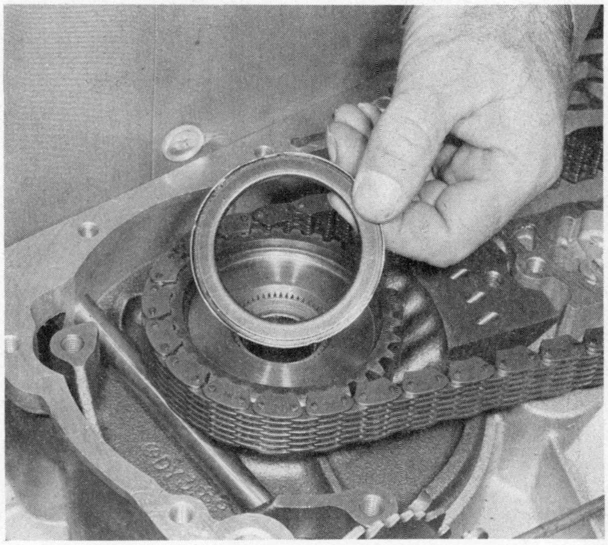

34. Remove the driven sprocket thrust bearing. When reassembling the transaxle, install the thrust bearing, as shown, on the drive gear. Thrust washer and bearing replacement are important.

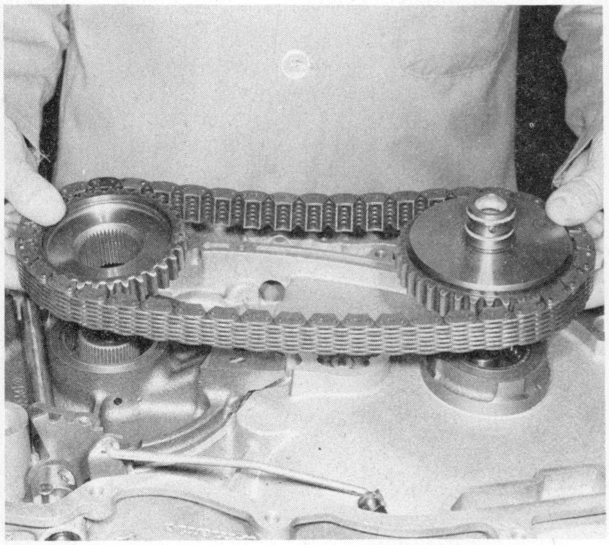

35. Jiggle the drive and driven sprocket assembly to loosen it. Lift the sprocket assembly up and out of the case with the link assembly, as shown. Installation is the reverse of removal.

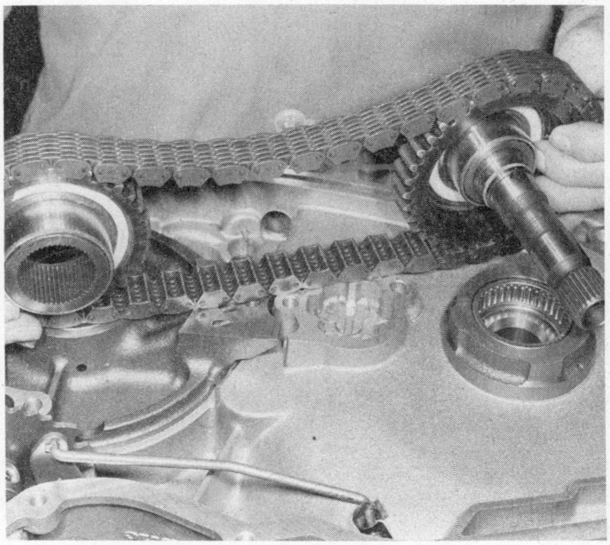

36. The thrust washers between the drive and driven sprocket and the support may remain in the case or come off with sprocket removal. In either case, remove both thrust washers and inspect them.

37. Rotate the detent lever toward the outside of the case. Use a pin punch to remove the roll pin in the shaft. This is the only position from which the pin can be removed.

38. Use needlenose pliers to remove the special nail holding the manual shaft to the case.

39. Remove the manual shaft from the transaxle case. Remove the detent lever, parking lock, and manual valve assembly. When reassembling, insert the shaft through the case and into the housing.

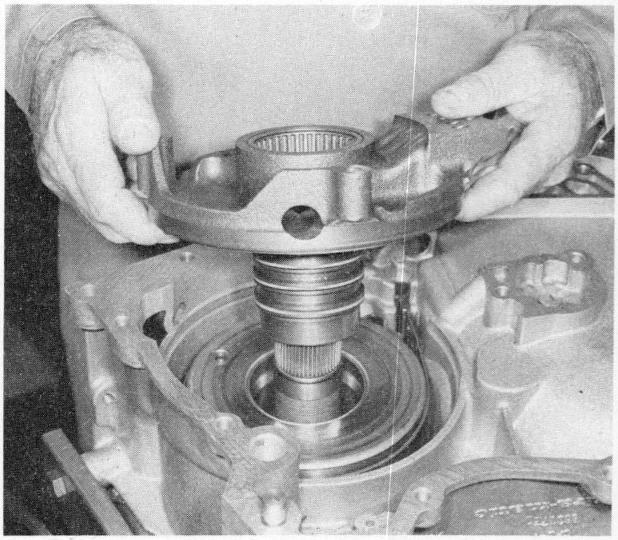

40. Remove the driven sprocket support. Remove the thrust washer between the support and the direct clutch housing. The thrust washer may come out with the support. Replace oil seal rings on support shaft.

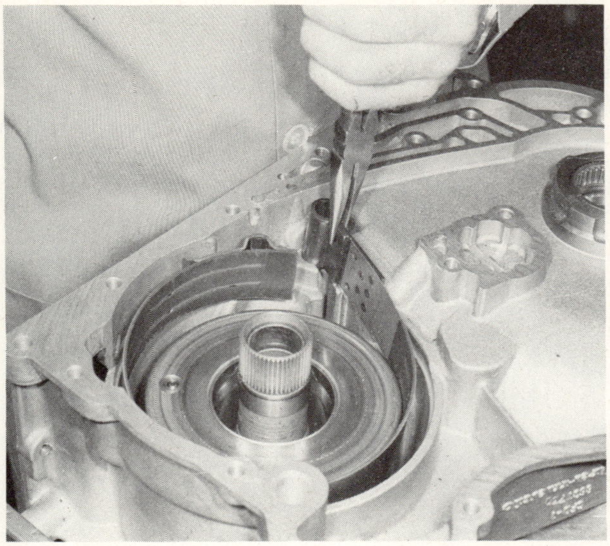

41. Remove the band anchor plug from the case. Use needlenose pliers to lift the intermediate band from the case. Check band surfaces. Replace band if any defects are found. Save band if still usable.

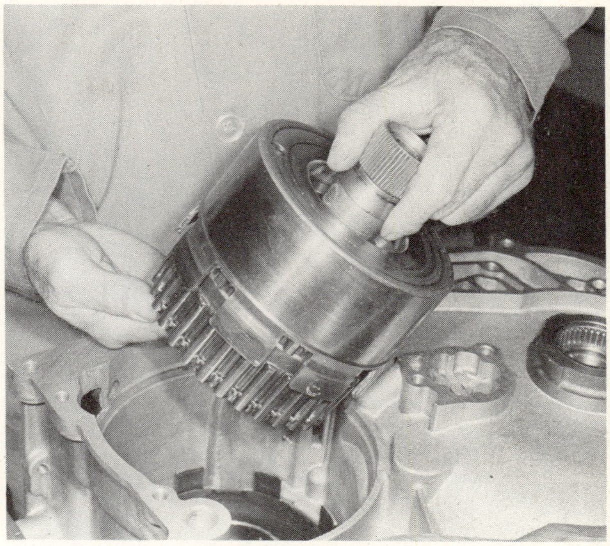

42. Hold the input shaft. Remove the direct and forward clutch assemblies. This unit is quite heavy, so use both hands.

43. Separate the direct and forward clutch assemblies by pulling them apart. You can work on the clutch assemblies as you remove them, or all at one time.

44. All clutches are serviced in essentially the same way. Pry the direct clutch snap ring from the clutch housing.

45. Turn the clutch housing upside down and tap gently to remove the backing plate and clutch pack. Separate steel plates from friction plates and inspect.

46. Another snap ring must be removed to gain access to the clutch housing piston. When replacing snap rings, make sure they snap into the proper groove all the way around.

47. Remove the apply ring and release the spring assembly. Check the springs for distortion and/or damage. If replacement is required, the entire assembly must be replaced as a unit.

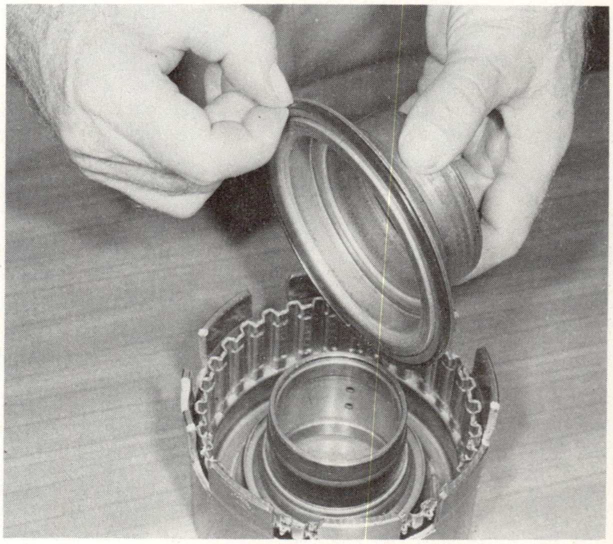

48. Turn the clutch housing upside down. Tap gently to free the direct clutch piston. Replace the inner and outer piston seals. Install a new housing center seal.

49. To reassemble the clutch, replace the piston, the spring assembly, and the snap ring. Alternate steel and friction clutch plates. Install the backing plate (flat side up) and the snap ring.

50. Continue transaxle disassembly by removing the thrust washer between the input ring and the input shaft. The notched side faces up.

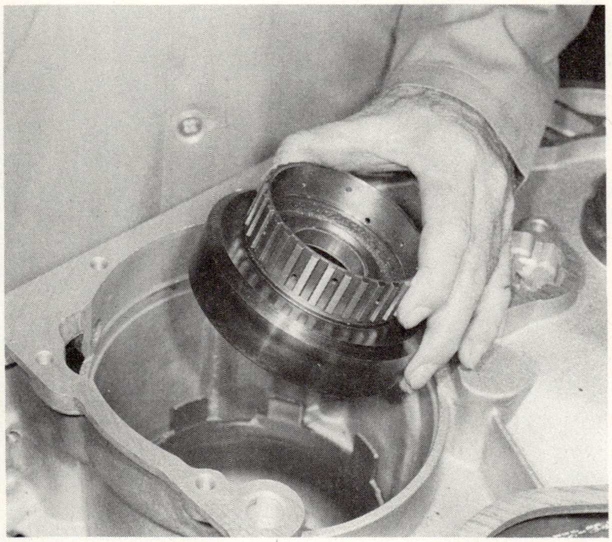

51. Remove the input ring gear from the case. Each part must be inspected and placed on a workbench in the order removed from the case for ease in reassembly.

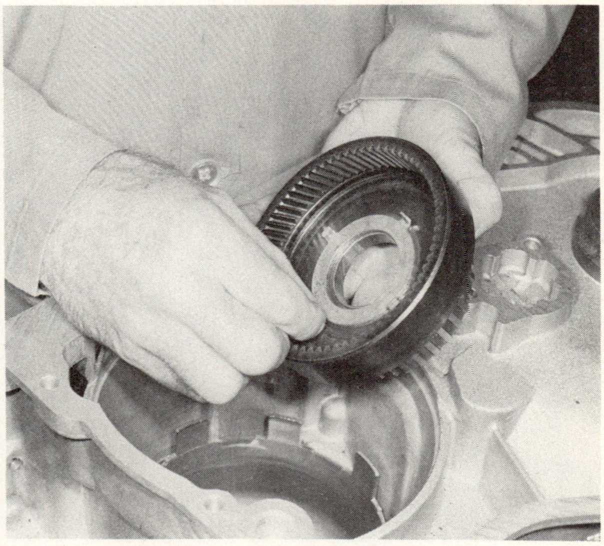

52. The thrust washer between the input carrier and the ring gear may be found on the underside of the ring gear. Otherwise, the thrust washer may remain with the carrier assembly.

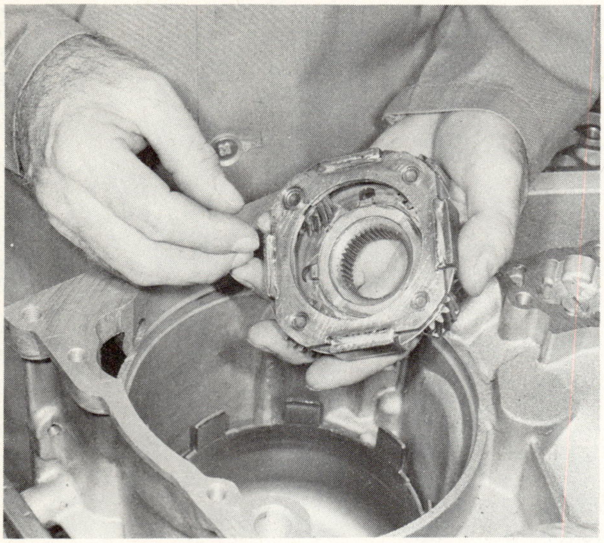

53. Remove the input carrier assembly from the case. Make sure that the thrust washer between the carrier and the sun gear is in position when reassembling the transaxle.

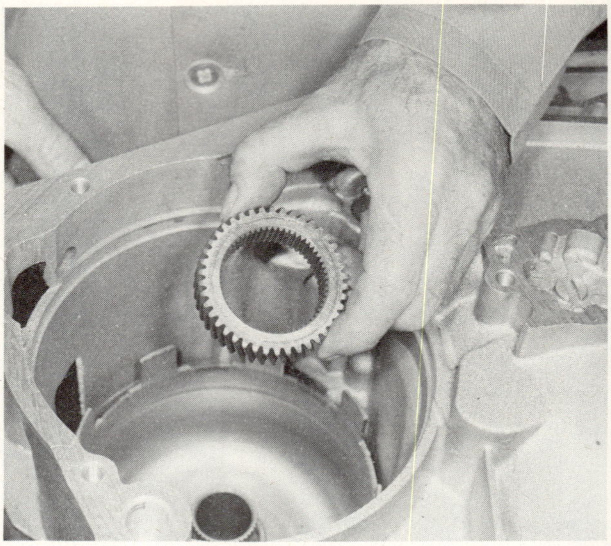

54. The input sun gear is the last part to be removed from the input drum. The sun gear can be replaced in the input drum with either side facing up.

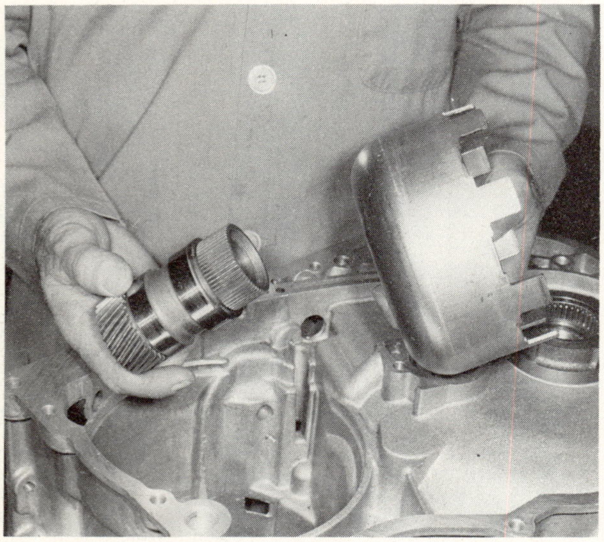

55. Remove the input drum and reaction sun gear from the transaxle case. The case is *not* empty at this point. There are still many parts to be removed. At this point, special tools are needed.

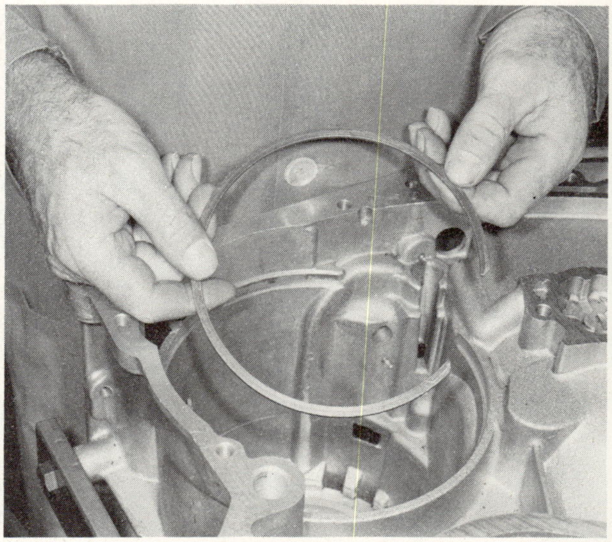

56. Remove the snap ring between the low/reverse clutch housing and case. Before further disassembly is possible, you'll need a special tool.

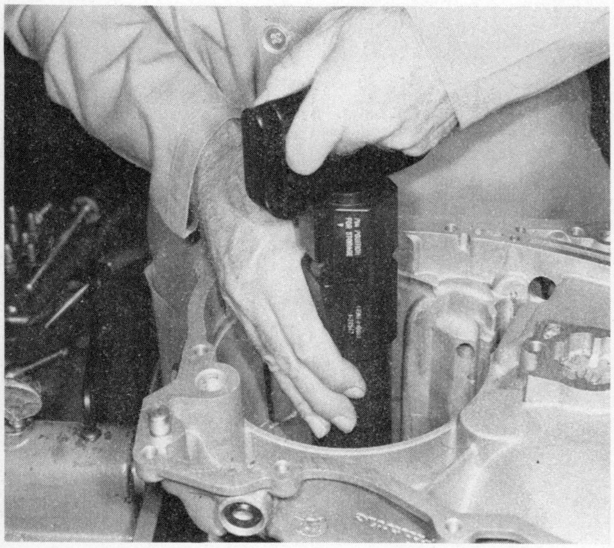

57. The special tool, GM tool J-28542 and adapter, is required to separate the clutch assembly from the reaction carrier. Do *not* attempt removal without this tool.

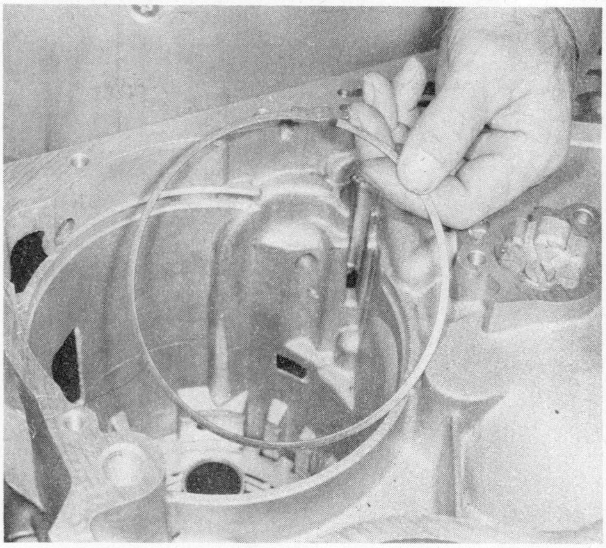

58. With the low/reverse clutch housing removed from the case, pry out and remove the spacer ring between the housing and the case. All snap rings and spacer rings must be inspected.

59. Hold the final-drive sun gear shaft and pull the reaction gearset out of the case.

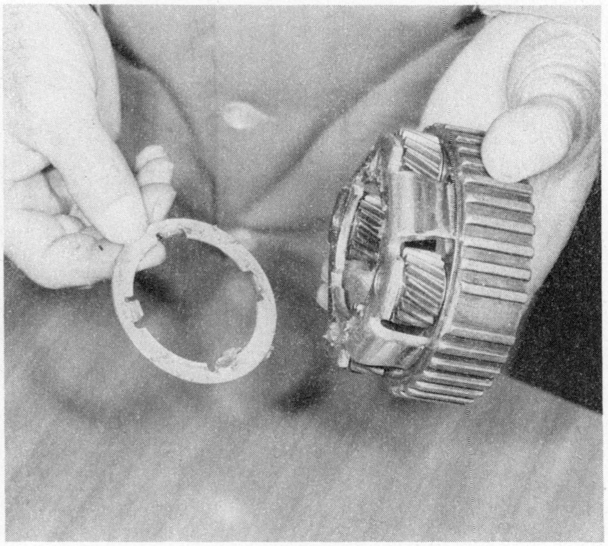

60. Remove the overrunning clutch and reaction carrier from the case. Separate the tanged thrust washer from the carrier. Then separate the ring gear from the end of the reaction carrier.

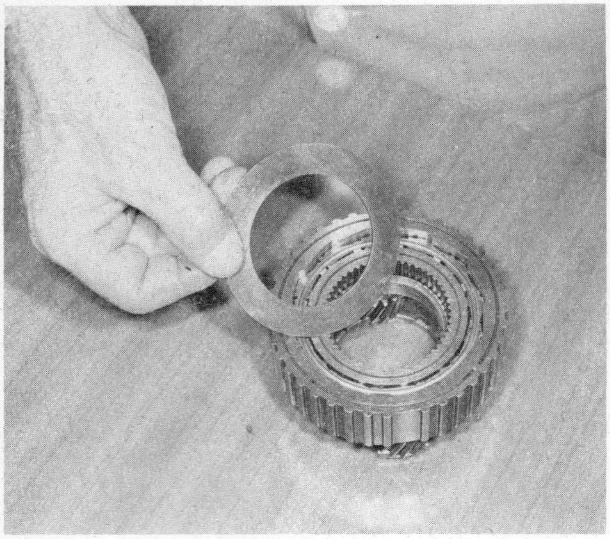

61. Another thrust washer is located on the overrunning clutch side of this assembly. Inspect the overrunning clutch for wear, damage, or faulty operation.

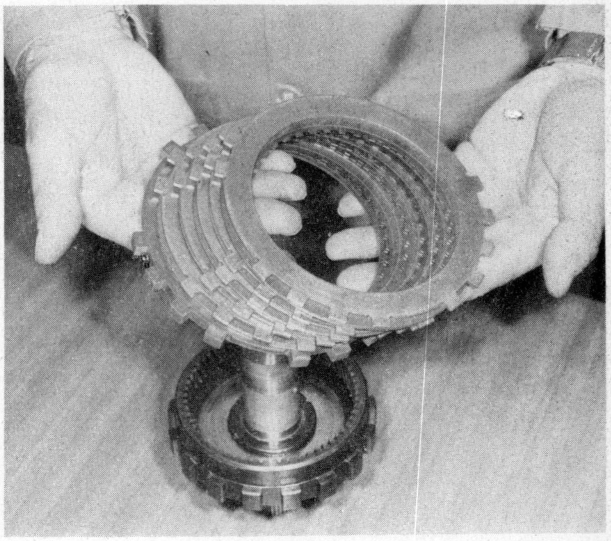

62. Remove the low/reverse clutch plates from the final-drive sun gear shaft. Inspect the frictions for wear and damage. Dip the plates in ATF before installation.

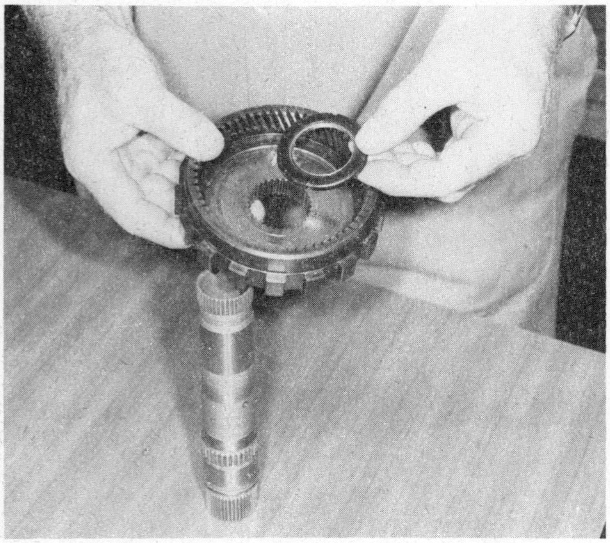

63. Remove the reaction ring gear and thrust bearing from the sun gear. All thrust bearings must be separated and inspected for wear.

64. Inspect the final-drive sun gear shaft carefully for wear or damage. Replace the oil seal rings with a new set.

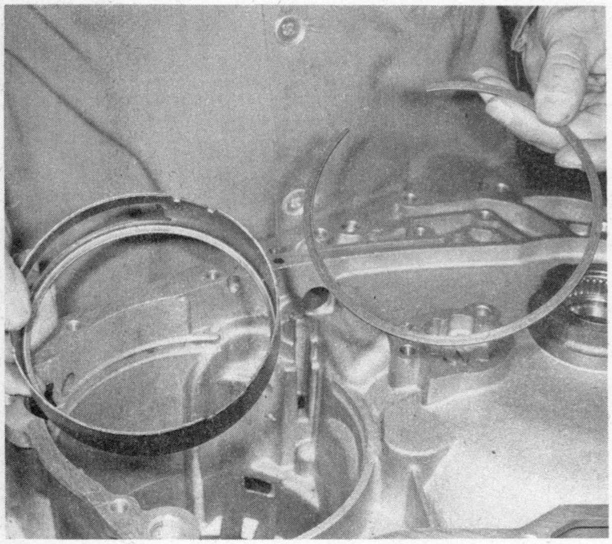

65. Pry out the snap ring from between the spacer and case. Remove the snap ring and the final-drive ring gear spacer from the case, as shown.

66. During reinstallation, the final-drive ring gear spacer and snap ring must be positioned properly. Place the spacer so that the parking pawl will freely enter the slot in the spacer, as shown.

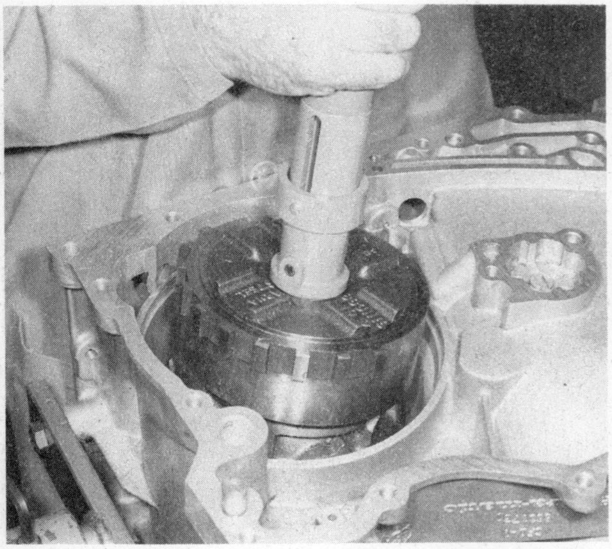

67. Another special GM tool (J-28545) is required to remove the final drive assembly from the transaxle case. Do *not* attempt to remove the final drive assembly without this tool.

68. Remove the selective thrust washer from between the differential final drive and the case. Also remove the final-drive thrust bearing from between the differential carrier and case.

69. Separate the final-drive ring gear from the differential case by rotating as you lift up. The differential is serviced as a separate unit (except for pinions).

70. Another thrust bearing is located inside the final drive unit. As with other thrust bearings, separate and check the bearing for wear and damage.

71. Remove the final-drive sun gear and the thrust bearing between the sun gear and differential carrier. The outside-diameter race of the thrust bearing must face against the carrier when reinstalled.

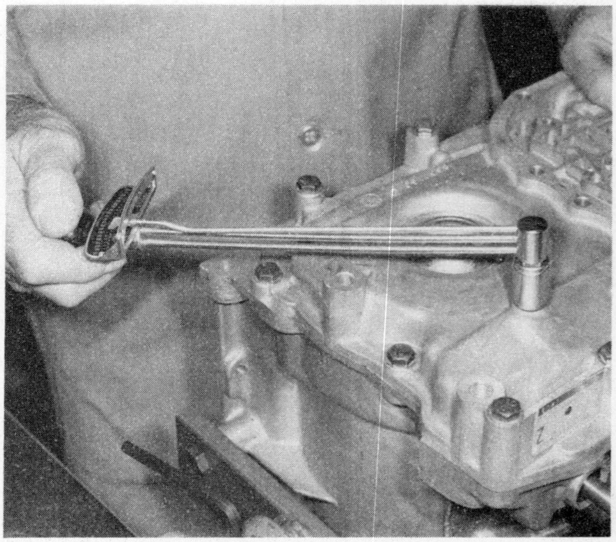

72. Reverse the procedure to reassemble the transaxle. Take note of special reassembly instructions already covered. Torque case cover bolts to specifications.

NAME ______________________ SECTION ______________ DATE ______________

Job Sheet 26-2

Overhauling the GM 125 Transaxle

Before you begin:

Read Topics 26.5 through 26.6.

Vehicle Make ______________________ Model ____________ Year ____________

Time Started ______________________ Time Finished ______________________

Job Completion Time ________________ Flat Rate Time ______________________

Tools Required

- Safety glasses
- Metric sockets
- Socket extensions, handle
- Speeder handle
- Metric open-end or combination wrenches
- Screwdrivers
- Screw extractors
- GM special tools for transaxle disassembly
- Dial indicator and base
- Long-nose pliers
- Ball-peen hammer
- Pin punch
- GM special tools to transaxle reassembly
- Torque wrench

References

Manufacturer's service manual ______________________________

Repair manual ______________________________

Service bulletin ______________________________

Procedure

1. Place the transaxle in a holding fixture. Remove the torque converter. Unbolt the governor cover and discard the O-ring seal.
2. Remove the thrust washer between the governor cover and the speedometer drive gear. Lift the governor and the drive gear assembly from the transaxle case.
3. Unbolt and remove the oil pan. Discard the old oil pan gasket. Remove any traces of the old gasket from the pan and mating flange.
4. Remove the oil strainer and discard.
5. Remove and discard the strainer assembly O-ring seal.
6. Unbolt the retaining brackets and remove the low/reverse oil pipe. Remove and discard the O-ring seal and backup ring from the bore.
7. Unbolt and remove the intermediate servo cover. Remove and discard the gasket.

8. Remove the intermediate servo assembly from the case. Install new oil seal rings. Refer to the manufacturer's service manual to check the band apply pin with the GM special tools.

Instructor Check ____________________

9. Remove the third accumulator check valve. If necessary, use a screw extractor to remove the cup plug that holds early production model valves in place.
10. Remove the oil pipe.

CAUTION: Do not bend the pipe during removal.

11. Remove the cup plug holding the low/reverse seal assembly from the case. Remove and discard the seal.
12. Unbolt and remove the dipstick stop and parking lock bracket.
13. Pull the actuator rod spring end from the case. Remove the rod from the attaching arm and case.
14. Rotate the final drive assembly until the C-ring can be seen in the access slot. Align the C-ring so that its open side faces the window. Push the ring from the output shaft with the GM special tool.
15. Remove the C-ring.
16. Remove the output shaft from the case.
17. Check the output shaft for burrs, scoring, or other damage.
18. Rotate the transaxle so that the case cover is facing upward. Remove the valve body cover and gasket. Discard the gasket. Clean the cover in fresh solvent. Clean the mating flange.
19. Remove the throttle lever and bracket assembly from the valve body.
20. Remove all valve body bolts, but leave the pump cover bolt attached. Note the positions of the shorter and longer bolts.
21. Remove the valve body from the transaxle case. Place the valve body (machined side up) on a workbench. Remove and discard the gasket between the valve body and the spacer plate.
22. Remove the remaining cover bolt. Lift off the cover. Make matchmarks on the pump vanes, housing, and vane holder for reassembly.

Instructor Check ____________________

23. Remove the #1 check ball in the spacer plate and keep separately.
24. Lift the spacer plate and gasket from the transaxle case. Remove and discard the bottom gaskets.
25. Remove the oil pump shaft from the transaxle case. Inspect the shaft for damage.
26. Find the five check balls in the valve body passages. Make notes or a drawing of where they belong. Remove the balls, and keep them separated from the #1 check ball.

27. Refer to the manufacturer's service manual to check input shaft end play.
In the space provided, write the manufacturer's specification and measured end play.

Manufacturer's Specified Input Shaft End Play .... ____________________

Measured End Play .... ____________________

Instructor Check: ____________________

28. Refer to the manufacturer's chart for the correct selective-fit snap ring to correct input shaft end play.
29. Remove the dial indicator, post, and GM special tool. Leave the special GM shaft-loading tool in the loaded (tightened) position.
30. Disconnect the manual valve rod from the manual valve.
31. Remove the two case cover screws and washers from the converter housing side. Then remove the remaining case cover screws. Dip the screws in ATF before reinstalling them, to prevent galling the aluminum threads.
32. Lift the case cover from the transaxle case. Turn the cover over and place it on a clean workbench. Save the 1–2 accumulator pin. Remove and discard the gasket.
33. Remove the 1–2 accumulator spring and the center cover gasket. Discard the gasket. Remove and discard the oil seal rings on the drive sprocket shaft.
34. Find and inspect the nylon drive sprocket thrust washer. If undamaged, save it in a clean container for reassembly.
35. Remove the 1–2 accumulator piston and replace the oil seal rings. Save the piston in a clean container for reassembly.
36. Remove the driven sprocket thrust bearing.
37. Remove the sprockets and chain from the case.
38. Remove both nylon thrust washers under the drive and driven sprockets. Inspect the washers for wear or damage. If undamaged, save them in a clean container for reassembly.

Instructor Check ____________________

39. Rotate the detent lever toward the outside of the case. Remove the roll pin in the detent shaft.
40. Remove and save the special nail holding the manual shaft to the case.
41. Remove the manual shaft from the transaxle case. Remove the detent lever, parking lock, and manual valve assembly.
42. Remove the driven sprocket support. Remove the thrust washer between the support and the direct clutch housing. Replace the oil seal rings on the support shaft.
43. Remove the band anchor plug from the case. Remove the intermediate band from the case. Inspect the band surface for defects. Save the band for reassembly if still good.
44. Hold the input shaft. Remove the direct and forward clutch assemblies.
45. Separate the direct and forward clutch assemblies by pulling them apart.
46. Pry the direct clutch snap ring from the clutch housing.

47. Remove the clutch assembly backing plate and clutch pack. Separate steel plates from friction plates, and inspect.
48. Wear safety glasses. Remove the snap ring holding the clutch housing piston.
49. Remove the apply ring and release the spring assembly. Check the springs for distortion and/or damage. If damaged, replace the entire assembly.
50. Remove the direct clutch piston. Replace the inner and outer piston seals. Install a new housing center seal.
51. Replace the clutch piston, the spring assembly, and the snap ring. Alternate steel and friction clutch plates. Install the backing plate (flat side up) and the snap ring.
52. Disassemble and reassemble forward clutch in the same way as direct clutch.
53. Remove the thrust washer between the input ring and the input shaft. The notched side faces up.
54. Remove the input ring gear from the case. Remove each part one at a time and make a note or drawing of the assembly order.
55. Find and inspect the thrust washer between the input carrier and the ring gear.
56. Remove the input carrier assembly from the case. Inspect the thrust washer between the carrier and the sun gear.
57. Remove the input sun gear from the input drum.
58. Refer to the manufacturer's service manual to check the measurement of the reaction sun gear to the input drum selective-fit snap ring.

In the spaces provided, write the manufacturer's specification and measured play.

Manufacturer's Specified End Play Measurement .... ____________________

Measured End Play .... ____________________

Instructor Check: ____________________

59. Refer to the manufacturer's service manual to check the measurement of the low and reverse clutch housing to the low overrunning clutch race.

In the spaces provided, write the manufacturer's specification and measured end play.

Manufacturer's Specified End Play Measurement .... ____________________

Measured End Play .... ____________________

Instructor Check: ____________________

60. Refer to the manufacturer's charts for the correct selective-fit snap rings to correct end play to specifications.
61. Remove the input drum and reaction sun gear from the transaxle case.
62. Remove the snap ring between the low/reverse clutch housing and case.
63. Use the GM special tool and adapter to separate the clutch assembly from the reaction carrier. Remove the clutch assembly from the case.

Instructor Check: ____________________

NAME ______________________ SECTION ____________ DATE ____________

64. Pry out and remove the spacer ring between the housing and the case. Inspect the snap ring for damage. Replace if any defects are found.
65. Hold the final-drive sun gear shaft and pull the reaction gearset out of the case.
66. Remove the overrunning clutch and reaction carrier from the case. Separate the tanged thrust washer from the carrier. Separate the ring gear from the end of the reaction carrier.
67. Remove and inspect the thrust washer on the overrunning clutch side of the reaction carrier assembly. Inspect the overrunning clutch for wear, damage, or faulty operation. Replace if any defects are found.
68. Remove the low/reverse clutch plates from the final-drive sun gear shaft. Inspect the friction plates for wear and damage. Dip the plates in ATF before installation.
69. Remove the reaction ring gear and thrust bearing from the sun gear. Separate all thrust bearings and inspected for wear. Replace if any defects are found.
70. Inspect the final-drive sun gear shaft carefully for wear or damage. Remove and discard the oil seal rings. Install new oil seal rings during reassembly.
71. Pry out the snap ring from between the spacer and case. Remove the snap ring and the final-drive ring gear spacer from the case.
72. Reinstall the final-drive ring gear spacer and snap ring. Place the spacer so that the parking pawl will freely enter the slot in the spacer.
73. Refer to the manufacturer's service manual to measure the final drive to case end play. In the spaces provided, write the manufacturer's specification and measured end play.

Manufacturer's Specified End Play .... ____________________

Measured End Play .... ____________________

Instructor Check: ____________________

74. Use the GM special tool to remove the final drive assembly from the transaxle case.
75. Refer to the manufacturer's chart for the correct selective-fit washer to correct end play.
76. Remove the selective-fit thrust washer from between the differential final drive and the case. Remove the final drive thrust bearing from between the differential carrier and case.
77. Separate the final-drive ring gear from the differential case.
78. Separate and check the thrust bearing inside the final drive unit for wear and damage. Replace if any defects are found.
79. Remove the final-drive sun gear and the thrust bearing between the sun gear and differential carrier. The outside-diameter race of the thrust bearing must face against the carrier when reinstalled.
80. Reverse the procedure to reassemble the transaxle. Refer to the manufacturer's service manual for specific instructions.

Job Completion Report

1. What difficulties, if any, were encountered during the transaxle overhaul?

____

____

____

____

____

____

2. What parts were replaced during overhaul? ______________________

____

____

____

____

____

____

INSTRUCTOR SIGNOFF ______________________ **DATE** ____________

26.7 TRANSAXLE INSTALLATION

To install the transaxle, the removal procedure is reversed. However, a few important points must be observed:

1. Install both axle shafts *after* the transaxle is in place.

2. When reinstalling the front suspension support assembly, follow the proper tightening sequence. See Figure 26-20. This maintains proper suspension alignment.

3. Check front suspension alignment after transaxle installation is complete.

4. Adjust the TV cable and shift linkage.

5. Fill the transaxle with the proper ATF.

6. Road test the vehicle to make sure all parts are operating properly.

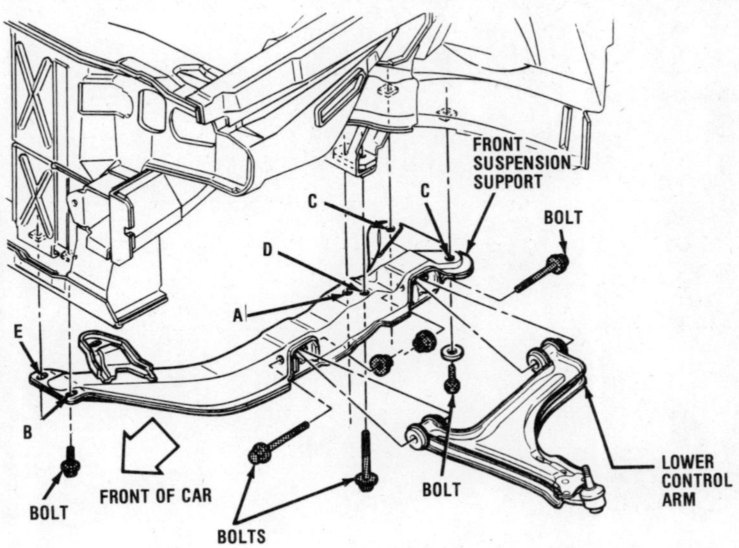

FRONT SUSPENSION SUPPORT ATTACHING BOLT/SCREW SEQUENCE:

1. INSTALL CENTER SCREW LOOSELY INTO HOLE [A].
2. INSTALL TIE BAR SCREW LOOSELY INTO SMALL SLOTTED OUTBOARD HOLE.
3. INSTALL BOTH BOLTS IN THE REAR HOLES AND TORQUE TO SPECIFICATIONS.
4. INSTALL AND TORQUE SECOND CENTER HOLE [D] BOLT/SCREW TO SPECIFICATIONS.
5. TORQUE CENTER HOLE [A] BOLT TO SPECIFICATIONS.
6. INSTALL OTHER TIE BAR BOLT (FRONT HOLE [E]) AND TORQUE TO SPECIFICATIONS.
7. TORQUE SECOND BOLT/SCREW TO SPECIFICATIONS IN FRONT HOLE [B].

Figure 26-20. Front suspension support assembly. (Oldsmobile Division-GMC)

NAME ______________________________ SECTION ________________ DATE ________________

Job Sheet 26-3

Installing the GM 125 Transaxle

Before you begin:

Read Topics 26.1 through 26.4 and 26.7.

Vehicle Make __________________________ Model ______________ Year ______________

Time Started __________________________ Time Finished __________________________

Job Completion Time ___________________ Flat Rate Time __________________________

Tools Required

- Hoist
- Safety glasses
- Special transaxle jack
- Metric sockets
- Socket extensions, universal joints, handle
- Open-end or combination wrenches
- Pliers
- Screwdrivers
- Special engine support fixture
- Torque wrench
- Brake bleeder
- Wheel-alignment rack and tools

References

Manufacturer's service manual ____

Repair manual ____

Service bulletin ____

Procedure

SAFETY CAUTION: ***Never*** **attempt to install the transaxle unless the vehicle is properly supported. Special hoists and transaxle jacks must be used. Trying to replace the transaxle using jackstands is dangerous. Serious damage and personal injury can result when using other than the recommended equipment.**

1. Put the transaxle on the special transaxle jack. Position the transaxle in the engine compartment and slide it toward the engine.
2. Replace one bolt to hold the transaxle to the engine.
3. Replace the screws holding the brake line and fuel line brackets to the left side of the underbody. Remove the plugs or caps and replace the brake and fuel lines.
4. Bleed the brake lines of air. Test the brakes for proper operation.

Instructor Check: ____________________

5. Replace the starter.
6. Reconnect the transaxle cooler lines.
7. Replace the bolts that hold the torque converter to the flexplate.
8. Replace the torque converter shield and screws.
9. Reconnect the transaxle strut at the transaxle.
10. Reconnect the speedometer cable.
11. Replace the stud holding the engine to the transaxle. Replace the nut holding the control cable bracket to the transaxle.
12. Remove the plugs in the transaxle bores and replace both axle shafts.
13. Follow the tightening sequence and replace the bolts holding the front right-hand suspension support assembly.
14. Follow the tightening sequence and replace the bolt holding the stabilizer bar to the lower right-hand control arm.
15. Repeat steps 13 and 14 at the left front wheel.
16. Assemble the ball joint to the control arm. Follow the tightening sequence and replace the castellated ball-joint nut. Torque the nut to the manufacturer's specifications.

Manufacturer's specified torque .... ____________________

Insert a new cotter pin and bend it into place.

17. Replace both front wheel and tire assemblies.
18. Lower the hoist.
19. Unlock the steering column.
20. Replace the upper bolts that hold the transaxle to the engine. Replace the bolt near the starter that holds the transaxle to the engine.
21. Reconnect and adjust the shift control linkage.
22. Replace the top transaxle mount and bracket assembly. Lower the engine slightly and replace the mounting assembly.
23. Remove the special engine support fixture.
24. Replace the hose on the air management valve.
25. Replace the wiring harness and bolt on the top of the transaxle.
26. Reconnect the TV cable to the transaxle rod.
27. Reconnect the TV cable at the carburetor. Adjust the TV cable.

Instructor Check: ____________________

28. Replace the carburetor air cleaner.
29. Remove the 1/4 x 2-inch-long bolt at the front right-hand engine mount.
30. Reconnect the negative battery cable to the transaxle.
31. Refill the transaxle with the correct fluid.
32. Check and/or realign the front end of the vehicle. Check the brakes for proper operation.

Instructor Check: ____________________

NAME ______________________ SECTION ____________ DATE ____________

33. Ask your instructor's permission to conduct a road test. If approved, road-test the vehicle and check for proper transmission operation in all gear positions. Recheck the fluid level after the road test.

SAFETY CAUTION: Drive in a safe manner. Jackrabbit starts, hard cornering, or sudden braking should be avoided. Driving conditions other than normal can worsen a problem, or create a new one. Carelessness during the test drive can result in an accident. Remember, you are responsible for someone else's vehicle. Drive only in a manner that is required to check or diagnose proper transmission operation.

Instructor Check: ____________________

Job Completion Report

1. What difficulties, if any, were encountered during transaxle installation?

____

____

____

____

____

____

INSTRUCTOR SIGNOFF ______________________ DATE ____________

NAME ______________________ SECTION ____________ DATE ____________

Certification Practice

Test 26

DIRECTIONS:

The following questions are similar to those used on mechanic certification tests. Answer each question by circling the letter of the correct choice.

1. The GM Turbo Hydra-Matic 125/125C includes all of the following EXCEPT
   A. one servo and band.
   B. two overrunning clutches.
   C. three multiple-disc clutches.
   D. two sprockets and a drive chain.

2. Which of the following statements about the GM 125/125C is correct?
   I. The oil pan alone does not hold enough fluid for proper transmission operation.
   II. The valve cover body is used as a reservoir to hold transmission fluid.
   A. I only B. II only C. Both I and II D. Neither I nor II

3. The tool shown
   A. is a transaxle support fixture.
   B. is an engine support fixture.
   C. will support the entire weight of the engine and transaxle.
   D. can be replaced with ordinary jackstands.

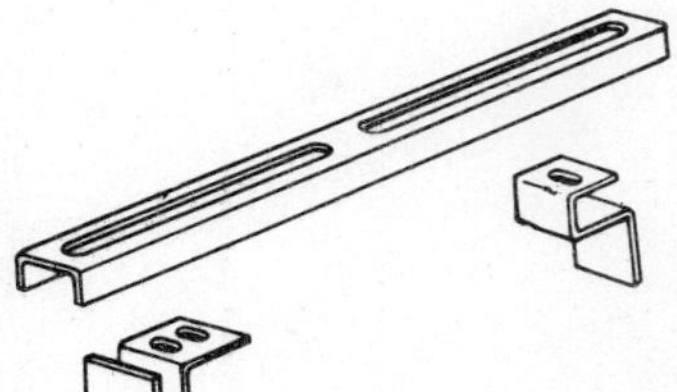

Question 3.
(Chevrolet Motor Division-GMC)

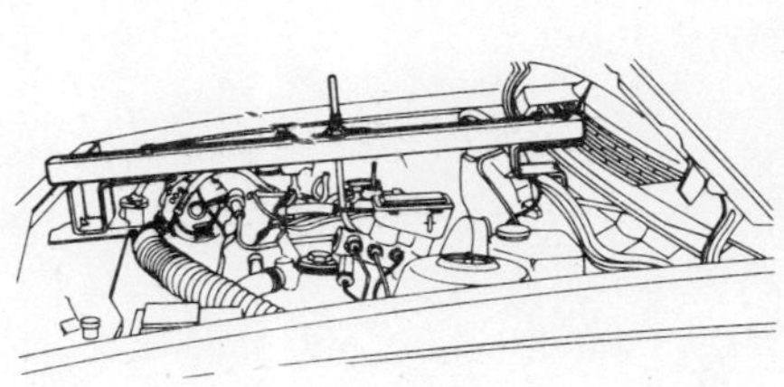

4. Mechanic A says the more important identification number on the 125/125C is the vehicle identification number.
   Mechanic B says the more important identification number on the GM 125/125C transaxle is the transaxle model code.
   Who is correct?
   A. A only B. B only C. Both A and B D. Neither A nor B

5. All of the following statements about removing the GM 125/125C transaxle are correct EXCEPT
   A. Parts of the front suspension must be removed.
   B. Parts of the brake system must be removed.
   C. The battery must be removed.
   D. The TV cable and shift linkage must be removed.

6. The GM 125/125C transaxle valve body has
   A. four check balls.
   B. five check balls.
   C. six check balls.
   D. seven check balls.

Question 7.

7. All of the following statements about the part shown are correct EXCEPT
   A. This part must be checked for burrs, scoring, or other damage.
   B. This part is the output shaft.
   C. A C-ring must be taken out to remove this part.
   D. During reassembly, the long, splined end goes into the transaxle case first.

8. The oil pump on the GM 125/125C is
   A. a rotor pump.
   B. a gear and crescent pump.
   C. a vane-type pump.
   D. none of the above.

9. Mechanic A says that the brake lines always must be bled during the transaxle installation procedure.

   Mechanic B says that the front end always must be realigned during the transaxle installation procedure.

   Who is correct?

   A. A only B. B only C. Both A and B D. Neither A nor B

10. Which of the following statements is correct?
    I. The TV cable and gearshift linkage must be adjusted during the transaxle installation procedure.
    II. The vehicle should be road tested after transaxle installation.

    A. I only B. II only C. Both I and II D. Neither I nor II

Glossary

A

accumulator A device that allows gradual engagement of servos by compressing a spring with some of the fluid pressure used to operate the servo.

air pressure test Applying air pressure to control valve body holes and passages leading to apply devices.

alignment The proper positioning of parts in relation to each other; also, the process of positioning parts properly.

apply device A hydraulically operated mechanism used to engage a clutch or band.

ATF *See* automatic transmission fluid.

automatic transmission A transmission that shifts itself automatically through forward gears as load and speed conditions change; the basic parts of an automatic transmission are the torque converter, planetary gearset, and control valve body.

automatic transmission fluid A specially compounded light oil used in automatic transmissions; two types of automatic transmission fluids are commonly used, Dexron II or Type F, depending on the type of friction materials used in the transmission clutches and bands.

axis The point or line around which a rotating part turns.

axle The shaft on which a wheel or gear rotates.

B

balance valve A spool valve that changes pressure.

ball bearing A bearing with spherical metal balls that rotate between races.

band adjustment *See* transmission band adjustment.

bead A thick line of material on a surface.

bearing An assembly used to hold, support, and allow a rotating part to turn smoothly.

bearing cone The cage and rollers in a roller bearing.

bearing cup *See* bearing race.

bearing puller A device with movable arms and a central, threaded shaft; used to remove bearings that have been pressed onto a shaft.

bearing race The smooth, round inner or outer metal part of a bearing against which the rollers or balls ride.

bellcrank A part that pivots to transfer motion at a right angle (90 degrees).

Belleville spring A concave disc with fingers pointed toward the center; a wavy disc that acts much like a diaphragm.

bellhousing The forward portion of the automatic transmission case that encloses the torque converter assembly.

bevel gear A gear whose teeth are cut at a slant.

bind To stick and hold; also, a place where sticking and holding occurs.

bleed To remove air bubbles from a fluid-filled system.

booster valve A balance valve with a pressure-regulated spring.

bore A cylindrical hole made in a part; also, the diameter of the hole.

boss A part that protrudes, or sticks out, from another part.

brake band *See* transmission band.

burr A sharp, thin ridge or edge on a piece of metal.

bushing A hollow cylinder-shaped metal lining; used as one form of bearing.

C

cable A metal wire that can move inside a flexible covering.

cage A thin sheetmetal enclosure for the rollers or balls of a bearing; also, the metal holder for planetary pinion gears.

capscrew A bolt or screw used to assemble the outer part of an assembly.

castellated nut A nut with projections on the outside face; used with a cotter pin to prevent the nut from unscrewing from a shaft.

casting porosity Holes in a casting caused by air bubbles in molten metal during the casting process.

centrifugal force An outward pull from the center of a rotating axis.

chamfer A beveled edge.

chassis The frame of the automobile.

check ball A ball that closes off orifices to change fluid flow; operates like a check valve.

check valve A valve that permits fluid flow in only one direction.

circlip A C-shaped snap ring that fits on the outside of a shaft.

clearance The space between two parts.

clip *See* retaining clip.

clutch assembly A friction-operated coupling device used to connect and disconnect driving and driven parts of a rotating mechanism.

clutch disc A metal disc with friction facings used in a clutch assembly.

clutch pack A clutch assembly consisting of multiple clutch discs separated by metal shims.

clutch plate *See* clutch disc.

coast mode A method used in test-driving procedures; the vehicle is allowed to coast down from a specific speed.

collapsible washer A washer that crushes when a nut is tightened against it.

collar *See* retaining collar.

column-shift linkage Shift linkage attached to the steering column of a vehicle.

control device A hydraulic, pneumatic, or electrical device used to control shift valves in an automatic transmission.

control valve Controls hydraulic pressure flow to operate other parts of the control valve body.

control valve body The part of an automatic transmission, consisting of multiple passages and valves, through which fluid is directed to control shifting and gear selection.

coolant The mixture of antifreeze and water circulated through the engine cooling system.

cotter pin A split metal pin driven through a hole in a shaft and then spread open; used to hold and attach parts.

crocus cloth A cloth, similar to a fine sandpaper, coated with a dark red material; used for polishing metal.

crossmember A part of the underbody structure of a vehicle that extends from one side of the frame to the other.

crossover blade *See* relay blade.

cruise mode A method used during test-driving procedures; the throttle is applied at all times, and engine speed is held at a predetermined rpm.

D

depth gauge A measuring instrument with a long, slender shaft, or bar; used to measure the distance to the bottom of a hole.

detent A device for positioning and holding a part in relation to another part.

diagnose To investigate the cause of a problem.

diagnosis A statement that identifies the cause of a problem.

dial indicator A precision measuring instrument; used to measure the distance a part moves or protrudes (sticks out).

diaphragm A thin rubber sheet that separates an area into different sections; used in modulator valves and vacuum-advance parts. Pressure acting within the parts moves the diaphragm in and out to change the pressure at each side of the diaphragm.

differential A gear assembly that allows the driving axles to rotate at different speeds when turning a corner.

differential case The part of the differential connected to the ring gear on which are mounted the differential pinion gears, pinion shaft, and side gears.

dipstick A metal rod extending into a lubricant supply; used to check the quantity of a lubricant.

direct drive The condition when the input and output shafts of a transmission turn at the same speed.

disengage To disconnect; used to describe the action when a clutch uncouples the power of the driving part from the driven part.

dog *See* driving lug.

dogged Connected by driving lugs that fit into grooves.

dowel A round pin or shaft; used to align or connect an attaching part.

downshift Automatic or manual shift to a lower gear.

drift A straight or tapered metal shaft; used with a hammer to drive shafts or pins through, or into, holes.

drive belt An engine belt that turns an accessory unit.

drive mode A method used in test-driving procedures; the throttle is depressed enough to accelerate the vehicle and apply engine torque.

drive nut A threaded nut, or plate, attached to a unit; used to connect the unit to another unit for turning purposes.

drive shell The housing that encloses a multiple-clutch assembly and may have a surface on which a transmission band engages.

drive stud A threaded shaft attached to a unit; used to connect the unit to another unit for turning purposes.

driveline The driveshaft and universal joints.

driven member The part of a rotating mechanism that receives power from a driving member. *See also* clutch.

driveshaft A steel tube or shaft that connects the transmission to the differential.

drivetrain The group of units that control and transfer power from the engine to the driving wheels; includes the transmission, driveline, differential, and driving axles.

driving axles The steel shafts that transfer differential torque to the driving wheels.

driving lug A metal tab on a part that fits into a groove; used to connect and turn a part.

driving member The part of a rotating mechanism that supplies power to a driven member. *See also* clutch, driven member.

driving shell *See* drive shell.

droplight An electrical light on a long extension cord.

E

emery cloth A cloth, similar to fine sandpaper, coated with a dark abrasive material; used for sanding and polishing metal.

end play The distance a shaft moves in and out of an assembly.

engage To connect; used to describe the action when a clutch couples a driving member of a rotating mechanism to the driven member.

epoxy A soft, plastic material that hardens when mixed with a chemical hardener.

extension housing A separate metal case on the end of a transmission.

F

feeler gauge A blade-shaped measuring gauge; used to measure distance between parts.

filler plug A pipe plug used to plug the opening to a lubricant supply.

finger-tight As tight as can be done by turning with the fingers.

firewall The metal wall separating the engine compartment from the passenger compartment.

flange A raised portion or rim around a part, to which other parts attach.

flat A part of a round shaft that has been machined flat.

flexplate A circular, flexible plate used to drive the torque converter; mounted in place of a flywheel.

float mode A method used during test-driving procedures; the throttle is backed off gradually and continually, and the vehicle is allowed to slow down in a controlled manner.

flow chart A map of the oil circuit system that shows fluid flow during certain conditions.

fluid coupling A liquid-filled device, consisting of an impeller and a turbine, that connects the power, or torque, from an engine to an automatic transmission. *See also* impeller, turbine, torque converter.

flywheel A heavy steel disc bolted to the rear of the engine that helps the engine to run more smoothly.

flywheel lock A tool used to keep a flywheel from turning during servicing.

free travel Movement of an operating part before it contacts another part.

friction disc *See* clutch disc.

friction plate *See* clutch disc.

frictions Clutch discs with friction facings.

front-wheel drive A powertrain system that provides engine torque to the front wheels.

FWD *See* front-wheel drive.

G

galling Wear caused by friction.

gate A guide containing a series of indentations, or notches; used to hold or guide another part.

gasket A shaped, soft material that forms a seal between parts; gaskets may be made of paper, cork, rubber, plastic, or soft metal.

gauge A device or instrument for measuring or testing; also, to measure or test.

gear A toothed wheel that meshes with another toothed wheel.

gear and crescent pump An oil pump with inner and outer driven gears; a crescent-shaped part controls fluid flow and prevents meshing of the two gears at one side.

gear ratio A number that indicates the relative turning speeds, or number of rotations, of a set of gears.

gearshift selector lever A lever attached through linkage to valves within an automatic transmission that determines which gear is selected.

gearshift linkage Linkage attached between the gearshift lever and the transmission shifting arms.

glazed Polished by heat and friction to a hard, smooth, shiny surface, unlike the original surface finish.

governor A mechanism that controls another device, depending on speed or rpm; it controls gearshifting depending on speed.

grommet A circular, protective ring inside an opening.

H

helical gear A gear with teeth cut at an angle to the axis of the gear.

high gear The forward gear in a transmission that produces maximum road speed.

hoist A hydraulically powered platform used to lift vehicles for inspection or repair.

hub The center of a circular part, around which the part pivots or rotates.

hydraulic Operated by fluid under pressure.

hydraulics The branch of science that deals with practical applications of liquid in motion.

hypoid A particular design of spiral bevel gear teeth that allows more than one tooth of a drive pinion gear to be in contact with the driven gear teeth; in a hypoid gearset, the center of the drive pinion gear is not aligned with the center of the driven gear.

I

idler gear An intermediate gear between two other gears.

idler shaft The shaft on which an idler gear turns.

ignition lock A linkage arrangement that prevents the steering wheel from turning when the transmission is shifted to Park and the ignition key is removed.

impact wrench A power tool for tightening or loosening bolts and nuts.

impeller The driven member of a torque converter, turned by engine power.

input shaft *See* transmission input shaft.

input shaft bearing The bearing in the transmission that supports the input shaft.

intermediate The middle assembly or system in an automatic transmission. For example, the intermediate clutch assembly would control second gear in a three-speed transmission.

J

jackstand A safety device meant to support weight safely during servicing procedures.

journal The area on a shaft contacted, and directly supported, by a bearing.

K

key A wedge-shaped locking device that holds two parts together.

kickdown band A transmission band that locks up a lower clutch assembly, such as the second, or intermediate, gear, and forces a downshift to overcome load or speed deficiencies.

kickdown linkage Linkage connected to the throttle linkage and the transmission shifting mechanisms; used to downshift the transmission into a lower gear for better acceleration as the throttle is quickly depressed.

knurl Machined ridges on a part; used to provide a friction surface for drive purposes.

L

lash The clearance, or play, measured between parts.

left-hand thread A thread that tightens by turning it to the left, or counterclockwise.

lever A bar-shaped part that transfers motion, and increases force.

linkage Parts that link, or connect, one assembly, or part, to another.

load External resistance to be overcome by a mechanism; the weight carried and the force of gravity (such as when climbing steep hills) determine the amount of load a vehicle must overcome.

lock ring A metal ring fitted into a groove that holds a part in position.

locknut A nut used to prevent a part or another nut from moving.

lockpin A steel pin driven through parts to hold them together.

lockscrew A screw that threads through a part and holds it securely in position.

lockup torque converter A torque converter with an additional clutch piston and drive devices to lock the impeller to the turbine at highway speeds; used to reduce converter slippage and increase fuel economy.

lockwasher A spring washer that grips when a nut or bolt is tightened against it; used to keep nuts and bolts from unscrewing.

longitudinal Mounted front-to-back.

lubricant Grease or oil.

lubricate To grease or oil an assembly.

lug A wheel nut or bolt.

M

mainline pressure Hydraulic pressure that has been regulated by the pressure regulator.

manual valve A valve shaft specially designed to direct fluid flow in the automatic transmission valve body, according to which quadrant position is selected.

mechanic's wire Bendable steel wire used to attach or fasten objects together.

mesh To fit together, as gear teeth.

metallic Made from metal.

micrometer A precision measuring tool with a fixed jaw and a movable jaw on a threaded shaft; used to measure thickness.

mill file A smooth file with grooves cut in only one direction.

mineral spirits Paint or varnish thinner.

mount An attachment for support.

multi-purpose lithium grease White chassis grease containing the element lithium.

multiple-disc clutch A clutch assembly using more than one clutch disc and intermediate metal pressure plates.

N

needle bearing A roller bearing with small, thin metal rollers.

neutral A transmission gear position that does not transmit power from the engine to the driveline.

neutral start switch An electrical switch that prevents the engine of a car with an automatic transmission from starting unless the transmission is shifted into park or neutral.

nominal Ideal or theoretical.

O

oil seal A ring-shaped assembly made of sheetmetal, rubber, or plastic, and a spring; used to prevent oil leaks around a shaft.

orbit To move in a circle around another part.

organic Made from plant or animal matter.

orifice An opening or hole.

O-ring A ring-shaped gasket, circular in cross section, that fits into a groove to form a seal.

overdrive A gear ratio that results in the output shaft of a transmission turning faster than the input shaft.

overrunning clutch A mechanism that allows rotating movement in only one direction.

P

pawl A pivoting finger-like part that engages a notch, or slots, on another part to control motion.

pedal A foot-operated lever.

penetrating oil A special, thin oil that penetrates rusted or tight threads and helps in removal of nuts and bolts.

pilot A shaft or stud used to guide one part into, or onto, another part; also, the machined area into which a pilot fits.

pilot bearing A bearing that supports and holds the end of the transmission input shaft in the flywheel.

pilot bushing A bushing that supports and holds the end of the transmission input shaft in the flywheel.

pinion gear A smaller gear that meshes with a larger gear.

piston A tightly sealed plunger that moves inside a hollow cylinder.

pivot A shaft or pin on which something turns; also, to turn around a shaft or pin.

planet carrier *See* planetary gear system.

planet gear *See* planetary gear system.

planetary gear system A gear system consisting of a central, or sun, gear, planet gears that can orbit around the sun gear and inside an internal ring gear, and a planet carrier with shafts around which the planet gears rotate; depending on which members are driven and held, multiple gear ratios and reverse rotation are possible.

play The distance a part in an assembly can be moved; also, looseness.

pneumatic Air-operated.

poppet valve A one-way valve that allows fluid to flow in one direction only.

powertrain The engine and drivetrain.

preload spring A part in a clutch assembly that applies pressure to a clutch pack to help coupling action.

press A device for exerting great pressure on parts; used for assembly and disassembly.

pressure Force exerted against a surface, divided by the surface area.

pressure plate A steel disc in a multiple-disc clutch assembly that presses against the clutches.

pressure regulator A valve that prevents hydraulic pressure from becoming excessive.

pressure test Attaching pressure gauges to a transmission and checking hydraulic pressures.

puller A tool used for pulling a part off, or out of, an assembly.

pulley A small wheel with a grooved rim; used to guide a cable.

pushrod A metal rod that pushes, or is pushed by, another part.

Q

quadrant An indicator that shows which gear of the automatic transmission has been selected. May be on the shift column, in the instrument cluster, or on a floor console.

R

race *See* bearing race.

range A choice of gears through which an automatic transmission may shift; first, second, and high gears make up a typical drive range.

reaction member The held member of a planetary gearset.

rear main bearing seal A seal at the rear of an engine that prevents engine lubricant from leaking.

rear-end torque Torque that causes the differential and axle housings to twist.

rear-wheel drive A powertrain system that provides engine torque to the rear wheels.

recess A groove or low area in a part.

relay blade A flat metal lever used in column-shift linkage.

reservoir A supply of liquid.

retainer A part that holds another part in position.

retaining clip A small metal part used to fasten or attach another part.

retaining collar A tight-fitting ring-shaped piece used to hold another part on a shaft.

ring gear A gear with a hollow center. *See also* planetary gear system.

roll pin A lockpin made of coiled spring steel.

roller bearing A bearing using cylindrical metal rollers that rotate between races.

roller clutch One type of overrunning clutch mechanism, with spherical rollers that can lock into depressions in the inner and outer races of the mechanism.

rotor pump An oil pump with inner and outer gears that are lobe-shaped.

rpm Revolutions per minute; used to measure the rotating speed of a shaft.

runout The distance a part or assembly juts, or moves, in or out from a specified position.

S

scoring A mark on a shaft or bearing, such as lines, grooves, scratches, or notches.

seal *See* oil seal.

sealant A thick liquid material that dries to form a seal between parts.

selector lever The lever the driver moves to select the gear into which the transmission is to be shifted.

semi-centrifugal clutch A clutch assembly in which centrifugal force is used to increase coupling action.

servo A device that changes hydraulic pressure into mechanical motion.

shaft A solid or hollow cylindrical bar used to support rotating pieces or to transmit torque through rotation.

shift lever *See* selector lever.

shift linkage The linkage that connects the selector lever to the shifting arms on the transmission.

shift valve Controls and times transmission shift points by directing hydraulic pressure to the apply devices.

shifting arm An arm on a transmission connected within the transmission to shifting valves, and by linkage outside the transmission to the selector lever.

shim A thin metal spacer of a specific thickness.

shim pack A group of shims between parts.

shoulder A raised edge on a shaft or inside an opening; used to limit how far a part may move.

sleeve A tubular part that fits over a shaft or into a hole.

slide hammer A metal rod with a tool at one end, and a heavy weight that slides on the rod to pull, or push, the tool.

slip joint A splined joint that can slide in and out.

slip yoke A yoke with a slip joint that slides back and forth on the transmission output shaft.

slippage A loss of power that occurs when the driven member of a rotating mechanism rotates slower than the driving member.

sludge A muddy-appearing deposit formed by contamination and chemical action on liquids; can clog openings and/or passages.

snap ring A springy split ring that fits into a groove on a shaft or in an opening; used to hold parts in place.

solenoid An electrical device that produces a pushing or pulling force through magnetic means.

solvent A chemical liquid that dissolves a solid; used for cleaning.

spacer A piece that fits between parts to maintain a distance between the parts.

spalling Chipping or flaking on a bearing or rotating part.

specifications Indications of correct measurements or procedures.

speedometer drive gear A gear on the transmission output shaft that turns the speedometer cable.
spiral bevel gear A gear cut with curved or spiral teeth.
splined Made with machined grooves into which a matching part fits.
spool valve A solid cylinder with varying size diameters to direct fluid in different directions and flow patterns.
sprag A flattened cylindrically shaped part, used as part of a sprag clutch.
sprag clutch One type of overrunning clutch, with sprags that can lock into depressions in the inner and outer races of the clutch mechanism.
sprocket Two are usually used on front-drive automatic transmissions to drive a chain that connects the torque converter assembly to the rest of the transmission assemblies.
spur gear A gear with straight teeth, cut so that the lines of the teeth all point toward the center of the gear.
stake To punch or chisel an indentation into two touching parts so that one of the two parts cannot turn against the other.
stall speed The maximum engine rpm that can be obtained by pressing the accelerator with the automatic transmission in gear and the brakes applied.
stall test A test made to determine the stall speed of a torque converter.
stator An intermediate part between the impeller and the turbine in a torque converter that redirects the fluid flow to increase coupling efficiency, multiply torque, and decrease slippage.
steels Clutch discs with steel, *not* friction, facings.
straight-cut gear A gear with teeth cut parallel to the gear axis; also called a spur gear.
straightedge A precision metal bar, ground so that the edge is in a very straight line; used in measuring and checking straightness.
stretching A condition, caused by overtightening, when a threaded fastener has become longer than it originally was.
stud A threaded, projecting shaft.
stylus The part of a dial indicator that contacts the part being measured.
suction gun A tool with a large hollow metal tube, a hose, and a piston plunger; used to suck out or force in lubricant.
sun gear *See* planetary gearset.
surface plate A specially ground precision metal plate that is flat across its entire surface; used to check the flatness of other parts.
swivel A device joining two parts so that one or both can pivot freely; also, to pivot freely.
synthetic Artificial or man-made.
syringe A tool with a rubber bulb and a tube; used to suck out or force in lubricant.

T

tachometer An instrument that measures the speed of a rotating part.
tap A tool for cutting threads in a cylindrical hole.
tension Pulling force.
throttle valve A balancing valve with a lever-regulated spring; controls valve-body action by pressure on the throttle pedal.
throttle-valve cable A cable that transfers throttle plate motion to automatic transmission shift control mechanisms.
thrust bearing A bearing that absorbs and controls the pressure exerted by a rotating part.
thrust block A block of material used like a thrust washer.
thrust washer A washer that absorbs and controls the pressure exerted by a rotating part.
timing marks Marks on a gearset that indicate how gear teeth should mesh.
tolerance The amount of variation that is allowed according to specifications.
torch A welding tool that burns gases to produce a high-temperature flame.
torque Twisting, or turning, force.
torque converter A fluid coupling with a stator, or stationary vane assembly, between the impeller and the turbine; used to couple and increase engine torque to an automatic transmission.
torque multiplication An increase in torque brought about through gear ratios or the action of a torque converter.
torque ratio A ratio that indicates the multiplication of torque through gear reduction.
torque steer A condition when a front-wheel-drive vehicle pulls to one side on acceleration, and to the opposite side on deceleration.

torque wrench A wrench that indicates how much torque is being applied during turning.

torquing-over The reaction that tends to rotate an assembly in a direction opposite to the direction a rotating shaft within the assembly is moving.

torsion spring A spring used to absorb shock and cushion engagement in a clutch assembly.

traction A gripping force; used to describe how tires grip the road.

transmission An assembly of shafts and gears inside a metal housing; used to control and transmit engine torque to the driveline.

transmission alignment pin A pilot shaft that screws into an upper transmission-attachment bolt hole.

transmission band A springy metal strap lined with friction material that can be clamped around a rotating drum to stop the drum from rotating.

transmission band adjustment To tighten or loosen the transmission band to eliminate slipping or dragging.

transmission cooler A heat exchanger used to cool automatic transmission fluid.

transmission input shaft The shaft through which engine power is transferred into a transmission.

transmission output shaft The shaft in the transmission connected to the driveline.

transverse Mounted side-to-side. *See also* longitudinal.

tripod joint A type of constant-velocity universal joint used at the transaxle end of drive axles on front-wheel-drive vehicles.

tripot joint *See* tripod joint.

troubleshooting The methods a mechanic uses to diagnose a problem; also, to diagnose a problem.

turbine The driven member of a torque converter, that transfers torque to the input shaft of an automatic transmission.

U

U-joint *See* Universal joint.

universal joint A movable joint, attached to the ends of a shaft, that transmits power smoothly as the shaft moves up and down or side to side.

upshift Automatic or manual shift to a higher gear.

V

vacuum lines Metal tubes or rubber hoses through which engine intake manifold vacuum is routed; used for control and operation of shifting in an automatic transmission.

valve A part that moves to open or close an opening or passage.

vane A blade on a torque converter impeller, turbine, stator, or oil pump.

vane-type pump An oil pump that circulates fluid by using vanes instead of gears.

varnish A coating of deposits that collects on parts; caused by oxidation and contamination of liquids; can cause binding.

vernier calipers A precision measuring instrument with a fixed and a sliding jaw, and a depth gauge; used to measure thickness and depth.

W

warpage Bending or twisting in a part that was originally straight.

wheel chocks Blocks to prevent the wheels of a vehicle from rolling.

white grease *See* multi-purpose lithium grease.

Index

A

Accelerator: linkage adjustments, 47; position, 160–161

Accumulator: function of, 107; inspection procedures, 132; parts, 107

Accumulator piston, 256

Air pressure: safety caution, 83, 92, 191, 225; tests, 225–226

American Motors Torque-Command transmission. *See* Chrysler TorqueFlite transmission

Apply devices, 105–108; defined, 105; fluid pressures in, 157; repairing, 113–116, 123–125, 131–134

ATF. *See* Automatic transmission fluid

Automatic transmission: all-metric, 282, 322; function of, 2; gear reduction in, 5; heavy duty, 105; hydraulic circuits in, 157–164; hydraulic flow in, 163; identification numbers (*see* Identification numbers); installation, 312–314; life of, 89–90; mechanical parts, 2; overdrive, overhauling, 281–302; refilling, 27–28; removal procedures, 239–242; safety caution, 17; servicing of, 17–20, 23–28; three-speed, 163–164; troubleshooting, 202–208, 213–215, 221, 225, 229–236; visual check of, 17–20. *See also names of parts and models of automatic transmissions*

Automatic transmission dipstick, 23–24; hole, 23; tube, 27

Automatic transmission fluid, 3, 153–154; adding, 27–28, 314; aeration, 23, 24; checking, 23, 24, 206; checking flow of, 95; color of, 24; cooling of (*see* Transmission cooler systems); DEXRON II, 154; draining from torque converter, 25–26; draining of, 24–25; external leaks, 20; filtering, 26; flow, 5, 8, 95; flow in torque converter, 66; flow redirection, 63; Ford Type CJ, 154; heating of, 89–90; level, 23–24, 27–28; odor, 24; reservoir, 4; safety caution, 24; screens, 26; temperature, 23; in turbine, 63; Type F, 154; viscosity, 153–154

Automatic transmission housing: casting porosity, 20; leaks, 18; mounts, checking, 18; seals, 18; visual check of, 18

Axle, 1, 12–13; bearing, 13; housing, 12–13; shaft, 12

Axle ratio, 11

B

Balance valves: with lever-regulated spring, 147–148; with pressure-regulated spring, 148

Ball check valves, 149

Belleville spring, 123

Bellhousing, inspection of, 70

Bench testing, defined, 77

Beveled gears, 10

Booster valve, 158–159

Borg-Warner A40 transmission, 365–368

Brake band, 104

Bushings: automatic transmission housing, 18; inspection, 133; transmission linkage, 18

C

Cable linkage, 18, 36, 44

Cables, checking, 19

Cap, defined, 91–92

Carrier, 11, 13; rotation of, 11, 12

Casting porosity: as cause of leaks, 18; repairing, 20

Check balls, 105; sizes, 191

Check valve, 105

Chrysler TorqueFlite transmission, 23, 47; band adjustments, 114; check valve, 47; cleaning and inspecting, 254–255; flexplate installation, 311–312; identification number, 250; installation, 312–314; location of parts, 251–253; model A-727, overhaul procedures, 255–272; removal procedures, 239–241; torque converter installation, 311–312

Clearance adjustments, multiple-clutch pack, 124

Clutch, 2; driving, 105; holding, 106; in manual transmissions, 2, 5

Clutch disc, 105–106; adjustment, 124; inspection, 123

Clutch drum, 105

Clutch pack, 105; clearance, 124; multiple-disc, 105–106, 123–124; snap ring adjustment, 124

Clutch plate. *See* Clutch disc

Column shift linkage, 18–19, 35, 37; rod-and-lever adjustments, 43–44. *See also* Shift linkage systems

Column shifter, 18

Companion flange, 9

Connecting rods, 4

Control devices, troubleshooting, 206–207

Control pressure, 158–160, 163

Control valve body, 2, 8, 157, 189–193; cleaning, 190–191; disassembly, 190–191; installation, 193, 272; internal parts, 189–190, 191; reassembly, 192; removal, 190, 255

Control valves, 145–150, 157

Converter housing, leakage from, 20

Converter pilot, visual inspection, 70

Converter-out fitting, 95

Coolant, engine, 90

Cooler system. *See* Transmission cooler system

D

Diagnosis. *See* Troubleshooting

Differential, 1, 4, 10–12; function of, 10, 11; gear reduction in, 5; location of, 10; movement of, 9; in transaxle-equipped vehicles, 371, 376

Differential case. *See* Carrier

Differential pinion gears, 10–11, 12

Differential ratio, 6

Differential ring gear, in transaxles, 376

Differential side gear, 11–12, 13

Dimples, removal of, 27

Dipstick. *See* Automatic transmission dipstick

Direct drive, 104, 108

Downshift valve, 161

Downshifting, 38, 48, 161, 214

Drain plugs, 24–25, 83

Drive nuts, 77

Drive oil. *See* Mainline pressure

Drive pinion gear, 10–11

Drive studs, 77

Driveline, 1, 8–9

Driven member, 5; of planetary gearset, 6

Driveshaft, 8–9

Drivetrain: in front-wheel-drive vehicles, 372–373; in rear-wheel-drive, 1–13; parts, 1–13

Driving axles, 1, 12–13; flange and stud assembly, 12–13

Driving clutch. *See* Clutch *headings*

Driving member, 5; of planetary gearset, 6

Driving wheels, 1; attachment to axles, 12–13; turning of, 10–12

Droplights, 17–18

E

Electrical connections: problems, 204; safety caution, 207; troubleshooting, 207–208

End play, 78; measurements, 253–254

Engine, 1; malfunctions, 202; removal, 239, 376; supporting, 376, 381; and transaxle, 371–373; "torquing over," 373

Engine coolant, 90

Extension housing, leakage from, 20

F

Filters, 26, 272

Firewall, 19

Flange, defined, 13

Flexplate, 62, 240–241; inspecting, 70; installation, 311–312; turning of, 65

Floor-shift linkage, 35, 37; cable, 37, 44, 374–375; rod-and-lever, 37, 43–44, 374–375

Floor shifter, 18

Floorpan, of front-wheel-drive vehicles, 372

Flow charts: defined, 167–168; key, 167; oil pressure and pressure regulation, 167–169; reading, 167–188; three-speed transmission, 163–164

Flow charts, General Motors 325 transmission, 170–175; intermediate range, 180–181; low range oil pressure, 182–183; part-throttle 3-2 downshift, 178–179; reverse, 184–185; transmission drive range—first gear, 172–173; transmission drive range—second gear, 174–175; transmission drive range—third gear, 176–177; transmission in neutral, 170–171

Fluid, automatic transmission. *See* Automatic transmission fluid

Fluid coupling. *See* Impeller; Turbine; Torque converter *headings*

Fluid flow, 5; control of, 8; redirection of, 63; in torque converter, 66

Fluid pressure. *See* Hydraulic pressure

Flushing: torque converter, 83; transmission cooler system, 95–96

Flywheel, 3

Ford AOD transmission, 281–302; parts, 281–284

Ford C–3/C–4/C–5 transmissions, 26, 345–347; C–4/C–5 band adjustment procedures, 114

Ford C–6 transmission, 353–356

Ford Jatco transmission, 359–362

Friction disc. *See* Clutch disc

Friction plate. *See* Clutch disc

Front suspension support assembly, installation, 415

Front-wheel-drive vehicles: floorpan, 372; transaxle, (*see* Transaxle); transmissions, 373

FWD. *See* Front-wheel-drive vehicles

G

Gaskets, oil pan, 27

Gate, defined, 37

Gear ratios, 6–7; calculation of, 7; defined, 6; meaning of, 7; overdrive, 7

Gear reduction, 1, 5, 108; and planetary gearset power flow, 108–109. *See also* Torque multiplication

Gear teeth, 5, 7

Gears: beveled, 10; helical-cut, 103; hypoid, 10–11; pinion, 5; planetary, 5–6; straight-cut, 103; torque measurement on, 4–5

Gearset, defined, 5

Gearshift lever adjustments, safety caution, 43, 44

Gearshift linkage, 36–37; cable, 18, 36; gate in, 37; pawl in, 37; rod-and-lever, 18, 36–37. *See also* Column-shift linkage; Floor-shift linkage

Gearshift linkage adjustments, 40, 43; cable systems, 44; rod-and-lever systems, 43; safety cautions, 43

Gearshift selector lever, 35–36; positions, 36, 39, 43–44

General Motors Turbo Hydra-Matic 125/125C transaxles, 379–384, 389–408; identification numbers, 381; installation procedure, 415; overhaul procedures, 389–408; parts of, 379–380; preparation for overhaul, 389; removal procedure, 381–384

General Motors Turbo Hydra-Matic 200/200C/200-4R transmissions, 321–322

General Motors Turbo Hydra-Matic 250/250C transmissions, 329–331

General Motors Turbo Hydra-Matic 325/325–4L transmissions, 325–326, 373; flow charts (*see* Flow charts, General Motors 325 transmission)

General Motors Turbo Hydra-Matic 350/350C transmissions, 329–335

General Motors Turbo Hydra-Matic 375/400/425 transmissions, 339–342

Governor plugs, 252

Governor pressure, 162–164

H

Heat, and automatic transmission life, 89–90

High gear, 7

Hoists, 17, 376

Holding clutch, 106

Housing. *See* Automatic transmission housing

Hydraulic circuits: defined, 145; elements of, 157; flow charts, 167–188 (*see also* Flow charts, General Motors 325 transmission)

Hydraulic force, 61, 143

Hydraulic pressure: and apply devices, 107; regulation of, 145–150; testing, 205. *See also* Hydraulic systems

Hydraulic systems, 141, 153–164; elements of, 145; force transfer in, 143; motion transfer in, 142; pressure in, 142, 144, 157; pressure regulation in, 145–150; pressure transfer in, 142; principles of, 141–150; repairing, 189–193; valves, 145–150

Hydraulics, defined, 2, 141

Hypoid gearset, 10–11

I

Identification numbers: Ford Jatco, 360; Ford C–3/C–4/C–5 transmissions, 346; Ford C–6 transmission, 354; Turbo Hydra-Matic 125/125C transaxles, 381; Turbo Hydra-Matic 125/125C transmissions, 381; Turbo Hydra-Matic 250/250C/ 350/350C/375B transmissions, 330; Turbo Hydra-Matic 375/400/425 transmissions, 340

Impeller, 3–4, 5, 61, 62; function, 3; rotation of, 62, 65–66

Input shaft, end play measurement, 253–254

Intercooler, 90

Interference tests, defined, 78

K

Kickdown linkage, 19, 37–38; adjustments, 40, 47–49; operation of, 37–38

Kickdown solenoid, 38

Kickdown switch, 48–49, 208

Kickdown valve, 48

L

Leakage: air pressure tests for, 225–226; checking for, 18, 28; external, repairing, 20; high-pressure, 24; from oil pan, 20; from torque converter, 20, 70; from transmission cooler system, 24, 91–92

Leakage testing tools, 83–84

Left side of vehicle, defined, 18

Lighting, importance of, 17

Line pressure. *See* Mainline pressure

Linkage adjustments, 204, 207

Linkage switch, 208

Linkage systems, 18–19. *See also* Shift linkage systems; Transmission linkage

Lockup clutch assembly, 61

Low gear, 7

M

Mainline pressure, 158–160, 163

Manual control valve, 158, 253; lever, 37, 43, 44

Manual transmissions, 2, 5

Manual valve. *See* Manual control valve

Matchmarks, 241

Multiple-clutch system, clearance, 124–125

Multiple-disc clutch assembly, 105–106; inspection procedures, 123–124

N

Needle bearings, 104

Neutral gear position, 104, 108

Neutral start switch, 38; adjustment, 43, 44, 50

O

O-rings, 134

Oil pan, 25, 157; as hydraulic system reservoir, 145; bolts and bolt holes, 27; cleaning, 25, 27; installation of, 27; leakage from, 20; removing, 24–25; valve body removal, 190–191

Oil pump, 65, 77, 78, 79, 155–156, 157; extension, 63; gear and crescent, 155; rotor, 155; TorqueFlite, 252; vane-type, 156

Output shaft: end play measurement, 253–254; locking of, 36

Overdrive gear ratio, 7, 104

Overdrive transmission, overhauling procedures, 281–302

Overhauling procedures: Chrysler TorqueFlite A-727 transmission, 255–272; Ford AOD transmission, 281–302; General Motors Turbo Hydra-Matic transaxle, 379–384, 389–408

Overrunning, defined, 64

Overrunning clutch, 61, 63–64, 107–108; function, 63, 107–108; inspection procedures, 73, 78, 132

P

Pascal, Blaise, 141

Pascal's Law, 141

Pawl, 37

Peening, defined, 20

Pinion gear, 5

Pinion shaft, 11, 12

Pistons, 4

Planet gears. *See* Planetary gearset

Planetary carrier, 6, 104

Planetary gearset, 2, 5–6, 103–104, 108–109; cleaning, 125; compound, 103, 108–109, 131; end play measurement, 253–254; Ford AOD, 282; gear ratios provided by, 104; and gear reduction, 108–109; input unit, 104; inspection, 125, 131; parts, 6; power flow, 108–109; reaction unit, 104; repairing, 113; single, 103, 104; TorqueFlite, 252; transmission of power in, 6

Plug, defined, 91

Poppet valves, 149

Powertrain: defined, 2; in front-wheel-drive vehicles, 372–374; in relation to shift linkage, 373–375

Pressure, fluid. *See* Hydraulic pressure

Pressure gauge, 205

Pressure plate: adjustment, 124; inspection, 123

Pressure regulating valves, 145, 158, 163

Pressure regulators, 253

Pressure testing, 205, 221–222; analysis of readings, 222

Primary pump. *See* Impeller

Pump. *See* Impeller; Oil pump

Q

Quadrant linkage, 36; adjusting system, 36–37, 40; parts, 36

R

Reaction shaft support, 252

Reaction unit, of planetary gearset, 104

Remote starter switch, 70

Reverse gear position, 104, 108, 230

Right side of vehicle, defined, 18

Ring and pinion. *See* Drive pinion gear; Ring gear

Ring gear, 6, 10–11, 104, 376. *See also* Planetary gearset

Road testing, 205, 213–215, 314; analysis of information, 215; problems and resolutions, 229–236; procedures, 213–214; safety caution, 28, 50, 213, 314

Rod-and-lever linkage, 18, 36–37, 374–375

Roller clutch, 64

Rotation, redirection of, 10

S

Screens, replacement of, 26

Seals: lip, 134; replacement of, 133–134; rings, 134; square-cut, 134

Separator plate, 157

Servos, 106–107; air pressure test, 225; in control valve body, 8; defined, 8; inspection procedures, 132; parts, 106

Shift lever, 27; in drive position, 8

Shift linkage systems, 35–40; adjustment of, 39–40; cable, 37, 44, 374–375; function, 35; in relation to powertrain, 373–375; rod-and-lever, 18, 36–37, 374–375; visual check of, 18. *See also* Gearshift linkage; Kickdown linkage; Quadrant linkage; Transmission linkage

Shift quadrant, 36

Shift valve, 159–160, 163, 253

Shuttle valve, 252

Slip joint, 9

Slip yoke. *See* Slip joint

Slippage, 3, 4, 205; of torque converter, 64–65

Sludge, defined, 95

Snap ring: clearance and adjustment, 124; replacement of, 133–134

Solenoid, 38

Spacer, clutch-pack, 125

Speedometer cable, removal of, 241

Spool valves, 146–147, 148, 149–150; multiland, 147

Sprag clutch, 64

Stall test, 73–74, 204–205, 213; diagnosis, 73–74; noises during, 74; procedures, 73; safety caution, 73

Stator, 3–4, 61, 63; function, 63; rotation of, 63–64

Stator overrunning clutch. *See* Overrunning clutch

Stator-to-impeller interference test, 78

Stator-to-turbine interference test, 78, 79

Steering ratio, 6

Studs, defined, 13

Sun gear, 6, 104

Switches, electrical, 208

Switching valves, 145, 148–150

Swivel lockscrew, 47–48

T

Temperature: automatic transmission fluid, 23; and automatic transmission life, 89–90

Throttle pressure, 160–161; and governor pressure, 163

Throttle valve: cable, 38; cable adjustment, 49–50; cable snaplock, 49; linkage adjustments, 47–48

Thrust bearings, 133

Thrust washers, 133

Torque: defined, 2, 4; driveline, 10; transfer of, 4, 10; units of measure of, 4. *See also* Torque multiplication

Torque converter, 3–4, 249; balance weights, 70; bench-testing, 77–79, 83–84; checking, 69–70, 73–74; draining, 25–26; end play, 78; fluid flow in, 66; flushing, 83; Ford AOD, 282; function, 61, 69; hub, 77; inspection, 240; installation, 311–312; with lockup mechanism, 64–65, 249–250; operation of, 65–66; parts, 61–66; removal with transmission, 240; rotation of oil pump gears by, 65; slippage, 64; stall test, 73–74, 204–205, 213; trouble indicators, 70; visual inspection, 70

Torque converter housing, 4, 62; rotation of, 65; visual inspection, 70

Torque converter leakage test, 83–84; safety cautions, 83–84; special tool for, 83

Torque multiplication, 4–5, 11; cause of, 3; planetary gears and, 108; when required, 66. *See also* Gear reduction

Torque wrench, 193

TorqueFlite transmission. *See* Chrysler TorqueFlite transmission

"Torquing-over," 373

Toyota. *See* Borg-Warner A40 transmission

Transaxle, 371–376, 379–384, 389–408; braces, 373–374; defined, 371; differences from/similarities to transmissions, 371–376; disassembly of, 376; installation procedure, 415; location and mounting position, 371–373; mounting procedure, 373–374; overhaul procedure, 389–408; preparation for overhaul, 389; removal, 376, 381–384

Transmission. *See* Automatic transmission; *and names of parts and models of automatic transmissions*

Transmission band, 106; adjustment procedures, 113–116; American Motors Torque Command, 114; Chrysler TorqueFlite, 114; controlled by servo, 106–107; Ford C-4/C-5, 114–115; General Motors Turbo Hydra-Matic 200/200C/200-4R band adjustment, 115–116; inspection procedures, 131; safety caution, 113

Transmission case: leakage from, 20; manual control valve arm, 37

Transmission cooler system, 24; external, 90; flow check, 95; flushing, 95–96; inspection of, 91; internal, 90; leakage, 24, 91–92; replacement of, 96

Transmission fluid. *See* Automatic transmission fluid

Transmission gear ratio, 6–7

Transmission input shaft, 253–254

Transmission jack, 239

Transmission linkage: column shift, 18–19, 35–37; floor shift, 18, 37, 43–44, 374–375; kickdown, 19, 37–38, 47, 48–49; visual check of, 17–18. *See also* Shift linkage systems

Transmission output shaft: endplay measurement, 253–254; locking of, 36

Transmission stand, 255

Transmission supports, 18, 242

Troubleshooting, 201–215; automatic transmission fluid, 206; basic transmission problems, 203–205; charts, 229–236; control devices, 206–207; control units, 203; defined, 201–202; determining problem, 202; electrical system, 204, 207–208; linkage, 204; pressure testing, 205, 221–222; road testing, 205, 213–215; stall testing, 73–74, 204–205, 213; transmission fluid, 203; vacuum control diaphragm, 206–207; vacuum lines, 203, 206

Trunnions, 9

Turbine, 3–4, 61, 62–63

Turbo Hydra-Matic. *See* General Motors Turbo Hydra-Matic *headings*

U

U-joints, *See* Universal joints

Universal joints, 8–9

Upshifting, problems and resolutions, 231

V

Vacuum control diaphragm, 203–204; checking, 206–207

Vacuum lines, 203, 206

Vacuum modulator. *See* Vacuum control diaphragm

Vacuum modulator pressure, 161

Vacuum pump, 207

Valve body. *See* Control valve body

Valves: balance, 147–148; booster, 158–159; control, 145–150, 157; downshift, 161; manual control, 158; poppet, 149; pressure regulator, 145, 158, 163; shift, 159–160; spool, 146–147, 148, 149–150; switching, 145, 148–150; throttle, 38, 160–161. *See also* Check valve

Vibration dampener weight, 313